HYDRODYNAMIC PROPULSION AND ITS OPTIMIZATION

FLUID MECHANICS AND ITS APPLICATIONS
Volume 27

Series Editor: R. MOREAU
MADYLAM
Ecole Nationale Supérieure d'Hydraulique de Grenoble
Boîte Postale 95
38402 Saint Martin d'Hères Cedex, France

Aims and Scope of the Series

The purpose of this series is to focus on subjects in which fluid mechanics plays a fundamental role.

As well as the more traditional applications of aeronautics, hydraulics, heat and mass transfer etc., books will be published dealing with topics which are currently in a state of rapid development, such as turbulence, suspensions and multiphase fluids, super and hypersonic flows and numerical modelling techniques.

It is a widely held view that it is the interdisciplinary subjects that will receive intense scientific attention, bringing them to the forefront of technological advancement. Fluids have the ability to transport matter and its properties as well as transmit force, therefore fluid mechanics is a subject that is particulary open to cross fertilisation with other sciences and disciplines of engineering. The subject of fluid mechanics will be highly relevant in domains such as chemical, metallurgical, biological and ecological engineering. This series is particularly open to such new multidisciplinary domains.

The median level of presentation is the first year graduate student. Some texts are monographs defining the current state of a field; others are accessible to final year undergraduates; but essentially the emphasis is on readability and clarity.

For a list of related mechanics titles, see final pages.

Hydrodynamic Propulsion and Its Optimization

Analytic Theory

by

J. A. SPARENBERG

Department of Mathematics,
University of Groningen,
The Netherlands

SPRINGER-SCIENCE+BUSINESS MEDIA, B.V.

A C.I.P. Catalogue record for this book is available from the Library of Congress

DOI 10.1007/978-94-017-1812-7

Printed on acid-free paper

To PAULA

FOREWORD

HYDRODYNAMIC PROPULSION AND ITS OPTIMIZATION

ANALYTIC THEORY

Hydrodynamic propulsion has been of major interest ever since craft took to the water. In the course of time, many attempts have been made to invent, develop, or to improve hydrodynamic propulsion devices. Remarkable achievements in this field were made essentially by experienced individuals, who were in need of reliable propulsion units such as paddle wheels, sculling devices, screw propellers, and of course, sails.

The problem of minimizing the amount of input energy for a prescribed effective output was first investigated seriously at the beginning of this century. In 1919, BETZ presented a paper on air-screw propellers with minimum consumption of energy which could be applied to ship-screw propellers also. Next, attempts were made to optimize hydrodynamic propulsion units. Ensuing investigations concerned the optimization of the hydrodynamic system: ship-propeller. The first simple theory of ship propulsion which was presented considered more or less only thrust augmentation, wake processing and modification of propeller characteristics when operating behind the ships hull. This theory has been little improved meanwhile and is still useful, particularly with regard to practical ship design and for evaluating results of ship model tests. However, this theory is not adequate for optimization procedures necessary for high-technology propulsion, particularly for ship propellers utilizing propulsion improving devices such as tip end plates or tip fins at the propeller blades, spoilers in front of the propeller, asymmetrical stern etc.

It is now time to develop the analytical tools suitable for optimizing hydrodynamic propulsion, and to define satisfactory procedures to determine the efficiency of the hydrodynamic system, the ship-propulsion unit, dependent on loading and on configurations of flow and geometry. The book "Hydrodynamic Propulsion and its Optimization - Analytic Theory" presents valuable mathematical tools, suitable for investigating and evaluating conventional and non-conventional propulsion. It closes a gap in the analytical treatment of ship propulsion and will be of significant value to experts and investigators in this field.

Professor Dr.-Ing. Helmut Schwanecke
Former Managing Director of the
Berlin Model Basin (VWS)

Contents

Preface

This is a monograph on a number of aspects of the analytical theory of hydrodynamic propulsion. However important they may be, numerical models which have to be handled almost directly with a computer, are not discussed here. The book has been written with technical propulsion systems in mind. Parts are much the same as, or are improved versions of the previous work by the same author, viz. Elements of Hydrodynamic Propulsion. The rest of the work, about half, is new – some of which will be referred to in this preface.

We assume the fluid to be incompressible which is admitted in conventional hydrodynamics. Furthermore, we assume the fluid to be inviscid. Of course it has to be realized that viscosity is important for many phenomena in real fluids, such as flow separation at a profile or the entrainment of fluid by a ship's hull. Another approximation which will be used often is that the problems will be treated by a linear or a semi-linear theory. In both cases the amount of vorticity shed by a wing or a propeller has to be sufficiently small, so that the square of its induced velocities can be neglected with respect to these velocities themselves. Hence it is necessary to evaluate the domain of validity of the results of these theories. At some places, however, we will discuss methods to take into account, be it approximately, the effects of viscosity and non-linearity.

In low Reynolds number flow, singular external forces and moments are very useful. It is one of the intentions of this work to promote the use of external force fields in the case of incompressible and inviscid fluids, as an expedient to generate velocity fields. For that purpose a general solution is given of the linearized equations of motion while an external force field is present. This solution, and that is new with respect to the previous work, is valid also in the region where the external forces operate or have operated. From this arises in a natural way the concept of the divergenceless dipole flow which can be considered as the building-stone of linearized divergenceless hydrodynamics. This dipole has inside of it a delta function of Dirac velocity field. An interesting feature of non-conservative external force fields is their ability to generate vorticity in an inviscid fluid. By this we have no need, in the discussion about the origin of vorticity in such a fluid, of a slight viscosity which afterwards is abandoned again.

In order to deal with a screw blade of which the generator line is curved in one way or another, a general linear theory is discussed for arbitrarily curved lifting surfaces.

These lifting surfaces are even allowed to change their shape and area while they move through the fluid. This theory shows that the Hadamard principle value of the integral, used to calculate the normal component of the velocities on the blades, is correct. This seems to have been recognized to date, only for flat wings or helicoidal screw blades with a straight generator line. Also a new addition is the proof that in the realm of a linear theory, the flow far behind any screw propeller is perpendicular to the helicoidal coordinate lines of the helicoidal coordinate system belonging to the propeller.

Another objective of this work is to discuss a linearized optimization theory for propellers or, more generally, for systems of lifting surfaces working in an incompressible and inviscid fluid. The theory applies to rather general prescribed force actions. Discussed is, among others, the quality number of a propeller which gives another type of information than the efficiency. It indicates how well a certain aspect of the geometry of the propeller is chosen. The lifting surfaces are assumed to form angles with the direction of the desired action, which are not small. An exception is the calculation of the maximum thrust of the sails of a yacht, sailing close to wind. This problem can be reformulated as a problem of optimum extraction of kinetic energy out of a fluid and in this way comes under the theory described here.

In the optimization theory the influence of the viscosity of water can be important with respect to the interpretation of the results. When we consider for instance a screw propeller with end plates, the efficiency of such a propeller depends on the dimension of the end plates. When they are too small, we lose unnecessarily much kinetic energy. When they are too large, the viscous losses become unnecessarily high. So a method will be discussed, based on an assumption about the viscous resistance of plates, to estimate the appropriate size of the end plates.

Some problems in relation to the existence or non-existence of optimum propulsion systems are discussed, mainly for the case of unsteady propulsion. It turns out that in some classes of admitted propellers, optimum propellers do exist and in other classes they do not. In the latter case it does not mean that the considered class cannot contain propellers which theoretically have a high efficiency. On the contrary, a non-existence proof can sometimes be based on the construction of a minimizing sequence of propellers in that class, for which the loss of kinetic energy per unit of time tends to zero. However, this will occur in general at the cost of wilder and wilder motions of the lifting surfaces, so that this sequence does not tend to an acceptable propeller. Hence the non-existence of an optimum propeller only means that we cannot construct an algorithm which yields, within the considered class, a propeller which has the minimum loss of kinetic energy.

For the proof of the existence of optimum propellers it seems that the abstract methods of functional analysis are unavoidable. We have to consider the lost kinetic energy per unit of time of a propeller as a functional on the space of motions of the lifting surfaces which form the admitted propellers. The functional and the space have to possess some properties which ensure that the functional assumes its minimum on at least one of the motions.

The choice of the subjects and the examples in this work reflects to some extent the field of interest of the author and his collaborators, it is not claimed that a complete survey of the theory of hydrodynamic propulsion is given. We do not discuss screw propellers acting in a wake, ducted propellers, contra-rotating screw propellers, cavitation and so on. However, it is hoped that the subjects treated here can be helpful towards procuring fundamental insight. The book is intended for readers who are interested in the application of mathematics to theoretical and practical problems in the theory of propulsion.

Finally the author is pleased to mention that former Ph.D. students at the Mathematics Department of the University of Groningen have in the course of time, contributed greatly to the contents of this monograph.

Groningen
J. A. Sparenberg

Acknowledgements

The author is indebted to Mr J.H. Bos of the Graphic- and Audio-Visual Department of the University of Groningen for making the drawings, The Dutch Technology Foundation (S.T.W.) for financial support (Projectnumber G.W.I.33.3197), and Kluwer Academic Publishers for offering the opportunity to prepare a revised edition of the previous work "Elements of Hydrodynamic Propulsion".

J.A. Sparenberg

Chapter 1

Basic Hydrodynamics

It is our intention to give a number of basic results which are useful for the development of the theory in subsequent chapters, but some of them are also of interest for their own sake. Examples of these are the divergenceless velocity dipole, the lifting surface theory and the orthogonality property of a helicoidal flow.

An important subject is the time-dependent force field acting at an inviscid and incompressible fluid, without the intermediary of a body. These considerations are mainly based on the linearized equations of motion, which will be solved explicitly. It is discussed that force fields which are conservative, hence which can be derived from a potential, are not of interest for a theory about propulsion. These fields only induce pressures and no velocities.

An extensive treatment is given of a rather general linear lifting surface theory. The deformable lifting surface is allowed to change its area by expanding or contracting.

The vorticity caused by an external force field or by a lifting surface can be divided into bound vorticity and free vorticity. We will show that such a denomination is subject to arbitrariness.

When a body moves through an inviscid and incompressible fluid it will induce velocities and pressures in the fluid. Hence the body will experience forces and moments caused by the integrated action of the pressures on its boundary. Inversely by the principle of action equals reaction, the body will exert forces on the fluid. Sometimes these force actions are accompanied by the shedding of vorticity as in the case of a lifting surface of finite span, sometimes there is no vortex shedding as in the case of the accelerated motion of a sphere, where we assume that no flow separation occurs.

It is shown that the work done by an external force field and by a moving flexible body in an incompressible and inviscid fluid can be found again as kinetic energy in the fluid. This holds for the non-linear as well as for the linear theory.

We conclude the chapter with the discussion of the roll up of vortex sheets in relation with the existence of a square root singularity of the intensity of the vorticity at an edge of the sheet. This shows why vorticity sheets of propulsors, which are optimized by a linear theory, still will roll up in reality although their induced velocity fields seem to indicate that these sheets will translate only (Chapter 5).

1.1. Representation of a Vector Field by Its Divergence and Its Rotation

A fundamental issue in the description of a vector field is its representation by its divergence and its rotation. Although this is well known we will give a discussion of the subject because then we have at our disposal some formulas which will be useful later on. We follow here the line of thought as given by Sijtsma in [60].

Consider, with respect to a Cartesian coordinate system (x, y, z), a vector field $\vec{v}$ which tends to zero at infinity. We will show first the unique dependence of such a vector field on its divergence and its rotation. Although we use the symbol $\vec{v}$, which in general will designate a velocity field, this section holds for any vector field, for instance a force field. We prescribe

$$\text{div}\,\vec{v} = \beta(x, y, z, t) \;\; ; \quad \text{rot}\,\vec{v} = \vec{\gamma}(x, y, z, t) \quad (\text{div}\,\vec{\gamma} = 0) \;\; , \tag{1.1.1}$$

where both β and $\vec{\gamma}$ have compact (finite and closed) supports, see below (D.1.1.). Suppose that besides $\vec{v}$ there exists another vector field $\vec{u}$ with the same divergence and rotation and which also tends to zero at infinity. Then for the vector field $\vec{w} = \vec{v} - \vec{u}$ we have

$$\text{div}\,\vec{w} = 0 \;\; , \qquad \text{rot}\,\vec{w} = 0 \;\; . \tag{1.1.2}$$

Using the identity

$$\Delta\vec{w} = \left(\frac{\partial^2}{\partial x^2} + \frac{\partial^2}{\partial y^2} + \frac{\partial^2}{\partial z^2}\right)\vec{w} = \text{grad div}\,\vec{w} - \text{rot rot}\,\vec{w} \;\; , \tag{1.1.3}$$

where the Laplace operator Δ is applied component wise to $\vec{w}$, we find by (1.1.2)

$$\Delta\vec{w} = 0 \;\; . \tag{1.1.4}$$

Because $\vec{w}$ tends to zero at infinity, it follows from (1.1.4)

$$\vec{w} = 0 \;\; , \qquad \vec{u} = \vec{v} \;\; . \tag{1.1.5}$$

This proves the unique dependence of a vector field, which tends to zero at infinity, on its divergence and rotation.

Next we express explicity $\vec{v}$ into β and $\vec{\gamma}$ (1.1.1) from which follows the existence of $\vec{v}$. Again by the identity (1.1.3) we have

$$\Delta\vec{v} = \text{grad}\,\beta - \text{rot}\,\vec{\gamma} \;\; , \tag{1.1.6}$$

which is a Poisson equation for $\vec{v}$. Using the convolution representation of the solution of (1.1.6), we obtain by (D.2.1) ... (D.2.6.)

$$\begin{aligned} \vec{v}(x, y, z, t) &= -\left(\frac{1}{4\pi R} * (\text{grad}\,\beta - \text{rot}\,\vec{\gamma}\,)\right) \\ &= \text{grad}\left(-\frac{1}{4\pi R} * \beta\right) + \text{rot}\left(\frac{1}{4\pi R} * \vec{\gamma}\right) \;\; , \\ R &= (x^2 + y^2 + z^2)^{1/2} \;\; , \end{aligned} \tag{1.1.7}$$

where the convolution $*$ (see (D.1.2)) is carried out component wise, when necessary. It is allowed to apply the convolution in (1.1.7) because we assumed that β and $\vec{\gamma}$ have compact supports. By applying the operations div and rot to (1.1.7) it is easily seen that we obtain (1.1.1).

Hence (1.1.7) gives the desired representation of the vector field $\vec{v}$ in terms of its divergence β and its rotation $\vec{\gamma}$. Elaboration of (1.1.7) yields

$$\begin{aligned}
\vec{v}(x,y,z,t) &= \text{grad}\,\Phi(x,y,z,t) + \text{rot}\,\vec{\Omega} \\
&= -\frac{1}{4\pi}\,\text{grad} \iiint\limits_{\mathbb{R}^3} \frac{\beta(\xi,\eta,\zeta,t)\,d\xi\,d\eta\,d\zeta}{\{(x-\xi)^2+(y-\eta)^2+(z-\zeta)^2\}^{1/2}} \\
&\quad + \frac{1}{4\pi}\,\text{rot} \iiint\limits_{\mathbb{R}^3} \frac{\vec{\gamma}(\xi,\eta,\zeta,t)\,d\xi\,d\eta\,d\zeta}{\{(x-\xi)^2+(y-\eta)^2+(z-\zeta)^2\}^{1/2}} \\
&= \frac{1}{4\pi} \iiint\limits_{\mathbb{R}^3} \frac{\beta(\xi,\eta,\zeta,t)\,\vec{R}\,d\xi\,d\eta\,d\zeta}{\{(x-\xi)^2+(y-\eta)^2+(z-\zeta)^2\}^{3/2}} \\
&\quad + \frac{1}{4\pi} \iiint\limits_{\mathbb{R}^3} \frac{\vec{\gamma}(\xi,\eta,\zeta,t)\times\vec{R}\,d\xi\,d\eta\,d\zeta}{\{(x-\xi)^2+(y-\eta)^2+(z-\zeta)^2\}^{3/2}}\,, \qquad (1.1.8)
\end{aligned}$$

where $\vec{R} = ((x-\xi),(y-\eta),(z-\zeta))$, $\times$ denotes the vector product and $\mathbb{R}^3$ is the whole space. The function Φ is the potential of that part of the field which is free of rotation, the vector function $\vec{\Omega}$ is the vector potential of that part of the field which is free of divergence.

It is not difficult to show that in any simply connected open region B in three-dimensional space, where rot $\vec{v} = 0$, there exists a potential function $\tilde{\Phi}(x,y,z,t)$ such that

$$\text{grad}\,\tilde{\Phi}(x,y,z,t) = \vec{v}(x,y,z,t)\ . \qquad (1.1.9)$$

A simply connected region is a region in which any closed contour can be contracted, while remaining within the region, into some point of that region. Choosing a point O as starting point in B, we can define the potential $\tilde{\Phi}$ at (x,y,z) by the line integral

$$\tilde{\Phi}(x,y,z,t) = \int\limits_{O}^{(x,y,z)} \vec{v}(\xi,\eta,\zeta,t)\cdot d\vec{s}\ , \qquad (1.1.10)$$

which is independent of the path along which we integrate from O towards (x,y,z), when this path remains in B, because there rot $\vec{v} = 0$. This potential $\tilde{\Phi}$ will in general not be the same as the potential Φ which occurs in (1.1.8), because now also velocities have to be taken into account which are induced by the vorticity outside the region B. When grad $\Phi = \vec{v}$, it follows that

$$\Delta\Phi = \text{div grad}\,\Phi = \text{div}\,\vec{v}\ . \qquad (1.1.11)$$

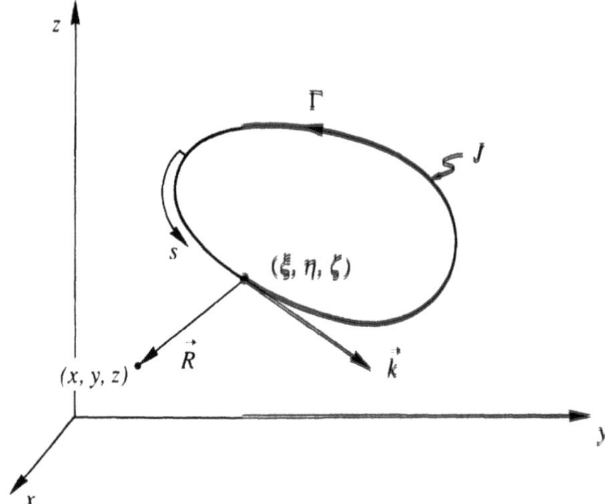

Fig. 1.1.1. Closed line J with concentrated vorticity $\vec{\Gamma}$.

Hence when div $\vec{v} = 0$, the potential satisfies

$$\Delta\Phi = 0 \ . \tag{1.1.12}$$

Now suppose that the divergence $\beta = 0$ and that the rotation of the vector field is concentrated at a closed line J, along which we have a length parameter s (Figure 1.1.1). In fact we consider the limit of a narrow tube around J, outside of which the rotation $\vec{\gamma}$ of the vector field $\vec{v}$ is zero. Inside this tube the rotation $\vec{\gamma}$ is non-zero, homogeneously distributed and "parallel" to J. When σ is the area of the cross section of this tube, we denote

$$\lim_{\sigma \to 0} \vec{\gamma} \cdot \sigma = \vec{\Gamma} \ . \tag{1.1.13}$$

The strength $\Gamma = |\vec{\Gamma}|$ of the concentrated vortex at J is called positive when the motion of the fluid around it is coupled with a right-hand screw to the positive direction of the parameter s along J. From the identity

$$\text{div}\, \vec{\gamma} = \text{div rot}\, \vec{v} = 0 \ , \tag{1.1.14}$$

we find that Γ is constant along J. Because $d\text{Vol} = \sigma ds$, we obtain from (1.1.8)

$$\vec{v} = \frac{1}{4\pi} \text{rot} \iiint_{\mathbf{R}^3} \frac{\vec{\gamma}(\xi,\eta,\zeta)\, \sigma\, ds}{\{(x-\xi)^2 + (y-\eta)^2 + (z-\zeta)^2\}^{1/2}} = \frac{1}{4\pi} \text{rot} \int_J \frac{\vec{\Gamma}\, ds}{R}$$

$$= \frac{1}{4\pi} \int_J \frac{\vec{\Gamma} \times \vec{R}}{R^3}\, ds = \frac{\Gamma}{4\pi} \int_J \frac{\vec{k} \times \vec{R}}{R^3}\, ds \ , \quad (x,y,z) \notin J \ , \tag{1.1.15}$$

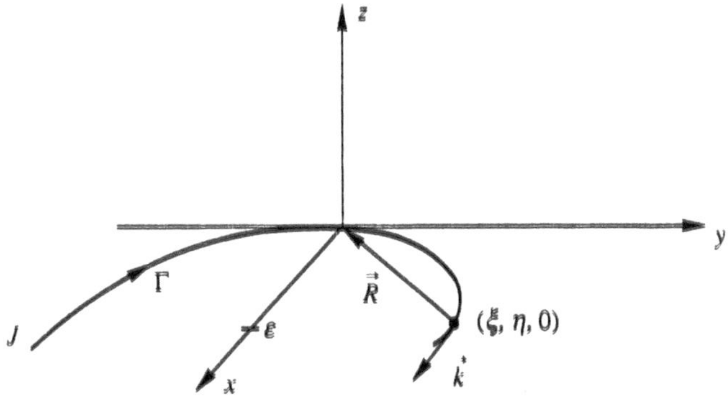

Fig. 1.1.2. Circular vortex line J of strength Γ.

where $\vec{R} = (x - \xi, y - \eta, z - \zeta)$, $R = |\vec{R}|$ and $\vec{k} = (d\xi/ds, d\eta/ds, d\zeta/ds)$ is the unit vector tangent to J.

Formula (1.1.15) is the well-known law of Biot and Savart for the velocity field induced by a vortex line.

In case we have a concentrated straight vortex of strength Γ, lying for instance along the z-axis with a right-hand screw in the positive z-direction, we find from (1.1.15) for its velocity field

$$v_x = \frac{-y\Gamma}{2\pi(x^2 + y^2)} , \qquad v_y = \frac{x\Gamma}{2\pi(x^2 + y^2)} ,$$

$$v_z = 0 , \qquad |\vec{v}| = \frac{\Gamma}{2\pi(x^2 + y^2)^{1/2}} . \tag{1.1.16}$$

We now show that (1.1.15) has a serious limitation. It makes no sense to calculate with this equation the velocities induced by a curved concentrated vortex, at the vortex line itself. As a simple example we take a circular vortex line J, of radius a, lying in the (x, y) plane and tangent to the y-axis at the origin O ($x = 0$, $y = 0$, $z = 0$) (Figure 1.1.2),

$$J: \quad x^2 + y^2 - 2ax = 0 , \quad z = 0 . \tag{1.1.17}$$

The velocity induced at O is in this case perpendicular to the (x, y) plane and in the negative z-direction (right-hand screw for the vortex). We only need to consider a part of J in the neighbourhood of O, say $0 \le x \le \varepsilon$, because parts of J which are at a finite distance of O, induce finite velocities.

For a point $(\xi, \eta, 0)$ at J, we have

$$\xi^2 + \eta^2 - 2a\xi = 0 , \quad \vec{k} = \left(\frac{\eta}{a}, \frac{(a - \xi)}{a}, 0\right) , \quad \vec{R} = (-\xi, -\eta, 0) ,$$

$$R^2 = 2a\xi , \quad \xi d\xi + \eta d\eta = a\, d\xi , \quad ds^2 = d\xi^2 + d\eta^2 = \frac{a^2\, d\xi^2}{(2a - \xi)\xi} . \tag{1.1.18}$$

Expressing everywhere η as a function of ξ, we find by (1.1.15) and (1.1.18) for the z-component of $\vec{v}$, induced by the part of J with $0 \le \xi \le \varepsilon$

$$v_z = -\frac{\Gamma}{4\pi\sqrt{2a}} \int_0^\varepsilon \frac{d\xi}{\xi(a-\xi)^{1/2}} , \tag{1.1.19}$$

where we used the fact that both halves of the considered part of J induce the same velocity at O. However, the integral (1.1.19) does not converge and is infinite. So a circular concentrated ring vortex induces infinite velocities at itself. In fact, (1.1.15) can be used only for points not "too close" to J. The same happens for more general curved vortex lines in space.

When we consider vorticity continuously distributed over a surface in space, then (1.1.15) can be used for calculating the induced velocities at the surface itself, but this has to be done carefully. This will be discussed further when we consider lifting surfaces.

1.2. Equations of Motion, Bernoulli's Equation, Boundary Condition

When in this section a frame of reference is needed, we use an inertial Cartesian coordinate system (x, y, z). We start from the Navier-Stokes equation, for an incompressible viscous fluid, which is derived in most textbooks on fluid dynamics, for instance [4]. This equation reads in vector notation

$$\rho \frac{d\vec{v}}{dt} = \rho \left\{ \frac{\partial \vec{v}}{\partial t} + (\vec{v} \cdot \text{grad})\vec{v} \right\} = \vec{F} - \text{grad}\, p + \mu \Delta \vec{v} , \tag{1.2.1}$$

where ρ is the density of the fluid, $\vec{v} = (v_x, v_y, v_z)$ the velocity field of the fluid, t is time, $\vec{F} = (F_x, F_y, F_z)$ the external force field per unit of volume, p the pressure, μ the coefficient of viscosity and Δ the Laplace operator (1.1.3).

We remark that here and in the following parts of this book, unless it is stated explicitly, we only consider fields of flow which are free of divergence,

$$\text{div}\, \vec{v} = 0 . \tag{1.2.2}$$

First we discuss the linearization of the equation of motion (1.2.1). We consider the case that an incoming parallel flow exists in the positive x-direction, with velocity U (Figure 1.2.1). Then we can write the velocity field $\vec{v}$ as

$$\vec{v} = (U, 0, 0) + \vec{w} = (U + w_x, w_y, w_z) , \tag{1.2.3}$$

where $\vec{w}$ is the disturbance of the main flow. We assume now that $\vec{F}$, $\vec{w}$, grad $\vec{w}$, and grad p are all of $O(\varepsilon)$, where ε is a small parameter with respect to which we linearize (1.2.1). When we substitute (1.2.3) into (1.2.1) and neglect quantities of $O(\varepsilon^2)$, Equation (1.2.1) assumes its linearized form

$$\rho \frac{\partial \vec{w}}{\partial t} + U \frac{\partial \vec{w}}{\partial x} = \vec{F} - \text{grad}\, p + \mu \Delta \vec{w} , \tag{1.2.4}$$

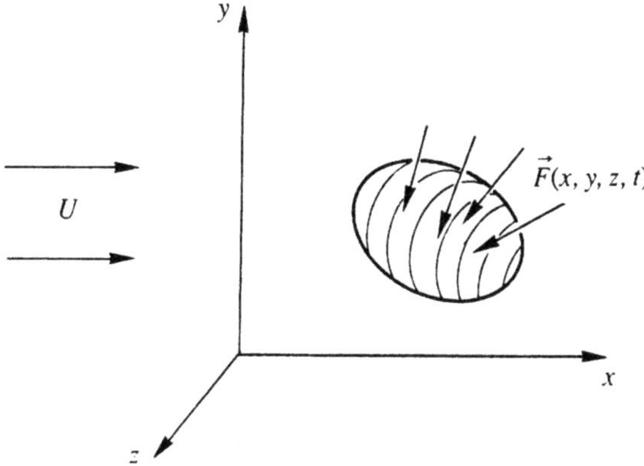

Fig. 1.2.1. Force field $\vec{F}$ in parallel flow $(U, 0, 0)$.

which is correct up to and including $O(\varepsilon)$. In case $U = 0$, then by (1.2.3) $\vec{v} = \vec{w}$ and (1.2.4) becomes

$$\rho \frac{\partial \vec{v}}{\partial t} = \vec{F} - \text{grad}\, p + \mu \Delta \vec{v} \ . \tag{1.2.5}$$

Next we will derive the unsteady Bernoulli-equation and its linearized forms. Here and in the following parts of the book, unless it is stated explicitly, we consider an inviscid fluid, hence we take $\mu = 0$. Using from Appendix C the formula (C.11), we can write (1.2.1) as

$$\frac{\partial \vec{v}}{\partial t} - \vec{v} \times \vec{\omega} = -\frac{1}{\rho}\, \text{grad}\, (p + \tfrac{1}{2}\, \rho\, |\vec{v}\,|^2) + \frac{1}{\rho}\, \vec{F} \ , \tag{1.2.6}$$

where $\vec{\omega} = \text{rot}\, \vec{v}$. To both sides of (1.2.6) we apply the operator rot, then we obtain

$$\frac{\partial \vec{\omega}}{\partial t} - \text{rot}\, (\vec{v} \times \vec{\omega}\,) = \frac{1}{\rho}\, \text{rot}\, \vec{F} \ . \tag{1.2.7}$$

When rot $\vec{F} \neq 0$ it follows from (1.2.7) that $\vec{\omega} \neq 0$ or $\partial \vec{\omega}/\partial t \neq 0$ or both, and hence that rotation or in other words vorticity is induced into the fluid. Inversely, when in a region B of space, during some period of time the flow is free of rotation, hence $\vec{\omega} = 0$ and $\partial \vec{\omega}/\partial t = 0$, then it follows from (1.2.7) that during that period rot $\vec{F} = 0$ in B. By (1.1.10) and the analogous formula for $\vec{F}$, we then have

$$\vec{v} = \text{grad}\, \Phi(x, y, z, t) \ , \qquad \vec{F} = \text{grad}\, K(x, y, z, t) \ , \tag{1.2.8}$$

where Φ is the velocity potential, which satisfies by (1.2.2)

$$\Delta \Phi = 0 \ , \tag{1.2.9}$$

and K is the potential of the force field. Substituting $\vec{\omega} = 0$ and (1.2.8) into (1.2.6) we find

$$\operatorname{grad}\left\{\frac{\partial\Phi}{\partial t} + \frac{1}{\rho}\,p + \tfrac{1}{2}\,|\vec{v}\,|^2 - \frac{1}{\rho}\,K\right\} = 0 \ . \tag{1.2.10}$$

Integrating (1.2.10) we obtain the unsteady Bernoulli equation

$$\frac{\partial\Phi}{\partial t} + \frac{1}{\rho}\,p + \tfrac{1}{2}\,|\vec{v}\,|^2 = \frac{1}{\rho}\,K + h(t) \ , \tag{1.2.11}$$

where $h(t)$ is an unknown function which is allowed to depend on t only.

Introducing an incoming parallel flow $(U,0,0)$ as in (1.2.3) and making the assumptions below (1.2.3), we find for the linearized form of (1.2.11)

$$\frac{\partial\Phi}{\partial t} + \frac{1}{\rho}\,p + Uw_x = \frac{1}{\rho}\,K + h(t) \ , \tag{1.2.12}$$

where we absorbed $\frac{1}{2}U^2$ in $h(t)$. For $U = 0$, (1.2.12) simplifies to

$$\frac{\partial\Phi}{\partial t} + \frac{1}{\rho}\,p = \frac{1}{\rho}\,K + h(t) \ . \tag{1.2.13}$$

Finally, we discuss in this section the boundary condition which the velocity field $\vec{v}$ of an inviscid fluid has to satisfy at an impermeable surface

$$H(x,y,z,t) = 0 \ , \tag{1.2.14}$$

which moves in one way or another through the fluid. Hence we ask for the condition which the velocity field has to satisfy in order that the fluid does not penetrate this surface. We assume again that the fluid comes from $x = -\infty$ as a homogeneous parallel flow with velocity U in the positive x-direction and we consider the disturbance field $\vec{w}$ (1.2.3), where now $\vec{w}$ need not to be small.

Consider $H(x,y,z,t)$ as a function defined in the whole space. Then each fluid "particle" perceives at the place where it is at a certain moment, some value of H. When the particle moves on, the rate of change of this value is given by

$$\begin{aligned}\frac{dH}{dt} &= \frac{\partial H}{\partial t} + \vec{v}\cdot\operatorname{grad} H\\ &= \frac{\partial H}{\partial t} + (U + w_x)\frac{\partial H}{\partial x} + w_y\frac{\partial H}{\partial y} + w_z\frac{\partial H}{\partial z} \ .\end{aligned} \tag{1.2.15}$$

However, a particle which moves along the surface has to observe the constant value $H = 0$, during its motion. This means that $dH/dt = 0$ or

$$\frac{\partial H}{\partial t} + (U + w_x)\frac{\partial H}{\partial x} + w_y\frac{\partial H}{\partial y} + w_z\frac{\partial H}{\partial z} = 0 \ ,$$

$$(x,y,z) \in H(x,y,z,t) = 0 \ , \tag{1.2.16}$$

which is the condition to be satisfied by the disturbance velocity field $\vec{w}$. In case $U = 0$, then $\vec{v} = \vec{w}$ and (1.2.16) simplifies to

$$\frac{\partial H}{\partial t} + v_x \frac{\partial H}{\partial x} + v_y \frac{\partial H}{\partial y} + v_z \frac{\partial H}{\partial z} = 0 \ , \quad (x, y, z) \in H(x, y, z, t) = 0 \ . (1.2.17)$$

In case we use cylindrical coordinates it follows from (1.2.15) and from (B.2.7) that (1.2.16) changes into

$$\frac{\partial H}{\partial t} + (U + w_x) \frac{\partial H}{\partial x} + w_r \frac{\partial H}{\partial r} + \frac{w_\varphi}{r} \frac{\partial H}{\partial \varphi} = 0 \ , \tag{1.2.18}$$

where w_x, w_r and w_φ are the physical components of the disturbance velocity.

If we consider a linearized theory, the conditions (1.2.16) or (1.2.17) can be adapted to such a theory in two respects. First with respect to the linearization of the condition itself and second with respect to the place where the condition is demanded. This will be discussed further whenever it is needed.

1.3. External Force Fields and Vorticity

We start from the equations which describe the motion of an inviscid and incompressible fluid with respect to the inertial Cartesian coordinate system (x, y, z). Then our basic equations become (1.2.1) ($\mu = 0$) and (1.2.2),

$$\frac{d\vec{v}}{dt} = \frac{\partial \vec{v}}{\partial t} + (\vec{v} \cdot \text{grad})\vec{v} = -\frac{1}{\rho} \text{grad}\, p + \frac{1}{\rho} \vec{F} \ , \tag{1.3.1}$$

$$\text{div}\, \vec{v} = 0 \ . \tag{1.3.2}$$

The way in which a force field is created does not come up for discussion at this stage, later on we will show how, in a linearized theory the action of a wing or of a screw blade can be represented by an external force field.

We assume in the following that the flow belonging to a force field exists and is uniquely determined by suitably chosen initial conditions. First consider a force field $\vec{F}(x, y, z, t)$ which satisfies rot $\vec{F} = 0$ and which comes into action at $t = 0$. Such a field can (1.2.8) be represented as

$$\vec{F} = \text{grad}\, K(x, y, z, t) \ , \qquad t > 0 \ . \tag{1.3.3}$$

Assume that for $t < 0$ the fluid is at rest hence $\vec{v}(x, y, z, t) = 0$, $t < 0$. Then we can satisfy (1.3.1) and (1.3.2) by

$$p = p_0 + K(x, y, z, t) \ , \qquad \vec{v} = 0 \ , \qquad t > 0 \ , \tag{1.3.4}$$

where p_0 is the already existing ambient pressure. Hence this force field will not induce any motion in the incompressible fluid, but only changes the pressure.

For instance, consider the case that we place at the boundary ∂B of a finite region B in a fluid at rest with pressure p_0, a normal external force field per unit of area, of constant strength C. The force potential (inside the region B can be taken equal to C, $K(x,y,z) = C$, $(x,y,z) \in B$ and $K(x,y,z) = 0$, $(x,y,z) \notin B$. Hence by (1.3.4) the pressure inside B is changed and becomes $p_0 + C$, while outside B the pressure remains p_0 and no velocities occur anywhere.

Another simple example is a pond with stagnant water. Then the potential of the external forces acting at the water particles is $K = -g\rho y$ where g is the acceleration of gravity and we find by (1.3.4) the hydrostatic pressure distribution.

Hence we arrive at the conclusion that for fluid flows only external force fields are of interest for which rot $\vec{F} \neq 0$ for some region of space.

Next consider a force field $\vec{F}_1(x,y,z,t)$ which starts to act upon a fluid at $t = 0$. The fluid has at $t = 0$ a prescribed initial velocity field and can be subjected to specified boundary conditions. For $t > 0$, the pressure field of this fluid will be denoted by $p_1(x,y,z,t)$ and the velocity field by $\vec{v}_1(x,y,z,t)$. Now consider another force field $\vec{F}_2(x,y,z,t)$ starting at $t = 0$ acting at a fluid with the same initial velocity field at $t = 0$ and with the same boundary conditions as in the previous case. Then the question can be posed, under which conditions for $t > 0$ is the velocity field $\vec{v}_2(x,y,z,t)$ belonging to $\vec{F}_2(x,y,z,t)$, the same as the velocity field $\vec{v}_1(x,y,z,t)$ belonging to $\vec{F}_1(x,y,z,t)$.

The answer to this question is of course closely related to (1.3.3) and (1.3.4). In fact, this will happen when there exists a function $\tilde{K}(x,y,z,t)$ such that

$$\vec{F}_1(x,y,z,t) - \vec{F}_2(x,y,z,t) = \text{grad}\,\tilde{K}(x,y,z,t) \ . \tag{1.3.5}$$

Then we can take for the pressure field p_2 belonging to $\vec{F}_2$

$$p_2(x,y,z,t) = p_1(x,y,z,t) - \tilde{K}(x,y,z,t) \ , \tag{1.3.6}$$

by which the right-hand sides of (1.3.1) in both cases become the same, so that we also have

$$\vec{v}_2(x,y,z,t) = \vec{v}_1(x,y,z,t) \ . \tag{1.3.7}$$

The equivalence relation (1.3.5) between two external force fields is important in the theory of actuator surfaces, which will be discussed in Sections 2.4 and 2.9.

External force fields lay the foundation for theories where vorticity is created in an inviscid and incompressible fluid. Indeed in domains where the rotation of the force field is non-zero, in a natural way rotation of the velocity field of an inviscid fluid, hence vorticity, is induced as follows from (1.2.7) and the text below this formula.

Consider a force field $\vec{F}(x,y,z,t)$ which is confined to a finite region of space B and which has a non-zero resultant or which has a non-zero moment about some point. Such a force field cannot be the gradient of a force potential function $K(x,y,z,t)$, this can be seen as follows.

Suppose $\vec{F} = \text{grad}\,K$ and non-zero only in the finite region B, then we show that both the resultant of $\vec{F}$ and the moment of $\vec{F}$ about any fixed point, are zero.

Consider the finite region V of space with $B \subset V$, of which the boundary ∂V has no points in common with the boundary ∂B of B. Then at ∂V the force potential $K =$ const. $= c$. When $\vec{n}$ is the outward unit normal at ∂V, we find for the resultant of $\vec{F}$,

$$\int_V \vec{F}\, d\text{Vol} = \int_V \text{grad}\, K\, d\text{Vol} = \int_{\partial V} K\vec{n}\, dS = c \int_{\partial V} \vec{n}\, dS = 0 \ . \tag{1.3.8}$$

The moment of $\vec{F}$ around the origin O of the coordinate system which by shifting B is a general point with respect to B, becomes

$$\int_V (\vec{r} \times \vec{F}\,)\, d\text{Vol} = \int_V (\vec{r} \times \text{grad}\, K)\, d\text{Vol}$$

$$= \int_{\partial V} K \cdot (\vec{r} \times \vec{n}\,)\, dS = c \int_{\partial V} (\vec{r} \times \vec{n}\,)\, dS = 0 \ . \tag{1.3.9}$$

The last equalities of (1.3.8) and (1.3.9) follow from the more general equality

$$\int_{\partial V} \{f_1(y,z)n_x + f_2(x,z)n_y + f_3(x,y)n_z\}\, dS = 0 \ , \tag{1.3.10}$$

which holds for "arbitrary" functions $f_i (i = 1, 2, 3)$ of the indicated arguments. Of course these functions are also allowed to depend on t.

From the foregoing follows the statement:

A force field which is confined to a finite region of space and which has a non-zero resultant, or which has a non-zero moment about some point of space, cannot be the gradient of a potential function, hence its rotation is non-zero and it induces velocities and vorticity in the fluid.

Consider a closed contour C in the fluid. At some moment of time we can calculate the circulation Γ of C which is defined by

$$\Gamma = \int_C \vec{v} \cdot \vec{ds} = \int_S \vec{\omega} \cdot \vec{n}\, dS \ , \tag{1.3.11}$$

where the first integral is taken along the contour C and the second one over some smooth surface S bounded by C, $\vec{ds}$ is a directed line element of length ds, $\vec{\omega} = \text{rot}\, \vec{v}$ and $\vec{n}$ is the unit normal at S. The direction of $\vec{n}$ is coupled with a right-hand screw to the direction of the length parameter s along C. The second equality of (1.3.11) is Stokes' formula. The contour C is allowed to move in one way or another with respect to our inertial coordinate system, this depends on the kind of theory we use.

First, we consider the non-linear theory of our inviscid and incompressible fluid, which is based on (1.3.1) and (1.3.2). For this theory we assume that the contour C is

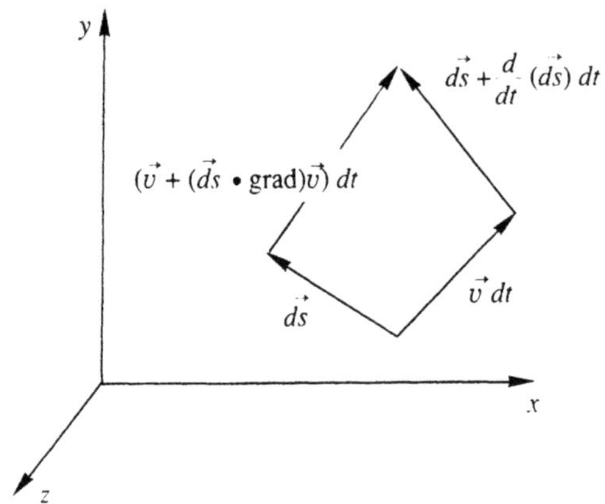

Fig. 1.3.1. Calculation of $d(\vec{ds})/dt$, for contour floating with the fluid.

coupled to the fluid particles, hence C is transported by the velocity field $\vec{v}$. Then we want to calculate the change of the circulation Γ per unit of time, hence we consider

$$\frac{d\Gamma}{dt} = \int_C \frac{d\vec{v}}{dt} \cdot \vec{ds} + \int_C \vec{v} \cdot \frac{d}{dt}(\vec{ds}) \ . \qquad (1.3.12)$$

By (1.3.1) the first integral of (1.3.12) can be written as

$$\frac{1}{\rho}\int_C (-\text{grad}\, p + \vec{F}) \cdot \vec{ds} = \frac{1}{\rho}\int_C \vec{F} \cdot \vec{ds} \ . \qquad (1.3.13)$$

Next we discuss the second integral of (1.3.12) and determine $d(\vec{ds})/dt$. From Figure 1.3.1 we find by summation of the vectors and dividing the result by dt

$$\frac{d}{dt}(\vec{ds}) = (\vec{ds} \cdot \text{grad})\vec{v} \ . \qquad (1.3.14)$$

Then the second integral of (1.3.13) becomes

$$\int_C \vec{v} \cdot \frac{d}{dt}(\vec{ds}) = \int_C \vec{v} \cdot (\vec{ds} \cdot \text{grad})\vec{v} = \tfrac{1}{2}\int_C \frac{d}{ds}(\vec{v}\,)^2\, ds = 0 \ . \qquad (1.3.15)$$

Hence we find by (1.3.13) and (1.3.15), for the contour C coupled to the fluid particles, with respect to the non-linear theory

$$\frac{d\Gamma}{dt} = \frac{1}{\rho}\int_C \vec{F} \cdot \vec{ds} = \frac{1}{\rho}\int_S \text{rot}\, \vec{F} \cdot \vec{n}\, dS \ . \qquad (1.3.16)$$

It follows that $d\Gamma/dt = 0$, or $\Gamma = \text{const.}$, as long as the contour C remains in a region where rot $\vec{F} = 0$.

Second, we consider the linearized theory. Then the force field is small of $O(\varepsilon)$, where ε is the small parameter with respect to which we linearize the equations. Further we assume that we have an incoming flow $(U, 0, 0)$. In this case, as we discussed in Section 1.2, we take $\vec{v} = (U, 0, 0) + \vec{w}$ and the equation of motion becomes (1.2.4) ($\mu = 0$)

$$\rho\left(\frac{\partial \vec{w}}{\partial t} + U\frac{\partial \vec{w}}{\partial x}\right) = \vec{F} - \text{grad}\, p \ , \tag{1.3.17}$$

for the disturbance velocity $\vec{w}$. Substituting $\vec{v} = (U, 0, 0) + \vec{w}$ into (1.3.11) we obtain

$$\Gamma = \int\limits_C \{(U, 0, 0) + \vec{w}\} \cdot \vec{ds} = \int\limits_C \vec{w} \cdot \vec{ds} \ . \tag{1.3.18}$$

From (1.3.18) we find

$$\frac{d\Gamma}{dt} = \int\limits_C \frac{d\vec{w}}{dt} \cdot \vec{ds} + \int\limits_C \vec{w} \cdot \frac{d}{dt}(\vec{ds}) \ . \tag{1.3.19}$$

By linearizing and using (1.3.17), we obtain for the first integral of (1.3.19)

$$\int\limits_C \frac{d}{dt}\vec{w} \cdot \vec{ds} = \int\limits_C \left\{\frac{\partial \vec{w}}{\partial t} + (U + w_x)\frac{\partial \vec{w}}{\partial x} + w_y\frac{\partial \vec{w}}{\partial y} + w_z\frac{\partial \vec{w}}{\partial z}\right\} \cdot \vec{ds}$$

$$\approx \int\limits_C \left\{\frac{\partial \vec{w}}{\partial t} + U\frac{\partial \vec{w}}{\partial x}\right\} \cdot \vec{ds} = \int\limits_C \vec{F} \cdot \vec{ds} \ . \tag{1.3.20}$$

For the second integral of (1.3.19) we find by using (1.3.14)

$$\int\limits_C \vec{w} \cdot (\vec{ds} \cdot \text{grad})\vec{v} = \int\limits_C \vec{w} \cdot (\vec{ds} \cdot \text{grad})\vec{w} = \tfrac{1}{2}\int\limits_C \frac{d}{ds}(\vec{w}\,)^2 ds = 0 \ , \tag{1.3.21}$$

where the last equality is exact, however the second and third integral of (1.3.21) could be neglected anyhow because they are $O(\varepsilon^2)$.

By (1.3.20) and (1.3.21) we arrive again at (1.3.16). We still have to give an interpretation to (1.3.20). It is used there that

$$\frac{d\vec{w}}{dt} \approx \frac{\partial \vec{w}}{\partial t} + U\frac{\partial \vec{w}}{\partial x} \ , \tag{1.3.22}$$

which means that we can consider now a contour C which is translating with the constant velocity U in the x-direction. This is a simplification with respect to the non-linear theory where C is coupled to the fluid particles.

Third, in the linearized case we can take $U = 0$ and then the equation of motion becomes (1.2.5) ($\mu = 0$) and (1.3.16) is valid, up to and including $O(\varepsilon)$, for a contour C which is fixed in space.

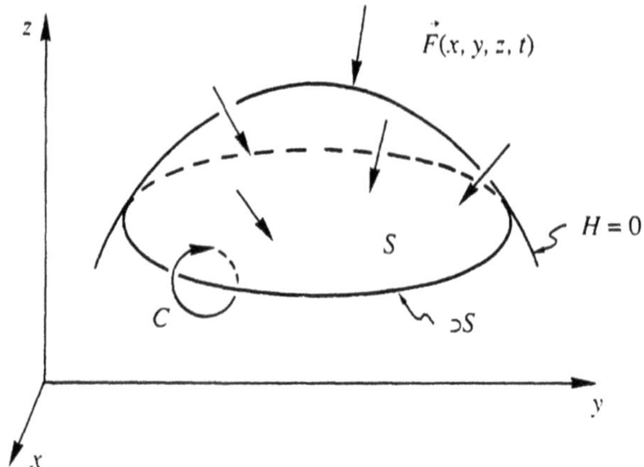

Fig. 1.3.2. Impulsive force field $\vec{F}x, y, z, t)$, perpendicular to $H = 0$.

Summarizing: (1.3.16) is valid in any one of the three mentioned cases, however in the first non-linear case the contour C floats with the fluid particles, in the second linearized case C translates with the velocity $(U, 0, 0)$ and in the third linearized case C is fixed in space.

We now give two examples of the use of Equation (1.3.16). Consider in the unbounded fluid a surface $H(x, y, z) = 0$ and from this surface a certain part S bounded by a contour ∂S (Figure 1.3.2). At S we apply an "impulsive" force field $\vec{F}(x, y, z, t)$ perpendicular to $H = 0$,

$$\vec{F} = f\,\vec{n}\,\delta(\sigma)\delta(t) \;, \tag{1.3.23}$$

where the constant f represents the strength of the field, $\vec{n}$ the unit normal at S, $\delta(\sigma)$ the Dirac delta function with respect to a length coordinate σ perpendicular to $H = 0$ which is zero at $H = 0$ and $\delta(t)$ the Dirac delta function with respect to time. Hence this impulsive force field is applied at $t = 0$. For $t < 0$ we assume the fluid to be at rest.

Next we consider a contour C coupled to the fluid particles, which encircles ∂S and which for simplicity cuts S perpendicularly. Then just after $t = 0$ we find by (1.3.16) for the circulation Γ of C

$$\Gamma = \frac{1}{\rho}\int_{-\varepsilon}^{\varepsilon}\left(\int_C \vec{F}\cdot d\vec{s}\right)dt = \frac{f}{\rho}\int_{-\varepsilon}^{\varepsilon}\left(\int_C \vec{n}\delta(\sigma)\cdot d\vec{s}\right)\delta(t)\,dt = \frac{f}{\rho}\;. \tag{1.3.24}$$

This circulation is the same for all contours C which encircle ∂S, however small they will be. For any other contour which does not encircle ∂S the circulation remains zero. This means that for values of t just after $t = 0$, there is a concentrated vortex along ∂S of strength f/ρ, which is induced by the impulsive force field $\vec{F}$. It is

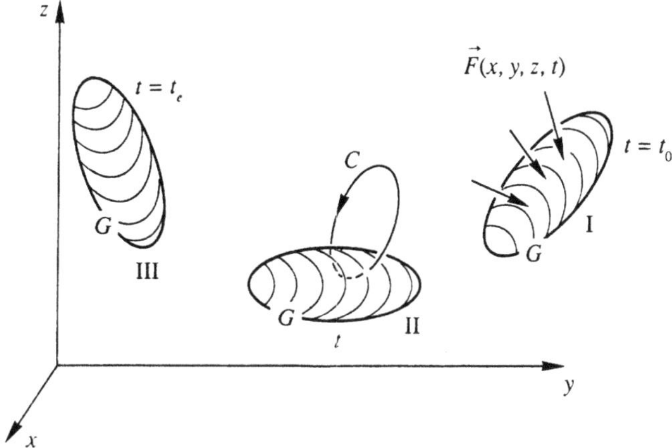

Fig. 1.3.3. Moving force field and fixed probing contour C.

clear that the fluid "under" S is pushed downwards, the fluid above S is sucked downwards and replenished around the edge ∂S by the pushed downwards fluid. In plain terms this replenishment yields the vortex. Directly after $t = 0$, this vortex and also the contours C with their circulation, float away by the velocity field induced by the vortex.

There is one difficulty with this example. At curved parts of ∂S, hence at curved parts of the concentrated vortex, the velocity induced by the concentrated vortex at itself becomes infinite as we discussed at the end of Section 1.1. Therefore, it is better to think of the delta function of Dirac $\delta(t)$ in (1.3.23) as a function of time which is spread out slightly around $t = 0$. Then the vortex is also spread out slightly and all velocities remain finite. Anyhow, the picture of the springing into existence of a more or less concentrated vortex in the neighbourhood of ∂S is correct.

Analogous considerations can be hold for the two linearized theories, where this difficulty does not occur, because there the vortex floats away with the velocity of the main stream U or remains at the place where it is formed.

As a second example of the use of (1.3.16), we consider the linearized case with $U = 0$. We investigate vorticity left behind in the fluid by a force field $\vec{F}(x, y, z, t)$, confined to a moving finite region G (Figure 1.3.3). Suppose that the circulation of the "probing" contour C, fixed in space, is zero when the force field is in position I ($t = t_0$). Then in general it becomes non-zero when G in position II is cut by C (1.3.16). When G moves to position III ($t = t_e$), it has no longer contact with C, hence the circulation of C remains constant and has the value

$$\Gamma = \frac{1}{\rho} \int_{t_0}^{t_e} \left(\int_C \vec{F} \cdot d\vec{s} \right) dt \ . \qquad (1.3.25)$$

In this way we can investigate, by using different probing contours C, where and of what strength, vorticity has been induced in the fluid by $\vec{F}(x, y, z, t)$.

1.4. Solution of the Linearized Equations of Motion

For the solution of the linearized equations of motion, we follow the line of thought as developed in [60]. There the theory is presented in a rigorous mathematical way by means of the theory of distributions. Here we will give a treatment which is more in the sense of classical applied mathematics, hence in first instance for sufficiently smooth functions.

We start from the linearized equation of motion (1.2.5) ($\mu = 0$) and the continuity equation,

$$\rho \frac{\partial \vec{v}}{\partial t} + \operatorname{grad} p = \vec{F} \ , \tag{1.4.1}$$

$$\operatorname{div} \vec{v} = 0 \ . \tag{1.4.2}$$

We assume that the external force field $\vec{F}(x, y, z, t)$ is at each moment confined to a finite region $G(t)$ of space and is "switched on" at a finite time ago $t = t_0$, while for $t < t_0$ the fluid is at rest.

Applying the operator div to both sides of (1.4.1) and using (1.4.2), we obtain

$$\Delta p = \operatorname{div} \vec{F} \ . \tag{1.4.3}$$

We assume, without reducing generality for our incompressible fluid, that $p = 0$ at infinity. Then the unique solution of this equation follows from (D.2.3) and (D.2.6), we find

$$p(x, y, z, t) = -\left(\frac{1}{4\pi R} * \operatorname{div} \vec{F}\right) = -\frac{1}{4\pi}\left(\operatorname{grad} \frac{1}{R} * \vec{F}\right)$$

$$= \frac{1}{4\pi} \int_{\mathbb{R}^3} \frac{\vec{F}(\xi, \eta, \zeta, t) \cdot \vec{R}}{R^3} \, d\text{Vol} \ , \qquad \vec{R} = (x - \xi, y - \eta, z - \zeta) \ , \tag{1.4.4}$$

where we used the second equality of (D.1.6). The convolution in (1.4.4) is allowed because $\vec{F}(x, y, z, t)$ is at each moment confined to a finite region $G(t)$ of space. By comparison of (1.4.4) with (D.2.7) we see that the pressure field is caused by a superposition of pressure dipoles. This pressure field is zero for $t < t_0$ and will be zero again when the force field is switched off.

We now apply the operator rot to both sides of (1.4.1), and integrate with respect to time, then we obtain

$$\operatorname{rot} \vec{v} = \frac{1}{\rho} \operatorname{rot} \int_{t_0}^{t} \vec{F}(x, y, z, \tau) \, d\tau \ . \tag{1.4.5}$$

Substitution of $\beta = \text{div}\, \vec{v} = 0$ and $\vec{\gamma} = \text{rot}\, \vec{v}$ from (1.4.5), into (1.1.8) and repeated application of (D.1.3), yields

$$\vec{v} = \frac{1}{4\pi\rho} \,\text{rot} \left(\frac{1}{R} * \text{rot} \int_{t_0}^{t} \vec{F}(x,y,z,\tau)\, d\tau \right)$$

$$= \frac{1}{4\pi\rho} \,\text{rot}\,\text{rot} \left(\frac{1}{R} * \int_{t_0}^{t} \vec{F}(x,y,z,\tau)\, d\tau \right) . \tag{1.4.6}$$

Using (C.12), (D.1.3) and (D.2.5) with (D.2.6) we can rewrite (1.4.6) as

$$\vec{v} = \frac{1}{4\pi\rho} \left[\text{grad}\,\text{div} \left(\frac{1}{R} * \int_{t_0}^{t} \vec{F}\, d\tau \right) - \Delta \left(\frac{1}{R} * \int_{t_0}^{t} \vec{F}\, d\tau \right) \right]$$

$$= \frac{1}{4\pi\rho} \,\text{grad} \left(\text{grad} \frac{1}{R} * \int_{t_0}^{t} \vec{F}\, d\tau \right) + \frac{1}{\rho} \int_{t_0}^{t} \vec{F}\, d\tau$$

$$= \text{grad} \left\{ -\frac{1}{4\pi\rho} \int_{t_0}^{t} \int_{\mathbb{R}^3} \frac{\vec{F}(\xi,\eta,\zeta,\tau) \cdot \vec{R}}{R^3} \,d\text{Vol}\, d\tau \right\} + \frac{1}{\rho} \int_{t_0}^{t} \vec{F}\, d\tau$$

$$= -\frac{1}{4\pi\rho} \int_{t_0}^{t} \int_{\mathbb{R}^3} \left\{ \frac{\vec{F}}{R^3} - 3 \frac{\vec{R} \cdot (\vec{F} \cdot \vec{R})}{R^5} \right\} d\text{Vol}\, d\tau + \frac{1}{\rho} \int_{t_0}^{t} \vec{F}\, d\tau \; , \tag{1.4.7}$$

where in the last integration over $\mathbb{R}^3$ the limit of a vanishing small excluded sphere has to be taken.

From the foregoing the following conclusions can be drawn about the relation in a linear theory, between the external force field and the pressures and the velocities, which are induced by it. Some of these conclusions have already been discussed in Section 1.3 but are repeated here for the sake of completeness.

The pressure field depends instantaneously on the force field, however only on div $\vec{F}$, as follows from the first equality in (1.4.4). The velocity field depends on the behaviour of the force field in the past, however only on rot $\vec{F}$, as follows from the first equality in (1.4.6). Hence a force field $\vec{F} = \text{grad}\, K(x,y,z,t)$, where K is some scalar function only induces pressures and no velocities. For a force field of the type $\vec{F} = \text{rot}\, \vec{W}(x,y,z,t)$ where $\vec{W}$ is called the vector potential of $\vec{F}$, we find by the first equality of (1.4.4) that it induces no pressures and then by (1.4.1) it follows that it induces velocities.

When a force field is "switched off" the pressure becomes zero everywhere, while the velocity field becomes independent of time. This zero pressure is in agreement with the linearized version (1.2.13) of the Bernouilli equation, because then $\partial\Phi/\partial t = 0$, $K = 0$ and $h(t) = 0$ when the pressure at infinity is assumed to be zero.

The last term at the right-hand side of (1.4.7) gives an important contribution to the velocity field; it is the representation of some non-classical velocity field as we will show in the next sections. This contribution is in general neither free of rotation, nor free of divergence. Its rotation is the rotation of the total velocity field. This rotation is left behind at regions where the force field has passed by, or is in the process of being formed at the region where the force field is present. Hence vorticity can be expected only at $\cup_\tau G(\tau)$, $\tau \in [t_0, t]$.

It follows that in linearized theory we do not have to make an extra assumption that vorticity remains where it is created. This follows simply from the solution of the linearized equations of motion.

The divergence of the last term of (1.4.7) is such that it annihilates the divergence of the first term of the last expression in (1.4.7), so the total flow is without divergence as we demanded by (1.4.2). To show this, we take the divergence of the first term at the right-hand side of (1.4.7). We find

$$\frac{1}{4\pi\rho}\,\mathrm{div}\left[\mathrm{grad\,div}\left(\frac{1}{R}*\int_{t_0}^{t}\vec{F}\,d\tau\right)\right]$$

$$=\frac{1}{4\pi\rho}\,\Delta\left(\frac{1}{R}*\mathrm{div}\int_{t_0}^{t}\vec{F}\,d\tau\right)=\frac{1}{4\pi\rho}\left(\Delta\frac{1}{R}*\mathrm{div}\int_{t_0}^{t}\vec{F}\,d\tau\right)$$

$$=-\frac{1}{\rho}\left(\delta(x)\delta(y)\delta(z)*\mathrm{div}\int_{t_0}^{t}\vec{F}\,d\tau\right)=-\frac{1}{\rho}\,\mathrm{div}\int_{t_0}^{t}\vec{F}\,d\tau\;, \tag{1.4.8}$$

which indeed is opposite to the divergence of the last term of (1.4.7).

Outside the region $\cup_\tau G(\tau)$, $\tau \in [t_0, t]$, in other words outside the region where the force field has passed or is present, it follows from (1.4.7) in agreement with Section 1.3 that the flow is free of vorticity. In that region it has the potential

$$\Phi(x,y,z,t)=-\frac{1}{4\pi\rho}\int_{t_0}^{t}\int_{G(t)}\frac{\vec{F}\cdot\vec{R}}{R^3}\,\mathrm{dVol}\,d\tau\;,$$

$$(x,y,z)\notin\cup_\tau\,G(\tau)\;,\quad \tau\in[t_0,t]\;. \tag{1.4.9}$$

In the next three sections we will give some examples of the application of (1.4.4) and (1.4.7) to a number of simple singular force fields. Then the particular meaning of the last term of (1.4.7) will become evident.

1.5. Singular Blow and the Divergenceless Dipole

We consider the case that the force field $\vec{F}=\vec{F}(x,y,z,t)$ is concentrated at the origin of the coordinate system (Figure 1.5.1), is infinitely strong and acting an infinitely

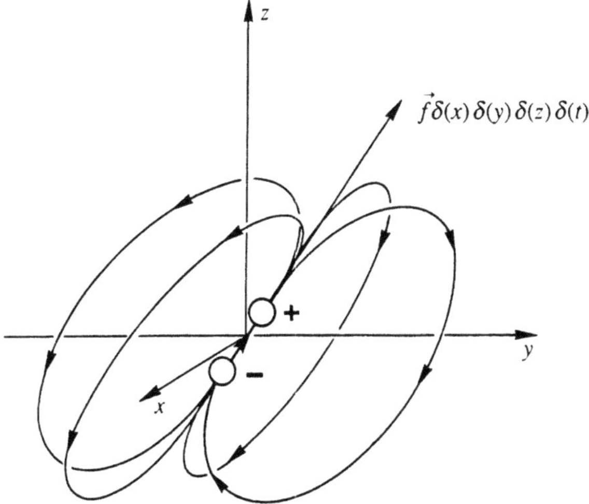

Fig. 1.5.1. Impression of the divergenceless dipole flow.

short time at $t = 0$. In formula we assume

$$\vec{F}(x, y, z, t) = \vec{f}\,\delta(x)\delta(y)\delta(z)\delta(t) \ , \tag{1.5.1}$$

where $\vec{f} = (f_1, f_2, f_3)$ is a constant vector which determines the strength of the blow and $\delta()$ is the delta function of Dirac. We assume that the formulas of the previous section can be used for this highly singular force field. That this is mathematically correct is proved in [60]. In applied mathematics this is "intuitively clear" by considering a series of smooth but more and more concentrated force fields.

The pressure field follows from (1.4.4)

$$\begin{aligned} p(x, y, z, t) &= \frac{1}{4\pi} \int\limits_G \frac{\vec{f} \cdot \vec{R}\,\delta(\xi)\delta(\eta)\delta(\zeta)\delta(t)}{\{(x-\xi)^2 + (y-\eta)^2 + (z-\zeta)^2\}^{3/2}}\, d\text{Vol} \\ &= \frac{1}{4\pi}\, \frac{\vec{f} \cdot \vec{R}\,\delta(t)}{(x^2 + y^2 + z^2)^{3/2}} \ , \qquad \vec{R} = (x, y, z) \ , \end{aligned} \tag{1.5.2}$$

where G is an open neighbourhood of the origin $O = (0, 0, 0)$. Hence the pressure field consists of one concentrated dipole, which is non-zero only for $t = 0$ and which is of "infinite strength".

Next we consider the induced velocities, by (1.4.7) where we take $t_0 < 0$. Then for $t > 0$ we find

$$\vec{v}(x, y, z, t) = -\frac{1}{4\pi\rho}\, \text{grad} \int\limits_{t_0}^{t} \left\{ \int\limits_G \frac{\vec{f} \cdot \vec{R}\,\delta(\xi)\delta(\eta)\delta(\zeta)\delta(\tau)\, d\text{Vol}}{\{(x-\xi)^2 + (y-\eta)^2 + (z-\zeta)^2\}^{3/2}} \right\} d\tau$$

$$+\frac{1}{\rho}\int_{t_0}^{t}\vec{f}\,\delta(x)\delta(y)\delta(z)\delta(\tau)\,d\tau$$

$$=-\frac{1}{4\pi\rho}\,\mathrm{grad}\left\{\frac{\vec{f}\cdot\vec{R}}{(x^2+y^2+z^2)^{3/2}}\right\}H(t)+\frac{1}{\rho}\,\vec{f}\,\delta(x)\delta(y)\delta(z)H(t)$$

$$=-\frac{1}{4\pi\rho}\left\{\frac{\vec{f}}{R^3}-3\frac{\vec{R}\cdot(\vec{f}\cdot\vec{R})}{R^5}\right\}H(t)+\frac{1}{\rho}\,\vec{f}\,\delta(x)\delta(y)\delta(z)H(t)\ , \quad (1.5.3)$$

where $H(t)$ is the Heaviside function; $H(t)=0, t<0$ and $H(t)=1, t>0$.

The first term at the right-hand side of (1.5.3) is the field of flow caused by a classical source-sink dipole, which is free of divergence except at the origin. The second term is a singular velocity field concentrated at the origin, which makes the total velocity field (1.5.3) free of divergence. Hence from minus the divergence of this second term follows the divergence of the classical source-sink dipole. This divergence becomes

$$-\mathrm{div}\left\{\frac{1}{\rho}\,\vec{f}\,\delta(x)\delta(y)\delta(z)H(t)\right\}$$

$$=-\frac{1}{\rho}(f_1\,\delta'(x)\delta(y)\delta(z)+f_2\,\delta(x)\delta'(y)\delta(z)+f_3\,\delta(x)\delta(y)\delta'(z))H(t)$$

$$=-\frac{1}{\rho}(\vec{f}\cdot\mathrm{grad})\,\delta(x)\delta(y)\delta(z)H(t)\ . \quad (1.5.4)$$

The mentioned second term represents a fluid transport from the sink of the dipole to the source of the dipole (Figure 1.5.1), hence in the opposite direction as a fluid is expected to flow. By this, the fluid which enters the sink is transported to the source, where it emerges again.

Although the divergence of $\vec{v}$ from (1.5.3) is zero everywhere, this is not the case with the rotation which is in agreement with the statement below (1.3.10). A simple calculation shows

$$\mathrm{rot}\,\vec{v}=\frac{1}{\rho}\{f_3\,\delta(x)\delta'(y)\delta(z)-f_2\,\delta(x)\delta(y)\delta'(z)\ ,f_1\,\delta(x)\delta(y)\delta'(z)$$

$$-f_3\,\delta'(x)\delta(y)\delta(z)\ ,f_2\,\delta'(x)\delta(y)\delta(z)-f_1\,\delta(x)\delta'(y)\delta(z)\}H(t)$$

$$=-\frac{1}{\rho}\left\{(\vec{f}\times\mathrm{grad})\,\delta(x)\delta(y)\delta(z)\right\}H(t)\ . \quad (1.5.5)$$

This vorticity is concentrated at the origin and is highly singular.

Because the divergence is zero everywhere and there is vorticity at the origin, we can also give an impression of the flow induced by the singular blow, by means of a vorticity consideration. For the sake of simplicity we take

$$\vec{f}=(0,0,f_3)\ , \quad (1.5.6)$$

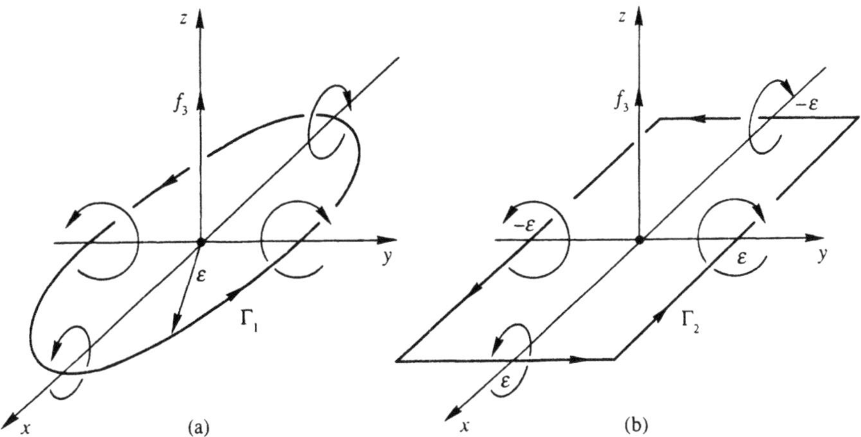

Fig. 1.5.2. Vortex approximations of the flow of a divergenceless dipole.

which in fact is only a question of choosing the coordinate system in an appropriate way. We spread the force (1.5.6) over a small region of area S in the (x, y) plane around the origin, hence we take

$$\vec{F}(x,y,z,t) = \frac{(0,0,f_3)}{S}\,\delta(z)\delta(t) \ , \qquad (x,y) \in S \ . \tag{1.5.7}$$

Then it follows from 1.3.24 and the text below that formula, that we have along the boundary ∂S of S, a vortex of strength $\Gamma_1 = f_3/\rho\, S$. For instance we can take for S a small circle of radius ε in the (x, y) plane around the origin. Then we have

$$\Gamma_1 = \frac{f_3}{\pi\rho\,\varepsilon^2} \ , \tag{1.5.8}$$

where in Figure 1.5.2 (a) this vortex is drawn. Its rotation direction is denoted by means of a right-hand screw. Then the velocity field (1.5.3) is induced by this circular vortex alone. The velocities can be calculated by means of the law of Biot and Savart (1.1.15) and by taken the limit $\varepsilon \to 0$.

In case we had taken for S a small square, with sides 2ε, we have to take for the strength Γ_2 of the vortex along the boundary of the square

$$\Gamma_2 = \frac{f_3}{4\rho\,\varepsilon^2} < \Gamma_1 \ . \tag{1.5.9}$$

The inequality follows from the fact that the area $4\varepsilon^2$ of the square is larger than the area $\pi\varepsilon^2$ of the circle.

The velocity field of the divergenceless dipole (1.5.3) can be considered as the building stone of linearized hydrodynamics with divergenceless flow fields. Indeed all such velocity fields can be represented by integration of (1.5.3), multiplied by a suitable weight function for the strength of these dipoles. These integrations can be carried out along a line, over a surface or in three-dimensional space.

1.6. Singular Force Moving through the Fluid

As in the preceeding section we consider a singular force field $\vec{F}(x, y, z, t)$ concentrated at a point, but now this point $\vec{Q}(t)$ is moving through the fluid, along a line L (Figure 1.6.1), which is assumed to be sufficiently smooth. The line L and the velocity $\vec{V}(t)$ of the point $\vec{Q}(t)$ will be prescribed by

$$L : \vec{Q}(t) = (\xi(t), \eta(t), \zeta(t)) \ ;$$
$$\vec{V}(t) = (\dot{\xi}(t), \dot{\eta}(t) \ , \dot{\zeta}(t)) \ , \quad V(t) = |\vec{V}(t)| \ , \tag{1.6.1}$$

where $\xi(t)$, $\eta(t)$ and $\zeta(t)$ are given functions of time and "$\cdot$" denotes here and later on differentiation with respect to time. We assume that $V(t) \geq a > 0$, for some fixed number a, which means that the point $\vec{Q}$ will not come to a standstill. Along L we have a length parameter s of which the positive direction is in the direction of motion of $\vec{Q}$.

The force field can be written as

$$\vec{F}(x, y, z, t) = \vec{f}(t) \ \delta(x - \xi(t))\delta(y - \eta(t))\delta(z - \zeta(t)) \ , \tag{1.6.2}$$

where $\vec{f}(t)$ is the singular force of magnitude $f(t) = |\vec{f}(t)|$. We suppose this force to be switched on at $t = t_0$, then $\vec{Q}$ is at $\vec{Q}(t_0)$, where we take $s = s_0$. For $t < t_0$ we assume the fluid to be at rest.

Substitution of (1.6.2) into (1.4.4) and (1.4.7) we find for the pressure and the velocity of the induced flow

$$p(x, y, z, t) = \frac{1}{4\pi} \frac{\vec{f} \cdot \vec{R}}{R^3} \ , \qquad \vec{R} = (x - \xi(t), y - \eta(t), z - \zeta(t)) \ , \tag{1.6.3}$$

$$\vec{v}(x, y, z, t) = -\frac{1}{4\pi\rho} \operatorname*{grad}_{x} \left\{ \int_{t_0}^{t} \frac{\vec{f}(\tau) \cdot \vec{R}}{R^3} \, d\tau \right\}$$
$$+ \frac{1}{\rho} \int_{t_0}^{t} \vec{f}(\tau) \ \delta(x - \xi(\tau))\delta(y - \eta(\tau))\delta(z - \zeta(\tau)) \, d\tau$$

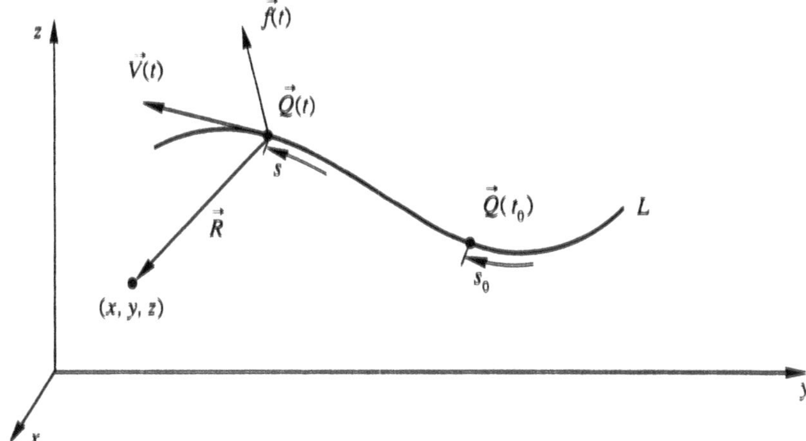

Fig. 1.6.1. The singular force $\vec{f}(t)$ moving along L.

$$= -\frac{1}{4\pi\rho} \int_{t_0}^{t} \left\{ \frac{\vec{f}(\tau)}{R^3} - 3\frac{\vec{R} \cdot (\vec{f}(\tau) \cdot \vec{R})}{R^5} \right\} d\tau$$

$$+ \frac{1}{\rho} \int_{t_0}^{t} \vec{f}(\tau)\, \delta(x - \xi(\tau))\, \delta(y - \eta(\tau))\, \delta(z - \zeta(\tau))\, d\tau \ , \quad (1.6.4)$$

respectively. The last integral in (1.6.4) is a singular contribution to the velocity field, which contains the rotation of the velocity field and which also makes the total flow free of divergence.

We will now use in the formulas (1.6.3) and (1.6.4), instead of the time, the length parameter s. When there is no risk of confusion we use the notation $m(t)$ or $m(s)$, which denotes that some quantity m can be considered to depend on t or on s. It does not mean that the dependence of m as a function of t or as a function of s is the same.

In principle, we can determine by (1.6.1) t as a function of s, $t = t(s)$ and the inverse function $s = s(t)$, where $\vec{Q}$ has at the time t the position s on L. Then we have at our disposal

$$\xi(s), \eta(s), \zeta(s), \vec{V}(s), \vec{f}(s), \vec{R}(s), dt = \frac{dt}{ds}\, ds = \frac{ds}{V(s)} \ . \quad (1.6.5)$$

Hence we find for (1.6.3) and for (1.6.4) when $\vec{Q}$ has arrived at s,

$$p(x, y, z, s) = \frac{1}{4\pi} \frac{\vec{f}(s) \cdot \vec{R}}{R^3} \ , \qquad s = s(t) \ , \quad (1.6.6)$$

$$\vec{v}(x,y,z,s)$$

$$= -\frac{1}{4\pi\rho} \operatorname*{grad}_{x} \left\{ \int_{s_0}^{s(t)} \frac{\vec{f}(\sigma)\cdot\vec{R}}{V(\sigma)\,R^3}\, d\sigma \right\}$$

$$+ \frac{1}{\rho} \int_{s_0}^{s(t)} \frac{\vec{f}(\sigma)}{V(\sigma)}\, \delta(x-\xi(\sigma))\, \delta(y-\eta(\sigma))\, \delta(z-\zeta(\sigma))\, d\sigma$$

$$= -\frac{1}{4\pi\rho} \int_{s_0}^{s(t)} \left\{ \frac{\vec{f}(\sigma)}{V(\sigma)R^3} - 3\frac{\vec{R}\cdot(\vec{f}(\sigma)\cdot\vec{R})}{V(\sigma)\,R^5} \right\} d\sigma$$

$$+ \frac{1}{\rho} \int_{s_0}^{s(t)} \frac{\vec{f}(\sigma)}{V(\sigma)}\, \delta(x-\xi(\sigma))\, \delta(y-\eta(\sigma))\, \delta(z-\zeta(\sigma))\, d\sigma \;, \tag{1.6.7}$$

where the integrations are along L.

From (1.6.7) and Section 1.5 it follows that the velocities can be described by means of an integral of divergenceless dipoles along L. The direction of these dipoles is in the direction of $\vec{f}$. Also it follows from Section 1.5 that we can induce approximately the velocity field by an integral of small flat closed vortex lines (Figure 1.5.2), of which the plane is perpendicular to $\vec{f}$ and which are around points of L where $\vec{f}$ has passed. The infinitesimal strength of such a vortex at the place σ at L which represents an element of length $d\sigma$ of L, follows from (1.5.8) when it is circular, it has the value

$$\frac{f(\sigma)\, d\sigma}{\pi\rho\, \varepsilon^2 V(\sigma)} \;, \tag{1.6.8}$$

when it is a square its strength follows from (1.5.9) as

$$\frac{f(\sigma)\, d\sigma}{4\rho\, \varepsilon^2 V(\sigma)} \;. \tag{1.6.9}$$

The extra factor $V(\sigma)$ in the denominators of (1.6.8) and (1.6.9) arises from the fact that in (1.6.7) we have $\vec{f}(\sigma)/V(\sigma)$ instead of $\vec{f}(\tau)$ as we had in (1.6.4). It shows that the faster the force moves along L, the smaller the representative vorticity has to be, hence the smaller the induced velocities become. This is the influence of the inertia of the fluid. From (1.6.8) or (1.6.9) we can, by an integration along L and using again the law of Biot and Savart (1.1.15), find an approximation of the velocity field (1.6.7) which tends to (1.6.7) for $\varepsilon \to 0$ in (1.6.8) or (1.6.9).

In each of the following two sections we discuss a special case of the formulas of this section. First we consider the case that the force $\vec{f}$ is tangent to L and second the case that $\vec{f}$ is perpendicular to L.

1.7. Singular Force Aligned with Its Velocity

We assume in this section that the force $\vec{f}(t)$ is tangent to L. Because we are interested in propulsion it seems natural to consider a force $\vec{f}(t)$ acting on the fluid in the negative s-direction. Then by "action equals reaction" the force on the system that causes $\vec{f}(t)$ is in the positive s-direction, hence it is a propulsive force in the direction of the velocity of the system. We assume again, as in the previous section, that the force $\vec{f}(t)$ is switched on at $s = s_0$ at the moment $t = t_0$ and that for $t < t_0$ the fluid is at rest.

First we consider the potential Φ of the velocity field outside of L, which by (1.6.7) has the form

$$\Phi(x,y,z,t) = -\frac{1}{4\pi\rho}\int_{s_0}^{s(t)} \frac{\vec{f}(\sigma)\cdot\vec{R}}{V(\sigma)\,R^3}\,d\sigma \ , \qquad (x,y,z) \notin L \ . \tag{1.7.1}$$

Because $\vec{f}(s)$ is tangent to L and $(d\xi/ds, d\eta/ds, d\zeta/ds)$ is the unit vector tangent to L, we can write the potential as

$$\Phi(x,y,z,t) = \frac{1}{4\pi\rho}\int_{s_0}^{s(t)} \frac{f(\sigma)}{V(\sigma)}\frac{d}{d\sigma}\left(\frac{1}{R}\right)\,d\sigma \ , \qquad (x,y,z) \notin L \ , \tag{1.7.2}$$

where we have to take $f(\sigma) < 0$, because $\vec{f}$ is in the negative s-direction. By partial integration we find from (1.7.2)

$$\Phi(x,y,z,t) = \frac{1}{4\pi\rho}\left\{ \left.\frac{f(\sigma)}{V(\sigma)R}\right|_{s_0}^{s(t)} - \int_{s_0}^{s(t)} \frac{1}{R}\frac{d}{d\sigma}\left(\frac{f(\sigma)}{V(\sigma)}\right)\,d\sigma \right\} \ ,$$

$$(x,y,z) \notin L \ . \tag{1.7.3}$$

Formula (1.7.3) has a simple interpretation. By (D.2.5) and (D.2.6) we see that the velocity potential

$$-\frac{1}{4\pi}\,\frac{1}{R} \ , \tag{1.7.4}$$

describes a source which produces a unit volume of fluid per unit of time. Hence (1.7.3), keeping in mind $f(\sigma) < 0$, can be considered as a source system which consists of three parts:

1. At $s = s_0$ we have a "starting source" of strength

$$\frac{f(s_0)}{\rho\,V(s_0)} \ , \tag{1.7.5}$$

2. At $s = s(t)$, which is the place of the propulsive force, we have a source of strength

$$\frac{-f(s(t))}{\rho\, V(s(t))} , \tag{1.7.6}$$

hence a sink. When at the place $s(t)$ the force is switched off, this sink will remain as an "ending sink".

3. Along L, from $s = s_0$ towards $s = s(t)$, we have a source-sink distribution of strength

$$\frac{1}{\rho}\frac{d}{ds}\left(\frac{f(s)}{V(s)}\right) , \tag{1.7.7}$$

per unit of length.

The velocity field outside of L can by (1.7.3) be written as

$$\vec{v}(x,y,z,t) = \text{grad}\, \Phi(x,y,z,t)$$

$$= \frac{1}{4\pi\rho}\left\{ \frac{-f(\sigma)\,\vec{R}}{V(\sigma)\,R^3}\Bigg|_{s_0}^{s(t)} + \int_{s_0}^{s(t)} \frac{\vec{R}}{R^3}\frac{d}{d\sigma}\left(\frac{f(\sigma)}{V(\sigma)}\right) d\sigma \right\} ,$$

$$(x,y,z) \notin L . \tag{1.7.8}$$

By the representation of the flow by means of a combination of a starting source, a sink at the place of the force and a source-sink distribution, we do not satisfy the continuity equation (1.4.2). However, as we stated already, this representation holds only for $(x,y,z,t) \notin L$ and we will show that the singular contribution to the velocity field, as given by the second integral at the right-hand side of (1.6.7) can repair this seeming shortcoming of the theory.

We look at the singular contribution for some arbitrary value $s = \tilde{s}$ at L. At this place we consider a small flat area S perpendicular to L (Figure 1.7.1) and calculate the flux through S by the mentioned singular part of the flow. We introduce at $s = \tilde{s}$ locally a new coordinate system $(s, \lambda_1, \lambda_2)$ which is rectangular and of which λ_1 and λ_2 are coordinates in S. Then we can write the contribution of the last integral of (1.6.7) as

$$\frac{1}{\rho}\int_{\tilde{s}-\varepsilon}^{\tilde{s}+\varepsilon} \frac{\vec{f}(\sigma)}{V(\sigma)}\,\delta(\sigma-\tilde{s})\delta(\lambda_1)\delta(\lambda_2)\, d\sigma , \tag{1.7.9}$$

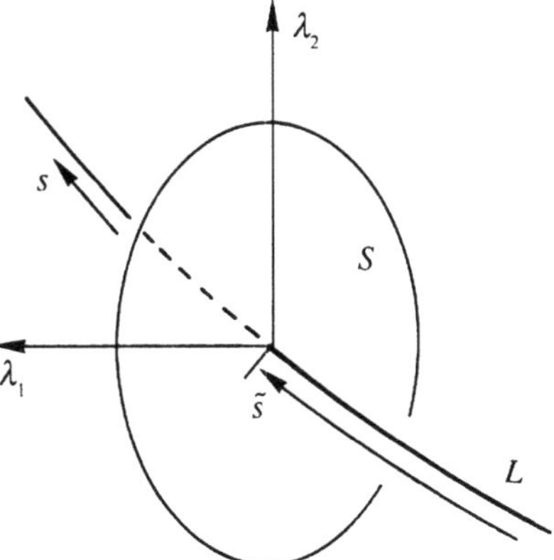

Fig. 1.7.1. Small area S, perpendicular to L.

where ε is some small positive number. Because $\vec{f}$ is perpendicular to S, the flux which is reckoned positive in the positive s-direction becomes

$$\frac{1}{\rho}\iint\limits_{S}\left\{\int\limits_{\tilde{s}-\varepsilon}^{\tilde{s}+\varepsilon}\frac{f(\sigma)}{V(\sigma)}\,\delta(\sigma-\tilde{s})\,d\sigma\right\}\delta(\lambda_1)\delta(\lambda_2)\,d\lambda_1\,d\lambda_2=\frac{1}{\rho}\,\frac{f(\tilde{s})}{V(\tilde{s})}\,, \qquad (1.7.10)$$

which is by $f(\tilde{s}) < 0$, in the negative s-direction. This is in agreement with the fact that $\vec{f}$ is a propulsive force and hence forms a backwards directed jet.

Equation (1.7.10) holds for arbitrarily small areas S, hence we have at L a jet of fluid of zero cross-section area, however with a non-zero volume transport per unit of time. Then the fluid velocity inside this jet has to be "infinite". The jet carries off the fluid from the sink (1.7.6) and feeds the source (1.7.5). It acts analogously with respect to the sources or sinks (1.7.7) distributed along L, because its flux changes in an appropriate way which is in correspondence with the strength of the distributed sources or sinks.

From (1.6.7) it is clear that the total flow pattern is the result of a string along L of divergenceless dipoles of the type (1.5.3), with their axes tangent to L. These are placed, loosely said, behind each other, the source of one "against the sink of the next", while they are of continuously varying strength. Then the combination of the sinks and sources of neighbouring dipoles give the source strength per unit of length given in (1.7.7). The dipoles at the two ends of the string, are "compensated only at one side" and by this give rise to the source (1.7.5) and the sink (1.7.6). Because the divergenceless dipoles are tangent to L, also their delta function velocity fields of the last term in (1.5.3) are placed in line and produce the jet along L.

Finally, we describe a vortex configuration which can induce the mentioned flow. To this end we return to (1.6.8) and the text above this formula. Then it is clear that we can approximate the flow by a tube along L of, for instance, circular vorticity of

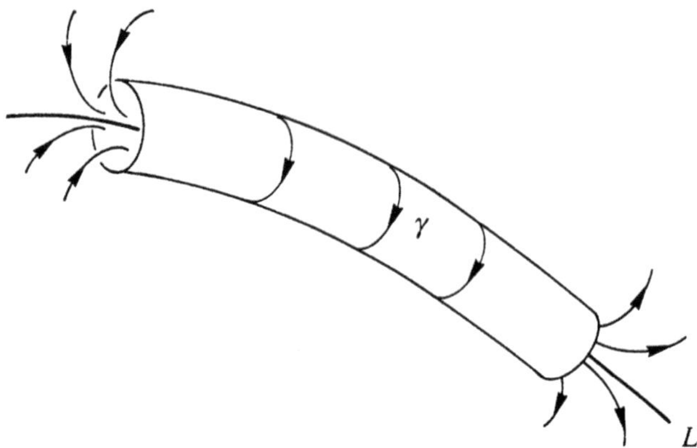

Fig. 1.7.2. Impression of the representation of the flow field by a vortex tube.

radius ε. This vorticity is perpendicular to L and its strength γ per unit of length in the s-direction is

$$\gamma(s) = \frac{f(s)}{\pi\rho\,\varepsilon^2 V(s)} \ . \tag{1.7.11}$$

In fact we have to take the limit $\varepsilon \to 0$ in order to induce the velocity field (1.7.8) outside L and the jet of which the flux equals (1.7.10) at L. We could also have taken a square cross section of the vortex tube by using (1.6.9) instead of (1.6.8) for the determination of the vorticity strength.

An impression of this representation of the velocity field is given in Figure 1.7.2. The direction of γ is given with a right-hand screw by an arrow, which agrees with the backwards directed slip stream of the propulsive force. From this representation it is also clear that the divergence of the flow is zero, because we only use vorticity.

1.8. Singular Force Perpendicular to Its Velocity

When the singular force is perpendicular to its velocity, no partial integration can be carried out, as the one from (1.7.2) towards (1.7.3). So we refer to (1.6.6) and (1.6.9) for the induced pressure and the velocity field. The flow is induced by a string of divergenceless dipoles along L, with their axes in the direction of $\vec{f}$, hence perpendicular to L.

We can, however, also for this case give an approximate description of the induced velocity field by means of vorticity. For the sake of simplicity we now take the representation of a divergenceless dipole by means of a small square vortex line (A_1, A_2, A_3, A_4), as is drawn in Figure 1.8.1 (a). The sides A_2A_3 and A_1A_4 are paralell to L. When the force moves with velocity V over a small distance $d\sigma \ll 2\varepsilon$ along L, the strength γ of the vorticity along the square (A_1, A_2, A_3, A_4) and which

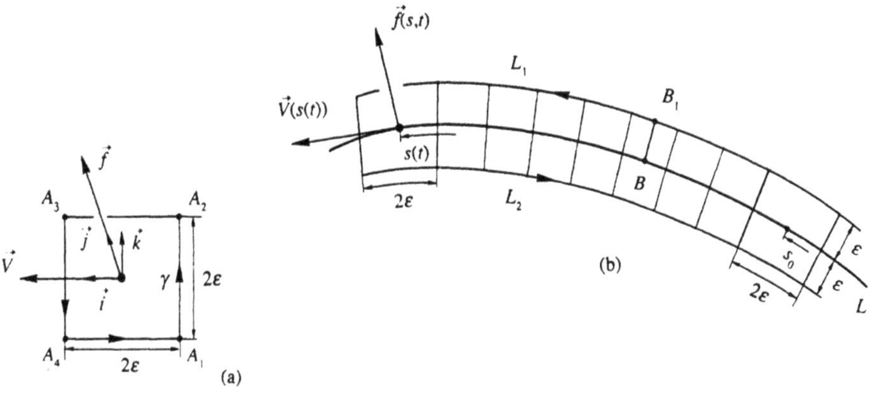

Fig. 1.8.1. (a) Square vortex line and unit vectors $\vec{\imath}, \vec{\jmath}, \vec{k}$. (b) Total vorticity induced by $\vec{f}$.

is left behind, is given by (1.6.9), hence

$$\gamma = \frac{f(\sigma)\, d\sigma}{4\rho\, \varepsilon^2 V(\sigma)} \ . \tag{1.8.1}$$

The rotation of the fluid around γ is given by an arrow with a right-hand screw in Figure 1.8.1 (a).

In Figure 1.8.1 (b) the resulting vorticity of the velocity field is drawn, when the force has moved from $s = s_0$ to $s = s(t)$. Five regions of vorticity can be distinguished:

(1) A neighbourhood of $s = s_0$ where the force is switched on. When the force moves from s_0 to $s_0 + 2\varepsilon$, the side A_1A_2 translates over the interval $(s_0 - \varepsilon, s_0 + \varepsilon)$ and sweeps a small square with sides equal to 2ε. Here by (1.8.1) vorticity is left there of approximate strength $f(s_0)/4\rho\, \varepsilon^2 V(s_0)$ per unit of length in the s-direction. For the total vorticity which remains behind in that square we then find

$$2\varepsilon \cdot \frac{f(s_0)}{4\rho\, \varepsilon^2 V(s_0)} = \frac{f(s_0)}{2\rho\, \varepsilon V(s_0)} \ , \tag{1.8.2}$$

which has a length 2ε, perpendicular to L. This is the "starting vorticity". Its rotation is coupled with a right-hand screw to the direction of $\vec{k}$.

(2) A neighbourhood of $s = s(t)$ where the moving force is present. When the force moves from $s(t) - 2\varepsilon$ to $s(t)$, the side A_3A_4 translates over the interval $(s(t) - \varepsilon, s(t) + \varepsilon)$ and sweeps also a small square with sides equal to 2ε. Hence vorticity is left there of approximate strength $f(s(t))/4\rho\, \varepsilon^2 V(s(t))$ per unit of length in the s-direction. For the total vorticity which remains in that square we obtain

$$2\varepsilon \cdot \frac{f(s(t))}{4\rho\,\varepsilon^2 V(s(t))} = \frac{f(s(t))}{2\rho\,\varepsilon V(s(t))}\,, \tag{1.8.3}$$

which has a length 2ε, perpendicular to L. This is the "bound vorticity" directly coupled to the instantaneous force action. Its rotation is connected with a right-hand screw to the direction of $-\vec{k}$. In essence, this is the well-known theorem of Joukowski. It is clear that when the force is switched off at time t, this vorticity remains and can be called "ending vorticity".

(3) The line L_1 passed through by the side A_2A_3 when f moves from s_0 to $s(t)$. Consider at L some point B with $s = \tilde{s}$ such that $s_0 + \varepsilon < \tilde{s} < s(t) - \varepsilon$ and at L_1 the point B_1, which is the projection of B on L_1. Then the strength of the vortex along L_1 at the point B_1 equals the sum of the vortex strengths of those A_2A_3 which contain B_1. Hence we find by taking in (1.8.1) $d\sigma = 2\varepsilon$, for this strength approximately

$$\frac{f(\tilde{s})}{2\rho\,\varepsilon V(\tilde{s})}\,, \tag{1.8.4}$$

which is the strength of the "tip vortex" lying along L_1.

(4) The line L_2, described by the side A_1A_4, analogously as in (3). Here we have a tip vortex of the same strength (1.8.4) but of opposite rotation.

(5) The region in between L_1 and L_2. Here we have distributed vorticity perpendicular to L, its strength is per unit of length in the s-direction. This vorticity we consider again for $s = \tilde{s}$. Its strength follows from the side A_3A_4 when the midpoint of the square $A_1A_2A_3A_4$ is at $\tilde{s} - \varepsilon$ and from the side A_1A_2 when the midpoint is at $\tilde{s} + \varepsilon$. Using (1.8.1) we find for the strength per unit of length of the distributed vorticity, hence by dividing the sum of the strengths of the mentioned two sides by $d\sigma$

$$\frac{1}{4\rho\,\varepsilon^2}\left\{\frac{f(\tilde{s}+\varepsilon)}{V(\tilde{s}+\varepsilon)} - \frac{f(\tilde{s}-\varepsilon)}{V(\tilde{s}-\varepsilon)}\right\} \approx \frac{1}{2\rho\,\varepsilon}\,\frac{d}{ds}\left\{\frac{f(s)}{V(s)}\right\}\bigg|_{s=\tilde{s}}. \tag{1.8.5}$$

This vorticity is shed by the unsteady behaviour of the bound vorticity of (2). It makes the total vorticity field free of divergence.

The rotational direction of the above-mentioned vorticities follows from Figure 1.8.1 (a) where, as we mentioned before, the vorticity with a right-hand screw is drawn. The sign of the distributed vorticity follows of course from the sign of the derivative in (1.8.5).

When we calculate the induced velocity at some point at a distance from L which is large with respect to ε, we can within the accuracy of the theory replace the vorticity at the regions (1) and (2) each by a concentrated vortex line of strength (1.8.2) or

(1.8.3) and of length 2ε. Again we have, in order to find the exact linearized velocity field, to take the limit $\varepsilon \to 0$.

Summarizing we have found the following: (1) a concentrated starting vortex line at $s = s_0$ of length 2ε and of strength (1.8.2); (2) a concentrated lifting line of length 2ε and of strength (1.8.3); (3) and (4) two tip vortices of opposite strength determined by (1.8.4) which are shed from the tips of the lifting line; (5) distributed vorticity in between the tip vorticities of strength (1.8.5), which is caused by the unsteady behaviour of the lifting line.

We now calculate by means of the law of Biot and Savart the induced velocity field. Because a vortex field is free of divergence, hence it can be considered to consist of "closed vortex lines", it follows that we can use the differential form of (1.1.15)

$$d\vec{v} = \frac{1}{4\pi}\Gamma\frac{\vec{l}\times\vec{R}}{R^3}\,d\sigma\ , \tag{1.8.6}$$

where $\vec{l}$ is a unit vector coupled to the vorticity by a right-hand screw and $d\sigma$ is the small length of the vortex.

The velocities induced by the starting vortex and the bound vortex are by (1.8.2) and (1.8.3),

$$\frac{1}{4\pi}\frac{f(s_0)}{2\rho\,\varepsilon V(s_0)}\frac{\vec{k}\times\vec{R}}{R^3}2\varepsilon = \frac{1}{4\pi}\frac{f(s_0)}{\rho\,V(s_0)}\frac{\vec{k}\times\vec{R}}{R^3}\ , \tag{1.8.7}$$

$$-\frac{1}{4\pi}\frac{f(s(t))}{\rho\,V(s(t))}\frac{\vec{k}\times\vec{R}}{R^3}\ , \tag{1.8.8}$$

respectively, where $\vec{R}$ has to be chosen in accordance with Figure (1.1.1).

The unit vectors along L_1 and L_2 have the form

$$\left(\vec{\imath}\pm\varepsilon\frac{d\vec{k}}{ds}\right)\cdot|\vec{\imath}\pm\varepsilon\frac{d\vec{k}}{ds}|^{-1}\ , \tag{1.8.9}$$

here and in the next formulas upper signs refer to L_1 and lower ones to L_2. Then the induced velocity by the vorticity along L_1 and L_2 becomes

$$\pm\frac{1}{4\pi\rho}\frac{1}{2\varepsilon}\int_{s_0}^{s(t)}\frac{f(s)}{V(s)}\left\{\frac{\left(\vec{\imath}\pm\varepsilon\frac{d\vec{k}}{ds}\right)}{|\vec{\imath}\pm\varepsilon\frac{d\vec{k}}{ds}|}\times\frac{(\vec{R}\mp\varepsilon\vec{k})}{|R\mp\varepsilon\vec{k}|^3}\right\}|\vec{\imath}\pm\varepsilon\frac{d\vec{k}}{ds}|\,ds\ . \tag{1.8.10}$$

Adding these two velocities and taking the limit $\varepsilon \to 0$, we obtain

$$\frac{1}{4\pi\rho}\int_{s_0}^{s(t)}\frac{f(s)}{V(s)}\frac{d}{d\lambda}\left\{\left(i+\lambda\frac{d\vec{k}}{ds}\right)\times\frac{(\vec{R}-\lambda\vec{k})}{|\vec{R}-\lambda\vec{k}|}\right\}\Big|_{\lambda=0}ds$$

$$= \frac{1}{4\pi\rho} \int_{s_0}^{s(t)} \frac{f(s)}{V(s)} \left[3\frac{(\vec{R}\cdot\vec{k})}{R^5} \cdot \vec{\imath} \times \vec{R} + \frac{1}{R^3} \frac{d\vec{k}}{ds} \times \vec{R} - \frac{1}{R^3} \vec{\imath} \times \vec{k} \right] ds \ . \qquad (1.8.11)$$

Finally, the velocity induced by the distributed vorticity (1.8.5) becomes

$$\frac{1}{4\pi\rho} \int_{s_0}^{s(t)} \frac{d}{ds} \left(\frac{f(s)}{V(s)} \right) \vec{k} \times \frac{\vec{R}}{R^3} \, ds \ . \qquad (1.8.12)$$

It can be shown that by adding the four parts (1.8.7), (1.8.8), (1.8.11) and (1.8.12) we find again the velocity field (1.6.7) without the singular parts, hence outside the line L. To this end a partial integration has to be carried out in (1.8.11) in order to get rid of the derivative of $\vec{k}$ with respect to s.

1.9. Reference Surface and Planform of Lifting Surface

In this section we consider a basic concept of linearized lifting surface theory, namely the reference surface. Herewith we can describe the type of motion which a lifting surface is allowed to perform, when we demand its induced velocities to be sufficiently small, so that a linearized theory will be valid.

We consider an inertial Cartesian coordinate system. The fluid comes in from $x = -\infty$ as a homogeneous parallel flow with velocity V in the positive x-direction. Suppose we have a flexible impermeable surface W of zero thickness and of finite dimensions. Then we ask first for the condition which ensures that W does not induce any disturbance velocity $\vec{w}$, when it moves through the fluid.

We assume W to be part of a geometrical surface

$$H(x, y, z, t) = 0 \ . \qquad (1.9.1)$$

In order that the fluid does not penetrate W we have the condition (1.2.16)

$$\frac{\partial H}{\partial t} + \vec{v} \cdot \operatorname{grad} H = \frac{\partial H}{\partial t} + (V + w_x)\frac{\partial H}{\partial x} + w_y \frac{\partial H}{\partial y} + w_z \frac{\partial H}{\partial z} = 0 \ ,$$
$$(x, y, z) \in H(x, y, z, t) = 0 \ . \qquad (1.9.2)$$

From (1.9.2) it follows that when W does not disturb the incoming flow, hence when $\vec{w} = (w_x, w_y, w_z) = 0$, then

$$\frac{\partial H}{\partial t} + V\frac{\partial H}{\partial x} = 0 \ . \qquad (1.9.3)$$

The general solution of this linear partial differential equation is

$$H = H(x - Vt, y, z) \ , \qquad (1.9.4)$$

where H is an "arbitrary" function of the three variables $x - Vt$, y and z. The meaning of (1.9.4) is that the surface $H = 0$ floats while it remains undeformed, with the velocity V in the positive x-direction. Hence such a surface has to be in rest with respect to the fluid. The flexible W has to glide along this surface. Then because W has no thickness, it is clear that it will not induce disturbance velocities.

Next we consider a lifting surface $\tilde{W}$ of finite dimension and zero thickness. In order that $\tilde{W}$ can develop a lift force or other force actions, it has to disturb the incoming parallel flow, in other words it has to induce disturbance velocities $\vec{w}$. When these velocities have to remain small, say $O(\varepsilon)$, it is clear from the above that it is necessary that $\tilde{W}$ moves in a close neighbourhood of some geometrical surface $H(x - Vt, y, z) = 0$, which is in rest with respect to the fluid. The close neighbourhood is not only with respect to the distances of the points $\tilde{W}$ from $H = 0$, but also the slopes and curvatures of $\tilde{W}$ are not allowed to deviate too much from those of $H = 0$. We will not give a rigorous mathematical definition of a close neighbourhood because from the technical point of view this will be sufficiently clear.

In other words $\tilde{W}$ has to remain in a close neighbourhood of its projection on $H(x - Vt, y, z) = 0$. This projection glides along $H = 0$ when $\tilde{W}$ is moving through the fluid, linking up with the first part of this section it will be denoted by W. Then we call the surface $H = 0$ the *reference surface* of $\tilde{W}$ and W is called the *planform* of $\tilde{W}$.

In the linearized theory we will in general satisfy the boundary conditions for the fluid velocity field not on the lifting surface $\tilde{W}$ itself, but on its planform W, which moves along the reference surface. Then we introduce errors of $O(\varepsilon^2)$, when ε is the small parameter of linearization, which limits the allowed deviations of the lifting surface from its reference surface. This magnitude of the errors is permitted because the linearized equations of motion are accurate only upto and including $O(\varepsilon)$. *From this it follows that the external forces exerted by the wing $\tilde{W}$ at the fluid can be considered to be present at the planform W of $\tilde{W}$.*

Suppose the lifting surface $\tilde{W}$ is represented as a part, of finite dimensions of the surface

$$\tilde{W} \subset G(x, y, z, t) = H(x - Vt, y, z) + \varepsilon h(x, y, z, t) = 0 \ , \tag{1.9.5}$$

where $\varepsilon h(x, y, z, t)$ determines the small deviations of $\tilde{W}$ from its planform $W \subset H = 0$. Then we have to replace H in (1.9.2) by $G = H + \varepsilon h$. Neglecting in the resulting equation quantities of $O(\varepsilon^2)$, we find for the linearized boundary condition

$$\vec{w} \cdot \operatorname{grad} H(x - Vt, y, z) = -\varepsilon \left(\frac{\partial h}{\partial t} + V \frac{\partial h}{\partial x} \right) \ ,$$
$$(x, y, z) \in W \subset H(x - Vt, y, z) = 0 \ . \tag{1.9.6}$$

A simple example of the foregoing is

$$H(x - Vt, y, z) = y \ , \qquad G(x, y, z, t) = y + \varepsilon h(x, z) \ . \tag{1.9.7}$$

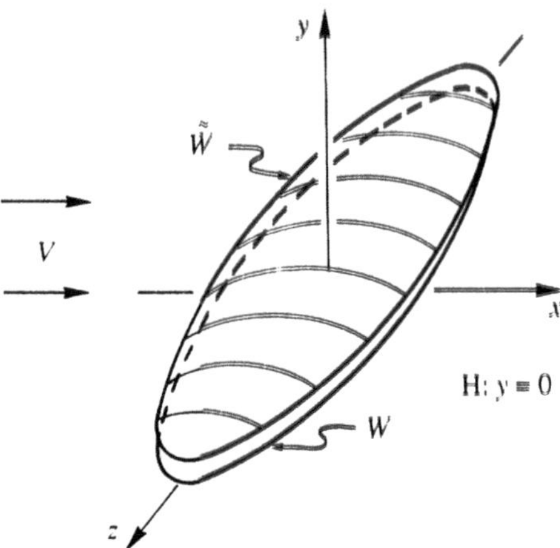

Fig. 1.9.1. Wing $\tilde{W}$ in parallel flow.

Then we have the steady case of a wing $\tilde{W}$ in a parallel flow (Figure 1.9.1). The boundary condition (1.9.6) assumes the form

$$w_y = -\varepsilon V \frac{\partial h}{\partial x} , \qquad (x, y, z) \in W \subset y = 0 , \tag{1.9.8}$$

which has to be satisfied at the projection W of $\tilde{W}$ on the reference surface $H = y = 0$.

An important case for later on will be a screw blade. Here we assume that the screw is rotating with fixed values of the x-coordinate. Hence the incoming velocity is V. This can be handled more easily in a cylindrical coordinate system (x, r, φ) (Appendix B). Using the representation of the gradient in cylindrical coordinates (B.2.7) and taking the physical components (below (B.1.10)) of the disturbance velocity $\vec{w} = (w_x, w_r, w_\varphi) = 0$, we find for the reference surface $H(x, r, \varphi, t) = 0$, that H has to satisfy

$$\frac{\partial H}{\partial t} + \vec{v} \cdot \operatorname{grad} H = \frac{\partial H}{\partial t} + V \frac{\partial H}{\partial x} = 0 . \tag{1.9.9}$$

The general solution of this equation is

$$H = H(x - Vt, r, \varphi) , \tag{1.9.10}$$

which could be expected, because also here $H = 0$ has to float with the incoming flow. Solving for $(x - Vt)$, we can write $H = 0$ as

$$x - Vt = M_1(r, \varphi) . \tag{1.9.11}$$

Another demand that we will put on the reference surface of a rigid screw blade is that the reference surface rotates around the x-axis, say with rotational velocity ω. Such a surface has the representation $H_2(x, r, \varphi - \omega t) = 0$ or

$$\varphi - \omega t = M_2(x, r) . \tag{1.9.12}$$

Because (1.9.11) and (1.9.12) represent the same surface we can eliminate t and find

$$\omega x - V\varphi = \omega M_1(r,\varphi) - VM_2(x,r) \ , \qquad \forall(x,r,\varphi) \ . \tag{1.9.13}$$

Hence $M_1(r,\varphi)$ has to be linear with respect to φ and $M_2(x,r)$ has to be linear with respect to x. Because the left-hand side of (1.9.13) is independent of r it is easily found that

$$M_1(r,\varphi) = -\frac{V}{\omega}\varphi - Vg(r) \ , \qquad M_2(x,r) = -\frac{\omega}{V}x - \omega g(r) \ , \tag{1.9.14}$$

where $g(r)$ is still an arbitrary function. Hence by (1.9.11) or (1.9.12), we find for the most general reference surface of a screw blade

$$H(x,r,\varphi,t) = \varphi + a(x - Vt) + g(r) = 0 \ , \qquad a = \frac{\omega}{V} \ . \tag{1.9.15}$$

Such a surface arises when we take some curved line $x = 0$, $\varphi = -g(r)$, rotate this line with the constant angular velocity ω in the negative φ-direction around the x-axis and translate it at the same time with the constant velocity V in the positive x-direction. The curved line is called the generator line of the helicoidal surface (1.9.15).

By the foregoing we can represent any rigid screw blade $\tilde{W}$ of zero thickness in the linear lifting surface theory as

$$G(x,r,\varphi,t) = \varphi + a(x - Vt) + g(r) + \varepsilon h(x,r) = 0 \ , \qquad a = \frac{\omega}{V} \ , \tag{1.9.16}$$

where $\varepsilon h(x,r)$ determines the small deviations of the blade from its reference surface. Here h needs to depend only on two coordinates x and r, as can easily be seen.

The linearized boundary condition for the disturbance velocity field $\vec{w}$, induced by the screw blade is found by replacing H in the first expression of (1.9.9) by G from (1.9.16) and neglecting terms of $O(\varepsilon^2)$, we obtain

$$aw_x + w_r g'(r) + \frac{w_\varphi}{r} = -\varepsilon V \frac{\partial h}{\partial x} \ . \tag{1.9.17}$$

This condition can be satisfied on the planform $W \subset H = 0$ of $\tilde{W}$.

1.10. Formulation of Lifting Surface Theory, Velocity Dipole Layer

The theory we develop is in first instance valid for a flexible lifting surface $\tilde{W}$ which moves in one way or another in the neighbourhood of a general reference surface H. By the foregoing section this reference surface is at rest with respect to the undisturbed fluid and we assume the fluid to be at rest with respect to the Cartesian coordinate (x, y, z). Our derivations follow the line of thought as developed in [60]. However, there the mathematical theory of distributions is rigorously used to

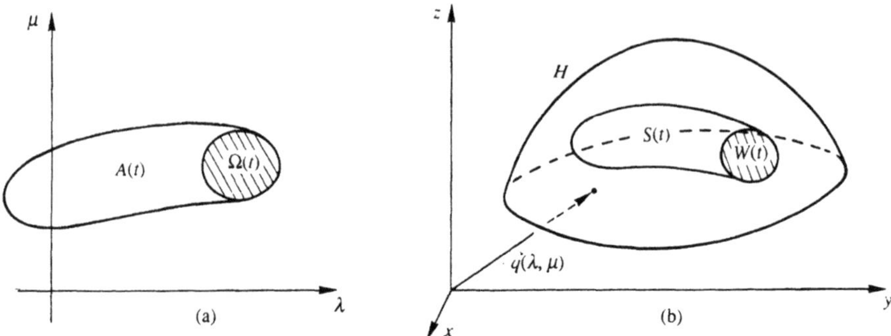

Fig. 1.10.1. (a) Parameter plane (λ, μ); (b) space with reference surface H and planform $W(t)$.

describe force fields concentrated at surfaces, here we use a more classical applied mathematics approach.

The reference surface H will be given in parameter form

$$H = \{\vec{q}(\lambda, \mu) = (q_1(\lambda, \mu), q_2(\lambda, \mu), q_3(\lambda, \mu))\} , \tag{1.10.1}$$

where $\vec{q}(\lambda, \mu)$ is some given vector-valued function of the parameters λ and μ, of which the components are assumed to depend analytically on λ and μ. These parameters form a curvilinear coordinate system on H and a Cartesian system in the parameter plane. For simplicity we assume that H does not intersect itself.

An element of area dS of H is given by

$$dS = Z(\lambda, \mu)\, d\lambda\, d\mu \ , \qquad Z(\lambda, \mu) = \left| \frac{\partial \vec{q}}{\partial \lambda} \times \frac{\partial \vec{q}}{\partial \mu} \right| \tag{1.10.2}$$

and we assume that a constant $c > 0$ exists such that

$$Z(\lambda, \mu) \geq c \ . \tag{1.10.3}$$

The unit normal $\vec{n}(\lambda, \mu)$ at the reference surface H can be written as

$$\vec{n}(\lambda, \mu) = \left(\frac{\partial \vec{q}}{\partial \lambda} \times \frac{\partial \vec{q}}{\partial \mu} \right) \cdot Z^{-1}(\lambda, \mu) \ . \tag{1.10.4}$$

The planform $W(t)$ which, as in the foregoing section, is the projection of the lifting surface $\tilde{W}$ at H, moves by definition along H. It is assumed to be the image of some given, possibly deforming finite area $\Omega(t)$ (Figure 1.10.1 (a)), moving in the parameter plane

$$W(t) = \{\vec{q}(\lambda, \mu) \ ; \ (\lambda, \mu) \in \Omega(t)\} \ . \tag{1.10.5}$$

We now consider the external force field per unit of area, which represents the action of the lifting surface $\tilde{W}$. This force field $\vec{F}(\lambda, \mu, t)$ can be assumed, as we discussed in the previous section, within the accuracy of the linearized theory, to act at the planform W and is normal to it.

$$\vec{F}(\lambda,\mu,t) = f(\lambda,\mu,t)\,\vec{n}(\lambda,\mu) \ , \qquad (\lambda,\mu) \in \Omega(t) \ ,$$

$$\vec{F}(\lambda,\mu,t) = 0 \ , \qquad (\lambda,\mu) \notin \Omega(t) \ . \tag{1.10.6}$$

The vorticity created by an external force field remains (Section 1.4), within the realm of a linear theory at the place where it is formed. So the region $S(t)$ where the vorticity of the force field (1.10.6) is present is at H and $S(t)$ can be represented by

$$S(t) = \cup_\tau \left\{W(\tau); t_0 \le \tau \le t\right\} \ , \tag{1.10.7}$$

where t_0 is the time at which the wing $\tilde{W}$ started to move or at which the external force field became active. We can also describe $S(t)$ by

$$S(t) = \{\vec{q}(\lambda,\mu) \ ; \ (\lambda,\mu) \in A(t) = \cup_\tau\{\Omega(\tau) \ ; \ t_0 \le \tau \le t\}\} \ , \tag{1.10.8}$$

where $A(t)$ is the region passed through by $\Omega(t)$.

Using (1.6.4), (1.10.2) and (1.10.6) and integrating, we find for the velocity field $\vec{v}(\vec{x},t)$ outside S

$$\vec{v}(\vec{x},t) = -\frac{1}{4\pi\rho} \operatorname*{grad}_x$$

$$\int_{t_0}^{t} \left\{ \iint_{\Omega(\tau)} \frac{f(\lambda,\mu,\tau)\vec{n}(\lambda,\mu)\cdot\vec{R}(\vec{x}-\vec{q}(\lambda,\mu))\,Z(\lambda,\mu)\,d\lambda\,d\mu}{R^3(\vec{x}-\vec{q}(\lambda,\mu))} \right\} d\tau \ . \tag{1.10.9}$$

In (1.10.9) we left the second term in the expressions of (1.6.4) out of consideration, because we only consider velocities outside S, hence outside the region where the force field has been acting or still acts. We introduce

$$g(\lambda,\mu,t) = \int_{t_0}^{t} f(\lambda,\mu,\tau)\,d\tau \ , \tag{1.10.10}$$

which is the total action upto and including t of the force field at a certain point $\vec{q}(\lambda,\mu)$ of S. It follows from Section 1.5 that this integral equals the strength of the pressure dipole layer at that place and at the moment t. Then $\vec{v}(\vec{x},t)$ becomes

$$\vec{v}(\vec{x},t) = -\frac{1}{4\pi\rho} \operatorname*{grad}_x$$

$$\iint_{A(t)} \frac{g(\lambda,\mu,t)\,\vec{n}(\lambda,\mu)\cdot\vec{R}(\vec{x}-\vec{q}\,(\lambda,\mu))\,Z(\lambda,\mu)}{R^3(\vec{x}-\vec{q}\,(\lambda,\mu))}\,d\lambda\,d\mu \ , \tag{1.10.11}$$

where now the integration with respect to λ and μ has to be carried out over the region $A(t)$ passed through by $\Omega(\tau)$, $t_0 \leq \tau \leq t$. It follows from Section 1.5 that (1.10.11) describes the velocity field outside S by means of a layer of classical velocity dipoles at $A(t)$.

Now we bring in (1.10.11) the operation $\underset{x}{\text{grad}}$ under the sign of integration and consider the resulting integrand which, by temporarily neglecting $g(\lambda, \mu, t)$, using (1.10.4) and an obvious reduction can be written as

$$- \underset{x}{\text{grad}} \left\{ \underset{x}{\text{grad}} \frac{1}{R(\vec{x} - \vec{q})} \cdot \left(\frac{\partial \vec{q}}{\partial \lambda} \times \frac{\partial \vec{q}}{\partial \mu} \right) \right\}. \tag{1.10.12}$$

Using the vector identity (C.13) with $\varphi = R^{-1}(\vec{x} - \vec{q})$, $\vec{a} = \partial \vec{q}/\partial \lambda$ and $\vec{b} = \partial \vec{q}/\partial \mu$, we can write (1.10.12) as

$$\begin{aligned}
&- \left(\frac{\partial \vec{q}}{\partial \mu} \cdot \underset{x}{\text{grad}} \right) \left\{ \underset{x}{\text{grad}} \frac{1}{R(\vec{x} - \vec{q})} \times \frac{\partial \vec{q}}{\partial \lambda} \right\} \\
&\quad + \left(\frac{\partial \vec{q}}{\partial \lambda} \cdot \underset{x}{\text{grad}} \right) \left\{ \underset{x}{\text{grad}} \frac{1}{R(\vec{x} - \vec{q})} \times \frac{\partial \vec{q}}{\partial \mu} \right\} \\
&= \left(\frac{\partial \vec{q}}{\partial \mu} \cdot \underset{q}{\text{grad}} \right) \left\{ \underset{x}{\text{grad}} \frac{1}{R(\vec{x} - \vec{q})} \times \frac{\partial \vec{q}}{\partial \lambda} \right\} \\
&\quad - \left(\frac{\partial \vec{q}}{\partial \lambda} \cdot \underset{q}{\text{grad}} \right) \left\{ \underset{x}{\text{grad}} \frac{1}{R(\vec{x} - \vec{q})} \times \frac{\partial \vec{q}}{\partial \mu} \right\} \\
&= \frac{\partial}{\partial \mu} \left\{ \underset{x}{\text{grad}} \frac{1}{R(\vec{x} - \vec{q})} \times \frac{\partial \vec{q}}{\partial \lambda} \right\} \\
&\quad - \frac{\partial}{\partial \lambda} \left\{ \underset{x}{\text{grad}} \frac{1}{R(\vec{x} - \vec{q})} \times \frac{\partial \vec{q}}{\partial \mu} \right\} .
\end{aligned} \tag{1.10.13}$$

In the second expression of (1.10.13) the operators $(\partial \vec{q}/\partial \lambda \cdot \underset{q}{\text{grad}})$ and $(\partial \vec{q}/\partial \mu \cdot \underset{q}{\text{grad}})$ are allowed to act also on $\partial \vec{q}/\partial \lambda$ and $\partial \vec{q}/\partial \mu$; this still gives the correct result because

$$\left(\frac{\partial \vec{q}}{\partial \mu} \cdot \underset{q}{\text{grad}} \right) \frac{\partial \vec{q}}{\partial \lambda} = \frac{\partial}{\partial \mu} \frac{\partial \vec{q}}{\partial \lambda} = \frac{\partial}{\partial \lambda} \frac{\partial \vec{q}}{\partial \mu} = \left(\frac{\partial \vec{q}}{\partial \lambda} \cdot \underset{q}{\text{grad}} \right) \frac{\partial \vec{q}}{\partial \mu} , \tag{1.10.14}$$

hence the extra terms which occur by the differentiations with respect to λ and μ cancel each other. By (1.10.13) we can write (1.10.11) as

$$\begin{aligned}
\vec{v}(\vec{x}, t) = \frac{1}{4\pi\rho} \iint\limits_{A(t)} g(\lambda, \mu, t) &\left[\frac{\partial}{\partial \lambda} \left\{ \underset{x}{\text{grad}} \frac{1}{R(\vec{x} - \vec{q})} \times \frac{\partial \vec{q}}{\partial \mu} \right\} \right. \\
&\left. - \frac{\partial}{\partial \mu} \left\{ \underset{x}{\text{grad}} \frac{1}{R(\vec{x} - \vec{q})} \times \frac{\partial \vec{q}}{\partial \lambda} \right\} \right] d\lambda \, d\mu .
\end{aligned} \tag{1.10.15}$$

We now choose some point $\vec{q}_s = \vec{q}\,(\lambda_s, \mu_s) \in$ interior $S(t)$ (1.10.7), hence $(\lambda_s, \mu_s) \in$ interior $A(t)$ and erect the unit normal $\vec{n}_s$ at that point

$$\vec{n}_s = \left(\frac{\partial \vec{q}}{\partial \lambda} \times \frac{\partial \vec{q}}{\partial \mu}\right)\Bigg|_{\substack{\lambda=\lambda_s\\ \mu=\mu_s}} \cdot Z^{-1}(\lambda_s, \mu_s) \ . \tag{1.10.16}$$

In general, when the index "s" is attached to a quantity which depends on λ and μ it means that this quantity is considered for the values $\lambda = \lambda_s$ and $\mu = \mu_s$. Consider a point $\vec{x}$ outside $S(t)$ and close to $\vec{q}_s$. Then we define a "pseudo normal component" v_n of $\vec{v}$ by

$$v_n(\vec{x}, t) = (\vec{v}(\vec{x}, t) \cdot \vec{n}_s) \ , \tag{1.10.17}$$

which does not have a physical meaning, only when $\vec{x} \to \vec{q}_s$, it will tend to the velocity component normal to S at $\vec{q}_s$.

Using (1.10.15), (1.10.17) and (C.2) we obtain

$$4\pi\rho\, v_n(\vec{x}, t) = \iint\limits_{A(t)} g(\lambda, \mu, t) \left[\frac{\partial}{\partial \lambda}\left\{\operatorname*{grad}_x \frac{1}{R(\vec{x} - \vec{q})} \cdot \left(\frac{\partial \vec{q}}{\partial \mu} \times \vec{n}_s\right)\right\}\right.$$
$$\left. - \frac{\partial}{\partial \mu}\left\{\operatorname*{grad}_x \frac{1}{R(\vec{x} - \vec{q})} \cdot \left(\frac{\partial \vec{q}}{\partial \lambda} \times \vec{n}_s\right)\right\}\right] d\lambda\, d\mu \ , \tag{1.10.18}$$

because $\vec{n}_s$ is a constant vector. In order to abbreviate this expression we introduce

$$\begin{aligned} H_1(\vec{x}, \vec{q}) &= \operatorname*{grad}_x \frac{1}{R(\vec{x} - \vec{q})} \cdot \left(\frac{\partial \vec{q}}{\partial \mu} \times \vec{n}_s\right) \\ &= \frac{-(\vec{x} - \vec{q})}{R^3(\vec{x} - \vec{q})} \cdot \left(\frac{\partial \vec{q}}{\partial \mu} \times \vec{n}_s\right) \ , \end{aligned} \tag{1.10.19}$$

$$\begin{aligned} H_2(\vec{x}, \vec{q}) &= -\operatorname*{grad}_x \frac{1}{R(\vec{x} - \vec{q})} \cdot \left(\frac{\partial \vec{q}}{\partial \lambda} \times \vec{n}_s\right) \\ &= \frac{(\vec{x} - \vec{q})}{R^3(\vec{x} - \vec{q})} \cdot \left(\frac{\partial \vec{q}}{\partial \lambda} \times \vec{n}_s\right) \ . \end{aligned} \tag{1.10.20}$$

Then (1.10.18) becomes

$$4\pi\rho\, v_n(\vec{x}, t) = \iint\limits_{A(t)} g(\lambda, \mu, t) \left\{\frac{\partial H_1}{\partial \lambda}(\vec{x}, \vec{q}) + \frac{\partial H_2}{\partial \mu}(\vec{x}, \vec{q})\right\} d\lambda\, d\mu \ . \tag{1.10.21}$$

1.11. Reformulation of Velocity Component Normal to S(t)

In this section, we derive a representation of the "normal" velocity component (1.10.21) which will be useful for taking the limit value $\vec{x} \to \vec{q}_s$.

We divide the region of integration $A(t)$ of (1.10.21) into two parts. First a "small" compact set B around the point (λ_s, μ_s) and second, the remaining part $A(t)\backslash B$. When the point $\vec{x}$ in (1.10.21) tends to $\vec{q}_s = \vec{q}(\lambda_s, \mu_s)$, then we have no difficulties with the second part $A(t)\backslash B$ of the region of integration, because there $\vec{x}$ does not come arbitrarily close to any point of it. So $R^{-3}(\vec{x}-\vec{q}_s)$ cannot increase unboundedly for that part of the integral, which we denote by I_1. Hence for I_1, the limit $\vec{x} \to \vec{q}_s$ exists. Therefore, we now direct our attention to the integration over B.

We write the integration over B in (1.10.21) as follows

$$\begin{aligned}
&\iint\limits_B g(\lambda, \mu, t) \left\{ \frac{\partial H_1}{\partial \lambda}(\vec{x}, \vec{q}) + \frac{\partial H_2}{\partial \mu}(\vec{x}, \vec{q}) \right\} d\lambda \, d\mu \\
&= \iint\limits_B \left\{ g - g_s - \frac{\partial g_s}{\partial \lambda} \cdot (\lambda - \lambda_s) - \frac{\partial g_s}{\partial \mu} \cdot (\mu - \mu_s) \right\} \\
&\quad \cdot \left\{ \frac{\partial H_1}{\partial \lambda}(\vec{x}, \vec{q}) + \frac{\partial H_2}{\partial \mu}(\vec{x}, \vec{q}) \right\} d\lambda \, d\mu \\
&\quad + g_s \iint\limits_B \left\{ \frac{\partial H_1}{\partial \lambda}(\vec{x}, \vec{q}) + \frac{\partial H_2}{\partial \mu}(\vec{x}, \vec{q}) \right\} d\lambda \, d\mu \\
&\quad + \frac{\partial g_s}{\partial \lambda} \iint\limits_B (\lambda - \lambda_s) \left\{ \frac{\partial H_1}{\partial \lambda}(\vec{x}, \vec{q}) + \frac{\partial H_2}{\partial \mu}(\vec{x}, \vec{q}) \right\} d\lambda \, d\mu \\
&\quad + \frac{\partial g_s}{\partial \mu} \iint\limits_B (\mu - \mu_s) \left\{ \frac{\partial H_1}{\partial \lambda}(\vec{x}, \vec{q}) + \frac{\partial H_2}{\partial \mu}(\vec{x}, \vec{q}) \right\} d\lambda \, d\mu \\
&= I_2 + I_3 + J_1 + J_2 \ , \qquad (1.11.1)
\end{aligned}$$

where in agreement with the remark below (1.10.16), $g_s = g(\lambda_s, \mu_s, t)$, $\partial g_s/\partial\lambda = \partial g/\partial\lambda(\lambda_s, \mu_s, t)$ and $\partial g_s/\partial\mu = \partial g/\partial\mu(\lambda_s, \mu_s, t)$ and where the different integrals are denoted by I_2, I_3, J_1 and J_2 which we will discuss successively.

Integral I_2 will not be rewritten in a different form, so we start with I_3,

$$\begin{aligned}
I_3 &= g_s \iint\limits_B \left\{ \frac{\partial H_1}{\partial \lambda}(\vec{x}, \vec{q}) + \frac{\partial H_2}{\partial \mu}(\vec{x}, \vec{q}) \right\} d\lambda \, d\mu \\
&= g_s \int\limits_{\partial B} \{ H_1(\vec{x}, \vec{q}) \, m_\lambda + H_2(\vec{x}, \vec{q}) \, m_\mu \} \, d\sigma \ , \qquad (1.11.2)
\end{aligned}$$

where $\vec{m} = (m_\lambda, m_\mu)$ is the outward normal of unit length at the boundary ∂B of B and $d\sigma$ is an element of length (Figure 1.11.1).

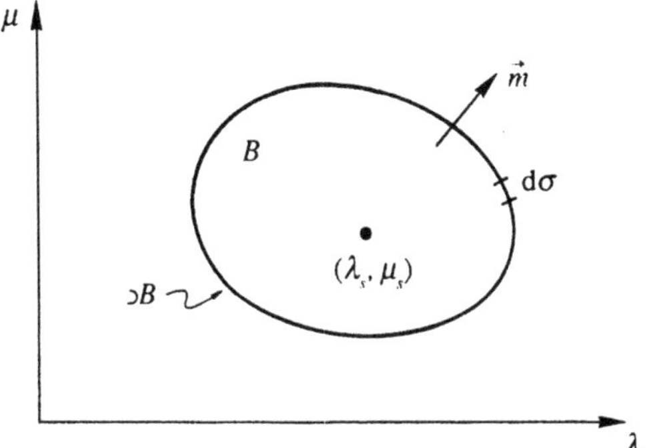

Fig. 1.11.1. The "small" compact region B around (λ_s, μ_s).

$$
\begin{aligned}
J_1 &= \frac{\partial g_s}{\partial \lambda} \iint_B (\lambda - \lambda_s) \left\{ \frac{\partial H_1}{\partial \lambda}(\vec{x}, \vec{q}) + \frac{\partial H_2}{\partial \mu}(\vec{x}, \vec{q}) \right\} d\lambda \, d\mu \\
&= \frac{\partial g_s}{\partial \lambda} \iint_B \left[\frac{\partial}{\partial \lambda} \{(\lambda - \lambda_s) H_1(\vec{x}, \vec{q})\} + \frac{\partial}{\partial \mu} \{(\lambda - \lambda_s) H_2(\vec{x}, \vec{q})\} \right] d\lambda \, d\mu \\
&\quad - \frac{\partial g_s}{\partial \lambda} \iint_B H_1(\vec{x}, \vec{q}) \, d\lambda \, d\mu \\
&= \frac{\partial g_s}{\partial \lambda} \int_{\partial B} (\lambda - \lambda_s) \{ H_1(\vec{x}, \vec{q}) \, m_\lambda + H_2(\vec{x}, \vec{q}) \, m_\mu \} \, d\sigma \\
&\quad - \frac{\partial g_s}{\partial \lambda} \iint_B H_1(\vec{x}, \vec{q}) \, d\lambda \, d\mu = I_4 + J_3 \ .
\end{aligned} \tag{1.11.3}
$$

Analogously, we find for J_2

$$
\begin{aligned}
J_2 &= \frac{\partial g_s}{\partial \mu} \iint_B (\mu - \mu_s) \left\{ \frac{\partial H_1}{\partial \lambda}(\vec{x}, \vec{q}) + \frac{\partial H_2}{\partial \mu}(\vec{x}, \vec{q}) \right\} d\lambda \, d\mu \\
&= \frac{\partial g_s}{\partial \mu} \int_{\partial B} (\mu - \mu_s) \{ (H_1(\vec{x}, \vec{q}) \, m_\lambda + H_2(\vec{x}, \vec{q}) \, m_\mu \} \, d\sigma \\
&\quad - \frac{\partial g_s}{\partial \mu} \iint_B H_2(\vec{x}, \vec{q}) \, d\lambda \, d\mu = I_5 + J_4 \ .
\end{aligned} \tag{1.11.4}
$$

Next we introduce

$$\tilde{H}_1(\vec{x},\vec{q}) = \operatorname*{grad}_x \frac{1}{R(\vec{x}-\vec{q})} \cdot \left(\frac{\partial \vec{q}}{\partial \mu} \times \vec{n}\right)$$
$$= -\frac{(\vec{x}-\vec{q})}{R^3(\vec{x}-\vec{q})} \cdot \left(\frac{\partial \vec{q}}{\partial \mu} \times \vec{n}\right) , \quad (1.11.5)$$

$$\tilde{H}_2(\vec{x},\vec{q}) = -\operatorname*{grad}_x \frac{1}{R(\vec{x}-\vec{q})} \cdot \left(\frac{\partial \vec{q}}{\partial \lambda} \times \vec{n}\right)$$
$$= \frac{(\vec{x}-\vec{q})}{R^3(\vec{x}-\vec{q})} \cdot \left(\frac{\partial \vec{q}}{\partial \lambda} \times \vec{n}\right) , \quad (1.11.6)$$

which are nearly the same definitions as in (1.10.19) and (1.10.20), only the fixed normal $\vec{n}_s = \vec{n}(\lambda_s, \mu_s)$ there, is replaced here by $\vec{n} = \vec{n}(\lambda, \mu)$.

We now write the last integral J_3 of (1.11.3) as

$$J_3 = -\frac{\partial g_s}{\partial \lambda} \iint_B H_1(\vec{x},\vec{q})\, d\lambda\, d\mu$$
$$= -\frac{\partial g_s}{\partial \lambda} \iint_B \{H_1(\vec{x},\vec{q}) - \tilde{H}_1(\vec{x},\vec{q})\}\, d\lambda\, d\mu - \frac{\partial g_s}{\partial \lambda} \iint_B \tilde{H}_1(\vec{x},\vec{q})\, d\lambda\, d\mu$$
$$= I_6 + J_5 , \quad (1.11.7)$$

and the last integral J_4 of (1.11.4) as

$$J_4 = -\frac{\partial g_s}{\partial \mu} \iint_B H_2(\vec{x},\vec{q})\, d\lambda\, d\mu$$
$$= -\frac{\partial g_s}{\partial \mu} \iint_B \{H_2(\vec{x},\vec{q}) - \tilde{H}_2(\vec{x},\vec{q})\}\, d\lambda\, d\mu - \frac{\partial g_s}{\partial \mu} \iint_B \tilde{H}_2(\vec{x},\vec{q})\, d\lambda\, d\mu$$
$$= I_7 + J_6 . \quad (1.11.8)$$

In order to make (1.11.7) and (1.11.8) more easy to handle, we rewrite (1.11.5) and (1.11.6) as follows. By (1.10.4) and (C.3)

$$\tilde{H}_1(\vec{x},\vec{q}) = \frac{1}{Z(\lambda,\mu)} \left[\operatorname*{grad}_x \frac{1}{R(\vec{x}-\vec{q})} \cdot \frac{\partial \vec{q}}{\partial \mu} \times \left\{ \frac{\partial \vec{q}}{\partial \lambda} \times \frac{\partial \vec{q}}{\partial \mu} \right\} \right]$$
$$= \frac{1}{Z(\lambda,\mu)} \left[\operatorname*{grad}_x \frac{1}{R(\vec{x}-\vec{q})} \cdot \left\{ \left|\frac{\partial \vec{q}}{\partial \mu}\right|^2 \frac{\partial \vec{q}}{\partial \lambda} - \left(\frac{\partial \vec{q}}{\partial \mu} \cdot \frac{\partial \vec{q}}{\partial \lambda}\right) \frac{\partial \vec{q}}{\partial \mu} \right\} \right]$$
$$= \frac{-1}{Z(\lambda,\mu)} \left\{ \left|\frac{\partial \vec{q}}{\partial \mu}\right|^2 \left(\operatorname*{grad}_q \frac{1}{R(\vec{x}-\vec{q})} \right) \cdot \frac{\partial \vec{q}}{\partial \lambda} \right.$$

$$- \left(\frac{\partial \vec{q}}{\partial \lambda} \cdot \frac{\partial \vec{q}}{\partial \mu} \right) \left(\operatorname{grad}_q \frac{1}{R(\vec{x} - \vec{q})} \right) \cdot \frac{\partial \vec{q}}{\partial \mu} \Bigg\}$$

$$= \frac{-1}{Z(\lambda,\mu)} \left\{ \left| \frac{\partial \vec{q}}{\partial \mu} \right|^2 \frac{\partial}{\partial \lambda} \frac{1}{R(\vec{x} - \vec{q})} - \left(\frac{\partial \vec{q}}{\partial \lambda} \cdot \frac{\partial \vec{q}}{\partial \mu} \right) \frac{\partial}{\partial \mu} \frac{1}{R(\vec{x} - \vec{q})} \right\}. \quad (1.11.9)$$

Analogously, we find

$$\tilde{H}_2(\vec{x}, \vec{q})$$

$$= \frac{-1}{Z(\lambda,\mu)} \left\{ \left| \frac{\partial \vec{q}}{\partial \lambda} \right|^2 \frac{\partial}{\partial \mu} \frac{1}{R(\vec{x} - \vec{q})} - \left(\frac{\partial \vec{q}}{\partial \lambda} \cdot \frac{\partial \vec{q}}{\partial \mu} \right) \frac{\partial}{\partial \lambda} \frac{1}{R(\vec{x} - \vec{q})} \right\}. \quad (1.11.10)$$

Then J_5 (1.11.7) and J_6 (1.11.8) become

$$J_5 = -\frac{\partial g_s}{\partial \lambda} \iint_B \tilde{H}_1(\vec{x}, \vec{q})\, d\lambda\, d\mu = \frac{\partial g_s}{\partial \lambda} \iint_B \frac{1}{Z(\lambda,\mu)}$$

$$\cdot \left\{ \left| \frac{\partial \vec{q}}{\partial \mu} \right|^2 \frac{\partial}{\partial \lambda} \frac{1}{R(\vec{x} - \vec{q})} - \left(\frac{\partial \vec{q}}{\partial \lambda} \cdot \frac{\partial \vec{q}}{\partial \mu} \right) \frac{\partial}{\partial \mu} \frac{1}{R(\vec{x} - \vec{q})} \right\} d\lambda\, d\mu$$

$$= \frac{\partial g_s}{\partial \lambda} \iint_B \left[\frac{\partial}{\partial \lambda} \left\{ \frac{1}{Z(\lambda,\mu)} \left| \frac{\partial \vec{q}}{\partial \mu} \right|^2 \frac{1}{R(\vec{x} - \vec{q})} \right\} \right.$$

$$\left. - \frac{\partial}{\partial \mu} \left\{ \frac{1}{Z(\lambda,\mu)} \left(\frac{\partial \vec{q}}{\partial \lambda} \cdot \frac{\partial \vec{q}}{\partial \mu} \right) \frac{1}{R(\vec{x} - \vec{q})} \right\} \right] d\lambda\, d\mu$$

$$- \frac{\partial g_s}{\partial \lambda} \iint_B \frac{1}{R(\vec{x} - \vec{q})} \left[\frac{\partial}{\partial \lambda} \left\{ \frac{1}{Z(\lambda,\mu)} \left| \frac{\partial \vec{q}}{\partial \mu} \right|^2 \right\} \right.$$

$$\left. - \frac{\partial}{\partial \mu} \left\{ \frac{1}{Z(\lambda,\mu)} \left(\frac{\partial \vec{q}}{\partial \lambda} \cdot \frac{\partial \vec{q}}{\partial \mu} \right) \right\} \right] d\lambda\, d\mu$$

$$= \frac{\partial g_s}{\partial \lambda} \int_{\partial B} \frac{1}{Z(\lambda,\mu)} \frac{1}{R(\vec{x} - \vec{q})} \left\{ \left| \frac{\partial q}{\partial \mu} \right|^2 m_\lambda - \left(\frac{\partial \vec{q}}{\partial \lambda} \cdot \frac{\partial \vec{q}}{\partial \mu} \right) m_\mu \right\} d\sigma + I_9$$

$$= I_8 + I_9\,, \quad (1.11.11)$$

$$J_6 = -\frac{\partial g_s}{\partial \mu} \iint_B \tilde{H}_2(\vec{x}, \vec{q})\, d\lambda\, d\mu$$

$$= -\frac{\partial g_s}{\partial \mu} \int_{\partial B} \frac{1}{Z(\lambda,\mu)} \frac{1}{R(\vec{x} - \vec{q})} \left\{ \left(\frac{\partial \vec{q}}{\partial \lambda} \cdot \frac{\partial \vec{q}}{\partial \mu} \right) m_\lambda - \left| \frac{\partial \vec{q}}{\partial \lambda} \right|^2 m_\mu \right\} d\sigma$$

$$+\frac{\partial g_s}{\partial \mu}\iint\limits_B \frac{1}{R(\vec{x}-\vec{q})}\left[\frac{\partial}{\partial \lambda}\left\{\frac{1}{Z(\lambda,\mu)}\left(\frac{\partial \vec{q}}{\partial \lambda}\cdot\frac{\partial \vec{q}}{\partial \mu}\right)\right\}\right.$$

$$\left.-\frac{\partial}{\partial \mu}\left\{\frac{1}{Z(\lambda,\mu)}\left|\frac{\partial \vec{q}}{\partial \lambda}\right|^2\right\}\right] d\lambda\, d\mu = I_{10}+I_{11} \ . \tag{1.11.12}$$

Hence we have the result that (1.10.21) changes into

$$4\pi\rho\, v_n(\vec{x},t)=\sum_{i=1}^{11} I_i \ . \tag{1.11.13}$$

1.12. Continuity of the Normal Velocity Component

The continuity of the normal velocity component across the dipole layer is a well-known fact and almost trivial from the point of view of the preservation of fluid, because we have no layers of pure sinks or sources on $S(t)$ (1.10.8). However, in proving this continuity, we obtain formulas necessary for the further development of the theory.

We now consider the convergence of $v_n(\vec{x},t)$ (1.11.13) for $\vec{x}\to\vec{q}_s$, in particular when $\vec{x}$ tends to $\vec{q}_s$ along the normal $\vec{n}_s$. Hence we have to consider the convergence of the integrals $I_1,\ldots,I_{11}$. This convergence is trivial for I_1 as we remarked in the second paragraph of Section 1.11. It is also trivial for those integrals which are along the boundary ∂B of B, hence for I_3, I_4, I_5, I_8, and I_{10}. We have to be careful with the remaining ones I_2, I_6, I_7, I_9, and I_{11}, which are 2-dimensional integrals over the interior of B. For these integrals the integrand becomes infinite for $\vec{x}\to\vec{q}_s$, because then $R(\vec{x}-\vec{q})\to 0$.

In the following we assume that the loading $f(\lambda,\mu,t)$ (1.10.6) of the lifting surface is sufficiently smooth. In fact, we assume that its time integral $g(\lambda,\mu,t)$ (1.10.10) has derivatives with respect to λ and μ which are continuous and which satisfy a Hölder condition. This condition states that for each finite compact subset B of $A(t)$, there exist $\alpha>0$ and $C\geq 0$ such that for all (λ_1,μ_1) and $(\lambda_2,\mu_2)\in B(t)$

$$\left|\frac{\partial g}{\partial \lambda}(\lambda_1,\mu_1)-\frac{\partial g}{\partial \lambda}(\lambda_2,\mu_2)\right|\leq C\,|(\lambda_1,\mu_1)-(\lambda_2,\mu_2)|^{\alpha} \ , \tag{1.12.1}$$

$$\left|\frac{\partial g}{\partial \mu}(\lambda_1,\mu_1)-\frac{\partial g}{\partial \mu}(\lambda_2,\mu_2)\right|\leq C\,|(\lambda_1,\mu_1)-(\lambda_2,\mu_2)|^{\alpha} \ , \tag{1.12.2}$$

where $|(\lambda_1,\mu_1)-(\lambda_2,\mu_2)|=\left\{(\lambda_1-\lambda_2)^2+(\mu_1-\mu_2)^2\right\}^{1/2}$.

We start with an inequality based on (1.10.3), there it is assumed that

$$Z(\lambda,\mu)=\left|\frac{\partial \vec{q}}{\partial \lambda}\times\frac{\partial \vec{q}}{\partial \mu}\right|\geq c>0 \ , \tag{1.12.3}$$

from this it follows that $\partial\vec{q}/\partial\lambda$ and $\partial\vec{q}/\partial\mu$ are nowhere parallel. Because we assumed that $\vec{q}(\lambda,\mu)$ depends analytically on λ and μ, it follows that in each compact set B of the (λ,μ) plane we have

$$\left|\frac{\partial\vec{q}}{\partial\lambda}(\lambda,\mu)\right| \le c_1 \ , \quad \left|\frac{\partial\vec{q}}{\partial\mu}(\lambda,\mu)\right| \le c_1 \ , \qquad (\lambda,\mu)\in B \ , \tag{1.12.4}$$

for some constant c_1. It follows from (1.12.3) and (1.12.4) that a constant c_2 exists so that

$$\left|\frac{\partial\vec{q}}{\partial\lambda}(\lambda,\mu)\right| \ge c_2 > 0 \ , \quad \left|\frac{\partial\vec{q}}{\partial\mu}(\lambda,\mu)\right| \ge c_2 > 0 \ , \qquad (\lambda,\mu)\in B \ . \tag{1.12.5}$$

For a small neighbourhood $\tilde{B}$ of the fixed vector $\vec{q}_s$ we have

$$\begin{aligned} |\vec{q}-\vec{q}_s| = {} & \frac{\partial\vec{q}_s}{\partial\lambda}\cdot(\lambda-\lambda_s) + \frac{\partial\vec{q}_s}{\partial\mu}\cdot(\mu-\mu_s) \\ & + O\left(|(\lambda,\mu)-(\lambda_s,\mu_s)|^2\right) \ . \end{aligned} \tag{1.12.6}$$

Because $\partial\vec{q}_s/\partial\lambda$ and $\partial\vec{q}_s/\partial\mu$ are not parallel, we find by (1.12.6)

$$R(\vec{q}_s-\vec{q}) = |\vec{q}-\vec{q}_s| \ge c_3\,|(\lambda,\mu)-(\lambda_s,\mu_s)| \ , \tag{1.12.7}$$

for some positive constant c_3. Now consider a point $\vec{x}$ at the normal $\vec{n}_s$ which is perpendicular to S at the point $\vec{q}_s$. Then it is clear that there exists a positive constant c_4 such that

$$R(\vec{x}-\vec{q}) \ge c_4\,|(\lambda,\mu)-(\lambda_s,\mu_s)| \ , \qquad (\lambda,\mu)\in\tilde{B} \tag{1.12.8}$$

which is the desired inequality.

First we discuss the continuity of I_2 (1.11.1)

$$\begin{aligned} I_2 = \iint\limits_B & \left\{ g - g_s - \frac{\partial g}{\partial\lambda}\cdot(\lambda-\lambda_s) - \frac{\partial g_s}{\partial\mu}\cdot(\mu-\mu_s) \right\} \\ & \cdot \left\{ \frac{\partial H_1}{\partial\lambda}(\vec{x},\vec{q}) + \frac{\partial H_2}{\partial\mu}(\vec{x},\vec{q}) \right\} \, d\lambda\, d\mu \ . \end{aligned} \tag{1.12.9}$$

So we have to estimate the two factors occurring in the integrand. Because

$$g(\lambda_2,\mu_2) - g(\lambda_1,\mu_1) = \int\limits_{\lambda_1}^{\lambda_2} \frac{\partial g}{\partial\lambda}(\lambda,\mu_1)\,d\lambda + \int\limits_{\mu_1}^{\mu_2} \frac{\partial g}{\partial\mu}(\lambda_2,\mu)\,d\mu \ , \tag{1.12.10}$$

and by using (1.12.1) and (1.12.2) we have

$$|g(\lambda_2,\mu_2) - g(\lambda_1,\mu_1) - \frac{\partial g}{\partial\lambda}(\lambda_1,\mu_1)(\lambda_2-\lambda_1) - \frac{\partial g}{\partial\mu}(\lambda_1,\mu_1)(\mu_2-\mu_1)|$$

$$\leq |\int_{\lambda_1}^{\lambda_2} \left\{ \frac{\partial g}{\partial \lambda}(\lambda, \mu_1) - \frac{\partial g}{\partial \lambda}(\lambda_1, \mu_1) \right\} d\lambda|$$

$$+ |\int_{\mu_1}^{\mu_2} \left\{ \frac{\partial g}{\partial \mu}(\lambda_2, \mu) - \frac{\partial g}{\partial \mu}(\lambda_1, \mu_1) \right\} d\mu|$$

$$\leq C| \int_{\lambda_1}^{\lambda_2} \max_{\lambda} |(\lambda, \mu_1) - (\lambda_1, \mu_1)|^{\alpha} d\lambda| + C| \int_{\mu_1}^{\mu_2} \max_{\mu} |(\lambda_2, \mu) - (\lambda_1, \mu_1)|^{\alpha}\, d\mu|$$

$$= C\left\{|(\lambda_2, \mu_1) - (\lambda_1, \mu_1)|^{\alpha} |\lambda_2 - \lambda_1| + |(\lambda_2, \mu_2) - (\lambda_1, \mu_1)|^{\alpha} |\mu_2 - \mu_1|\right\}$$

$$\leq \tilde{C}|(\lambda_2, \mu_2) - (\lambda_1, \mu_1)|^{1+\alpha} \,, \tag{1.12.11}$$

for some constant $\tilde{C}$. Taking $(\lambda_1, \mu_1) = (\lambda_s, \mu_s)$ and $(\lambda_2, \mu_2 = (\lambda, \mu)$, we have found an estimate for the first factor of the integrand of I_2 (1.12.9). Next we consider the second factor.

From (1.10.19) we have

$$H_1(\vec{x}, \vec{q}) = \frac{-(\vec{x} - \vec{q})}{R^3(\vec{x} - \vec{q})} \cdot \left(\frac{\partial \vec{q}}{\partial \mu} \times \vec{n}_s \right) \,. \tag{1.12.12}$$

Hence by differentiating with respect to λ and using (1.12.7) we find

$$\begin{aligned} \left| \frac{\partial H_1}{\partial \lambda}(\vec{x}, \vec{q}) \right| = & \left| \frac{\partial \vec{q}}{\partial \lambda} \cdot \frac{1}{R^3(\vec{x} - \vec{q})} \left(\frac{\partial \vec{q}}{\partial \mu} \times \vec{n}_s \right) \right. \\ & - (\vec{x} - \vec{q}) \cdot \left\{ \frac{3(\vec{x} - \vec{q})}{R^5(\vec{x} - \vec{q})} \cdot \frac{\partial \vec{q}}{\partial \lambda} \right\} \left(\frac{\partial \vec{q}}{\partial \mu} \times \vec{n}_s \right) \\ & \left. - \frac{(\vec{x} - \vec{q})}{R^3(\vec{x} - \vec{q})} \cdot \frac{\partial}{\partial \lambda} \left(\frac{\partial \vec{q}}{\partial \mu} \times \vec{n}_s \right) \right| \\ & \leq C\, |(\lambda, \mu) - (\lambda_s, \mu_s)|^{-3} \end{aligned} \tag{1.12.13}$$

for some constant C. Analogously, we have

$$\left| \frac{\partial H_2}{\partial \mu}(\vec{x}, \vec{q}) \right| \leq C\, |(\lambda, \mu) - (\lambda_s, \mu_s)|^{-3} \,. \tag{1.12.14}$$

Then by (1.12.11), (1.12.13) and (1.12.14) we have for the integrand of I_2 the estimate

$$\left| \left\{ g - g_s - \frac{\partial g_s}{\partial \lambda} \cdot (\lambda - \lambda_s) - \frac{\partial g_s}{\partial \mu} \cdot (\mu - \mu_s) \right\} \cdot \left\{ \frac{\partial H_1}{\partial \lambda}(\vec{x}, \vec{q}) + \frac{\partial H_1}{\partial \mu}(\vec{x}, \vec{q}) \right\} \right|$$

$$\leq C\, |(\lambda, \mu) - (\lambda_s, \mu_s)|^{-2+\alpha} \,, \qquad \alpha > 0 \,. \tag{1.12.15}$$

The right-hand side of (1.12.15) is integrable over B, hence the limit of I_2 for $\vec{x} \to \vec{q}_s$ exists by the dominated convergence theorem [75].

Subsequently, we consider I_6 (1.11.7)

$$I_6 = -\frac{\partial g_s}{\partial \lambda} \iint_B \left\{ H_1(\vec{x},\vec{q}) - \tilde{H}_1(\vec{x},\vec{q}) \right\} d\lambda \, d\mu \ . \tag{1.12.16}$$

The integrand of this integral can be estimated as follows. By (1.10.19) and (1.11.5) we have

$$|H_1(\vec{x},\vec{q}) - \tilde{H}_1(\vec{x},\vec{q})| = | - \frac{(\vec{x}-\vec{q})}{R^3(\vec{x}-\vec{q})} \cdot \frac{\partial \vec{q}}{\partial \mu} \times (\vec{n}_s - \vec{n})| \ . \tag{1.12.17}$$

Because $\vec{q}(\lambda,\mu)$ depends analytically on λ and μ we have by (1.10.4) for a sufficiently small neighbourhood of (λ_s, μ_s)

$$|\vec{n}_s - \vec{n}(\lambda,\mu)| \leq C|(\lambda_s,\mu_s) - (\lambda,\mu)| \ , \tag{1.12.18}$$

which then also holds for B, possibly with another C. So by (1.12.17) we draw the conclusion that

$$|H_1(\vec{x},\vec{q}) - \tilde{H}_1(\vec{x},\vec{q})| \leq C\,|(\lambda_s,\mu_s) - (\lambda,\mu)|^{-1} \ . \tag{1.12.19}$$

Hence again by the dominated convergence theorem the limit of I_6 exists for $\vec{x} \to \vec{q}_s$.

In an analogous way it can be shown that the remaining integrals I_7, I_9 and I_{11} converge for $\vec{x} \to \vec{q}_s$. Herewith we can calculate the normal component $v_n(\vec{q}_s, t)$ of the velocity field at the planform of the lifting surface by

$$v_n(\vec{q}_s, t) = \lim_{\vec{x} \to \vec{q}_s} v_n(\vec{x}, t) \ , \tag{1.12.20}$$

where $\vec{x} \to \vec{q}_s$ along the normal $\vec{n}_s$. However, it is clear, but mathematically somewhat more complicated, that the limit (1.12.20) also exists when $\vec{x} \to \vec{q}_s$ along a line which forms an angle with the normal $\vec{n}_s$, which is smaller than 90°.

The limit (1.12.20) is independent of the side of S from which $\vec{x}$ approaches $\vec{q}_s$, because the integrals $I_1, \ldots, I_{11}$ converge to fixed values. This agrees with the preservation of fluid.

1.13. Simplification of the Normal Velocity Component

We will now simplify (1.11.13) in which appear the integrals $I_1, \ldots, I_{11}$. This can be done by making a special choice for the compact region B. First we consider a series of regions B_β of which the area tends to zero when $\beta \to 0$. Then it follows that the contributions of the convergent 2-dimensional integrals, namely I_2, I_6, I_7, I_9 and I_{11}, which we discussed in the previous section, tend to zero with $\beta \to 0$. So there remains

$$4\pi\rho\, v_n(\vec{q}_s, t) = \lim_{\beta \to 0} \{I_1 + I_3 + I_4 + I_5 + I_8 + I_{10}\} \Big|_{\vec{x}=\vec{q}_s} \ . \tag{1.13.1}$$

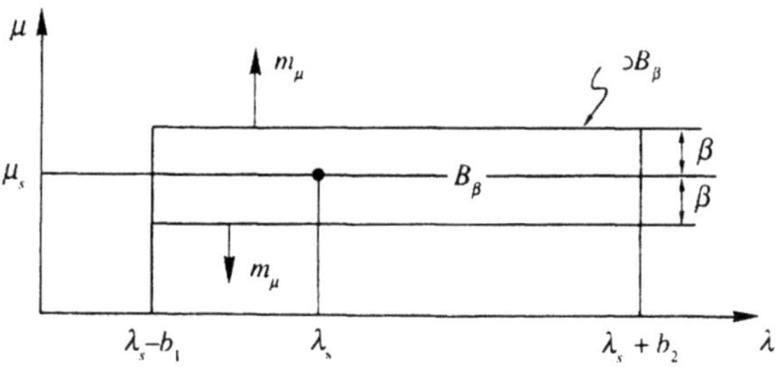

Fig. 1.13.1. Choice of region B_β.

The region B_β which we will use is drawn in Figure 1.13.1 and is defined by

$$B_\beta(b_1,b_2) = \{(\lambda,\mu) \ ; \ \lambda_s - b_1 \le \lambda \le \lambda_s + b_2 \ , \\ \mu_s - \beta \le \mu \le \mu_s + \beta\} \ , \tag{1.13.2}$$

for some arbitrary numbers $b_1 > 0$ and $b_2 > 0$. In the integrals of (1.13.1) the contributions of the terms with m_λ are zero for the long sides of B_β and tend to zero with $\beta \to 0$ for the short sides of B_β. Then it follows that we can write

$$4\pi\rho\, v_n(\vec{q}_s,t) = \lim_{\beta\to 0}\Bigg[\iint\limits_{A(t)\backslash B_\beta} g(\lambda,\mu,t)\left\{\frac{\partial H_1}{\partial\lambda}(\vec{q}_s,\vec{q}) + \frac{\partial H_2}{\partial\mu}(\vec{q}_s,\vec{q})\right\} d\lambda\, d\mu$$

$$+ g_s \int\limits_{\partial B_\beta} H_2(\vec{q}_s,\vec{q})\, m_\mu\, d\sigma + \frac{\partial g_s}{\partial\lambda}\int\limits_{\partial B_\beta}(\lambda-\lambda_s)H_2(\vec{q}_s,\vec{q})\, m_\mu\, d\sigma$$

$$+ \frac{\partial g_s}{\partial\mu}\int\limits_{\partial B_\beta}(\mu-\mu_s)H_2(\vec{q}_s,\vec{q})\, m_\mu\, d\sigma$$

$$- \frac{\partial g_s}{\partial\lambda}\int\limits_{\partial B_\beta}\frac{1}{Z(\lambda,\mu)}\cdot\frac{1}{R(\vec{q}_s-\vec{q})}\left(\frac{\partial\vec{q}}{\partial\lambda}\cdot\frac{\partial\vec{q}}{\partial\mu}\right) m_\mu\, d\sigma$$

$$+ \frac{\partial g_s}{\partial\mu}\int\limits_{\partial B_\beta}\frac{1}{Z(\lambda,\mu)}\cdot\frac{1}{R(\vec{q}_s-\vec{q})}\left|\frac{\partial\vec{q}}{\partial\lambda}\right|^2 m_\mu\, d\sigma\Bigg] \ . \tag{1.13.3}$$

Now we will show that in the limit $\beta \to 0$, all but the first two integrals in (1.13.3) tend to zero, hence in the limit only I_1 and I_3 contribute to $v_n(\vec{q}_s,t)$. It is clear that we have to consider for the five line integrals only a small neighbourhood of (λ_s,μ_s). This is because at each finite fixed distance of (λ_s,μ_s) the integrands at both lines $\mu = \mu_s - \beta$ and at $\mu = \mu_s + \beta$ are continuous and by the reverse of sign of m_μ at these lines, their contributions cancel for $\beta \to 0$. A possible contribution can only

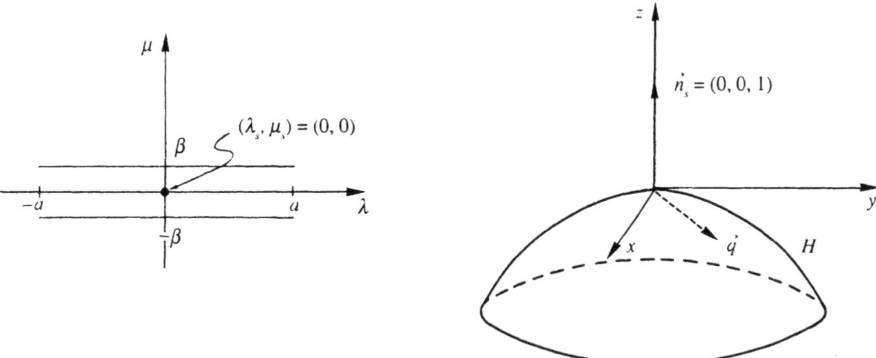

Fig. 1.13.2. Simplified configurations of parameter system (λ, μ) and coordinate system (x, y, z).

occur by the singularity of the integrands for $(\lambda, \mu) \to (\lambda_s, \mu_s)$. So we can choose a sufficiently small but fixed number a and consider integrands of the type

$$\lim_{\beta \to 0} \left\{ \int_{-a}^{a} G(\lambda, \mu = \mu_s + \beta)\, d\lambda - \int_{-a}^{a} G(\lambda, \mu = \mu_s - \beta)\, d\lambda \right\} , \qquad (1.13.4)$$

where $G(\lambda, \mu)$ represents any of the integrands of the line integrals in (1.13.3).

We can, without loosing essential information, simplify the calculations by assuming that the Cartesian coordinate system (x, y, z) is placed with its origin at the point $\vec{q}_s = \vec{q}(\lambda_s, \mu_s)$ at which we calculate the normal velocity component $v_n(\vec{q}_s, t)$. This is done in such a way that the (x, y) plane becomes tangent to S which is part of the surface H (Figure 1.10.1). Finally, we assume that the (λ, μ) coordinate system is translated such that its origin coincides with the original point (λ_s, μ_s), hence we assume that $(\lambda_s, \mu_s) = (0, 0)$ and $\vec{q}_s = (0, 0, 0)$. Then we obtain the configuration as is drawn in Figure 1.13.2. In the final result we have to consider again the original general configuration.

Because $\vec{q}(\lambda, \mu)$ depends analytically on λ and μ, we can expand the integrands of the line integrals of (1.13.3) in the neighbourhood of $(\lambda_s, \mu_s) = (0, 0)$. We write

$$\begin{aligned} \vec{q}(\lambda, \mu) = (&\alpha_1 \lambda + \alpha_2 \mu + O(\nu^2), \beta_1 \lambda + \beta_2 \mu + O(\nu^2) , \\ &\gamma_1 \lambda^2 + \gamma_2 \lambda \mu + \gamma_3 \mu^2 + O(\nu^3)) , \end{aligned} \qquad (1.13.5)$$

where $\nu^2 = \lambda^2 + \mu^2$ and we introduced the coefficients $\alpha_1, \ldots, \gamma_3$.

By (1.13.5) the unit normal $\vec{n}_s$ (1.10.4) becomes

$$\vec{n}_s = \left(\frac{\partial \vec{q}_s}{\partial \lambda} \times \frac{\partial \vec{q}_s}{\partial \mu} \right) \cdot \left| \frac{\partial \vec{q}_s}{\partial \lambda} \times \frac{\partial \vec{q}_s}{\partial \mu} \right|^{-1}$$

$$= \frac{(\alpha_1, \beta_1, 0) \times (\alpha_2, \beta_2, 0)}{|(\alpha_1, \beta_1, 0) \times (\alpha_2, \beta_2, 0)|} = \frac{(0, 0, \alpha_1\beta_2 - \alpha_2\beta_1)}{|\alpha_1\beta_2 - \alpha_2\beta_1|} . \tag{1.13.6}$$

Hence we have

$$\vec{n}_s = (0,0,1) , \qquad (\alpha_1\beta_2 - \alpha_2\beta_1) > 0 ;$$
$$\vec{n}_s = (0,0,-1) , \qquad (\alpha_1\beta_2 - \alpha_2\beta_1) < 0 . \tag{1.13.7}$$

In the first instance, we take $(\alpha_1\beta_2 - \alpha_2\beta_1) > 0$, hence $\vec{n}_s = (0,0,1)$ as is drawn in Figure 1.13.2. The other possibility $(\alpha_1\beta_2 - \alpha_2\beta_1) < 0$, with $\vec{n}_s = (0,0,-1)$ can be considered analogously and will be discussed shortly at the end of this section.

In the denominator of the integrands of (1.13.3) occurs some power of $R(\vec{q}_s - \vec{q})$, we have by (1.13.5)

$$R^2(\vec{q}_s - \vec{q}) = |\vec{q}|^2 = \{(\alpha_1\lambda + \alpha_2\mu)^2 + (\beta_1\lambda + \beta_2\mu)^2 + O(\nu^3)\}$$
$$= \{s_1\lambda^2 + s_2\lambda\mu + s_3\mu^2 + O(\nu^3)\} , \quad \nu^2 = \lambda^2 + \mu^2 , \tag{1.13.8}$$

where we introduced s_1, s_2 and s_3. We remark that $(s_1\lambda^2 + s_2\lambda\mu + s_3\mu^2)$, because it is the lowest order part of $|\vec{q}|^2$, is strictly positive, which means that this expression becomes zero if and only if $\lambda = \mu = 0$.

In this way we can expand $Z(\lambda, \mu)$ (1.10.3), $H_2(\vec{q}_s, \vec{q})$ (1.10.20), etc., which occur in the integrands of (1.13.3). Then we arrive at line integrals of the type

$$J_{mk} = \int_{-a}^{a} \frac{\lambda^m \, d\lambda}{(s_1\lambda^2 + s_2\lambda\mu + s_3\mu^2)^{k+1/2}} . \tag{1.13.9}$$

These integrals can be calculated in closed form, by recursive relations with respect to m and k, for $\mu = \beta$ and for $\mu = -\beta$ (for instance [23], p. 39). Then the results can be expanded with respect to β and the limit $\beta \to 0$ of the combinations as given in (1.13.4), can be taken. These are cumbersome manipulations, but simple and straightforward. We find that the contributions of the last four integrals in (1.13.3) tend to zero with $\beta \to 0$. The first and second integral in (1.13.3) give non-zero contributions. We direct our attention to the second one.

For this integral, again for the configuration of Figure 1.13.2, we find by (1.10.20)

$$\lim_{\beta \to 0} g_s \int_{\partial B_\beta} H_2(\vec{q}_s = \vec{o}, \vec{q}) \, m_\mu \, d\sigma$$

$$= \lim_{\beta \to 0} g_s \left[- \int_{-a}^{a} \frac{\{p_1\beta + O(\nu^2)\} \, d\lambda}{\{s_1\lambda^2 + s_2\beta\lambda + s_3\beta^2 + O(\nu^3)\}^{3/2}} \right.$$

$$\left. - \int_{-a}^{a} \frac{\{p_1\beta + O(\nu^2)\} \, d\lambda}{\{s_1\lambda^2 - s_2\beta\lambda + s_3\beta^2 + O(\nu^3)\}^{3/2}} \right]$$

$$= -\frac{16\, g_s p_1 \sqrt{s_1}}{(4s_1 s_3 - s_2^2)} \cdot \frac{1}{\beta} + O(\beta) \; ; \qquad p_1 = (\alpha_2\beta_1 - \alpha_1\beta_2) \; . \tag{1.13.10}$$

We remark that this result only depends on the lowest order terms of the numerator and denominator of the integrands in (1.13.10).

Now we return to the general case in which $(\lambda_s, \mu_s) \neq (0,0)$ is some point in the parameter plane and $\vec{q}_s \neq (0,0,0)$ is some point in space. By the definition of $H_2(\vec{x}, \vec{q})$ (1.10.20) we find for the second integral at the right-hand side of (1.13.3)

$$\lim_{\beta \to 0} g_s \int\limits_{\partial B_\beta} H_2(\vec{q}_s, \vec{q})\, m_\mu \, d\sigma$$

$$= \lim_{\beta \to 0} g_s \Bigg[\int\limits_{\lambda_s - a}^{\lambda_s + a} \frac{(\vec{q}_s - \vec{q})}{R^3(\vec{q}_s - \vec{q})} \cdot \left(\frac{\partial \vec{q}}{\partial \lambda} \times \vec{n}_s \right) \Bigg|_{\mu = \mu_s + \beta} d\lambda$$

$$- \int\limits_{\lambda_s - a}^{\lambda_s + a} \frac{(\vec{q}_s - \vec{q})}{R^3(\vec{q}_s - \vec{q})} \cdot \left(\frac{\partial \vec{q}}{\partial \lambda} \times \vec{n}_s \right) \Bigg|_{\mu = \mu_s - \beta} d\lambda \Bigg] \; . \tag{1.13.11}$$

Because in the previous special case the result (1.13.10) depends only on the lowest order terms, this will happen also for the general case, which originates from the special one by changing the positions of the reference systems (x, y, z) and (λ, μ). So we have to determine for the general case the lowest order terms of the numerator and the denominator of the integrands in (1.13.11).

First we consider the numerator, which by (1.10.4) can be written as

$$(\vec{q}_s - \vec{q}) \cdot \left\{ \frac{\partial \vec{q}}{\partial \lambda} \times \left(\frac{\partial \vec{q}_s}{\partial \lambda} \times \frac{\partial \vec{q}_s}{\partial \mu} \right) Z^{-1}(\lambda_s, \mu_s) \right\} \; . \tag{1.13.12}$$

Because it is assumed that a is fixed but small, we can introduce into (1.13.12)

$$\vec{q}_s - \vec{q} = -\frac{\partial \vec{q}_s}{\partial \lambda} \cdot (\lambda - \lambda_s) - \frac{\partial \vec{q}_s}{\partial \mu} \cdot (\mu - \mu_s) + O(\nu^2) \; , \tag{1.13.13}$$

$$\frac{\partial \vec{q}}{\partial \lambda} = \frac{\partial \vec{q}_s}{\partial \lambda} + \frac{\partial^2 \vec{q}_s}{\partial \lambda^2} \cdot (\lambda - \lambda_s) + \frac{\partial^2 \vec{q}_s}{\partial \lambda \partial \mu} \cdot (\mu - \mu_s) + O(\nu^2) \; . \tag{1.13.14}$$

Then by using (C.3) we find for (1.13.12)

$$\left\{ \left| \frac{\partial \vec{q}_s}{\partial \lambda} \right|^2 \left| \frac{\partial q_s}{\partial \mu} \right|^2 - \left(\frac{\partial \vec{q}_s}{\partial \lambda} \cdot \frac{\partial \vec{q}_s}{\partial \mu} \right)^2 \right\} Z^{-1}(\lambda_s, \mu_s)(\mu - \mu_s) \; . \tag{1.13.15}$$

Hence we find for the corresponding value of p_1 (1.13.10) in the general case

$$p_1 = -\left\{ \left| \frac{\partial q_s}{\partial \lambda} \right|^2 \left| \frac{\partial \vec{q}_s}{\partial \mu} \right|^2 - \left(\frac{\partial \vec{q}_s}{\partial \lambda} \cdot \frac{\partial \vec{q}_s}{\partial \mu} \right)^2 \right\} Z^{-1}(\lambda_s, \mu_s) \; . \tag{1.13.16}$$

Next we need for the general case the values of s_1, s_2 and s_3 which occur in (1.13.8). These can be determined from

$$|\vec{q}_s - \vec{q}|^2 = \left\{ \frac{\partial \vec{q}_s}{\partial \lambda} \cdot (\lambda - \lambda_s) + \frac{\partial \vec{q}_s}{\partial \mu} \cdot (\mu - \mu_s) + O(\nu^2) \right\}^2$$

$$= \{ s_1 \cdot (\lambda - \lambda_s)^2 + s_2 \cdot (\lambda - \lambda_s)(\mu - \mu_s)$$

$$+ s_3 \cdot (\mu - \mu_s)^2 + O(\nu^3) \} . \qquad (1.13.17)$$

Hence we find

$$s_1 = \left| \frac{\partial \vec{q}_s}{\partial \lambda} \right|^2 , \quad s_2 = 2 \left(\frac{\partial \vec{q}_s}{\partial \lambda} \cdot \frac{\partial \vec{q}_s}{\partial \mu} \right) , \quad s_3 = \left| \frac{\partial \vec{q}_s}{\partial \mu} \right|^2 . \qquad (1.13.18)$$

Substitution of (1.13.16) and (1.13.18) into (1.13.10) yields

$$\lim_{\beta \to 0} g_s \int_{\partial B} H_2(\vec{q}_s, \vec{q}) \, m_\mu \, d\sigma = \frac{4 g_s |\partial \vec{q}_s / \partial \lambda|}{|\partial \vec{q}_s / \partial \lambda \times \partial \vec{q}_s / \partial \mu|} \cdot \frac{1}{\beta} + O(\varepsilon)$$

$$= c_0 g_s \cdot \frac{1}{\beta} + O(\beta) \ , \qquad (1.13.19)$$

where we introduced c_0. This results holds for the general case.

There is a seeming discrepancy between (1.13.10) and (1.13.19). In the general result (1.13.19) there are only, apart from g_s, positive quantities, while in (1.13.10) the value of $p_1 = (\alpha_2\beta_1 - \alpha_1\beta_2)$ can be positive or negative. In our derivation of the special case (1.13.10) we assumed $(\alpha_2\beta_1 - \alpha_1\beta_2) > 0$, by which $\vec{n}_s = (0, 0, 1)$ (1.13.7). If we had assumed $(\alpha_2\beta_1 - \alpha_1\beta_2) < 0$, then the normal $\vec{n}_s$ becomes $(0, 0, -1)$. By this the vector product in the definition of $H_2(\vec{x}, \vec{q})$ will deliver instead of the factor $(\alpha_2\beta_1 - \alpha_1\beta_2)$ a factor $-(\alpha_2\beta_1 - \alpha_1\beta_2)$ which again is positive. So (1.13.10) is not in contradiction with (1.13.19).

By the foregoing we find that we can write (1.13.3) as

$$4\pi\rho \, v_n(\vec{q}_s, t)$$

$$= \lim_{\beta \to 0} \left[\iint_{A(t) \setminus B_\beta} g(\lambda, \mu, t) \left\{ \frac{\partial H_1}{\partial \lambda}(\vec{q}_s, \vec{q}) + \frac{\partial H_2}{\partial \mu}(\vec{q}_s, \vec{q}) \right\} d\lambda \, d\mu + c_0 g_s \frac{1}{\beta} \right] ,$$

$$c_0 = \frac{4 |\partial \vec{q}_s / \partial \lambda|}{|\partial \vec{q}_s / \partial \lambda \times \partial \vec{q}_s / \partial \mu|} . \qquad (1.13.20)$$

1.14. "Stationary" Lifting Surface Theory

We now consider a special case of the foregoing theory. Suppose that $\Omega(t)$ (Figure 1.10.1), which represents in the (λ, μ) plane the planform $W(t)$ of the lifting

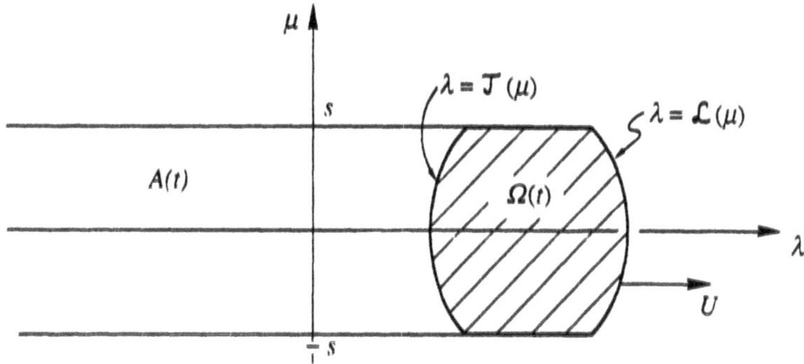

Fig. 1.14.1. (λ, μ) plane with translating region $\Omega(t)$ of fixed shape.

surface, has a fixed shape and translates with a constant velocity U in the λ-direction. Then it can be described by

$$\Omega(t) = \{(\lambda, \mu) : \mathcal{T}(\mu) + Ut \leq \lambda \leq \mathcal{L}(\mu) + Ut \ , \ -s \leq \mu \leq s\} \ , \tag{1.14.1}$$

where the function $\mathcal{L}(\mu)$ determines the "leading edge" of $\Omega(t)$, $\mathcal{T}(\mu)$ its "trailing edge" and s its "span" (Figure 1.14.1). We assume that the force field started at time $t = -\infty$, which is necessary for a stationary theory, then we have

$$A(t) = \{(\lambda, \mu) : -\infty < \lambda \leq \mathcal{L}(\mu) + Ut \ , \ -s \leq \mu \leq s\} \ . \tag{1.14.2}$$

The leading edge and the trailing edge of the planform $W(t)$ of the lifting surface are defined by

$$\text{Leading edge } W(t) = \{\vec{q}(\lambda, \mu) : \lambda = \mathcal{L}(\mu) + Ut \ , \ -s \leq \mu \leq s\} \ , \tag{1.14.3}$$

$$\text{Trailing edge } W(t) = \{\vec{q}(\lambda, \mu) : \lambda = \mathcal{T}(\mu) + Ut \ , \ -s \leq \mu \leq s\} \ , \tag{1.14.4}$$

respectively. However, we have to show that at time t the lines in space given in (1.14.3) and (1.14.4) are indeed the physical leading edge and the physical trailing edge of the planform $W(t)$. This can be done as follows.

Consider at the line (1.14.3) a point $\vec{x}^*$ corresponding to the parameter value $\mu = \mu^*$

$$\vec{x}^*(t) = \vec{q}(\mathcal{L}(\mu^*) + Ut, \mu^*) \ . \tag{1.14.5}$$

At the time $t + dt$ this point has moved to

$$\vec{x}^*(t + dt) = \vec{q}(\mathcal{L}(\mu^*) + Ut + Udt, \mu^*) \ . \tag{1.14.6}$$

Because $\mathcal{L}(\mu^*) + Ut + Udt > \mathcal{L}(\mu^*) + Ut$ it follows from (1.14.1) that $\mathcal{L}(\mu^*) + Ut + U\,dt \notin \Omega(t)$, or

$$\vec{x}^*(t + dt) \notin W(t) \ . \tag{1.14.7}$$

However, by definition $\vec{x}^*(t+dt) \in W(t+dt)$, in other words the point $\vec{x}^*(t+dt)$ is at the moment t outside W and at the moment $t+dt$ on the boundary of W, hence $\vec{x}^*(t)$ (1.14.5) is situated at the leading edge of W.

In the same way we can show that the points of the line in space given by (1.14.4) belong to the trailing edge of $W(t)$.

There still can be points with $\mu = s$ and $\mu = -s$, which are also on the boundary $\partial\Omega(t)$ of $\Omega(t)$. These points can be considered as points lying at the leading edge, when the parts of $\partial\Omega$ with $\mu = s$ and $\mu = -s$ are slightly inclined and directed to a point at the λ-axis far in front of $\Omega(t)$. These points can also be considered to lie at the trailing edge when the mentioned parts of $\partial\Omega$ are directed to a point at the λ-axis far behind $\Omega(t)$. It means that the "tips" of $W(t)$ slide along the lines $\vec{q}(\lambda, \pm s)$ on H (Figure 1.10.1), just as the tips of $\Omega(t)$ slide along the lines $\mu = \pm s$ in the (λ, μ) plane.

Further, we assume in the formulation of the "stationary" lifting surface theory that the intensity $f(\lambda, \mu, t)$ (1.10.6) of the force field translates with $\Omega(t)$. Therefore, we introduce a local coordinate system $(\tilde{\lambda}, \tilde{\mu})$ moving with $\Omega(t)$, of which the coordinates $(\tilde{\lambda}, \tilde{\mu})$ are related to the original coordinates (λ, μ) by

$$\tilde{\lambda} = \lambda - Ut \ , \qquad \tilde{\mu} = \mu \ . \tag{1.14.8}$$

The region $\Omega(t)$ with respect to this local system $(\tilde{\lambda}, \tilde{\mu})$ will be denoted by $\tilde{\Omega}$. Also we will need later a region $\tilde{B}_\beta$ instead of the region B_β, which is defined by

$$\tilde{B}_\beta = \{(\tilde{\lambda}, \tilde{\mu}) \in \tilde{\Omega} \ , \ |\tilde{\mu} - \tilde{\mu}_s| \leq \beta\} \ , \qquad \tilde{\mu}_s = \mu_s \ . \tag{1.14.9}$$

We now prescribe the intensity of the normal forces at $\tilde{\Omega}$ by $\tilde{f}(\tilde{\lambda}, \tilde{\mu})$, with

$$\tilde{f}(\tilde{\lambda}, \tilde{\mu}) = 0 \ , \qquad (\tilde{\lambda}, \tilde{\mu}) \notin \tilde{\Omega} \ . \tag{1.14.10}$$

Then we find for the original force intensity $f(\lambda, \mu, t)$ the more special form

$$f(\lambda, \mu, t) = \tilde{f}(\lambda - Ut, \mu) \ . \tag{1.14.11}$$

We remark that the theory which we describe in this section needs not to be a stationary theory for the planform $W(t)$. We only asked that the shape and the velocity U of $\Omega(t)$ are fixed and that the force intensity $\tilde{f}(\tilde{\lambda}, \tilde{\mu})$ is independent of time. Of course, the shape of $W(t)$ and its loading will depend on time for an arbitrary chosen reference surface H determined by some vector-valued function $\vec{q} = \vec{q}(\lambda, \mu)$. So the theory of this section is somewhat more general than the classical stationary theory for $W(t)$, however it will be seen that the stationary theory for $W(t)$ follows simply from this one by a suitable choice of $\vec{q}(\lambda, \mu)$.

Now we calculate at a point on the planform $W(t)$ with local coordinates $(\tilde{\lambda}_s, \tilde{\mu}_s)$, the normal velocity which, by the foregoing paragraph, can still be dependent on t and which is denoted by $\tilde{v}_n(\tilde{\lambda}_s, \tilde{\mu}_s, t)$. It follows from (1.13.20), (1.10.10), (1.14.8) and (1.14.11)

$$4\pi\rho\, v_n(\vec{q}_s,t)$$

$$= \lim_{\beta\to 0}\Bigg[\iint\limits_{A(t)\setminus B_\beta}\int\limits_{-\infty}^{t}\tilde{f}(\lambda-U\tau,\mu)\,d\tau\,K(\tilde{\lambda}_s+Ut,\tilde{\mu}_s,\lambda,\mu)\,d\lambda\,d\mu$$

$$+\frac{c_0}{\beta}\int\limits_{-\infty}^{t}\tilde{f}(\tilde{\lambda}_s+Ut-U\tau,\tilde{\mu}_s)\,d\tau\Bigg]\ , \tag{1.14.12}$$

where the kernel function K follows by (1.14.8) from

$$K(\lambda_s,\mu_s,\lambda,\mu)=\left\{\frac{\partial H_1}{\partial\lambda}(\vec{q}_s,\vec{q})+\frac{\partial H_2}{\partial\mu}(\vec{q}_s,\vec{q})\right\}\ . \tag{1.14.13}$$

The first integral of (1.14.12) will be rewritten as follows

$$\int\limits_{-\infty}^{t}\iint\limits_{A(t)\setminus B_\beta}\tilde{f}(\lambda-U\tau,\mu)\,K(\tilde{\lambda}_s+Ut,\tilde{\mu}_s,\lambda,\mu)\,d\lambda\,d\mu\,d\tau$$

$$=\int\limits_{-\infty}^{t}\iint\limits_{A(t-\tau)\setminus B_\beta(b_1+U\tau,b_2+U\tau)}\tilde{f}(\tilde{\lambda},\tilde{\mu})\,K(\tilde{\lambda}_s+Ut,\tilde{\mu}_s,\tilde{\lambda}+U\tau,\tilde{\mu})\,d\tilde{\lambda}\,d\tilde{\mu}\,d\tau$$

$$=\int\limits_{-\infty}^{t}\iint\limits_{\tilde{\Omega}\setminus\tilde{B}_\beta}\tilde{f}(\tilde{\lambda},\tilde{\mu})K(\tilde{\lambda}_s+Ut,\tilde{\mu}_s,\tilde{\lambda}+U\tau,\tilde{\mu})\,d\tilde{\lambda}\,d\tilde{\mu}\,d\tau$$

$$=\iint\limits_{\tilde{\Omega}\setminus\tilde{B}_\beta}f(\tilde{\lambda},\tilde{\mu})\left\{\int\limits_{-\infty}^{t}K(\tilde{\lambda}_s+Ut,\tilde{\mu}_s,\tilde{\lambda}+U\tau,\tilde{\mu})\,d\tau\right\}d\tilde{\lambda}\,d\tilde{\mu}\ . \tag{1.14.14}$$

The second equality is correct in case the limit $\beta\to 0$ is taken and the last equality is correct because we used $\tilde{B}_\beta$ (1.14.9) which stretches upto the leading edge of $\tilde{\Omega}$. By this it is prevented that in the time integral over K, which appears in the last expression of (1.14.14), points of integration $(\tilde{\lambda}+U\tau,\tilde{\mu})$ become equal to $(\tilde{\lambda}_s+Ut,\tilde{\mu}_s)$ for some value of τ with $-\infty<\tau\le t$. When this could happen, the time integral over K would obtain a non-integrable singularity in the $(\tilde{\lambda},\tilde{\mu})$ domain. In other words, by this it is not possible for the strings of velocity dipoles behind the elementary forces $f(\tilde{\lambda},\tilde{\mu})d\tilde{\lambda}d\tilde{\mu}$ (first paragraph of Section 1.8) to pass through the point of consideration $(\tilde{\lambda}_s,\tilde{\mu}_s)$. That $\tilde{B}_\beta$ also stretches upto the trailing edge of $\tilde{\Omega}$ is not essential for the convergence of the integral over $(\tilde{\lambda},\tilde{\mu})$, however $\tilde{B}_\beta$ has to include $(\tilde{\lambda}_s,\tilde{\mu}_s)$.

The second integral of (1.14.12) can also be written more simply by changing the integration variable τ to the local coordinate $\tilde{\lambda}$

$$\tilde{\lambda} = \tilde{\lambda}_s + Ut - U\tau \ , \qquad d\tilde{\lambda} = -U d\tau \ . \tag{1.14.15}$$

Then the second expression at the right-hand side of (1.14.12) becomes

$$\frac{c_0}{U\beta} \int\limits_{\tilde{\lambda}_s}^{\mathcal{L}(\tilde{\mu}_s)} \tilde{f}(\tilde{\lambda}, \tilde{\mu}_s) \, d\tilde{\lambda} \tag{1.14.16}$$

because $\tilde{f}(\tilde{\lambda}, \tilde{\mu}_s) = 0$ for $\tilde{\lambda} > \mathcal{L}(\tilde{\mu}_s)$.

By (1.14.14) and (1.14.16) we find for (1.14.12)

$$4\pi\rho \, \tilde{v}_n(\vec{q}_s, t)$$

$$= \lim_{\varepsilon \to 0} \left[\iint\limits_{\tilde{\Omega} \backslash \tilde{B}_\beta} \tilde{f}(\tilde{\lambda}, \tilde{\mu}) \tilde{K}(\tilde{\lambda}_s, \tilde{\mu}_s, \tilde{\lambda}, \tilde{\mu}) \, d\tilde{\lambda} \, d\tilde{\mu} + \frac{c_0}{U\beta} \int\limits_{\tilde{\lambda}_s}^{\mathcal{L}(\tilde{\mu}_s)} \tilde{f}(\tilde{\lambda}, \tilde{\mu}_s) \, d\tilde{\lambda} \right] , \tag{1.14.17}$$

where

$$\tilde{K}(\tilde{\lambda}_s, \tilde{\mu}_s, \tilde{\lambda}, \tilde{\mu}) = \int\limits_{-\infty}^{t} K(\tilde{\lambda}_s + Ut, \tilde{\mu}_s, \tilde{\lambda} + U\tau, \tilde{\mu}) \, d\tau \tag{1.14.18}$$

and K is given in (1.14.13).

When $\vec{q}(\lambda, \mu)$ is such that (see also below (1.14.11)), indeed a stationary lifting surface $W(t)$ in the physical sense occurs of which the loading is independent of time, then also $\tilde{v}_n$ becomes independent of time. Then we can take $t = 0$ in (1.14.17) and in (1.14.18). In this case, the kernel function $\tilde{K}$ simplifies to $\tilde{K}_0$ which is given by

$$\tilde{K}_0(\tilde{\lambda}_s, \tilde{\mu}_s, \lambda, \mu) = \int\limits_{-\infty}^{0} K(\tilde{\lambda}_s, \tilde{\mu}_s, \tilde{\lambda} + U\tau, \tilde{\mu}) \, d\tau \ , \tag{1.14.19}$$

where again K is given by (1.14.13).

The limit procedure which is applied to the general unsteady case (1.13.20) or to the steady case (1.14.17) is called the Hadamard principal value of the two-dimensional integral over the $(\tilde{\lambda}, \tilde{\mu})$ domain. It is not difficult to show by a series expansion that the integrand has a singularity of the order $(\tilde{\mu}_s - \tilde{\mu})^{-2}$ which is not integrable in the μ-direction. In Section 3.4 we will discuss the meaning of the Hadamard principal value in terms of the vorticity on the blade of a screw propeller.

Finally, we make a short remark about the periodic unsteady theory for a lifting surface. Then instead of (1.14.11) we can assume

$$f(\lambda, \mu, t) = \tilde{f}(\lambda - Ut, \mu) \, e^{i\omega t} \ , \tag{1.14.21}$$

and substitute this into (1.13.20) in the same way as we did before. It is clear that then also the Hadamard principal value will occur. We do not pursue this any further, but refer for practical calculations of unsteady screw propellers to [8].

1.15. Forces and Moments Exerted on a Fluid by a Moving Body

We will discuss here the force actions exerted *on* a fluid by a moving body. This body is allowed to change its shape and its volume. Our discussion will be restricted to the derivation of some formulas needed in the next sections.

First we consider a body $B(t)$ of finite extent, moving in an inviscid and incompressible fluid. In this fluid we have a Cartesian coordinate system (x, y, z) with respect to which the fluid at infinity is at rest. We assume that there is no external force field acting at the fluid and that no vorticity is shed from the body into the fluid. Hence the motion of the fluid can be assumed to be irrotational and its velocity field $\vec{v} = (v_x, v_y, v_z)$ can be derived from a potential $\Phi = \Phi(x, y, z, t)$ (1.1.10) at all points of space outside the body, $\vec{v} = \text{grad}\ \Phi$. Because div $\vec{v} = 0$ we have $\Delta\Phi = 0$.

Consider around B a geometrical surface $\tilde{H}$ which is coupled to the fluid particles, hence it floats with the fluid. To the specified amount of fluid in the region $\tilde{\Omega}$ between $\tilde{H}$ and the boundary ∂B of B we can apply the theorem of momentum. This theorem states:

> *The resultant force exerted on an amount of fluid equals the change of momentum per unit of time, of that amount of fluid.*

For the use of this theorem, however, it seems more convenient to replace $\tilde{H}$ by a geometrical surface H fixed in space, around B. Then we consider the region Ω bounded by H and ∂B. But now we have to add to the rate of change of momentum of the fluid in Ω, the flux of momentum leaving Ω through H and to subtract the incoming flux. For H we take a sphere with sufficiently large radius R_H, which has its centre in the neighbourhood of B (Figure 1.15.1).

The momentum $\vec{I}(t) = \big(I_x(t), I_y(t), I_z(t)\big)$ of the fluid in the region Ω is

$$\vec{I} = \rho \int\limits_{\Omega} \text{grad}\ \Phi\ d\text{Vol} = \rho \int\limits_{\partial B+H} \Phi\, \vec{n}\, dS\ , \tag{1.15.1}$$

where the unit normal $\vec{n}$ points out of the region Ω and dS is an element of area. We want to calculate the resultant force $\vec{F} = \vec{F}(t)$ exerted on the fluid by the body B during its motion. We also introduce the force $\vec{F}_H = \vec{F}_H(t)$ exerted on the fluid inside H by the fluid outside H. Then by (1.15.1) we can write the theorem of momentum as

$$\vec{F} + \vec{F}_H = \rho \frac{d}{dt} \int\limits_{\partial B+H} \Phi\, \vec{n}\, dS + \rho \int\limits_{H} \vec{v} \cdot (\vec{v} \cdot \vec{n})\, dS\ , \tag{1.15.2}$$

where the last term is the momentum flux through H.

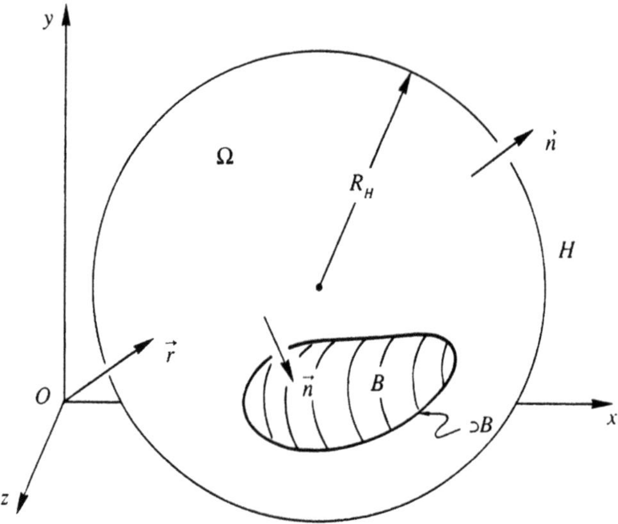

Fig. 1.15.1. Body B with control surface H.

The force $\vec{F}_H$ can be written as

$$\vec{F}_H = -\int_H p\,\vec{n}\,dS \ . \tag{1.15.3}$$

We now use Bernoulli's equation (1.2.11) with $K = 0$ and $h(t) = p_\infty = \text{const.}$ These choices can be made because there is no external force field acting at the fluid and at infinity $\vec{v} = 0$ and $p = p_\infty$. Hence (1.2.11) can be written as

$$p = p_\infty - \tfrac{1}{2}\rho\,|\vec{v}|^2 - \rho\,\frac{\partial \Phi}{\partial t} \ . \tag{1.15.4}$$

Substitution of (1.15.4) into (1.15.3) yields

$$\vec{F}_H = \rho \int_H \left(\tfrac{1}{2}\,|\vec{v}|^2 + \frac{\partial \Phi}{\partial t} \right) \vec{n}\,dS \ , \tag{1.15.5}$$

where by (1.3.10) the contribution of p_∞ is zero.

Because H is at rest with respect to our coordinate system, we have

$$\int_H \frac{\partial \Phi}{\partial t}\,\vec{n}\,dS = \frac{d}{dt}\int_H \Phi\,\vec{n}\,dS \ . \tag{1.15.6}$$

Substitution of (1.15.5) and (1.15.6) into (1.15.2) yields

$$\vec{F} = \rho\,\frac{d}{dt}\int_{\partial B} \Phi\,\vec{n}\,dS - \rho \int_H \left\{ \tfrac{1}{2}|\vec{v}|^2\,\vec{n} - \vec{v}\cdot(\vec{v}\cdot\vec{n}) \right\}\,dS \ . \tag{1.15.7}$$

Now we consider the limit of (1.15.7) when the radius R_H of H tends to infinity. At large distances the velocities induced by B tend to zero at least as fast as R_H^{-2}. This happens when B changes its volume, hence acts as a source or sink at large distances, otherwise the induced velocities tend to zero even more quickly. Anyhow the contribution of the integral over H in (1.15.7) tends to zero for $R_H \to \infty$. This means that

$$\vec{F} = \rho \frac{d}{dt} \int_{\partial B(t)} \Phi \, \vec{n} \, dS \ . \tag{1.15.8}$$

Next we determine the moment $\vec{M} = \vec{M}(t)$, exerted by the body B on the fluid. This moment will be calculated with respect to the origin O of the coordinate system. We apply the theorem:

The resultant moment about O of the forces exerted on an amount of fluid, equals the change of the moment of momentum about O per unit of time, of that amount of fluid.

The moment of momentum of the fluid in Ω, about O is

$$\rho \int_{\Omega} (\vec{r} \times \vec{v}) \, d\text{Vol} = \rho \int_{\partial B + H} \Phi \cdot (\vec{r} \times \vec{n}) \, dS \ , \tag{1.15.9}$$

where the equality follows by $\vec{v} = \text{grad}\,\Phi$ and partial integration. We now take the centre of the sphere H at O, which is, by moving B, an arbitrary point with respect to B. Because the hydrodynamic forces exerted at the fluid inside H by the fluid outside H, are perpendicular to H, their moment about O is zero. By this and by using (1.15.9) the last-mentioned theorem assumes the form

$$\vec{M} = \rho \frac{d}{dt} \int_{\partial B + H} \Phi \cdot (\vec{r} \times \vec{n}) \, dS + \rho \int_{H} (\vec{r} \times \vec{v})(\vec{v} \cdot \vec{n}) \, dS \ , \tag{1.15.10}$$

the last term is the flux of moment of momentum through the fixed geometrical surface H. Because of the choice of the centre of H at O, we have at H that the vector product $\vec{r} \times \vec{n} = 0$, hence in the first integral of (1.15.10) the integration over H gives no contribution. When the radius R_H of H tends to infinity, the contribution of the second integral of (1.15.10) tends to zero, because from the discussion below (1.15.7) it follows that its integrand tends to zero at least as fast as R_H^{-3}. Hence we find

$$\vec{M} = \rho \frac{d}{dt} \int_{\partial B} \Phi \cdot (\vec{r} \times \vec{n}) \, dS \ . \tag{1.15.11}$$

Next we consider the case of a two-sided infinitely long cylinder B with generators parallel to the z-axis. This cylinder is allowed to move arbitrarily and to change its shape and its "volume", but such that its generators remain parallel to the z-axis.

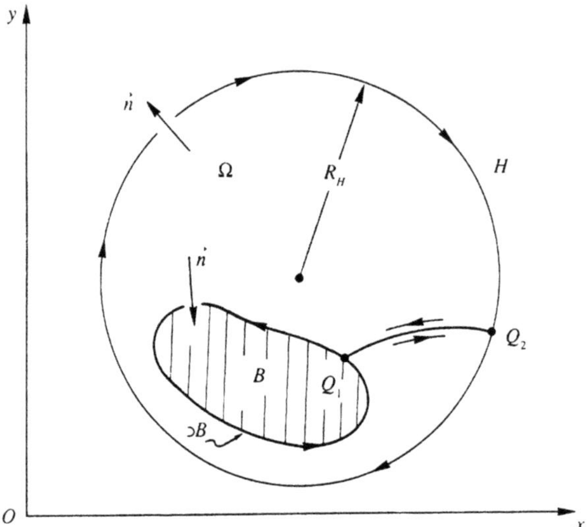

Fig. 1.15.2. Cross section of cylinder B and control surface H.

Then the induced flow depends only on x and y. Also, we assume that B starts to move at $t = t_0$ and that for $t < t_0$ the fluid is at rest. This is not a restriction of generality because the field of flow of our incompressible and inviscid fluid, when no vorticity is shed, depends only on the instantaneous motion of B. In the following we assume $t > t_0$.

Again we assume the fluid to be at rest with respect to the coordinate system (x, y) at large distances from B, that there is no external force field acting at the fluid and, as we mentioned already, that no vorticity is shed from the cylinder into the fluid. We surround the cylinder B by a circular cylindrical control surface H with radius R_H and fixed in space (Figure 1.15.2). Our considerations will be given for a slab of space of unit width in the z-direction and we consider H and the boundary ∂B of B as lines in the (x, y) plane. The 2-dimensional region bounded by these lines is denoted by Ω. An essential difference with the previous 3-dimensional case is that here Ω is doubly connected.

The momentum $\vec{I} = (I_x, I_y)$ of the fluid in the region Ω (taken of unit width) is

$$\vec{I}(t) = \rho \iint\limits_{\Omega} \operatorname{grad} \Phi \, dx \, dy \ , \tag{1.15.12}$$

where Φ is the velocity potential. Because Ω is doubly connected, the function Φ can be "multivalued". We make Ω simply connected by introducing a cut (Q_1, Q_2) from ∂B towards H.

By partial integration we change the 2-dimensional integral in (1.15.12) into an integral along the contour L, which consists of H, ∂B and (Q_1, Q_2) in the indicated directions. We find

$$(I_x, I_y) = \rho \left\{ -\int_L \Phi \, dy, \int_L \Phi \, dx \right\} = \rho \int_L (y, -x) \, d\Phi \ . \tag{1.15.13}$$

We define as the circulation of the cylinder B, the circulation Γ of the contour ∂B, when $\vec{v}$ of (1.3.11) is the velocity just outside ∂B. This circulation has to remain constant during the motion of B, for the following reason. Consider also the circulation Γ_2 of the contour H which, for this intermediate discussion, is assumed to float with the fluid particles. Then by (1.3.16) $d\Gamma_2/dt = 0$, because there is no external force field. However, when Γ of ∂B changes with time, the circulation of the previously introduced contour L has to change with time. Hence, by the expression with $\vec{\omega}$ in (1.3.11), there has to occur rotation in the fluid, which for $t < t_0$ was not present and which can only arise from B. However, this was supposed not to happen, hence Γ is constant. Now we go on with the main line of reasoning where again H is fixed in space.

Because it follows from the above that the difference of Φ at both sides of (Q_1, Q_2) is constant, the contributions of both sides of (Q_1, Q_2) cancel in the last expression of (1.15.13). When i is the imaginary unit, we can write (1.15.13) as

$$I = (I_x + iI_y) = -i\rho \int_{\partial B + H} \zeta \, d\Phi \ , \tag{1.15.14}$$

where $\zeta = (x + iy)$.

The resultant of the force exerted by B is denoted by $F = (F_x + iF_y)$. The resultant force on the fluid inside the fixed H by the fluid outside H is given by

$$F_H = -i \int_H p \, d\zeta \ . \tag{1.15.15}$$

Now we apply again the theorem of momentum. Then we have to consider as in the previous 3-dimensional case, the momentum flux through H. Also here it is easily seen that the contribution of this flux to the force $\vec{F}$ tends to zero for $R_H \to \infty$. Hence we can write

$$F_H + F \approx -i\rho \frac{d}{dt} \int_{\partial B + H} \zeta \, d\Phi \ , \tag{1.15.16}$$

where $\approx$ means that the momentum flux through H is left out of consideration.

Because H is fixed in space, we have

$$\frac{d}{dt} \int_H \zeta \, d\Phi = \int_H \zeta \, d\frac{\partial \Phi}{\partial t} = -\int_H \frac{\partial \Phi}{\partial t} \, d\zeta \ . \tag{1.15.17}$$

The latter equality can be checked as follows. The potential Φ can change by a certain amount by encircling B along H. This amount is independent of time because, as we discussed, the circulation around B is constant. Hence $\partial\Phi/\partial t$ assumes the same value at H after encircling B. From this it follows by partial integration that (1.15.17) is correct.

Next we rewrite (1.15.15) by substitution of (1.15.4) and substitute this result together with (1.15.17) into (1.15.16). Then we find for $R_H \to \infty$

$$F = (F_x + iF_y) = -i\rho \frac{d}{dt} \int_{\partial B} \zeta \, d\Phi \ . \tag{1.15.18}$$

We repeat that the above denoted forces are exerted on the fluid, so by the principle of action equals reaction, minus these force actions are exerted on the body.

1.16. Force Actions Exerted by a Body and Shed Vorticity

In this section we will investigate whether a body (screw propeller, wing, etc.), which exerts a force action at a fluid, has to shed vorticity into the fluid.

Consider a body B of finite extent moving through an incompressible and inviscid fluid, which is at rest at infinity with respect to our Cartesian coordinate system. The body will move with a mean velocity U in the positive x-direction, while its motion is periodic with time period σ. Hence it repeats its motion after each covered distance

$$b = U\sigma \ , \tag{1.16.1}$$

in the x-direction. When we assume that no vorticity is shed, we will prove that B cannot exert a mean force on the fluid.

In this case (1.15.8) is valid. The mean value of $\vec{F}(t)$ over one period σ of time becomes

$$\frac{1}{\sigma} \int_t^{t+\sigma} \vec{F}(\tau)\, d\tau = \frac{\rho}{\sigma} \Bigg\{ \int_{\partial B(t+\sigma)} \Phi(x+b, y, z, t+\sigma)\, \vec{n}(x+b, y, z, t+\sigma)\, dS - \int_{\partial B(t)} \Phi(x, y, z, t)\, \vec{n}(x, y, z, t)\, dS \Bigg\} \ . \tag{1.16.2}$$

The velocities of the fluid at times t and $t+\sigma$ are the same for points (x, y, z) and $(x+b, y, z)$. Hence the difference of the potential Φ at corresponding points and times can only be a constant c. The normals at ∂B are the same for t and $t+\sigma$, at corresponding points. Then by (1.15.8) and (1.3.10) we find

$$\frac{1}{\sigma} \int_t^{t+\sigma} \vec{F}(\tau)\, d\tau = \frac{\rho\, c}{\sigma} \int_{\partial B} \vec{n}\, dS = 0 \ . \tag{1.16.3}$$

From (1.16.13) we find:

A body of finite extent, moving periodically in the way as we described, cannot exert a force with a non-zero mean value, without shedding vorticity.

Inversely, by the principal of action equals reaction, such a body cannot experience a mean force exerted by the fluid. It follows that a thrust producing screw propeller or a lift producing wing, have to shed vorticity into the fluid. Because vorticity is coupled to velocities, we find that a propeller or a wing can function only by spilling kinetic energy into the fluid. The less kinetic energy is spilled, the better they are designed.

Next we consider the moment, with respect to the origin O, of the force actions exerted by a body on the fluid. The mean value with respect to time over one period, of the x-component of this moment becomes

$$\frac{\vec{e}_x\rho}{\sigma}\int_t^{t+\sigma}\vec{M}(\tau)\,d\tau = \frac{\vec{e}_x\rho}{\sigma}\Bigg[\int_{\partial B(t+\sigma)}\Phi(x+b,y,z,t+\sigma)$$

$$\cdot\{(\vec{r}(x,y,z)+b\vec{e}_x)\times\vec{n}(x+b,y,z,t+\sigma)\}\,dS$$

$$-\int_{\partial B(t)}\Phi(x,y,z,t)\{\vec{r}(x,y,z)\times\vec{n}(x,y,z,t)\}\,dS\Bigg]\ , \tag{1.16.4}$$

where $\vec{e}_x$ is the unit vector in the x-direction. Using the arguments below (1.16.2), we find again by (1.3.10)

$$\frac{\vec{e}_x\rho}{\sigma}\int_t^{t+\sigma}\vec{M}(\tau)\,d\tau = \frac{\rho\,c}{\sigma}\int_{\partial B}(y\,n_z - z\,n_y)\,dS = 0\ . \tag{1.16.5}$$

From (1.16.5) we find

A body of finite extent, moving periodically in the way as we described, cannot exert a moment about the x-axis with a non-zero mean value, without shedding vorticity into the fluid.

Hence a screw propeller has to shed vorticity, not only because it delivers a thrust but also because it exerts a moment around its axis.

The 2-dimensional case is different from the 3-dimensional one, as follows from our considerations in the previous section. It can be considered as 3-dimensional, while the velocities are independent of the z-coordinate, hence they do not tend to zero at infinity. For instance, a two-sided infinitely long wing can have a lift force per unit of span without shedding vorticity. In fact, it sheds free tip vorticity at infinity, outside the field of vision.

For the time-dependent force $F = (F_x + iF_y)$ per unit of span exerted by the flexible contour ∂B on the fluid, we use (1.15.18). Then the mean value of F over one period of time σ, can be written as

$$\frac{1}{\sigma}\int_t^{t+\sigma} F(\tau)\,d\tau = \frac{1}{\sigma}\int_t^{t+\sigma} (F_x(\tau)+iF_y(\tau))\,d\tau$$

$$= \frac{-i\rho}{\sigma}\left\{ \int_{\partial B(t+\sigma)} (\zeta+b)\,d\Phi(x,y,t+\sigma) - \int_{\partial B(t)} \zeta\,d\Phi(x,y,t) \right\}$$

$$= \frac{-i\rho\,b}{\sigma}\int_{\partial B(t)} d\Phi(x,y,t) = -i\rho\,U\Gamma\ , \tag{1.16.6}$$

where Γ is the circulation around the 2-dimensional body B. From (1.16.6) we find

$$\frac{1}{\sigma}\int_t^{t+\sigma} F_x(\tau)\,d\tau = 0\ , \qquad \frac{1}{\sigma}\int_t^{t+\sigma} F_y(\tau)\,d\tau = -\rho\,U\Gamma\ , \tag{1.16.7}$$

where we have to remember that F_x and F_y are forces per unit of length exerted by the body on the fluid. The mean value of the lift per unit of span, experienced by B becomes

$$\rho\,U\Gamma\ . \tag{1.16.8}$$

Hence we have the result:

A 2-dimensional body, moving periodically as we described, cannot exert on the fluid a non-zero mean force in the direction of its motion, without shedding vorticity into the fluid.

1.17. Work Done by External Force Field and Moving Body

In this section we will show that the work done by a moving flexible body and an external force field, in an incompressible and inviscid fluid, can be found again as kinetic energy of that fluid. This holds for the non-linear theory as well as for the linear theory.

Consider an incompressible and inviscid fluid which is at rest at infinity with respect to a Cartesian coordinate system (x,y,z). First we derive an expression for the change of the kinetic energe E of the fluid when an external force field and a moving body are present. We start with the non-linear case. The non-linear equations of motion (1.2.1) (with $\mu = 0$) and the equation of mass conservation (1.2.2) are

$$\rho\,\frac{d\vec{v}}{dt} = \rho\,\frac{\partial\vec{v}}{\partial t} + \rho\,(\vec{v}\cdot\mathrm{grad})\,\vec{v} = -\mathrm{grad}\,p + \vec{F}\ , \tag{1.17.1}$$

$$\mathrm{div}\,\vec{v} = 0\ . \tag{1.17.2}$$

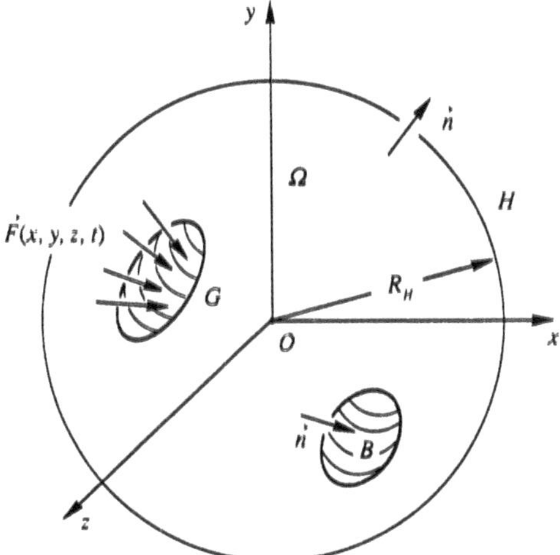

Fig. 1.17.1. Force field $\vec{F}(x, y, z, t)$ and body B.

The finite region for which $\vec{F} \neq 0$ is denoted by $G(t)$. The moving body $B(t)$ is allowed to change its shape and its volume but we assume that it does not shed vorticity. We assume that the body and the region $G(t)$ stay away from each other. Both the body and the region $G(t)$ are surrounded by a large spherical geometrical control surface H of radius R_H, which is fixed in space. The region inside H with the exception of B is called Ω (Figure 1.17.1).

From (1.17.1) it follows

$$\rho \int_\Omega \vec{v} \cdot \left(\frac{\partial \vec{v}}{\partial t} + v_x \frac{\partial \vec{v}}{\partial x} + v_y \frac{\partial \vec{v}}{\partial y} + v_z \frac{\partial \vec{v}}{\partial z} \right) d\text{Vol}$$

$$= - \int_\Omega \vec{v} \cdot \text{grad}\, p \, d\text{Vol} + \int_G \vec{v} \cdot \vec{F} \, d\text{Vol} \ . \tag{1.17.3}$$

By using (1.17.2), the left-hand side of this equation can be reduced to

$$\frac{\rho}{2} \int_\Omega \left\{ \frac{\partial \vec{v}^{\,2}}{\partial t} + \text{div}\,(\vec{v}^{\,2} \cdot \vec{v}) \right\} d\text{Vol} \ . \tag{1.17.4}$$

The first part of this integral is rewritten as

$$\frac{\rho}{2} \int_\Omega \frac{\partial \vec{v}^{\,2}}{\partial t} \, d\text{Vol} = \frac{\rho}{2} \frac{d}{dt} \int_\Omega \vec{v}^{\,2} \, d\text{Vol} - \frac{\rho}{2} \int_{\partial B} \vec{v}^{\,2} \cdot (\vec{v} \cdot \vec{n}) \, dS$$

$$= \frac{dE}{dt} - \frac{\rho}{2} \int_{\partial B} \vec{v}^{\,2} \cdot (\vec{v} \cdot \vec{n}) \, dS \ , \tag{1.17.5}$$

where E is the kinetic energy of the fluid inside H and $\vec{n}$ is the unit normal at the boundary ∂B of B, pointing out of Ω. The second part of the integral (1.17.4) can be written as

$$\frac{\rho}{2}\int_{\Omega} \text{div}\,(\vec{v}^{\,2}\cdot\vec{v})\,d\text{Vol} = \frac{\rho}{2}\int_{\partial B+H} \vec{v}^{\,2}\cdot(\vec{v}\cdot\vec{n})\,d\text{Vol}\ . \tag{1.17.6}$$

Hence (1.17.4) becomes

$$\frac{dE}{dt} + \frac{\rho}{2}\int_{H} \vec{v}^{\,2}\cdot(\vec{v}\cdot\vec{n})\,dS\ . \tag{1.17.7}$$

The first term at the right-hand side of (1.17.3) can be brought into the form

$$-\int_{\Omega} \text{div}\,p\vec{v}\,d\text{Vol} = -\int_{\partial B+H} p\cdot(\vec{v}\cdot\vec{n})\,dS\ . \tag{1.17.8}$$

It is easily seen that the integrations over H tend to zero for $R_H \to \infty$. Hence by (1.17.7) and (1.17.8) we find from (1.17.3)

$$\frac{dE}{dt} = -\int_{\partial B} p\cdot(\vec{v}\cdot\vec{n})\,dS + \int_{G} \vec{v}\cdot\vec{F}\,d\text{Vol}\ . \tag{1.17.9}$$

The first term at the right-hand side is the work done per unit of time by the moving body $B(t)$, it is the integral of the scalar product of the infinitesimal forces $-p\,\vec{n}\,dS$ with the velocity $\vec{v}$ at their point of application. Analogously, the second term at the right-hand side is the work done per unit of time by the infinitesimal forces $\vec{F}\,d\text{Vol}$.

In case that $\vec{F}$ is the gradient of a scalar function K, hence $\vec{F} = \text{grad}\,K$ and when the body is absent, (1.17.9) changes into

$$\frac{dE}{dt} = \int_{\Omega} \vec{v}\cdot\text{grad}\,K\,d\text{Vol} = \int_{\Omega} \text{div}\,(K\vec{v})\,d\text{Vol} = \int_{H} K\cdot(\vec{v}\cdot\vec{n})\,dS\ . \tag{1.17.10}$$

Because the force field is supposed to be zero outside of the finite region $G(t)$, we have $K = \text{const.}$ at H for R_H sufficiently large. Then because $\text{div}\,\vec{v} = 0$, we find from (1.17.10)

$$\frac{dE}{dt} = 0\ . \tag{1.17.11}$$

This is in agreement with the fact that a conservative force field does not induce a velocity field, see below (1.3.4).

We now discuss shortly the linearized theory. Then all velocities have to be sufficiently small, say of $O(\varepsilon)$, where ε is a small parameter. First, we assume that $|\vec{F}(x,y,z,t)| = O(\varepsilon)$. Second, the velocities induced by the body have to be $O(\varepsilon)$. This means that the normal component of the velocities $\vec{v}^{\,*}$ of points of the boundary

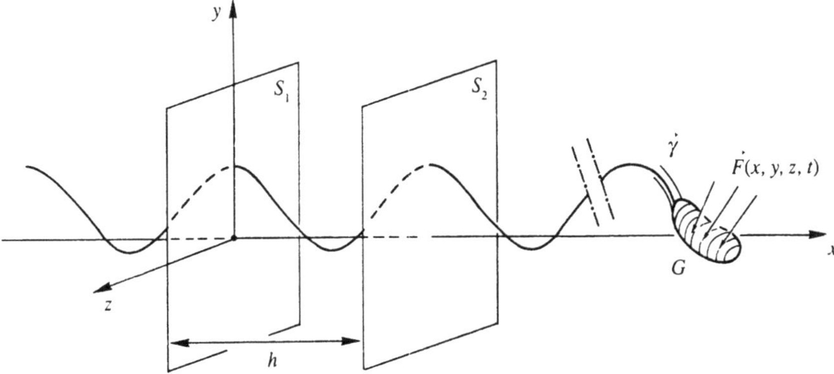

Fig. 1.17.2. Periodically moving force field $\vec{F}(x, y, z, t)$.

$\partial B(t)$ of the body $B(t)$ have to be small, in formula $\vec{n} \cdot \vec{v}^* = O(\varepsilon)$. When these conditions are fulfilled, the potential Φ will be of $O(\varepsilon)$, when an appropriate possibly time-dependent "constant" is subtracted and hence by (1.2.11) grad $p = O(\varepsilon)$. Then it follows that in the previous formulas of this section, we have to retain only the lowest order ($O(\varepsilon^2)$) terms and we arrive again at (1.17.9) ... (1.17.11). *Hence it follows that also in the approximative linear theory the energy is preserved.*

We consider the case of a periodically moving external force field $\vec{F}(x, y, z, t)$, with a mean velocity U in the direction of the positive x-axis, which started at $x = -\infty$. The length period of the motion will be h. We assume the linearized equations of motion to be valid. Then the created free vorticity $\vec{\gamma}$ behind the force field stays where it has been formed (Section 1.4). Hence in the linearized theory we have behind the force field a periodic system of vortices upto $x = -\infty$. It is clear that when we assume that $\vec{F}(x, y, z, t)$ has been switched on not infinitely long ago but very long ago, the velocity field in the neighbourhood of the force region $G(t)$ is nearly periodic. Then the increase of kinetic energy of the fluid during one period of motion is nearly the kinetic energy between two unbounded planes S_1 and S_2 (Figure 1.17.2) parallel to the (y, z) plane, at a distance h apart and situated halfway the most left point of the shed vorticity and the force field. However, in this case our previous reasoning is valid, it means that this increase in kinetic energy equals the work done by $\vec{F}(x, y, z, t)$ over one period.

From this it follows that in the purely periodic case, in which the force field is switched on at $t = -\infty$ hence at $x = -\infty$, the work done by $\vec{F}(x, y, z, t)$ per period equals the kinetic energy between two planes S_1 and S_2 perpendicular to the mean direction of motion of $\vec{F}(x, y, z, t)$, a period apart and situated "infinitely" far behind $\vec{F}(x, y, z, t)$.

1.18. Vorticity of a Lifting Surface and Induced Resistance, Linear Theory

In this section we will introduce two types of vorticity, the bound vorticity and the free vorticity, belonging to a lifting surface of finite span, and discuss the induced resistance of the lifting surface. The fluid is inviscid and incompressible and is at infinity at rest with respect to a Cartesian coordinate system (x, y, z). We use the linearized theory and consider for the sake of simplicity a wing of zero thickness which is nearly parallel to and in the neighbourhood of the plane $H : y = c_1$. Hence this plane is the reference plane of the wing.

The wing moves with a constant velocity U in the positive x-direction and started infinitely long ago at $x = -\infty$. The considerations for this stationary wing can be easily extended to more general situations. In linearized theory it is allowed, as we discussed in Section 1.9, to replace the wing by its planform W which in this case is flat and which is part of the reference surface $H : y = c_1$. We also have at the planform a local coordinate system $(\tilde{x}, \tilde{z})$ which is related to the Cartesian system by $\tilde{x} = x - Ut$, $\tilde{z} = z$. We assume that there is a pressure difference between upper (+) and lower (-) side of the wing denoted by $[p]_-^+(\tilde{x}, \tilde{z}) = p^+(\tilde{x}, \tilde{z}) - p^-(\tilde{x}, \tilde{z})$. Then we can represent the wing by the external force field

$$\vec{F}(x - Ut, y, z) = [p]_-^+(x - Ut, z)\,\delta(y - c_1)\vec{e}_y \;, \tag{1.18.1}$$

where $\delta(y - c_1)$ is the delta function of Dirac and $\vec{e}_y$ is the unit vector in the y-direction. It is also possible to say that the external force field is concentrated at W, is perpendicular to W and has the intensity $F_y(\tilde{x}, \tilde{z}) = [p]_-^+(\tilde{x}, \tilde{z})$ per unit of area. For a more or less realistic wing, which experiences a lift force upwards, we have $[p]_-^+ < 0$ and the equivalent external force field is in the negative y-direction.

We will detect by the use of probing contours, which were introduced in Figure 1.3.3, the vorticity which belongs to this wing. First we consider the rectangular contour A (Figure 1.18.1) of which the sides are parallel to the x- and the y-axis and of which the angular points are A_1, A_2, A_3 and A_4. The position of A is fixed and is such that when W moves on from position I, W is cut by the side A_1A_2 of A along the chord W_1W_2 of length $|W_1W_2|$. We take $|A_2A_3| > |W_1W_2|$. It follows from the first equality of (1.3.16) that as long as the wing is cut by A_1A_2 the circulation Γ_A around A satisfies

$$\frac{d\Gamma_A}{dt} = \frac{1}{\rho}\int\limits_A \vec{F}\cdot d\vec{s} = \frac{[p]_-^+}{\rho} \;. \tag{1.18.2}$$

Now we assume that W has reached position II, where W_1W_2 is entirely inside A. Because $\Gamma_A(I) = 0$ when W is in position I, we find for position II

$$\Gamma_A(II) = \frac{1}{U\rho}\int\limits_{W_1}^{W_2} [p]_-^+(\tilde{x}, \tilde{z})\,d\tilde{x} \;. \tag{1.18.3}$$

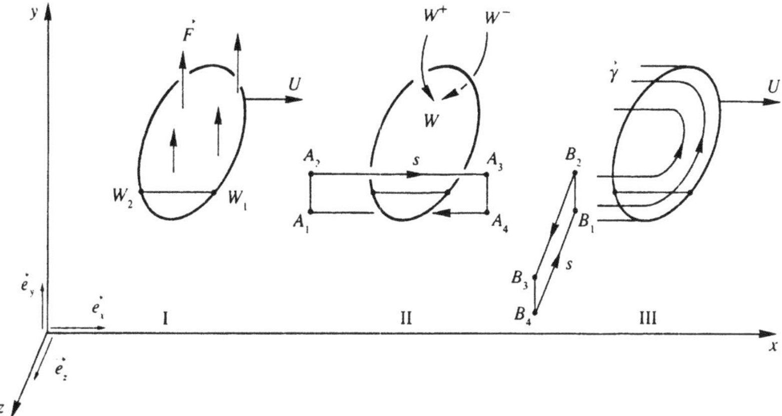

Fig. 1.18.1. Wing with probing contours A and B.

When W arrives at position III, $\Gamma_A = 0$ again because when A_3A_4 cuts W, the direction of the integration variable s along A is in the opposite direction compared to A_1A_2. This means that behind our stationary wing there is no shed vorticity which has a non-zero mean component perpendicular to contour A. When the force field $\vec{F}$ is with respect to the wing time dependent, it is easily seen that this is not true in general. The circulation Γ_A (1.18.3) of contour A, hence in case the wing is in position II, is called the circulation of the profile W_1W_2 of W. The circulation is equal to the integral over W_1W_2 of the vorticity component in the negative z-direction (right-hand screw). We will call this vorticity *bound* vorticity because it is able to sustain a pressure difference between the two sides of the wing.

Next we consider contour B, of which the sides are parallel to the y- and z-axis and of which the angular points are B_1, B_2, B_3 and B_4. We assume the position of B_1B_2 such that when W moves from position II to position III, W is cut by B_1B_2 along the chord W_1W_2 and that B_2B_3 is long enough so that B_3B_4 does not cut W.

As long as W is cut by B_1B_2, (1.18.2) is valid when we replace $d\Gamma_A/dt$ by $d\Gamma_B/dt$ and analogously as before when W has passed from position II to position III, its circulation is changed from 0 to the value (1.18.3). When W moves on, Γ_B does not change anymore, hence behind the wing we have vorticity parallel to the x-axis. It follows that the total vorticity in the negative z-direction (right-hand screw) which is equal to the circulation of the profile W_1W_2 or to Γ_A (1.18.3), is also passing through contour B and is left there behind. This is in agreement with the fact that the 2-dimensional vorticity field $\vec{\gamma}(x, z)$ is free of divergence. The vorticity component in the positive x-direction (right-hand screw) which is detected by contour B, we call *free* vorticity even when it is at the planform W itself. It does not contribute to the lift forces, but makes the total vorticity free of divergence.

We want to refine the probing contours. First we make from contour A the side

A_2A_3 very short, then both A_1A_2 and A_3A_4 cut W at the same time when W is passing. Now contour A gives information how the bound vorticity is divided along the chord W_1W_2. It is not difficult to show that the bound vorticity component γ_z (right-hand screw in positive z-direction) has the value

$$\gamma_z = -\frac{[p]_-^+(\tilde{x}, \tilde{z})}{\rho U} \tag{1.18.4}$$

per unit of length in the x-direction. The relation between the external force field (1.18.1) and the bound vorticity (1.18.4) is locally nothing but the law of Joukowski for the force production by bound vorticity, see also the remark below (1.8.3).

Next we take B_2B_3 from contour B very short, such that both B_1B_2 and B_3B_4 are cut by W when W passes from position II to position III. Then it is easily found that the strength γ_x of the free vorticity follows from (1.18.3) and has the strength

$$\gamma_x = \frac{1}{U\rho}\frac{d}{d\tilde{z}}\int_{W_1(\tilde{z})}^{W_2(\tilde{z})} [p]_-^+(\tilde{x}, \tilde{z})\, d\tilde{x} = \frac{1}{U\rho}\frac{d}{d\tilde{z}}\int_{W_1(\tilde{z})}^{W_2(\tilde{z})} F_y(\tilde{x}, \tilde{z})\, d\tilde{z} \tag{1.18.5}$$

per unit of length in the z-direction. This free vorticity is coupled with a right-hand screw to the positive x-direction, when $[p]_-^+ > 0$. In case of a more realistic wing when $[p]_-^+ < 0$, then all signs of the vorticity are reversed.

We assumed in the foregoing the loading of the wing to be known. The question arises what are the profiles needed to induce this loading. We do not discuss this problem here but refer to Chapter 3, where the analogous problem is discussed for the blades of the screw propeller.

Using the results of the previous section we can calculate the induced resistance R_i of the wing. This resistance can be calculated in two ways. First by calculating the work done per unit of time by the force field $\vec{F} = (0, F_y, 0)$. Here we assume for simplicity that F_y is continuous at the closure of W, hence including the boundary of W. Because we have no work performing body B, but only the external force field which represents the wing, we have by (1.17.9) that this work equals the change of kinetic energy of the fluid per unit of time and is equal to

$$\frac{dE}{dt} = \int_G \vec{F} \cdot \vec{v}\, d\text{Vol} = \int_W F_y \cdot v_y\, d\tilde{x}\, d\tilde{z} = UR_i \ . \tag{1.18.6}$$

When F_y is $O(\varepsilon)$, then also v_y is $O(\varepsilon)$ and hence R_i is $O(\varepsilon^2)$.

Another way of looking at (1.18.6) is as follows. We divide both sides of the last equality by U, then we can interpret v_y/U as the local angle of incidence of the wing. The reaction forces at the wing are perpendicular to the wing surface, hence they have a small component of $O(\varepsilon^2)$ in the x-direction, which represents the local resistance which has to be integrated over W.

As is well known (1.18.6) is not valid when $F_y = [p]_-^+$ has a square root singularity at the leading edge of W, because then we have to take into account a suction force of

$O(\varepsilon^2)$ at the leading edge which has to be subtracted from the value given in (1.18.6). We return to this in Section 1.21.

The second way of calculating the induced resistance R_i is by calculating dE/dt (1.18.6) from the kinetic energy between two planes perpendicular to the x-axis and a distance U apart (Figure 1.17.2). The velocities here are induced by the two-sided infinite strip of free vorticity behind the wing. In this way however a possible suction force is taken into account correctly and we have $R_i = U^{-1}\, dE/dt$.

Next we make a remark about the free vorticity shed by an unsteady lifting surface. As an example we consider the following event. The wing comes from $x = -\infty$ in the same way as before and has upto and including position II the steady load $[p]_-^+(\tilde{x}, \tilde{z})$. Hence in position II the circulation of contour A is again equal to $\Gamma_A(II)$ (1.18.3). But now by some trick, we suddenly make the load of W equal to zero directly after the arrival at position II and keep $[p]_-^+(\tilde{x}, \tilde{z}) = 0$, while W moves on to position III. Then by (1.18.2) there is no change anymore of the circulation of contour A, or Γ_A keeps the value (1.18.3). It follows that, as contrasted with the previously considered steady case, free vorticity has been left behind by W which has a component in the z-direction. In the same way as before, this free vorticity can be analyzed more closely by taking a contour A with a very short side A_2A_3. Periodically moving lifting surfaces can be treated analogously.

Next we discuss the concept of bound vorticity somewhat more closely by means of a linearized 2-dimensional flow. We consider in the neighbourhood of the x-axis a profile (a, b) without thickness, of chord length c and moving in the positive x-direction with a constant velocity U (Figure 1.18.2). The profile experiences a pressure jump $[p]_-^+(s) < 0$ along its chord, where s is a length parameter along the projection of the profile at the x-axis, $0 \leq s \leq c$. Hence we can replace the profile by an external force field with only a non-zero y-component

$$F_y(s) = [p]_-^+(s) \;, \tag{1.18.7}$$

acting at the part (a, b) of the x-axis. Then analogous as before, we have bound vorticity along the profile of strength

$$\gamma(s) = \frac{-F_y(s)}{\rho\, U} \tag{1.18.8}$$

per unit of length in the x-direction. We call $\gamma(s)ds$ the amount of vorticity bound to the local force $F_y(s)ds$. This denomination is coupled to the possibility of giving a meaning to "a force $F_y(s)ds$ acting at the fluid for a given value of s and moving with the velocity U." When however we consider a profile of time-dependent length $c = c(t)$ and of which the forces are time dependent $F_y(s, t)ds$, then this meaning is lost. The velocity of a point of the profile or of the elementary force $F_y(s, t)ds$ can be chosen at will.

This holds even for the simple case of Figure 1.18.2 and we will show the arbitrariness of the concept bound vorticity with the help of this case. For simplicity we

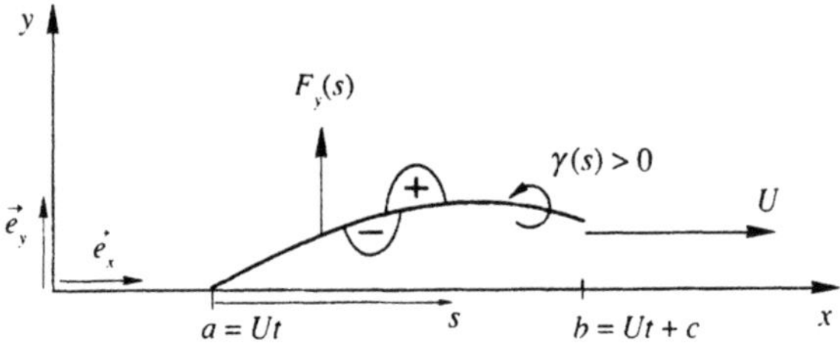

Fig. 1.18.2. Stationary motion of profile (a, b).

assume a constant pressure jump over the profile hence

$$\gamma = \frac{-F_y}{\rho U} = \frac{-[p]_-^+}{\rho U} = \text{const.} \tag{1.18.9}$$

We consider a row of time-dependent, concentrated elementary forces along the entire x-axis $(-\infty < x < \infty)$ at intervals of length ds. These forces are parallel to the y-axis and move with the velocity $V > U$ in the positive x-direction. Behind the profile the strength of these forces is zero. When they catch up with the trailing edge of the profile, they are switched on to the desired strength $F_y ds = [p]_-^+ \, ds$. Then their concentrated bound vorticity has the strength

$$\tilde{\gamma}\, ds = \frac{-F_y ds}{\rho V} \tag{1.18.10}$$

while the starting vortices $-\tilde{\gamma} ds$ remain at the place where they are formed. This is analogous to (1.8.2) where the length of the starting vortex is ε. Behind the profile these latter vortices give rise to a free vortex sheet of strength

$$\frac{F_y(V-U)}{\rho VU}\,, \tag{1.18.11}$$

per unit of length in the x-direction, when they are again spread out evenly.

Then the elementary forces move along the profile towards the leading edge and create a bound vortex layer of strength

$$\tilde{\gamma} = \frac{-F_y}{\rho V} \tag{1.18.12}$$

which is different from (1.18.9).

Next the elementary forces reach the leading edge and when they pass this edge, are switched off and leave behind at that place their vorticity as ending vorticity (below (1.8.3)). This ending vorticity is then passed by the profile at which it creates a free vortex layer of strength per unit of length

$$\frac{-F_y(V-U)}{\rho VU}\,. \tag{1.18.13}$$

Hence at the profile we have two vortex layers of strength (1.18.12) and (1.18.13), which together form a layer of strength

$$\frac{-F_y}{\rho V} - \frac{F_y(V-U)}{\rho VU} = \frac{-F_y}{\rho U} , \tag{1.18.14}$$

which equals the strength of the bound vorticity in the first approach (1.18.8). Behind the trailing edge we have the two layers of strength (1.18.11) and (1.18.13), which cancel each other, as it has to be in comparison with our first approach.

From this it follows that although the bound and the free vorticity differs in both approaches, their flow fields are the same because the law of Biot and Savart makes no distinction between bound and free vorticity. When it is obvious we will use the words "bound vorticity" of for instance a screw propeller blade, in the sense as was used in the first approach (1.18.8).

The foregoing can be generalized to arbitrarily flexible lifting surfaces which are allowed to expand or to contract, as we formulated in Section 1.10. The velocities of its points are of no interest and can be chosen within certain limits at will. The same holds for the velocities of the time-dependent elementary forces which represent the lifting surfaces, only the vorticity created by them is of importance.

The description given here will be useful when we consider the optimization of flexible wings.

Finally, we discuss in this section a method to give an impression of the distribution of the vorticity of a lifting surface. We assume the vorticity to be situated at the planform W of the lifting surface and on that part of its reference surface H which has been passed through by W. Because the 2-dimensional vorticity field is free of divergence, we can define a "vorticity stream function" $\Psi = \Psi(\xi, \eta)$ where ξ and η are coordinates at H. These coordinates are curvilinear when we consider for instance a screw blade. The calculation of Ψ is as follows.

For two points (ξ_1, η_1) and (ξ_2, η_2) we have

$$\Psi(\xi_2, \eta_2) - \Psi(\xi_1, \eta_1) = \int_{(\xi_1,\eta_1)}^{(\xi_2,\eta_2)} \gamma_n \, ds , \tag{1.18.15}$$

where γ_n is the component of the vorticity perpendicular to the length element ds of an arbitrary line connecting the two points.

When we prescribe at one point of H the value Ψ, it is fixed everywhere at H by (1.18.15). So we can take $\Psi = 0$ at some point of H in front of the leading edge of W. Then Ψ becomes zero everywhere at the boundary of the vorticity at H, hence in particular at the leading edge of W. By plotting the level lines of Ψ, we find the shape of the vortex lines at H. In case the difference of the values of Ψ at these lines is constant, then the distance of these lines gives insight in the strength of the vorticity field. When the lines are close, this strength is large and inversely. The function Ψ can also be interpreted as the strength of the dipole layer at H because it is equal to the jump of the velocity potential of the flow. An application of this method is, for instance, given in Figure 6.5.4.

1.19. Bound Vortex "Ending" at Plate of Finite Dimensions

Sometimes the statement is heard: a bound vortex can end at a rigid plate. This seems to be in contradiction with the law that a vortex field is free of divergence (div rot $\vec{v} \equiv 0$).

Bound vorticity "ending" at a rigid plate is important in ship propulsion. There exist screw propellers with end plates at the tips of their blades. More common is the ducted propeller, where the tips of the blades move closely to the inner side of the duct.

In order to simplify the problem, without loosing its essence, we consider the case of a flat plate of finite extent and zero thickness. The plate W (Figure 1.19.1) coincides with part of the (x, y) plane and contains the origin. The fluid in which it is embedded has a velocity U in the positive x-direction. Concentrated at the z-axis for $z \geq a > 0$ we have an external force field

$$\vec{F}(x,y,z) = F^*(x,y)\,\vec{e}_y = F \cdot \delta(x)\delta(y)\,\vec{e}_y \ , \qquad 0 < a \leq z \ ,$$

$$\vec{F}(x,y,z) = 0 \ , \qquad z < a \ , \tag{1.19.1}$$

where $\vec{e}_y$ is the unit vector in the y-direction and we take F = const. > 0.

The z-axis for $z \geq a$ represents a wing or a screw blade, W represents an end plate or a part of a duct. We will consider what happens in the limit $a \to 0$.

We use the linearized theory and assume the fluid to be incompressible and inviscid. Then the vorticity belonging to the force field can be detected by probing contours such as A and B which float with constant velocity U in the positive x-direction. First we consider contour A of which the sides are parallel to the y- and z-axis and have length $2l$ ($l < a$). The midpoint of A is at the line $(y = 0, z = a)$. At time $t = 0$ the x-coordinate of A equals $\tilde{x} < 0$. This means that only side A_4A_1 will cut the external force region. By integrating (1.3.16) with respect to time, we find for the circulation of A, after it has drifted to positive values of x,

$$\Gamma_A = \int \left\{ \int_{-l}^{l} F^*(\tilde{x} + Ut, y)\,\vec{e}_y \cdot (dy\,\vec{e}_y) \right\} dt$$

$$= \frac{F}{\rho} \int \left\{ \int_{-l}^{l} \delta(\tilde{x} + Ut)\,\delta(y)\,dy \right\} dt = \frac{F}{\rho U} \ . \tag{1.19.2}$$

This circulation does not change anymore when A floats downstream. Also the circulation Γ_A does not depend on the length l of the sides of A, these can be chosen as short as we like. This means that there is a concentrated free vortex along the line $(x > 0, y = 0, z = a)$. Analogously, the bound vortex along the z-axis for $a < z$ is detected by means of contour B. The strength of both vortices is equal and has the value

$$\Gamma = \frac{F}{\rho U} \ , \tag{1.19.3}$$

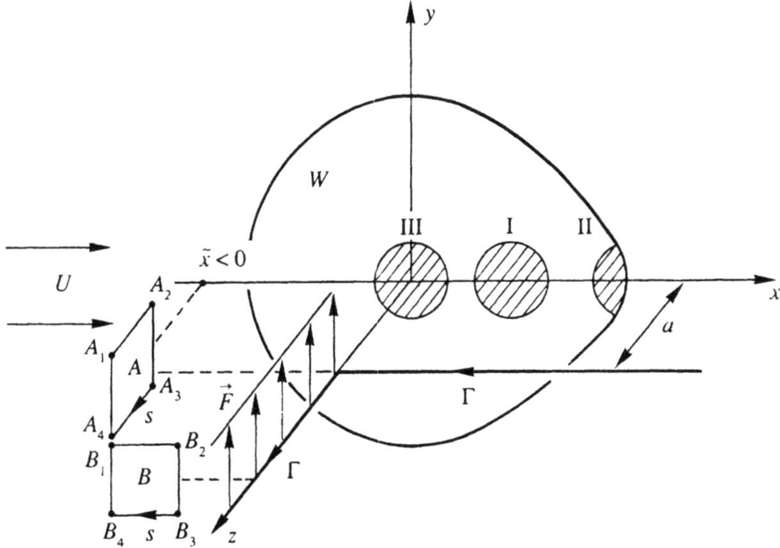

Fig. 1.19.1. Bound vortex ending closely to a rigid plate.

where the indicated directions of Γ in Figure 1.19.1 are coupled with a right-hand screw to the locally induced velocities.

We now discuss the question, what happens when $a \to 0$, that is when the free vortex tends to the positive x-axis and hence will coincide with the plate W. We consider this limit separately for three small characteristic regions I, II and III which are all cut by the x-axis. Region I is somewhat in between the origin and the trailing edge of W, region II is partly bounded by the trailing edge and region III contains the origin.

First region I. When a is sufficiently small we have predominantly the situation of a two-sided infinitely long free vortex of strength Γ in the neighbourhood of the unbounded (x, y) plane. The occurring velocity field is simple. At points with $z > 0$ the influence of the plate is the same as the influence of a vortex which is the image of the free vortex under discussion (Figure 1.19.2 (a)). By this the tangential velocity $\tilde{v} = \tilde{v}(y)$ is known (1.1.16), $\tilde{v} = -\Gamma a/\pi(a^2 + y^2)$. Behind the unbounded plate ($z < 0$) the velocity induced by the original free vortex is zero. Hence by the definition of a vortex layer the vorticity which can represent the plate is

$$\tilde{\gamma}(y) = \frac{\Gamma a}{\pi a\,(a^2 + y^2)}\,, \tag{1.19.4}$$

which is parallel to the x-axis and coupled with a right-hand screw to the positive x-direction (Figure 1.19.2 (b)). We note that the total amount of vorticity $\tilde{\gamma}$ at the plate is

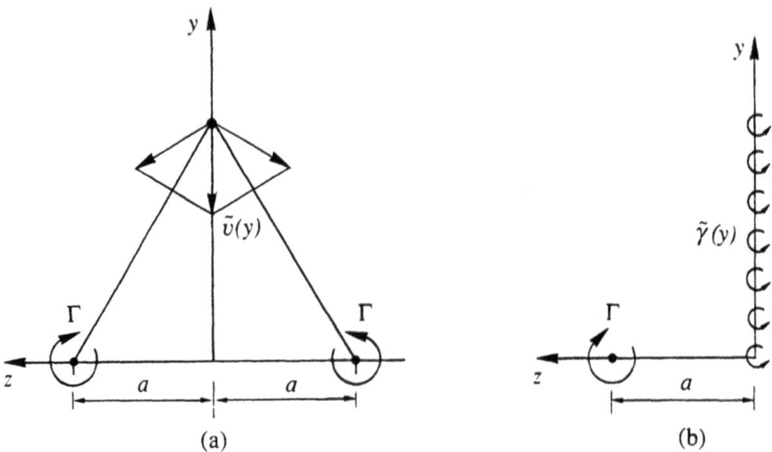

Fig. 1.19.2. Cross section of region I, $a \to 0$.

$$\int_{-\infty}^{\infty} \tilde{\gamma}(y)\, dy = \frac{\Gamma a}{\pi} \int_{-\infty}^{\infty} \frac{dy}{(a^2 + y^2)} = \Gamma \ . \tag{1.19.5}$$

In the neighbourhood of the small region I the velocities induced by the free vortex, the distributed vorticity $\tilde{\gamma}(y)$ and the bound vortex along the z-axis, are all perpendicular to the main stream $(U, 0, 0)$. Hence by the linearized version of Bernoulli's theorem (1.2.12) it follows that the pressure in the fluid equals the undisturbed pressure p_∞ at infinity.

Second we consider region II, which for sufficiently small values of a behaves for the dominant free vorticity as a half-infinite plate with an oblique trailing edge (Figure 1.19.3). The solution for this problem is simple: the vorticity is the same as we found for region I. We only have to take away that part of the unbounded plate used in the discussion of region I, which is downstream of the trailing edge, however we leave the vortex system $\tilde{\gamma}(y)$ intact. This is correct because upstream of the trailing edge we satisfy the boundary condition of a vanishing normal velocity component at the plate and we satisfy the Kutta condition at trailing edge. This latter statement is true because as we already remarked, the pressures at region I and hence now also at region II are undisturbed and equal to p_∞, hence we have no pressure jump at the trailing edge.

For the two regions I and II we consider the limit $a \to 0$. It follows from (1.19.4) that $\tilde{\gamma}(y) \to 0$ for each fixed $y \neq 0$, however as follows from (1.19.5) the total strength of $\tilde{\gamma}$ remains equal to Γ. Hence for $a \to 0$ the concentrated free vortex is annihilated by the vorticity $\tilde{\gamma}(y)$ at W as well as behind the trailing edge. So we expect that in the limit $a = 0$, no free concentrated vorticity will exist anymore.

Finally, we describe the vorticity for region III in the limit case $a = 0$, hence the dominant vorticity in the neighbourhood of the origin O, where the concentrated

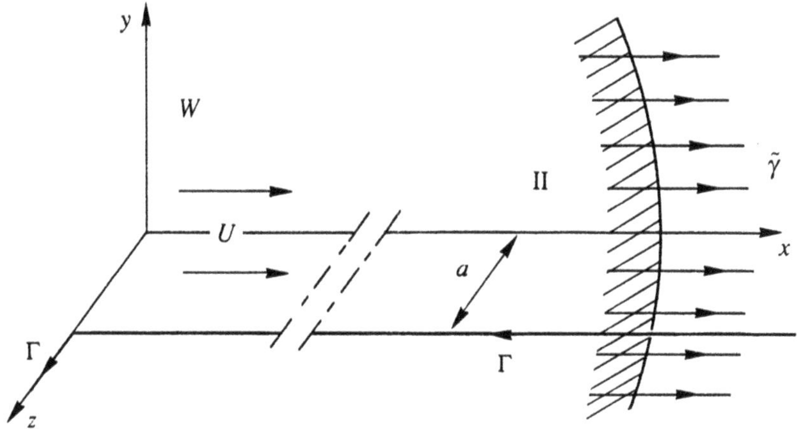

Fig. 1.19.3. Region II with trailing edge.

bound vortex along the z-axis meets the plate W. Then we neglect the influence of the boundary of the plate and consider a vortex ending at an unbounded plate. In this case we have an axisymmetric problem of which the solution is as follows. At the plate we have radial vorticity of strength

$$\frac{\Gamma}{2\pi\,(x^2+y^2)^{1/2}} \tag{1.19.6}$$

per unit of length along the circle with radius $(x^2+y^2)^{1/2}$ (Figure 1.19.4). Then it is seen that the total, 3-dimensional vorticity field is without divergence and it can be proved simply by symmetry considerations that the component of the induced velocity normal to the plate is zero. Hence the boundary condition for the velocity field is satisfied.

We will go one step further and consider the component of the induced velocity just behind the plate. The velocity, induced by the half-infinite bound vortex, at the plate is tangent to the circles around O and has the magnitude $\Gamma/4\pi(x^2+y^2)^{1/2}$. The velocity just before and just behind the plate, induced by the radial vorticity, is also tangent to the circles around O and has a jump across the plate of magnitude $\Gamma/2\pi(x^2+y^2)^{1/2}$. It is easily seen that just behind the plate ($z=-0$) half of this velocity jump annihilates the first-mentioned velocity induced by the bound vortex. So we find that the velocity induced by the bound vortex and the radial vorticity, is zero behind the plane $z=0$ when this plane is considered to be unbounded.

When we consider the situation of a finite end plate (Figure 1.19.1) for the case $a=0$, it is plausible from the foregoing that the spreading of the vorticity is qualitatively as given in Figure 1.19.5. Because we have to satisfy the Kutta condition at the trailing edge, hence no pressure jump is allowed at this edge, the vorticity at the plate has to meet the trailing edge, parallel to the direction of the incoming main flow, hence parallel to the x-axis.

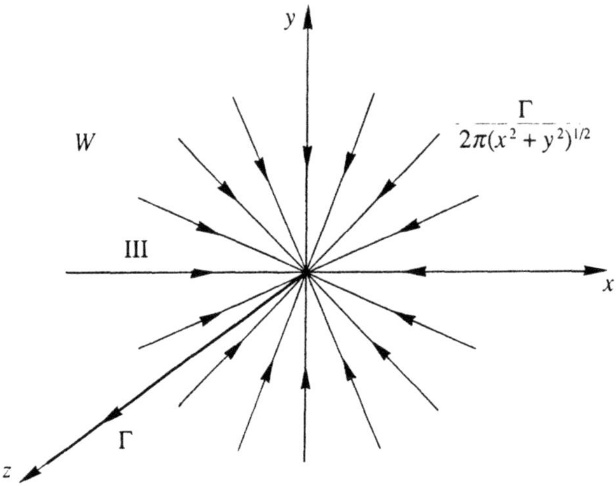

Fig. 1.19.4. The dominant vorticity at region III, $a = 0$.

Finally, we observe an important feature of the influence of an end plate. First consider the configuration drawn in Figure (1.19.1) with $a > 0$, hence when the bound vortex ends away from the plate. Then we have parallel to the x-axis, a concentrated half-infinite free vortex. Such a vortex has theoretically an infinite amount of kinetic energy per unit of length around it. In fact, when we are looking far behind the plate, we find for this kinetic energy per unit of length

$$\frac{\rho}{2}\int_{-\infty}^{\infty}\int_{-\infty}^{\infty} \vec{v}^{\,2}\,dx\,dy = \frac{\rho}{2}\int_{-\infty}^{\infty}\int_{-\infty}^{\infty} \frac{\Gamma^2}{4\pi^2(x^2+y^2)}\,dx\,dy$$

$$= \frac{\rho\,\Gamma^2}{4\pi}\int_{0}^{\infty}\frac{dr}{r} = \frac{\rho\,\Gamma^2}{4\pi}\ln r\,\Big|_0^{\infty} \;. \tag{1.19.7}$$

This result becomes infinite for $r = \infty$ as well as for $r = 0$ and both boundaries of integration yield a positive infinite contribution. So the kinetic energy around such a vortex is infinite at "large distances" as well as "directly around" the vortex.

When we have a lifting line or a concentrated bound vortex of finite span along the z-axis, then we have two concentrated trailing vortices of opposite sign. Because now infinitely far behind the lifting line, for large values of r the velocity becomes $O(r^{-2})$, the kinetic energy is "finite for $r \to \infty$". However, the kinetic energy in the close neighbourhood of each concentrated vortex remains infinite.

Hence when instead of an incoming flow we consider the external force field of (1.19.1) moving with the velocity U through a fluid at rest, the kinetic energy of this fluid increases per unit of time, by an infinite amount. This means that the induced resistance of the force field for $a > 0$ is infinite. However, when $a = 0$ the

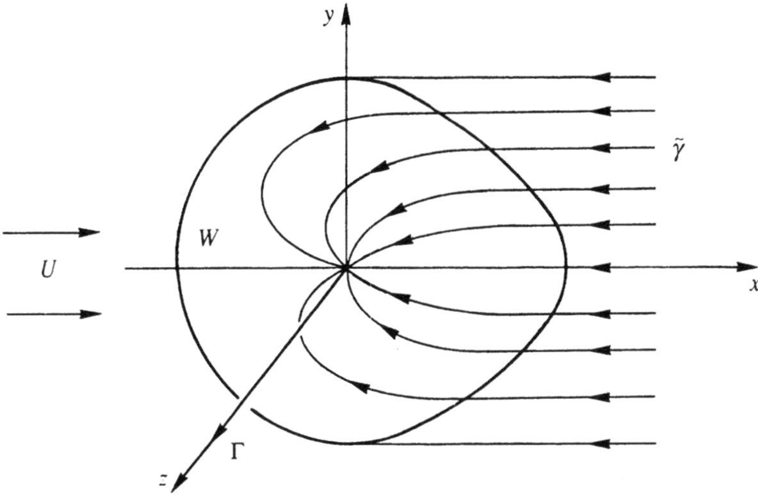

Fig. 1.19.5. Impression of a concentrated vortex ending at a plate of finite dimensions.

concentrated bound vortex connected to the force field ends at the plate and the shed vorticity starting at the trailing edge of the plate has no concentrated part. Then the energy losses per unit of time and hence the induced resistance are finite.

It follows that end plates (or a duct) can have a favourable influence, anyhow from the point of view of potential theory, on the induced resistance of wings or screw propeller blades with loaded tips.

The larger the span in y-direction of the end plate is taken, the better the shed vorticity can be spread and the smaller the induced resistance can be. However, in a real fluid the end plate will experience viscous resistance which also causes energy loss. So in practical cases these two opposing properties of the spreading of shed free vorticity has to be carefully weighed against each other. This will be discussed in Section 6.6.

1.20. Stream Function in Curvilinear Coordinates and Orthogonality Property of Flow behind a Screw Propeller

We will consider a flow in 3-dimensional space which is described with respect to a general curvilinear coordinate system, hence this system does not need to be locally orthogonal, which is important for the theory of the screw propeller. The flow is assumed to be irrotational and to depend only on two of the three coordinates. Our aim is to show that in this case a stream function can be defined and also that under certain conditions the velocity field of the flow is orthogonal to that coordinate direction on which the flow does not depend [64].

We use the notation of Appendix B. The Cartesian coordinate system x^i (i = 1,

2, 3) is embedded in an unbounded space filled with an incompressible and inviscid fluid. In this space we have also a curvilinear coordinate system ξ^i $(i = 1, 2, 3)$ related to the Cartesian system by

$$x^i = x^i(\xi^j) \ . \tag{1.20.1}$$

The velocity field with respect to the system ξ^i is denoted (B.1.9) by

$$\vec{v} = \vec{v}^{\,i}\vec{g}_i = v_i\,\vec{g}^{\,i} \ , \tag{1.20.2}$$

where $\vec{g}_i$ and $\vec{g}^{\,i}$ are the co- and contravariant base vectors, respectively.

In the following we assume that $g_{ij} = \vec{g}_i \cdot \vec{g}_j$ and v^i are independent of ξ^3. Under this assumption we can write the continuity equation of the flow as (B.1.19)

$$\operatorname{div}\vec{v} = \frac{1}{\sqrt{g}}\left\{\frac{\partial}{\partial\xi^1}(\sqrt{g}\,v^1) + \frac{\partial}{\partial\xi^2}(\sqrt{g}\,v^2)\right\} \ , \tag{1.20.3}$$

where $g = |g_{ij}|$, (B.1.7).

The continuity equation (1.20.3) can be satisfied by taking

$$\vec{v} = \operatorname{rot}\vec{B} \ , \tag{1.20.4}$$

where $\vec{B}$ is called the vector potential of $\vec{v}$. We take $\vec{B}$ in the form

$$\vec{B} = B_i\,\vec{g}^{\,i} \ , \quad B_1 = \chi(\xi^1,\xi^2) \ , \quad B_2 = 0 \ , \quad B_3 = \Psi(\xi^1,\xi^2) \ , \tag{1.20.5}$$

χ and Ψ are functions to be determined in relation to the problem under consideration. Then the contravariant components of $\vec{v}$ become (B.1.20)

$$v^1 = \frac{1}{\sqrt{g}}\frac{\partial\Psi}{\partial\xi^2} \ , \qquad v^2 = -\frac{1}{\sqrt{g}}\frac{\partial\Psi}{\partial\xi^1} \ , \qquad v^3 = -\frac{1}{\sqrt{g}}\frac{\partial\chi}{\partial\xi^2} \ . \tag{1.20.6}$$

The choice $B_2 = 0$ is not a restriction of generality. The reason is that $B_2(\xi^1,\xi^2)$ would only contribute to v^3, however this contribution can also be supplied by a suitable adaptation of $\chi(\xi^1,\xi^2)$.

Next we introduce in space a surface H of finite area, by

$$H : \xi^1 = f(u^1) \ , \ \xi^2 = h(u^1) \ , \ \xi^3 = u^2 \ ; \ u^\alpha_a < u^\alpha < u^\alpha_b \ , (\alpha = 1,2) \ , \tag{1.20.7}$$

where f and h are chosen functions, u^1 and u^2 are parameters which form a 2-dimensional coordinate system on H and u^α_a and u^α_b determine the intervals for the values of u^α. The vector of position $\vec{R}$ of a point (ξ^1,ξ^2,ξ^3) of H depends on u^1 and u^2. Hence the two covariant base vectors $\vec{a}_\alpha$ $(\alpha = 1, 2)$ of the coordinate system (u^1, u^2) become by (1.20.7), analogously to (B.1.6),

$$\vec{a}_1 = a^i_1\,\vec{g}_i = \frac{\partial\vec{R}}{\partial u^1} = \frac{\partial\vec{R}}{\partial\xi^i}\frac{\partial\xi^i}{\partial u^1} = \frac{df}{du^1}\cdot\vec{g}_1 + \frac{dh}{du^1}\cdot\vec{g}_2 \ , \tag{1.20.8}$$

$$\vec{a}_2 = a_2^i\, \vec{g}_i = \frac{\partial \vec{R}}{\partial u^2} = \frac{\partial \vec{R}}{\partial \xi^i} \frac{\partial \xi^i}{\partial u^2} = \vec{g}_3 \;, \tag{1.20.9}$$

which are tangent to the coordinate directions at H. It follows from (1.20.5) and (1.20.7) that at H the functions χ and Ψ are functions of u^1 alone.

The flux F through H becomes

$$F = \int_H \vec{v} \cdot \vec{n}\, dS = \int_H \vec{v} \cdot \vec{n}\, |\vec{a}_1\, du^1 \times \vec{a}_2\, du^2| \;, \tag{1.20.10}$$

where $\vec{n}$ is the unit normal at H and dS is an element of area. The unit normal can be written as

$$\vec{n} = \frac{\vec{a}_1 \times \vec{a}_2}{|\vec{a}_1 \times \vec{a}_2|} = \varepsilon_{klm} \frac{a_1^k\, a_2^l\, \vec{g}^{\,m}}{|\vec{a}_1 \times \vec{a}_2|} \;, \tag{1.20.11}$$

where ε_{klm} is the permutation tensor defined in (B.1.16). That (1.20.11) is correct follows from the fact that in the Cartesian coordinate system the formula reduces to the well-known vector product and hence, because it is written in tensor notation, it is valid for all coordinate systems.

Substitution of (1.20.11) into (1.20.10) yields for the flux

$$F = \int_{u_a^2}^{u_b^2} \int_{u_a^1}^{u_b^1} \left(\frac{\partial \Psi}{\partial \xi^1} \frac{\partial \xi^1}{\partial u^1} + \frac{\partial \Psi}{\partial \xi^2} \frac{\partial \xi^2}{\partial u^1} \right) du^1\, du^2$$

$$= \int_{u_a^2}^{u_b^2} \int_{u_a^1}^{u_b^1} \frac{d\Psi}{du^1}\, du^1\, du^2 = \left(u_b^2 - u_a^2\right) \left[\Psi(u_b^1) - \Psi(u_a^1)\right] \;. \tag{1.20.12}$$

It is seen that the component v^3 does not contribute to F because $\vec{a}_2 = \vec{g}_3$ (1.20.9) is tangent to H. From (1.20.12) it follows that Ψ can be interpreted as a stream function because, when u_b^2 and u_a^2 are kept constant, the flux F depends only on the difference of Ψ for the values u_b^1 and u_a^1. Hence Ψ = const. yields a stream tube in 3-dimensional space.

We consider fields of flow of which the vorticity is along the ξ^3 coordinate lines. Hence rot $\vec{v} = \vec{\gamma}(\xi^1, \xi^2) = \gamma^3(\xi^1, \xi^2)\vec{g}_3$ which satisfies the condition div $\vec{\gamma} = 0$ (B.1.19). Then we find for the contravariant components of rot $\vec{v}$

$$\frac{1}{\sqrt{g}} \left(\frac{\partial v_3}{\partial \xi^2} - \frac{\partial v_2}{\partial \xi^3} \right) = 0 \;, \tag{1.20.13}$$

$$\frac{1}{\sqrt{g}} \left(\frac{\partial v_1}{\partial \xi^3} - \frac{\partial v_3}{\partial \xi^1} \right) = 0 \;, \tag{1.20.14}$$

$$\frac{1}{\sqrt{g}}\left(\frac{\partial v_2}{\partial \xi^1} - \frac{\partial v_1}{\partial \xi^2}\right) = \gamma^3(\xi^1, \xi^2) \ . \tag{1.20.15}$$

Because $\vec{v}$ is independent of ξ^3 it follows from (1.20.13) and (1.20.14) that v_3 is independent of ξ^1 and ξ^2, hence

$$v_3 = g_{i3}\, v^i = C \ , \tag{1.20.16}$$

where C is some constant. Then by the first two equalities of (1.20.6) and by (1.20.16) we can, instead of the last equality of (1.20.6), obtain for v^3 the expression

$$v^3 = \frac{1}{\sqrt{g}\, g_{33}}\left\{-g_{13}\frac{\partial \Psi}{\partial \xi^2} + g_{23}\frac{\partial \Psi}{\partial \xi^1} + C\sqrt{g}\right\} \ , \tag{1.20.17}$$

which is a consequence of rot $\vec{v} = \gamma^3(\xi^1, \xi^2)\, \vec{g}_3$.

From (1.20.16) and (B.1.7) it follows that

$$\vec{v} \cdot \vec{g}_3 = v_j \vec{g}^{\,j} \cdot \vec{g}_3 = v_3 = C \ . \tag{1.20.18}$$

Hence we have found the important property that the velocity field is perpendicular to the ξ^3 coordinate direction in case for some problem $C = 0$.

We still have to use (1.20.15), which can be written as

$$\frac{\partial}{\partial \xi^1}\left(g_{i2}\, v^i\right) - \frac{\partial}{\partial \xi^2}\left(g_{j1}\, v^i\right) = \sqrt{g}\, \gamma^3(\xi^1, \xi^2) \ . \tag{1.20.19}$$

Substitution of the first two equations of (1.20.6) and of (1.20.17) yields for Ψ the equation

$$\begin{aligned} &\frac{\partial}{\partial \xi^1}\left[\frac{1}{\sqrt{g}}\left(-g_{22} + \frac{(g_{23})^2}{g_{33}}\right)\frac{\partial \Psi}{\partial \xi^1} + \frac{1}{\sqrt{g}}\left(g_{12} - \frac{g_{13}g_{23}}{g_{33}}\right)\frac{\partial \Psi}{\partial \xi^2} + \frac{g_{23}}{g_{33}}C\right] \\ &\quad - \frac{\partial}{\partial \xi^2}\left[\frac{1}{\sqrt{g}}\left(-g_{12} + \frac{g_{13}g_{23}}{g_{33}}\right)\frac{\partial \Psi}{\partial \xi^1}\right. \\ &\quad \left. + \frac{1}{\sqrt{g}}\left(g_{11} - \frac{(g_{13})^2}{g_{33}}\right)\frac{\partial \Psi}{\partial \xi^2} + \frac{g_{13}}{g_{33}}C\right] = \sqrt{g}\, \gamma^3(\xi^1, \xi^2) \end{aligned} \tag{1.20.20}$$

in which C is still unknown. This constant can be determined by (1.20.18) from the behaviour of the flow at infinity or from the desired flux through a surface.

When we have a Cartesian coordinate system, we obtain from (1.20.6) and (1.20.20) ($x = \xi^1, y = \xi^2, z = \xi^3$)

$$v_x = \frac{\partial \Psi}{\partial y} \ , \qquad v_y = -\frac{\partial \Psi}{\partial x} \ , \tag{1.20.21}$$

$$\frac{\partial^2 \Psi}{\partial x^2} + \frac{\partial^2 \Psi}{\partial y^2} = -\gamma^3(x, y) \ , \tag{1.20.22}$$

where v_x and v_y are automatically the physical components of the velocity field (below (B.1.10)). In (1.20.22) we used still the notation γ^3 in an obvious way, as we will do also in (1.20.24) and (1.20.26).

For a cylindrical coordinate system (Appendix B.2) ($x = \xi^1, r = \xi^2, \varphi = \xi^3$), we find for the physical components of the velocity v_x, v_r and for the differential equation for Ψ

$$v_x = \frac{1}{r}\frac{\partial \Psi}{\partial r} , \qquad v_r = -\frac{1}{r}\frac{\partial \Psi}{\partial x} , \tag{1.20.23}$$

$$\frac{\partial^2 \Psi}{\partial x^2} + r\frac{\partial}{\partial r}\left(\frac{1}{r}\frac{\partial \Psi}{\partial r}\right) = -r^2\gamma^3(x,r) . \tag{1.20.24}$$

For a helicoidal coordinate system (Appendix B.2) ($\tilde{\rho} = \xi^1, \zeta = \xi^2, \sigma = \xi^3$) we obtain the formulas

$$v_{\tilde{\rho}} = \frac{a^2}{\tilde{\rho}}\frac{\partial \Psi}{\partial \zeta} , \qquad v_\sigma = -\frac{a^2}{\tilde{\rho}}\frac{\partial \Psi}{\partial \tilde{\rho}} , \tag{1.20.25}$$

$$\tilde{\rho}\frac{\partial}{\partial \tilde{\rho}}\left(\frac{\tilde{\rho}}{(1+\tilde{\rho}^{\,2})}\frac{\partial \Psi}{\partial \tilde{\rho}}\right) + \frac{\partial^2 \Psi}{\partial \zeta^2} = \frac{2\tilde{\rho}^{\,2}C}{a(1+\tilde{\rho}^{\,2})^2} - \frac{\tilde{\rho}^{\,2}}{a^4}\gamma^3(\tilde{\rho},\zeta) . \tag{1.20.26}$$

In this case it follows from (B.2.13) that $|\vec{g}_3| = O(\tilde{\rho})$ for $\tilde{\rho} \to \infty$. Because by (1.20.18) $\vec{v} \cdot \vec{g}_3$ is a constant C, it is sufficient that $|\vec{v}| = O(\tilde{\rho}^{-1-\varepsilon})$ ($\varepsilon > 0$) for $\tilde{\rho} \to \infty$, in order that C and hence $\vec{v} \cdot \vec{g}_3$ are zero. This happens to be the case when the total circulation of the vorticity $\vec{\gamma}(\tilde{\rho}, \zeta)$ is zero. For simplicity we can assume here that this vorticity is confined to a cylinder of finite radius. Then it is found easily that $|\vec{v}| = O(\tilde{\rho}^{-2})$ for $\tilde{\rho} \to \infty$.

However the free vorticity left behind by a screw propeller described by a linear theory has the desired properties. This means that: *far behind a screw propeller in the linearised theory the induced velocities are orthogonal to the helicoidal lines of the helicoidal coordinate system which belongs to the propeller.* The orthogonality property is important in considerations about optimum shed vorticity (Section 6.3).

1.21. Suction Force at Leading Edge of Lifting Surface

We consider a thin profile under an angle of incidence. When no flow separation occurs, the flow resembles the pattern drawn in Figure 1.21.1 (a). We consider two points A and B in the neighbourhood of the nose of the profile, one at the upper side of the profile and one at its lower side. Between these two points, at the nose a low pressure region occurs caused by the strong curvature of the stream. The fluid particles behave like mass points which fly around the nose, kept in their orbit by a string on which they exert a "centrifugal force". This low pressure results in a force K in the x-direction at the nose. When the profile becomes thinner, the

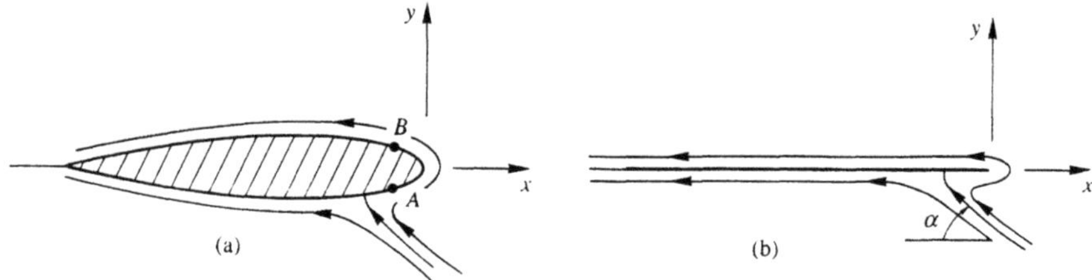

Fig. 1.21.1. Flow around nose of a profile.

frontal area of the nose diminishes, however the curvature at that place and the local velocity increase and hence the pressure becomes lower. In the limit when the profile becomes a plate of zero thickness (Figure 1.21.1 (b)), the frontal area is zero and the pressure has become infinitely low in such a way that still a suction force is exerted at the leading edge in the positive x-direction. Thrust production by means of suction forces does not seem to be very reliable in practice because flow separation in the neighbourhood of the leading edge of a wing or screw blade can disturb this phenomenon rather easily.

In the following we calculate the limit value of the suction force for the case of a 2-dimensional strip of zero thickness as in Figure 1.21.1 (b), our considerations refer to a unit of length in the spanwise direction of the strip. It is clear that the suction force has to be caused, in this case, by infinite velocities just in front of the leading edge of the plate. Hence we can neglect for the calculation of it the vorticity at a "finite distance" of the leading edge, because this induces only finite velocities at that place. For this reason we consider the following problem. Along the x-axis we assume vorticity of strength

$$\gamma(x,t) = \frac{Q(t)}{(-x)^{1/2}} \ , \qquad -a \le x \le 0 \ ,$$

$$\gamma(x,t) = 0 \ , \qquad x < -a \qquad \text{or} \qquad x > 0, \quad a > 0 \ . \tag{1.21.1}$$

This is the characteristic singular behaviour of the vorticity in the neighbourhood of the nose, represented by O, of a profile of zero thickness (for instance Section 4.4 and 4.5). The strength of this singular behaviour, determined by $Q(t)$, is allowed to be time dependent.

First we state the velocities and the potential belonging to a two-sided, infinitely long, straight concentrated vortex (1.1.16). The vortex is perpendicular to the (x,y) plane (Figure 1.21.2), cuts this plane at $(x,y) = (\xi,0)$ and is of strength Γ. Its velocity field is given by

$$v_x = -\frac{\Gamma}{2\pi}\frac{y}{\{(x-\xi)^2+y^2\}} \ , \qquad v_y = \frac{\Gamma}{2\pi}\frac{(x-\xi)}{\{(x-\xi)^2+y^2\}} \ , \tag{1.21.2}$$

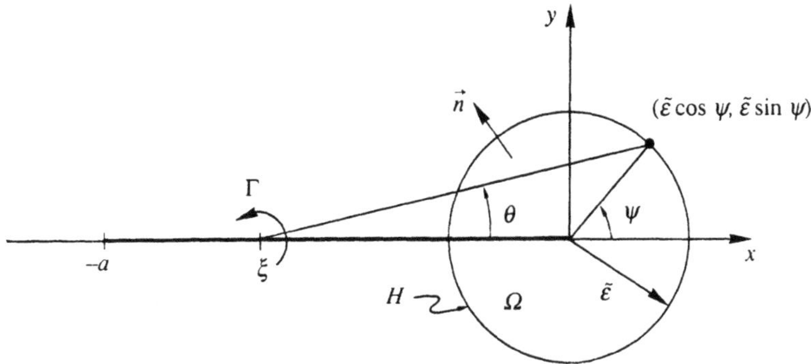

Fig. 1.21.2. Coordinate system at leading edge.

and its potential by

$$\tilde{\Phi} = \frac{\Gamma\theta}{2\pi} , \tag{1.21.3}$$

where θ is the angle denoted in Figure 1.21.2. That this potential is "infinitely many valued" causes no difficulties in our derivation.

By superposition we find from (1.21.2) for the x-component of the velocity induced by the vortex layer (1.21.1) at the point $x = \tilde{\varepsilon}\cos\psi$, $y = \tilde{\varepsilon}\sin\psi$,

$$\begin{aligned} v_x &= -\frac{Q(t)}{2\pi}\int_{-a}^{0} \frac{1}{(-\xi)^{1/2}}\,\frac{\tilde{\varepsilon}\sin\psi}{\{(-\xi+\tilde{\varepsilon}\cos\psi)^2+\tilde{\varepsilon}^2\sin^2\psi\}}\,d\xi \\ &= -\frac{Q(t)\sin\psi}{2\pi\,\tilde{\varepsilon}^{1/2}}\int_{0}^{a/\tilde{\varepsilon}} \frac{1}{\eta^{1/2}(\eta^2+2\eta\cos\psi+1)}\,d\eta\ . \end{aligned} \tag{1.21.4}$$

Later on we want to take the limit $\tilde{\varepsilon} \to 0$, by this the upper bound of the integral tends to ∞, the resulting integral is known ([23] II, page 31). Because this integral is absolutely convergent for each fixed value of ψ with $-\pi < \psi < \pi$, we can write (1.21.4) as

$$v_x = -\frac{Q(t)}{2(2\tilde{\varepsilon})^{1/2}}\left\{\frac{\sin\psi}{(1+\cos\psi)^{1/2}} + o(\varepsilon)\right\} , \qquad -\pi < \psi < \pi , \tag{1.21.5}$$

where $o(\tilde{\varepsilon})$ means a quantity with $o(\tilde{\varepsilon})/\tilde{\varepsilon} \to 0$ for $\tilde{\varepsilon} \to 0$. Analogously, we find for the velocity component in the y-direction

$$v_y = \frac{Q(t)}{2(2\tilde{\varepsilon})^{1/2}}\left\{(1+\cos\psi)^{1/2} + o(\tilde{\varepsilon})\right\} , \qquad -\pi < \psi < \pi . \tag{1.21.6}$$

From (1.21.5) and (1.21.6) we find for the velocity component v_n in the direction of the normal $\vec{n}$ at the control circle H of radius $\tilde{\varepsilon}$ around the nose O of the profile

$$v_n = (v_x \cos\psi + v_y \sin\psi)$$
$$= \frac{Q(t)}{2(2\tilde{\varepsilon})^{1/2}} \left\{ \frac{\sin\psi}{(1+\cos\psi)^{1/2}} + o(\tilde{\varepsilon}) \right\} , \qquad -\pi < \psi < \pi . \tag{1.21.7}$$

It follows from (1.21.5), (1.21.6) and (1.21.7) that these results are easily continuously extended to the closed interval $-\pi \le \psi \le \pi$.

Next we apply the theorem of momentum to the fluid in the small circular region Ω around the nose O of the profile. This is done in the same way as in Section 1.15. The resulting force acting on the fluid inside H equals the change of momentum per unit of time of the fluid occupying the region Ω, to which we have to add the flux of momentum entering or leaving Ω through H.

First we consider the force exerted at Ω by the fluid outside H. The pressure follows from Bernoulli's equation (1.2.11) with $K = 0$ because there is no external force field in the fluid

$$p = \rho \left\{ h(t) - \tfrac{1}{2} |\vec{v}|^2 - \frac{\partial \Phi}{\partial t} \right\} . \tag{1.21.8}$$

The possibly time-dependent $h(t)$ cannot give a resultant force by integrating around H. Hence we have to calculate $|\vec{v}|^2$ and $\partial\Phi/\partial t$. From (1.21.5) and (1.21.6) it follows

$$|\vec{v}|^2 = v_x^2 + v_y^2 = \frac{Q^2(t)}{4\tilde{\varepsilon}} + \frac{o(\tilde{\varepsilon})}{\tilde{\varepsilon}} . \tag{1.21.9}$$

The first term of (1.21.9) gives no contribution to the pressure integral around H, because it is only time dependent and the second term becomes too small for $\tilde{\varepsilon} \to 0$. The potential of the vortex layer has by (1.21.3) the value

$$\Phi(x, y, z, t) = \frac{Q(t)}{2\pi} \int_{-a}^{0} \frac{1}{(-\xi)^{1/2}} \, \theta(x, y, \xi) \, d\xi \tag{1.21.10}$$

which remains bounded for all values of x and y in a neighbourhood of the origin. Hence also $\partial\Phi/\partial t$ remains bounded and cannot contribute to the integral of the pressure forces at H, for $\tilde{\varepsilon} \to 0$.

Next we consider the flux of momentum through H. This becomes in the x-direction and in the y-direction

$$\rho \int_{-\pi}^{\pi} v_x \, v_n \, \tilde{\varepsilon} \, d\psi = -\frac{\rho \, \pi \, Q^2(t)}{4} + o(\tilde{\varepsilon}) , \tag{1.21.11}$$

$$\rho \int_{-\pi}^{\pi} v_y \, v_n \, \tilde{\varepsilon} \, d\psi = o(\tilde{\varepsilon}) , \tag{1.21.12}$$

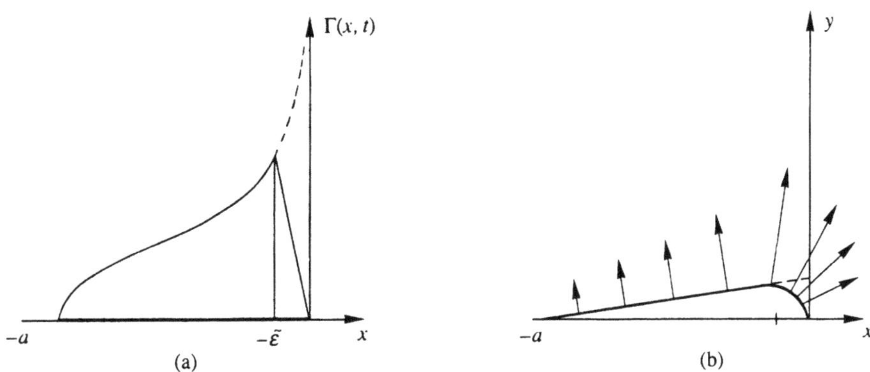

Fig. 1.21.3. Multilated vorticity and profile belonging to it.

respectively.

Finally, we check the momentum of the fluid inside Ω. Because the velocity components (1.21.5) and (1.21.6) are $O\left((x^2+y^2)^{-1/4}\right)$ for $x^2+y^2 \to 0$, and the area of Ω is $O(\tilde{\varepsilon})^2$, it follows that the momentum inside Ω as well as its time derivative tend to zero with $\tilde{\varepsilon} \to 0$.

From the foregoing follows the desired result. On the fluid inside the small circle with radius $\tilde{\varepsilon}$ around O acts a force of which the x- and y-component are given by (1.21.11) and (1.21.12), respectively. Hence by the principle action equals reaction and taking the limit $\tilde{\varepsilon} \to 0$, we find at the nose of the profile a suction force per unit of span K in the positive x-direction, given by

$$K = \frac{\rho\pi}{4} Q^2(t) \ . \tag{1.21.13}$$

This force occurs typically when the vorticity of the profile in the neighbourhood of the leading edge has a singular behaviour as given in (1.21.1). When the singularity is less strong, say of order $(-x)^{-1/2+\varepsilon}$, $\varepsilon > 0$, then no suction force occurs. This happens when the curvature or camber of the profile is chosen in such a way that the loading of the profile becomes sufficiently smooth near the leading edge. We will discuss this later on in Section 6.5 in relation with the design of ship screws.

Now we give another, much more simple reasoning to show the theoretical existence of a suction force at a sharp leading edge. Consider a flat, two-sided, infinite strip of zero thickness which moves with a constant velocity U and with a constant angle of incidence in the positive x-direction. Such a wing does not leave behind vorticity and hence by the last paragraph of Section 1.16.8, the x-component of the time-independent force experienced by the profile has to be zero. However, the pressure jump over the profile, when the leading edge and the trailing edge are left out of consideration, yields a force which forms an angle α with the y-axis. At the trailing edge the vorticity has no infinite singularity (Kutta condition), hence no singular force can occur there. Then the only possibility to get the resultant force at the inner part of the profile perpendicular to the x-axis is by a suction force at the leading edge.

We conclude this section with an heuristic argument, which can be made rigorous, about the origin of the suction force. In (1.21.1) we considered vorticity with a square root singularity. We now multilate this vorticity as is drawn in Figure 1.21.3 (a). From $x = -\tilde{\varepsilon}$ to $x = 0$, the vorticity tends to zero, for instance linearly. Then it is not difficult to show that the profile which belongs to this vorticity, is curved very steeply downwards in the neighbourhood of the leading edge (Figure 1.21.3 (b)). Hence the force action of the fluid on the profile gets a horizontal component which changes, in the limit $\tilde{\varepsilon} \to 0$, into the suction force.

This indicates that the suction force is not a new concept for profiles of zero thickness, but has the same origin as the other horizontal components of the forces induced by the pressure jump across the profile. Only part of these has shifted to the leading edge.

1.22. About the Roll-Up of Free Vortex Sheets

In this section we will indicate the difficulties which arise in a non-linear theory when we consider the evolution of the configuration of the shed free vorticity behind a lifting surface. This evolution is caused by the induced velocities of the vorticity on the wings or the screw blades and by the induced velocities of the shed free vorticity itself. The effect is important for heavily loaded screw propellers where the roll-up of the tip vorticity will influence the velocity induced at the blades, because each blade passes rather closely the free vorticity shed by the foregoing blade.

Problems of vortex sheet evolution in relation to more or less complicated 3-dimensional flow behind lifting surfaces are for instance discussed by Hoeymakers [26]. In general, the vortex sheets are represented by a number of discrete singular vortex lines which are transported by the local velocity field.

More simple is the roll-up of 2-dimensional vortex sheets such as a two-sided infinite strip of finite width of which the vorticity is parallel to the edges of the strip. An important case is when the intensity of the vorticity has a square root singularity near the edges of the strip. This occurs in the linearized theory, behind a wing with least induced resistance. This 2-dimensional problem is discussed for instance by Krasny [42]. When such a vortex sheet deforms in the course of time, it gives insight in the 3-dimensional evolution of the shed free vorticity behind a wing. When the wing has the velocity of advance U, we can expect that the shape of the 2-dimensional sheet at time t resembles the shape of the cross section of the 3-dimensional sheet at a distance Ut behind the wing.

Because here we will show only the fundamental aspects of the roll-up process, we discuss a still more "simple" 2-dimensional problem, namely the strip of half-infinite width which has only one edge. So we leave in first instance out of consideration the mutual interaction of the edges of a sheet of finite width.

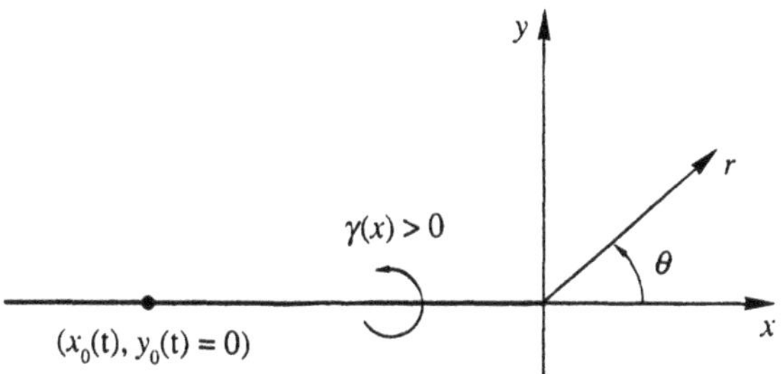

Fig. 1.22.1. Half infinite vortex sheet along the negative x-axis, $t < 0$.

We follow the beautiful formulation of the roll-up of a vortex sheet as given by Birkhoff [6], which is used by Pullin for the half-infinite vortex sheet [56].

Consider a free vortex sheet (Figure 1.22.1) which at times $t < 0$ is stretched along the negative x-axis, while the vorticity lines are perpendicular to the (x, y) plane. The intensity of the vorticity $\gamma(x)$ for $t < 0$ is assumed to be

$$\gamma(x) = \frac{a}{|x|^{1/2}} \ , \quad x < 0 \ ; \qquad \gamma(x) = 0 \ , \quad x > 0 \ ; \qquad t < 0 \ . \tag{1.22.1}$$

We introduce the complex velocity potential which belongs to the vorticity field (1.22.1)

$$\begin{aligned} W(z,t) &= \Phi(x,y) + i\Psi(x,y) = -iaz^{1/2} \\ &= -iar^{1/2}\left(\cos\frac{\theta}{2} + i\sin\frac{\theta}{2}\right) \ , \qquad z = x + iy \ , \quad t < 0 \ , \end{aligned} \tag{1.22.2}$$

where r and θ are denoted in Figure 1.22.1, i is the imaginary unit, $\Phi(x, y)$ is the velocity potential of the flow

$$\Phi(x,y) = ar^{1/2}\sin\frac{\theta}{2} \ , \qquad v_x = \frac{\partial\Phi}{\partial x} \ , \qquad v_y = \frac{\partial\Phi}{\partial y} \ , \tag{1.22.3}$$

and $\Psi(x, y)$ is the stream function

$$\Psi(x,y) = -ar^{1/2}\cos\frac{\theta}{2} \ , \qquad v_x = \frac{\partial\Psi}{\partial y} \ , \qquad v_y = -\frac{\partial\Psi}{\partial x} \ . \tag{1.22.4}$$

Hence we have

$$\frac{dW}{dz} = \frac{\partial\Phi}{\partial x} + i\frac{\partial\Psi}{\partial x} = v_x - iv_y = \overline{v_x + iv_y} \ . \tag{1.22.5}$$

We remark that this velocity field can be considered as a flow around a rigid impermeable plate of zero thickness which stretches along the negative x-axis. In connection with the previous section we find that at the origin $(x, y) = (0, 0)$ a

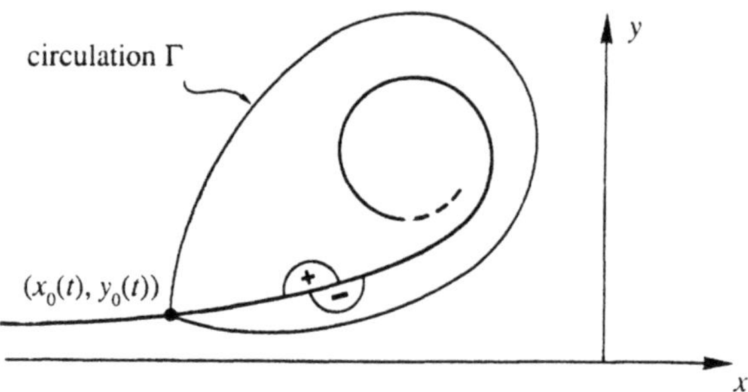

Fig. 1.22.2. Impression of half-infinite vortex sheet after release, at some time $t > 0$.

suction force K per unit of length of the edge of the plate, is acting which has the strength

$$K = \frac{\rho\,\pi}{4}\,a^2\,. \tag{1.22.6}$$

When we release at $t = 0$ the vortex sheet by annihilating the rigid plate, no singular suction force can be sustained anymore, the sheet cannot remain in its simple rectilinear shape and will start to roll up (Figure 1.22.2).

We emphasize that this evolution of the shape of the half-infinite sheet at the first moment $t = 0$, is not a result of the velocities induced by the sheet on itself. These velocities are for all $x < 0$ tangent to the sheet, but the deformation of the sheet is initiated by the disappearance of the possibility to sustain the suction force. Analogous situations arise in the case when we optimize, by means of a linear theory, the vorticity of a screw propeller or any other propulsor (see for instance Section 6.2).

Now we consider at the sheet for $t < 0$ some point $(x_0, y_0 = 0)$ (Figure 1.22.1), then the amount Γ of vorticity between this point and the origin $(0, 0)$ has the value

$$\Gamma = \int_{x_0}^{0} \gamma(x)\,dx = 2a\,|x_0|^{1/2}\ , \qquad t < 0\ . \tag{1.22.7}$$

When we follow the position $\big(x_0(t), y_0(t)\big)$ of this point, after release of the vortex sheet, it is clear that the amount of vorticity between this point and the "edge" of the sheet, has to remain constant. Hence the point $(x_0(t), y_0(t))$ can be characterized at the deformed sheet by this value of Γ. So any point of the deforming vortex sheet can be denoted as

$$z_0(\Gamma, t) = \big(x_0(\Gamma, t) + iy_0(\Gamma, t)\big)\ , \qquad 0 \le \Gamma < \infty\ , \quad t > 0\ . \tag{1.22.8}$$

Next we determine at $t > 0$ the velocity induced by the vortex sheet at some point z outside the sheet. By (1.1.16) it is easily seen that

$$v_x(z,t) - iv_y(z,t) = \frac{1}{2\pi i}\int_0^{\infty} \frac{d\Gamma^*}{(z - z_0(\Gamma^*, t))}$$

$$= \frac{1}{2\pi i} \int\limits_{\mathcal{L}} \frac{(\partial z_0/\partial\Gamma^*)^{-1}\, d\sigma}{(z-\sigma)} \;, \tag{1.22.9}$$

where we introduced a new variable of integration $\sigma = z_0(\Gamma^*, t)$ instead of Γ^* and $\mathcal{L}$ is the line of integration in the complex plane which belongs to σ. The last integral is a Cauchy integral (A.1.3) which assumes different values (A.1.6), (A.1.7) at the + and − side of the vortex sheet. Hence when $z \to z_0(\Gamma, t)$ from + or − side, then

$$v_x(x_0, y_0) - i v_y(x_0, y_0) = \pm\frac{1}{2}\left(\frac{\partial z_0}{\partial\Gamma}(\Gamma, t)\right)^{-1} + \frac{1}{2\pi i} \oint\limits_{\mathcal{L}} \frac{(\partial z_0/\partial\Gamma^*)^{-1}\, d\sigma}{(z_0(\Gamma, t) - \sigma)}$$

$$= \pm\frac{1}{2}\left(\frac{\partial z_0}{\partial\Gamma}(\Gamma, t)\right)^{-1} + \frac{1}{2\pi i} \oint\limits_0^\infty \frac{d\Gamma^*}{(z_0(\Gamma, t) - z_0(\Gamma^*, t))} \;, \tag{1.22.10}$$

where $\oint$ is the Cauchy principal value of the integral. The difference of the velocity at both sides of the sheet is just a consequence of the vorticity at the sheet.

For the transport of the point under consideration $z_0(\Gamma, t)$ we have to take the mean value of the velocity at both sides of the sheet. Then we find by (1.22.10)

$$\frac{\partial \bar{z}_0(\Gamma, t)}{\partial t} = \frac{1}{2\pi i} \oint\limits_0^\infty \frac{d\Gamma^*}{(z_0(\Gamma, t) - z_0(\Gamma^*, t))} \;. \tag{1.22.11}$$

This is an integro-differential equation for the function $z_0(\Gamma, t)$ which determines the points, characterized by some value of Γ, of the sheet as a function of time. For this equation we have the starting condition (1.22.7)

$$\lim_{t\to 0} z_0(\Gamma, t) = x_0(\Gamma, 0) = -\frac{\Gamma^2}{4a^2} \;. \tag{1.22.12}$$

We remark that in the problem given by (1.22.11) and (1.22.12) the unknown function $z_0(\Gamma, t)$ depends on two independent variables Γ and t. In the following we describe, by means of dimensional analysis, a method to introduce a self-similar solution, so that only one independent variable remains.

The place z_0 of an elementary vortex of the sheet is a function of Γ and t, but depends also on a (1.22.1) and the density ρ, so we can write

$$z_0 = f(\rho, a, \Gamma, t) \;. \tag{1.22.13}$$

When $[\,l\,]$ is the dimension of length, $[m]$ of mass and $[\,t\,]$ of time, we find for the dimensions of z_0, ρ, a and Γ

$$[z_0] = [\,l\,] \;; \qquad [\rho] = [\,l\,]^{-3}[m] \;,$$

$$[a] = [\,l\,]^{3/2}[\,t\,]^{-1} \;, \qquad [\Gamma] = [\,l\,]^2[\,t\,]^{-1} \;. \tag{1.22.14}$$

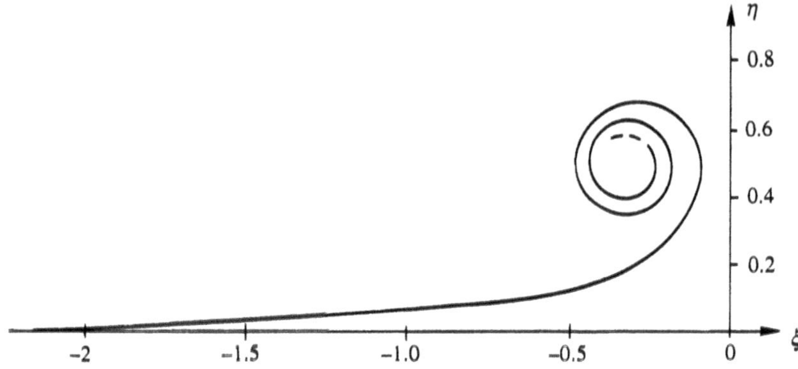

Fig. 1.22.3. Self-similar, half-infinite vortex sheet shape, $h = \xi + i\eta$ (from [56]).

Then applying the abridged version of the dimensional analysis as described in the last paragraph of Appendix E, we have to solve β_1, β_2, β_3 and β_4 from

$$[\,l\,] = ([\,l\,]^{-3}[m])^{\beta_1} \cdot ([\,l\,]^{3/2}[\,t\,]^{-1})^{\beta_2} \cdot ([\,l\,]^{2}[\,t\,]^{-1})^{\beta_3} \cdot ([\,t\,])^{\beta_4} \ . \tag{1.22.15}$$

Equating the powers of $[\,l\,]$, $[m]$ and $[\,t\,]$ at both sides of the equality sign, we find the equations

$$1 = -3\beta_1 + \frac{3}{2}\beta_2 + 2\beta_3 \ ; \qquad 0 = \beta_1 \ ; \qquad 0 = -\beta_2 - \beta_3 + \beta_4 \ . \tag{1.22.16}$$

Hence we can choose

$$\beta_1 = 0 \ , \qquad \beta_2 = \frac{2}{3} - \frac{4}{3}\beta_3 \ , \qquad \beta_4 = \frac{2}{3} - \frac{1}{3}\beta_3 \ . \tag{1.22.17}$$

Then we find that (1.22.13) has to have the form

$$z_0 = a^{2/3}t^{2/3}h(\Gamma a^{-4/3}t^{-1/3}) = (at)^{2/3}h(\lambda) \ , \tag{1.22.18}$$

where h is an unknown function of $\lambda = \Gamma a^{-4/3}t^{-1/3}$. Substitution of (1.22.18) into (1.22.11) yields the new integro-differential equation

$$\frac{1}{3}\left(2\bar{h}(\lambda) - \lambda\frac{d\bar{h}(\lambda)}{dt}\right) = \frac{1}{2\pi i}\oint_0^\infty \frac{d\lambda^*}{(h(\lambda) - h(\lambda^*))} \ , \tag{1.22.19}$$

for the unknown function $h(\lambda)$ which depends only on one variable $\lambda = \Gamma a^{-4/3}t^{-1/3}$.

For $t \to 0$ or $\lambda \to \infty$ we have by (1.22.12)

$$\lim_{t\to 0} \ (at)^{2/3} \ h(\Gamma a^{-4/3}t^{-1/3}) = \frac{-\Gamma}{4a^2} \ , \tag{1.22.20}$$

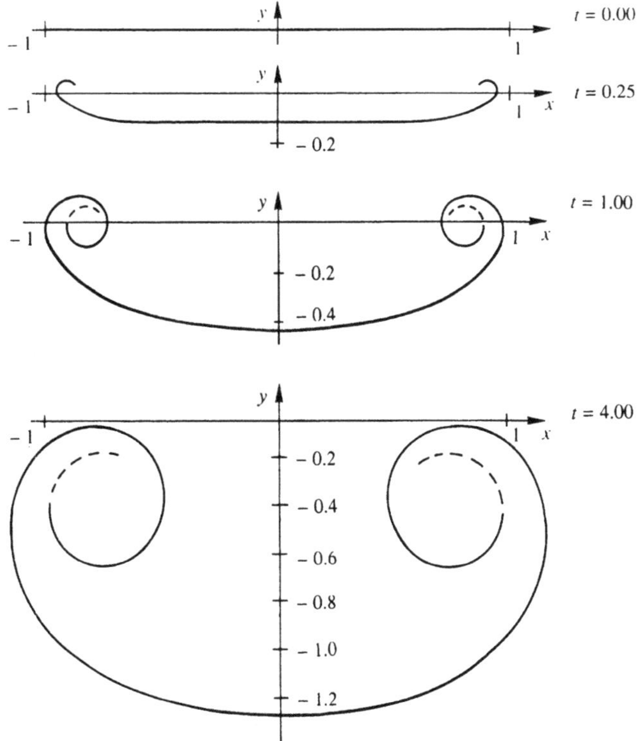

Fig. 1.22.4. Evolution of vortex sheet of finite width (from [42]).

hence

$$\lim_{\lambda \to \infty} h(\lambda) = -\frac{1}{4}\lambda^2. \tag{1.22.21}$$

The problem is now to solve (1.22.19) under the asymptotic condition (1.22.21). This has been carried out in [56]. We do not discuss the numerical process but give in Figure 1.22.3 the shape of the self-similar vortex sheet in the complex $h = \xi + i\eta$ plane. The inner windings of the spiral are only drawn upto a finite number.

We also give the drawing for the 2-dimensional roll-up of a vortex sheet of finite width, as calculated in [42] by numerical means. The free vorticity has the intensity

$$\gamma(x) = \frac{x}{(1 - x^2)^{1/2}}, \qquad |x| < 1 , \quad t = 0 . \tag{1.22.22}$$

This vorticity corresponds to the intensity of a vortex sheet behind a wing with elliptic loading, hence with least induced resistance (see Section 5.6). The evolution of the strip is drawn in Figure 1.22.4.

Here an analogous phenomenon occurs as for the half-infinite sheet. Now the velocity v_y induced at the strip for $t = 0$ by its own vorticity (1.22.22) is downwards and constant,

$$v_y(x,0) = \frac{1}{2\pi} \oint_{-1}^{1} \frac{\xi}{(1-\xi^2)^{1/2}} \frac{d\xi}{(x-\xi)} = -\frac{1}{2} \,, \qquad |x| < 1 \,. \tag{1.22.23}$$

So at first sight one expects the rectilinear strip to translate downwards without changing its shape. However, also here, just as in the case of the half-infinite sheet, we have a square root singularity of the vorticity at the edges of the strip, which induces suction forces. When the vortex sheet is released, these suction forces have to disappear and this initiates the roll-up of the strip.

Chapter 2

The Actuator Surface

An actuator surface can be defined as a 2-dimensional geometric region in a fluid at which a discontinuity in any flow property can occur [27]. We confine ourselves to a more restricted definition. We will use the term actuator surface for a 2-dimensional region in a fluid on which an external force field is concentrated. At this surface we have to admit a possible discontinuity of the pressure or of the tangential component of the velocity or of both.

According to this definition a lifting surface is a special type of actuator surface. Perpendicular to it we have an external force field which creates a pressure jump over the surface. The flow is tangential to it and has a discontinuity corresponding with the vorticity at the surface.

The actuator surfaces we discuss in this chapter are passed through by the fluid particles. For instance a circular region perpendicular to the main stream at which the force field represents the propulsive action of a screw propeller. This type of actuator surface is generally called an actuator disk. It can be used when the complicated details of the pressure field and of the flow field, induced by a screw propeller is not so much of interest and only a global knowledge of these fields will be sufficient.

We start this chapter with the linearized theory of an actuator disk with a normal load which is constant with respect to time. This theory is directly based on the linearized theory of external force fields as discussed in Sections 1.3–1.7. After this we discuss the vorticity shed by the disk.

Two applications are given of linearized actuator disk theory. First, we consider the interaction of a propeller and a body. This sheds some light upon the fundamental aspects of the complicated phenomenon called thrust deduction. It is shown that also the opposite effect, thrust augmentation can occur, although this does not seem to be of practical importance. Second, we consider the influence of a duct on an actuator disk of which the loading is periodically dependent on time. This shows in principle the efficiency increasing effect which a duct or any impermeable surface aligned with the mainstream, can have on a time-dependent propulsive force action.

The chapter is concluded with a general treatment of the non-linear theory of the actuator disk with a time-independent load. The complicated flow pattern which possibly can occur at the edge of the disk is discussed. This latter problem is still open for discussion and offers a challenge for theoretical investigation.

2.1. Linearized Actuator Disk Theory

We will consider an external force field $\vec{F}(x, y, z, t)$, parallel to the x-axis, which is concentrated at the translating flat plane $x = Ut$ ($U > 0$) (Figure 2.1.1). The field is confined to the disk region B of the plane, it is switched on at $t = t_0$ and switched off at $t = t_e$, hence $t_e > t_0$. We will take $\vec{F}(x, y, z, t)$ of the form

$$\vec{F}(x,y,z,t) = \left(-f(y,z)\delta(x-Ut), 0, 0\right)\{H(t-t_0) - H(t-t_e)\} = O(\varepsilon) \ ,$$
$$f(y,z) = 0 \ , \qquad (y,z) \notin B \ , \tag{2.1.1}$$

where δ is the delta function of Dirac and $H(t)$ is the Heaviside function

$$H(t) = 0 \ , \ t < 0 \ ; \qquad H(t) = 1 \ , \ t > 0 \ . \tag{2.1.2}$$

Because we consider a linearized theory we have to take, as we denoted in (2.1.1), $\vec{F}$ sufficiently small, say of $O(\varepsilon)$. In (2.1.1) we specialized to the case that the intensity $f(y, z)$ of the force field is independent of time. In order to be in concurrence with the theme of this book, we suppose $f(y, z) > 0$, then by the minus sign in (2.1.1) $\vec{F}$ acts in the negative x-direction and the reaction is a propulsive force T at the disk B. This propulsive force or thrust has the value

$$T = \iint_B f(y,z)\, dy\, dz \ . \tag{2.1.3}$$

The pressure field induced by the force field follows from (1.4.4). Because the force field is defined for all points of the 3-dimensional space, we can integrate over the whole space

$$p(x,y,z,t) = \frac{1}{4\pi}\iiint_{-\infty}^{\infty} \frac{\vec{F}(\xi,\eta,\zeta,t)\cdot\vec{R}}{R^3}\, d\xi\, d\eta\, d\zeta$$
$$= -\frac{1}{4\pi}\iint_B \frac{f(\eta,\zeta)(x-Ut)\, d\eta\, d\zeta}{\{(x+Ut)^2 + (y-\eta)^2 + (z-\zeta)^2\}^{3/2}}$$
$$\cdot\{H(t-t_0) - H(t-t_e)\} \ , \tag{2.1.4}$$

which is the field of a pressure dipole layer at the disk for $t_0 < t < t_e$. For $t < t_0$ $p(x, y, z, t)$ is zero, which is not surprising, however $p(x, y, z, t)$ is also zero for $t > t_e$ when the force field is switched off, but still a velocity field will be present. The reason for this is that we consider a linearized theory so the velocities are of

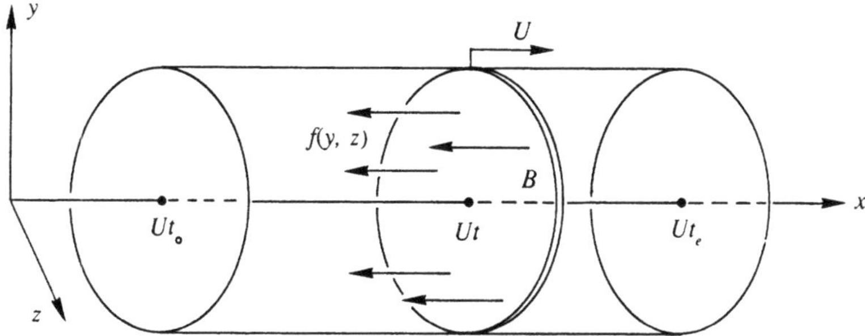

Fig. 2.1.1. Translating actuator disk.

$O(\varepsilon)$, hence their influence on the pressure is of $O(\varepsilon^2)$ (1.2.11), which is neglected with respect to the $O(\varepsilon)$ of (2.1.4).

Next we consider the velocity field induced by the disk. We use the general formulas (1.4.7) and consider the last but one expression of it. Because the force field is also defined for all values of time, we can integrate over the whole space and with respect to time over the interval $(-\infty, t)$. So we have

$$v(x,y,z,t) = \text{grad}\left\{ -\frac{1}{4\pi\rho}\int_{-\infty}^{t}\iiint_{-\infty}^{\infty}\frac{\vec{F}(\xi,\eta,\zeta,t)\cdot\vec{R}}{R^3}\,d\xi\,d\eta\,d\zeta\,d\tau\right\}$$

$$+\frac{1}{\rho}\int_{-\infty}^{t}\vec{F}\,d\tau = \text{grad}\,I_1(x,y,z,t) + \vec{I_2}(x,y,z,t)\ , \qquad (2.1.5)$$

where we introduced the abbreviations I_1 and $\vec{I_2}$.

Substituting (2.1.1) into (2.1.5), we find for $I_1(x,y,z,t)$ after changing the order of integration

$$I_1 = \frac{1}{4\pi\rho}\iiint_{-\infty}^{\infty}\frac{f(\eta,\zeta)(x-\xi)}{R^3}$$

$$\cdot\left[\int_{-\infty}^{t}\delta(\xi-U\tau)\{H(\tau-t_0)-H(\tau-t_e)\}\,d\tau\right]d\xi\,d\eta\,d\zeta\ . \qquad (2.1.6)$$

First we consider separately the integration with respect to τ. For $t < t_0$ the integrand is zero, hence the integral is zero. For $t_0 < t < t_e$ we obtain

$$\int_{t_0}^{t}\delta(\xi-U\tau)\,d\tau = -\frac{1}{U}\int_{\xi-Ut_0}^{\xi-Ut}\delta(\lambda)\,d\lambda$$

$$= \frac{1}{U}\{H(\xi - Ut_0) - H(\xi - Ut)\} \ . \tag{2.1.7}$$

For $t > t_e$ the original integration with respect to τ becomes independent of t and its value follows from (2.1.7) by replacing t by t_e. In addition (2.1.7) is zero for $\xi < Ut_0$, it is U^{-1} for $Ut_0 < \xi < Ut$ and again zero for $\xi > Ut$.

So when we substitute (2.1.7) into (2.1.6) we obtain for $t_o < t < t_e$

$$\begin{aligned} I_1 &= \frac{1}{4\pi\rho U}\iint\limits_{-\infty}^{\infty} f(\eta,\zeta)\left[\int\limits_{Ut_0}^{Ut} \frac{(x-\xi)\,d\xi}{\{(x-\xi)^2+(y-\eta)^2+(z-\zeta)^2\}^{3/2}}\right] d\eta\, d\zeta \\ &= \frac{1}{4\pi\rho U}\iint\limits_{B} f(\eta,\zeta)\left[\frac{1}{\{(x-Ut)^2+(y-\eta)^2+(z-\zeta)^2\}^{1/2}}\right. \\ &\quad \left. - \frac{1}{\{(x-Ut_o)^2+(y-\eta)^2+(z-\zeta)^2\}^{1/2}}\right] d\eta\, d\zeta \ . \end{aligned} \tag{2.1.8}$$

Resuming the above results we have for $I_1(x,y,z,t)$

$$\begin{aligned} &t < t_0 \ , \quad I_1(x,y,z,t) = 0 \ ; \quad t_0 < t < t_e \ , \quad I_1(x,y,z,t) = (2.1.8) \ ; \\ &t_e < t \ , \quad I_1(x,y,z,t) = (2.1.8) \text{ with } t = t_e \text{ (independent of time)} \ . \end{aligned} \tag{2.1.9}$$

Next we consider $\vec{I}_2(x,y,z,t)$

$$\vec{I}_2 = \left(\frac{-f(y,z)}{\rho}\int\limits_{-\infty}^{t} \delta(x-Ut)\,\{H(\tau-t_0)-H(\tau-t_e)\}\ d\tau, 0, 0\right) . \tag{2.1.10}$$

It follows that

$$\begin{aligned} &\vec{I}_2(x,y,z,t) = (0,0,0) \quad \text{for} \quad x < Ut_0 \quad \text{or} \quad x > U\cdot\text{minimum}\,(t,t_e) \ , \\ &\vec{I}_2(x,y,z,t) = \left(\frac{-f(y,z)}{\rho U}, 0, 0\right) \\ &\qquad\qquad \text{for} \quad t > t_0 \quad \text{and} \quad Ut_0 < x < Ut \le Ut_e \ . \end{aligned} \tag{2.1.11}$$

From the above it follows that the velocity field induced by the actuator disk in operation ($t_0 < t < t_e$), can be represented by three parts. First, the velocity field caused by a layer of sinks at the disk. This follows from the first term between brackets in the last expression of (2.1.8). The strength per unit of area, of these sinks is $f(y,z)/\rho U$. Second, the velocity field caused by a layer of starting sources at $x = Ut_0$. This follows from the second term between brackets in the last expression of (2.1.8). The strength of these sources is also $f(y,z)/\rho U$. Third, in the region

$$Ut_0 < x < Ut \le Ut_e \ , \qquad (y,z) \in B \ , \tag{2.1.12}$$

we have the parallel velocity field $\vec{I}_2$ as follows from (2.1.11)

$$\left(\frac{-f(y,z)}{\rho U}, 0, 0\right) . \tag{2.1.13}$$

It is easy to check that the total velocity field is continuous at the actuator disk and at the layer of starting sources. The jumps of the normal velocity components at the sink and source layers, are compensated by the jumps of (2.1.13) or (2.1.11) at those layers. Hence the divergence of the flow is zero everywhere.

For $t > t_e$ the actuator disk does not exist anymore and the sinks which were present at the disk remain in the fluid as ending sinks at $x = Ut_e$, $(y, z) \in B$. Also the starting source layer at $x = Ut_0$ and the parallel velocity field are still present. Then the velocity field has become independent of time.

Because the velocity field is continuous at the actuator disk and at the layer of starting sources ($x = Ut_0$), vorticity can occur only inside the region (2.1.12) passed through by the disk and at the cylindrical part of the boundary of this region. This vorticity can be calculated a priori. By (1.3.25) the circulation Γ of any contour C can be determined. Taking C "infinitely" small and in a flat plane, we can find by (1.3.11) the component normal to that plane of the local vorticity $\vec{\omega}$. So taking different small contours C we can at any place, calculate the vorticity induced by the moving force field. Then, because the divergence of the flow is zero, it follows from Section 1.1 that the velocity field is determined by the vorticity field alone. Hence another way of calculating the velocity field induced by the moving disk is by applying the law of Biot and Savart (1.1.15) to the vorticity field caused by the disk.

It is also of interest to consider these results from the point of view of Section 1.7, where the singular force aligned with its velocity is discussed. The solution found in that section can also be used as a Green-function for the actuator disk problem, this is left to the reader.

When the actuator disk started its action a long time ago, say at $t_0 = -\infty$, we can easily calculate the rate of change of the momentum of the fluid, which has to be equal to the resultant external force exerted by the disk at the fluid (see Section 1.15, fourth paragraph). Also we can easily calculate the rate of change of the kinetic energy of the fluid from which follows the efficiency η of the disk.

First we consider the momentum. When $t_0 = -\infty$, it follows that far behind the disk, the contribution of grad $I_1(x, y, z, t)$ (2.1.8) to the velocity field has disappeared. Hence far behind the disk we have only the flow (2.1.11), which is called the slip stream. So there is added to the momentum of the fluid each unit of time, the momentum of a part of the slip stream of length U. By (2.1.13) the rate of change of momentum becomes

$$U\rho \cdot \iint\limits_B \left(\frac{-f(y,z)}{\rho U}, 0, 0\right) dy\, dz = \iint\limits_B (-f(y,z), 0, 0)\, dy\, dz = -T \ , \tag{2.1.14}$$

where T is the thrust of the disk (2.1.3). We remark that the rate of change of momentum of the fluid always has to be equal to $-T$, also when t_0 is finite, only then it is more complicated to carry out a numerical calculation.

Next we consider the rate of change of the kinetic energy of the fluid for the case that $t_0 = -\infty$. Each unit of time there is added, in the same way as for the momentum, the kinetic energy of a part of the slip stream of length U, which we denote by E

$$E = U \cdot \tfrac{1}{2}\rho \iint_B \left\{ \frac{f(y,z)}{\rho U} \right\}^2 dy\, dz = \frac{1}{2\rho U} \iint_B f^2(y,z)\, dy\, dz \ . \tag{2.1.15}$$

From (2.1.15) follows the efficiency η of the actuator disk, which is defined as the quotient of the useful work per unit of time UT and the total work per unit of time $UT + E$

$$\eta = \frac{UT}{UT + E} = 1 - O(\varepsilon) \ . \tag{2.1.16}$$

The latter equality follows from the fact that $T = O(\varepsilon)$ and $E = O(\varepsilon)^2$. Here we used the result of Section 1.17, that the work done by the force field can be retraced as the kinetic energy of the fluid also in the case of a linearized theory.

In contrast to the momentum consideration, the work per unit of time of the external force field differs in the cases t_0 being finite or infinite. Namely when t_0 is finite, the disk experiences the velocities induced by the starting source layer at $x = Ut_0$. These velocities reduce the normal component of the total velocity field at the disk. So in the early stages of the operation of the disk, the work done per unit of time by the external forces will be smaller than E (2.1.15). The efficiency η (2.1.16) is the efficiency of the disk when its action has become stationary.

It is of interest to determine by means of the floating probing contours of Section 1.3, the shed free vorticity behind the disk. When the intensity of the force field $f(x,y)$ is a continuous function at B with non-zero values at the boundary ∂B of B, then it can be seen that behind ∂B there is a concentrated vortex layer while behind the interior of B the vorticity is continuously distributed in space.

2.2. Vorticity of the Linearized Actuator Disk

The most simple special case of the actuator disk, described in the previous section, arises when $t_0 = -\infty$ and when the load at the disk is independent of y and z, hence in (2.1.1) $f(y,z) = f = \text{const.} > 0$. Also we assume for convenience that the region B, where the external force field is acting, is circular and has the radius b (Figure 2.2.1).

Instead of a coordinate system which is at rest with respect to the undisturbed fluid (Figure 2.1.1) we use here a coordinate system which is at rest with respect to the disk. Then we have, in order to obtain the total velocity field, to add to the induced velocities the incoming velocity $(-U, 0, 0)$ and we get the situation of Figure 2.2.1 (a).

In this case the only vorticity occurs at the half-infinite cylinder $x < 0, y^2 + z^2 = b^2$, hence at the boundary of the slip stream, and it is parallel to the (y, z) plane. Its

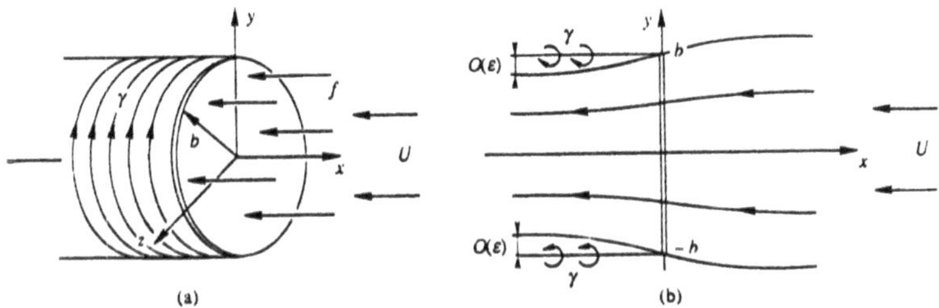

Fig. 2.2.1. Circular actuator disk with constant normal load in parallel flow.

strength γ can be determined besides by means of probing contours, also from the difference of the velocity just outside the slip stream and just inside the slip stream. By (2.1.13) with $f(x, y) = f$ = const., we find

$$\gamma = \frac{f}{\rho U} , \tag{2.2.1}$$

per unit of length in the x-direction. It is drawn with a right-hand screw in Figure 2.2.1 (a).

As mentioned before the disturbance velocity field can be described in two different ways. First at the disk we have a layer of sinks of strength $f/\rho U$ (2.1.8) (with $t_0 = -\infty$) and behind the disk we have to add the parallel flow in the negative x-direction of magnitude $f/\rho U$. Second, by applying the law of Biot and Savart to the circular vorticity of strength (2.2.1).

A cross section of the flow is given in Figure 2.2.1 (b). It is clear that this picture is only correct up to and including $O(\varepsilon)$. The free vorticity which in the linearized theory is at the half-infinite cylinder with radius b (see above (1.4.8) in italics), in reality is transported by the fluid. Hence this vorticity lies in fact at the stream tube which passes through the edge of the disk.

The question arises if we can give a representation of the disk under consideration, which is more related to a screw propeller. We introduce a cylindrical coordinate system (x, r, φ) as drawn in Figure 2.2.2. Consider a straight bound vortex OA, of length b and lying in the y, z plane. The end O coincides with the origin of the coordinate system and the vortex rotates with angular velocity ω. The strength Γ, which is independent of r, of this vortex is coupled with a right-hand screw to the positive r-direction. From O starts a free vortex of strength Γ, stretching along the negative x-axis and from the other end A starts a free vortex of the same strength along the helicoidal line

$$\varphi + \omega t - ax = 0 , \quad r = b , \quad a = \omega/U . \tag{2.2.2}$$

These two free vortices are connected by the starting vortex O_2A_2, which was shed long ago at the beginning of the process and which makes the vortex field free of divergence.

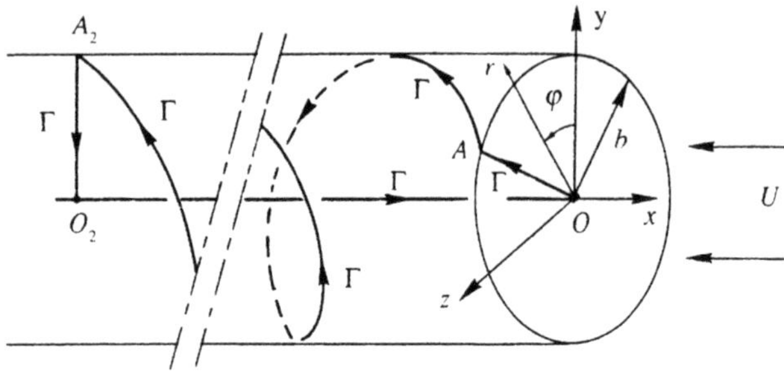

Fig. 2.2.2. Rotating bound vortex OA.

The velocity of a point of OA at a distance r of O, relative to the fluid is $(U, 0, \omega r)$, where the components are the physical components in the x, r and φ-direction (see below Formula (B.1.10)). In the following always physical components are used. Hence by the theorem of Joukowski (below (1.8.3)) the force per unit of length acting at the fluid at that point is

$$\rho\,(-\omega r\Gamma, 0, U\Gamma)\ . \tag{2.2.3}$$

The opposite force acts at the rotating vortex, hence the rotating bound vortex acts as a propeller.

Now suppose that ω increases indefinitely and Γ decreases so that $\omega\Gamma$ remains constant. Then several limits have to be considered. First, the rotating vortex at the disk, the free vortex along the negative x-axis and the starting vortex disappear. Second, the strength of the free helicoidal vortex tends to zero, inversely proportional to ω, but its number of windings per unit of length along the negative x-axis increases in proportion to ω. So there arises a layer of circular vortices of which the strength per unit of length in the x-direction assumes the value

$$\Gamma \cdot \frac{\omega}{2\pi} \cdot \frac{1}{U} = \frac{\Gamma\omega}{2\pi U}\ , \tag{2.2.4}$$

coupled with a right-hand screw to the negative φ-direction. Third, the force action exerted at the fluid becomes perpendicular to the disk ($x = 0, 0 \le r \le b$) and its mean value per unit of area at a radius r becomes

$$\frac{\rho\,\omega r\Gamma\,dr}{2\pi r\,dr} = \frac{\rho\,\omega\Gamma}{2\pi}\ , \tag{2.2.5}$$

which is independent of r. This force action is in the negative x-direction. If we suppose

$$\lim_{\omega\to\infty} \omega\Gamma = 2\pi f/\rho\ , \tag{2.2.6}$$

we obtain the actuator disk of Figure 2.2.1.

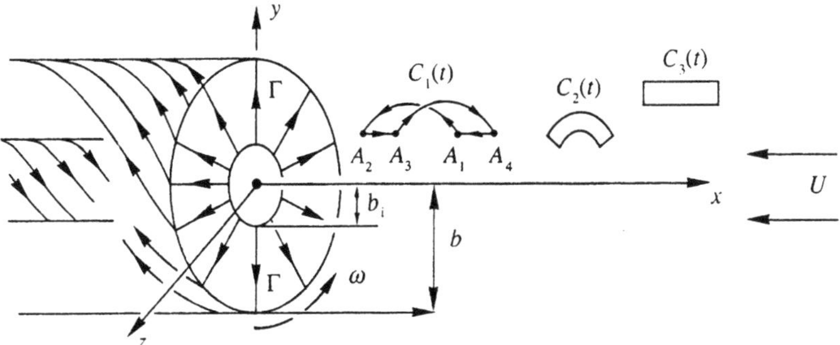

Fig. 2.2.3. Bound vortex system with finite rotational velocity ω.

The efficiency η of the disk in this case, hence with $f(x,y) = f =$ const. and $t_0 = -\infty$ follows from (2.1.15) and (2.1.16)

$$\eta = \frac{UT}{UT+E} = \left(1 + \frac{T}{2\rho\, U^2 S}\right)^{-1} = (1 - O(\varepsilon)) \ , \qquad (2.2.7)$$

where $S = \pi b^2$ is the area of the disk. Later on (Section 5.9) it will be shown that this efficiency is, with respect to linearized theory the lowest upperbound for the efficiency of all possible propellers with thrust T, "working area" S, mean velocity of advance U and acting in an inviscid and incompressible fluid. Therefore this special actuator disk is often called an "ideal propeller".

In case of a real screw propeller we have behind the propeller besides a jet directed backwards, also a rotation of the fluid in the jet around the axis of the propeller. We now discuss shortly an actuator disk, with a more general type of loading, which also shows this behaviour. The rotation of the fluid can be induced by admitting at the disk besides forces parallel to the x-axis, also tangential forces per unit of area. We assume in this case the disk region (Figure 2.2.3) to be given by

$$x = 0 \ , \quad 0 < b_i \le r \le b \ , \quad 0 \le \varphi < 2\pi \ . \qquad (2.2.8)$$

The loading, physical (x, r, φ) components, per unit of area is taken as

$$\vec{F}(x,r) = \left(f\,\delta(x), 0, \frac{h}{r}\,\delta(x)\right) \ , \quad b_i \le r \le b \ , \qquad (2.2.9)$$

where f and h are constants. In order to determine the vorticity which is induced by this force field, we will now for a change of method use probing contours (Section 1.3) which float with the fluid. First a contour $C_1(t)$ of which the sides (A_4, A_1) and (A_2, A_3) are parallel to the x-axis. The sides (A_1, A_2) and (A_3, A_4) are parallel with the plane $x = 0$, they are at a constant radius r and have the length l. The position of (A_1, A_2) is supposed to be given by $x = -Ut$.

When the contour $C_1(t)$ passes at $t = 0$, with its side (A_1, A_2) through the disk, then by (1.3.16) only the force component in the φ-direction of (2.2.9) can give a contribution to the change of circulation $\Gamma_1(t)$ of $C_1(t)$. In the position of $C_1(t)$ that (A_1, A_2) has just passed through the disk and (A_3, A_4) is still in front of the disk, we find by (1.3.16)

$$\Gamma_1 = \int_{-\tilde{\varepsilon}}^{\tilde{\varepsilon}} \left\{ \frac{1}{\rho} \int_{C(t)} \vec{F}(x,r) \cdot \vec{ds} \right\} dt = \frac{1}{\rho} l \int_{-\tilde{\varepsilon}}^{\tilde{\varepsilon}} F_\varphi(-Ut, r)\, dt$$

$$= \frac{l}{\rho} \cdot \frac{h}{r} \cdot \int_{-\tilde{\varepsilon}}^{\tilde{\varepsilon}} \delta(-Ut)\, dt = \frac{lh}{\rho\, rU} , \tag{2.2.10}$$

where $\tilde{\varepsilon} > 0$ is some sufficiently small number. Hence the vorticity inside $C_1(t)$ becomes

$$h/\rho\, rU , \tag{2.2.11}$$

per unit of length in the φ-direction. This radial vorticity has to be situated at the disk, because we can take the sides (A_4, A_1) and (A_2, A_3) as short as we want. Its rotational direction is coupled with a right-hand screw to the positive r-direction. When the side (A_3, A_4) has passed also through the disk, the circulation Γ_1 is again zero.

In the same way we can show by means of the floating contours $C_2(t)$ and $C_3(t)$ that downstream of this actuator disk only vorticity occurs at the cylindrical surfaces

$$x < 0, \qquad r = b \quad \text{or} \quad r = b_i \ . \tag{2.2.12}$$

The contour $C_2(t)$ is parallel with the plane $x = 0$, its longer sides are circular arcs around the x-axis, while its short sides are radial. The contour $C_3(t)$ is in a meridian plane, its longer sides are parallel to the x-axis and its short sides are parallel to the plane $x = 0$. By means of C_2 we can detect the x-component of the vorticity and by C_3 we can detect the φ-component.

It turns out that at the two boundaries of the slip stream (2.2.12), there is helicoidal vorticity. This free vorticity floats downstream with the velocity U in the negative x-direction and is such that it can be considered to be shed by the radial vorticity (2.2.11) at the disk, when this is considered as bound vorticity which rotates with an angular velocity

$$\omega = fU/h \ . \tag{2.2.13}$$

Now we can couple the beginning of the previous reasoning to its end. It is easily seen by the theorem of Joukowski (below (1.8.3)) that the originally prescribed load of the disk (2.2.9) is found again from the force action caused by the mentioned rotating bound vorticity (2.2.11) at the disk.

We remark that there is no free vorticity inside the slip stream of the above discussed actuator disk because of the special choice of the load (2.2.9). Another type of load can fill the whole slip stream with vorticity.

When we consider actuator disks with axisymmetric loading placed in a uniform parallel flow, the stream tube through the edge of the disk will also be axisymmetric. When the normal component of the loading at the disk is in the downstream direction of the incoming flow, there will be a contraction of this stream tube (Figure 2.2.1 (b)). We consider the helicoidal vorticity behind the disk which in the general case fills the whole slip stream region. The axial flow in the stream is caused by the free vorticity component which is parallel to the plane $x = 0$ and which forms circular vortex rings. When we are at the disk we have only a half-infinite region with these vorticity rings, however when we are infinitely far downstream of the disk, we have a two-sided infinite region with these rings. Hence at the disk the induced axial velocity is half of the induced axial velocity at infinity. This holds for each radius inside the slip stream. From this follows a property of the contraction of the mentioned stream tube through the edge of the disk. Suppose the radius of the disk is b and the radius of the stream tube infinitely far upstream at $x = \infty$ is $b + \Delta b$ (Δb of $O(\varepsilon)$), then the radius of the stream tube infinitely far downstream has to be $b - \Delta b$. This will in general no longer be true for a heavily loaded disk, described by a non-linear theory.

The problem of the actuator disk, discussed by a linear theory, can be posed in two ways. The first one is, as we discussed in Section 2.1, by prescribing explicitly the external force field at the disk. The second one is, by prescribing the bound vorticity rotating or moving in some way with velocities of $O(\varepsilon^0)$ in the plane of the disk. These two methods are in our linearized theory practically equivalent. From the forces, the vorticity can be found explicitly and inversely. This is no longer possible in a non-linear theory, where from the prescribed moving bound vorticity first by solving the problem the induced velocities have to be calculated from which then follows by Joukowski's theorem the forces at the bound vorticity. The reason for this is that in a non-linear theory the induced velocities and the velocities of the bound vorticity are of the same order of magnitude ($O(\varepsilon^0)$).

Another complicated problem arises when at the disk a rotating homogeneously distributed material structure is defined, which represents for instance a large number of rotating blades of a screw propeller. Then the vorticity at the disk has to be determined such that the flow which arises, satisfies certain conditions imposed by the moving structure. For such a problem we refer to Hess [24] who calculated the forces on a quickly rotating boomerang of prescribed shape by means of a linear theory.

2.3. Thrust Deduction and Thrust Augmentation

Consider a ship of which the screw propeller is placed in the neighbourhood of the stern and that travels with a certain constant speed. Then it is well known that the propulsive force in the propeller shaft is in general larger than the force needed to tow the ship, without screw propeller, with the same speed as before [45]. One of the causes is that by the action of the propeller a low pressure region is created at the stern which induces a force in the backward direction, which phenomenon is called

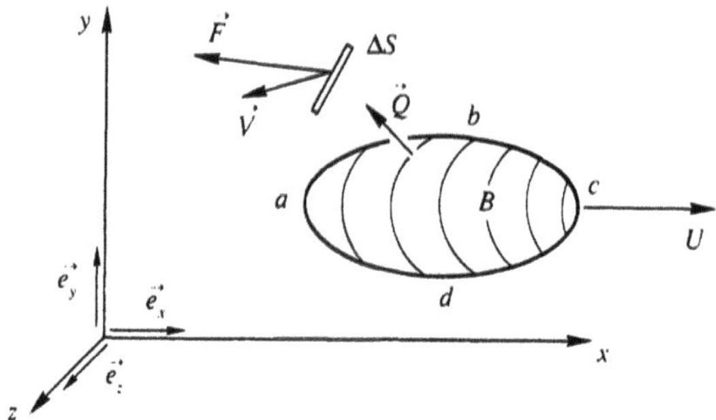

Fig. 2.3.1. Body B with nearby actuator surface ΔS.

thrust deduction. We will show by using a semi-linear theory, that such an effect can be predicted qualitatively from the interaction of an actuator disk and a body. Also it will be found that the inverse phenomenon can occur when the screw is acting at other places with respect to the body, then we have thrust augmentation.

A body B is translating with velocity U in the positive x-direction of a Cartesian coordinate system (x, y, z), through an unbounded inviscid and incompressible fluid. The fluid is assumed to be at rest at infinity with respect to the coordinate system. The body has finite dimensions, hence its disturbance velocity field $\vec{V}(x, y, z, t)$ is $O(\varepsilon^0)$. In the neighbourhood of the body we consider a small actuator surface ΔS by which an external force $\vec{F}$ of $O(\varepsilon)$ per unit of area is exerted on the fluid (Figure 2.3.1).

It is most easy to consider the actuator surface ΔS, translating at a fixed relative position with respect to the body B without any mechanical contact with B. We discuss the force $\vec{Q}$ of $O(\varepsilon)$ exerted at B by ΔS. This force, as we mentioned before, is caused by the pressures induced by ΔS. In "reality" the actuator surface is coupled to B and then also the reaction force $-\vec{F}\Delta S$ is acting at B as a force with a propulsive component in the x-direction.

We now use formula (1.17.9) which reads in this case

$$\frac{dE}{dt} = -\int_{\partial B} p \cdot (\vec{v} \cdot \vec{n}) \, dS + \int_{\Delta S} \vec{v} \cdot \vec{F} \, dS \ . \tag{2.3.1}$$

The total velocity $\vec{v}$ of the fluid can be split into two parts

$$\vec{v} = \vec{V} + \vec{v}_a \ , \tag{2.3.2}$$

where $\vec{v}_a$ is of $O(\varepsilon)$ and it is the change caused by the presence of ΔS, of the original velocity field $\vec{V}$. Also the total pressure field can be divided into a part P of $O(\varepsilon^0)$ and a part p_a of $O(\varepsilon)$ induced by the presence of ΔS,

$$p = P + p_a \ . \tag{2.3.3}$$

The condition for the total velocity field $\vec{v}$ at the boundary ∂B of B, reads

$$\vec{v} \cdot \vec{n} = U\vec{e}_x \cdot \vec{n} \ . \tag{2.3.4}$$

Hence (2.3.1) can be written as

$$\frac{dE}{dt} = -U\vec{e}_x \cdot \int\limits_{\partial B} (P + p_a)\, \vec{n}\, dS + \int\limits_{\Delta S} (\vec{V} + \vec{v}_a) \cdot \vec{F}\, dS \ . \tag{2.3.5}$$

Because the resistance of a rigid body translating with a constant velocity in an inviscid and incompressible medium is zero, we have

$$-U\vec{e}_x \cdot \int\limits_{\partial B} (P + p_a)\, \vec{n}\, dS = -U\vec{e}_x \cdot \int\limits_{\partial B} p_a\, \vec{n}\, dS = -U\vec{e}_x \cdot \vec{Q} \ . \tag{2.3.6}$$

Hence (2.3.5) changes into

$$\frac{dE}{dt} = \int\limits_{\Delta S} (\vec{V} + \vec{v}_a) \cdot \vec{F}\, dS - U\vec{e}_x \cdot \vec{Q} \ . \tag{2.3.7}$$

The left-hand side of (2.3.7) is the rate of change of the kinetic energy E in the fluid. Infinitely far behind B we have only disturbance velocities of $O(\varepsilon)$ caused by the free vorticity shed by the actuator surface ΔS, because the velocity field $\vec{V}$ of $O(\varepsilon^0)$ induced by B is zero at infinity. Hence the rate of change of E is $O(\varepsilon^2)$. Comparing in (2.3.7) quantities of $O(\varepsilon)$ we find

$$\vec{e}_x \cdot \vec{Q} = \frac{\Delta S}{U}\, \vec{V} \cdot \vec{F} \ , \tag{2.3.8}$$

where we have approximated the integration over ΔS, which is allowed when ΔS is sufficiently small.

In case that $\vec{F}$ is a propulsive force, hence predominantly in the negative x-direction, it follows from (2.3.8) that $\vec{e}_x \cdot \vec{Q}$ will be negative for positions of ΔS where the x-component of $\vec{V}$ is positive, then we have thrust deduction. For instance at the positions a and c of Figure 2.3.1. Inversely we have thrust augmentation when ΔS is at the positions b or d where the x-component of $\vec{V}$ will be negative. Remember that in case the actuator disk is connected mechanically to the body B the total force on B in the positive x-direction becomes

$$\vec{e}_x \cdot (-\vec{F}\Delta S + \vec{Q}) = -\Delta S\, \vec{F} \cdot \left(\vec{e}_x - \frac{\vec{V}}{U} \right) \ . \tag{2.3.9}$$

If we consider a larger actuator surface we have to carry out an integration over its area.

Loosely stated we can give the following slightly different version of the results. If a lightly loaded propeller is placed in a region where the fluid velocities with respect to the propeller are mainly lower than the translational velocity U of the propelled body, linearized potential theory predicts thrust deduction. Inversely, when the propeller is placed in a region where these velocities are mainly larger than U, we have thrust augmentation. This result is in agreement with the expected influence of the pressure field of the propeller on the body.

For papers which discuss the interaction of a propeller and a hull in a more realistic, be it approximate, way we refer to for instance [19].

2.4. Unsteady Actuator Disk with Duct

In [15] Dickmann considers several kinds of unsteady propulsion, such as sculling propulsion by wings or by an oscillating piston in a cylinder of finite length, which induces a periodic jet. In order to obtain insight in the action of these propellers, he started with the unsteady actuator disk and discussed several features of its efficiency.

One of the interesting subjects of the hydrodynamics of propulsion is the influence of a rigid surface in the neighbourhood of a propulsion system, on the efficiency of it. Here we will show by means of a linear theory, that the efficiency of an unsteady actuator disk can be increased by surrounding it by a duct. We follow Klaren [40].

Consider a Cartesian coordinate system (x, y, z) in an unbounded inviscid and incompressible fluid, which flows with a velocity U in the negative x-direction. Also we will use a cylindrical coordinate system (x, r, φ) (Figure 2.4.1). The unsteady actuator disk is represented by the external force field

$$\vec{F}(x, r, \varphi, t) = -(f_1 + f_2\, e^{i\omega t})\, \delta(x)\, \vec{e}_x \ , \quad r \leq a \ , \tag{2.4.1}$$

and $\vec{F} = 0$ for $r > a$, where f_1 and f_2 are, because the theory is linear, constants of $O(\varepsilon)$, $\delta(x)$ is the delta function of Dirac, a the radius of the disk and $\vec{e}_x$ the unit vector in the x-direction.

In Figure 2.4.1 we have drawn the actuator disk at $x = 0$, with a rigid and impermeable cylindrical duct which has the same radius a. Hence we assume that there is no gap between disk and duct. The length of the duct is L and it stretches from $x = 0$ upto $x = L$. So the disk is placed at the trailing edge of the duct. We

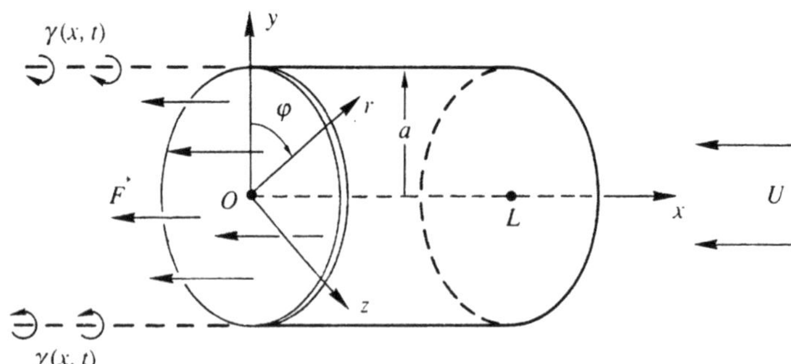

Fig. 2.4.1. Actuator disk and duct in parallel flow U.

will now show that this represents the general configuration of the disk inside the duct. In other words, the velocities induced by the disk in the presence of the duct are independent of the position of the disk in the duct, in case there is no gap between them.

We compare two circular actuator surfaces I and II, of which we have drawn the cross sections in Figure 2.4.2 (a) and (b). I is the disk of which the force field is defined in (2.4.1). II consists of two parts. First a cylinder B of radius a of which the intersections with the plane $z = 0$ are given by (B_1, B_2) and (B_4, B_3). The length of this cylinder is b, hence it stretches from $x = 0$ upto $x = b$. Normal to this cylinder we place the same load as the one of I. Second, we add to the cylinder at $x = b$ the same actuator disk as I. Together this cylinder and the translated disk form the actuator surface II. The statement is now that the actuator systems I and II induce the same velocity field.

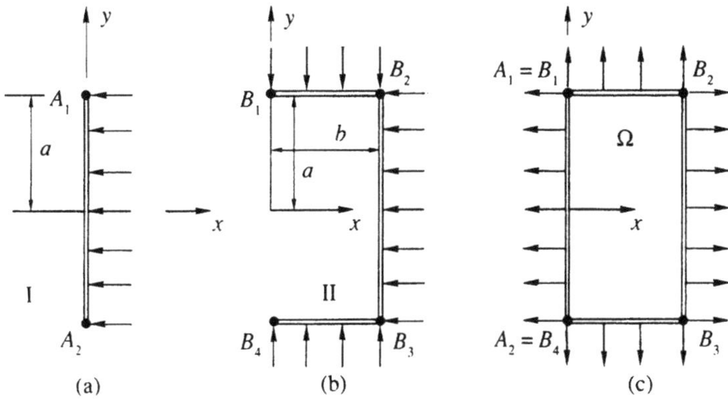

Fig. 2.4.2. Cross-section of two equivalent actuator surfaces.

The proof of this follows from (1.3.5) and (1.3.7). In fact we consider the difference $(\vec{F}_I - \vec{F}_{II})$ of the two force fields I and II, as is drawn in Figure 2.4.2 (c). Then we can write

$$\vec{F}_I - \vec{F}_{II} = \text{grad}\, K(x, y, z, t) \ , \tag{2.4.2}$$

where $K(x, y, z, t)$ is a function which has the value $|\vec{F}_I| = |\vec{F}_{II}|$ inside the region Ω and is zero outside this region. Hence its value has at each moment a constant jump at the boundary $\partial\Omega$ of Ω. Then its gradient yields the force field (2.4.2). Now we know from the text belonging to (1.3.5) and (1.3.7) that this force field does not induce any disturbance velocities. From this it follows that the velocity fields caused by disk I and actuator surface II have to be the same.

However, when we put II inside the duct, the force field at the cylinder B of II cannot induce any velocities because it lies against the innerside of the duct. Hence only the translated disk I which is now at a distance b from the trailing edge of the

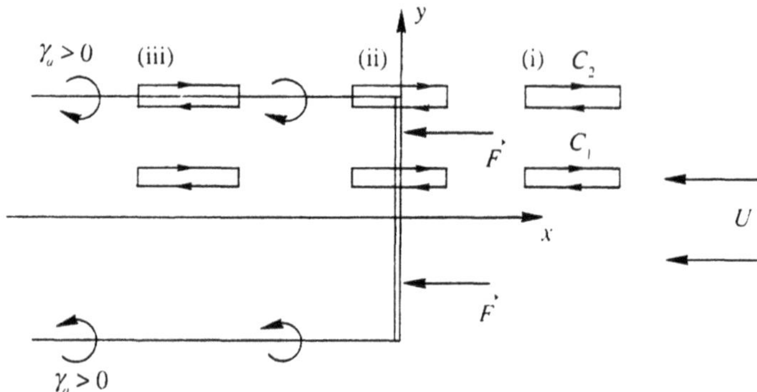

Fig. 2.4.3. Cross-section of disk, probing contours C_1 and C_2 floating with the main stream.

duct induces the velocity field, which by the foregoing has to be the same as when it was placed at the trailing edge at $x = 0$.

This result holds only when $0 \leq b \leq L$, otherwise we have to add to the disk the normal load at that part of the cylinder B which is outside the duct. Moreover, it is essential that actuator disk and duct have the same diameter, otherwise the normal load at the cylinder B induces also a velocity field.

First we consider the vorticity shed by the time periodic action of the actuator disk (2.4.1) alone, hence without the influence of the duct. We use narrow probing contours C_1 and C_2 in a meridian plane with sides parallel to the x-axis and y-axis, floating downstream with the velocity U in the negative x-direction. The length of these contours is l and their width is $\tilde{\varepsilon}$, $\tilde{\varepsilon} \ll l$. The contour C_1 passes through the actuator disk for values of $r < a$, while C_2 meets the boundary of the disk at $r = a$ (Figure 2.4.3).

In position (i), the circulations Γ_1 and Γ_2 of C_1 and C_2 respectively are zero. In position (ii) we have by (1.3.16)

$$\frac{d\Gamma_1}{dt} = \frac{1}{\rho} \int_{C_1} \vec{F} \cdot \vec{ds} = 0 \ ,$$

$$\frac{d\Gamma_2}{dt} = \frac{1}{\rho} \int_{C_2} \vec{F} \cdot \vec{ds} = \frac{1}{\rho} \left(f_1 + f_2 \, e^{i\omega t} \right) . \tag{2.4.3}$$

We now assume the length l of the contours so small that $\vec{F}$ nearly does not change when the contours pass the disk. This is important for C_2. We find when C_1 and C_2 are just downstream of the disk

$$\Gamma_1 = 0 \ , \qquad \Gamma_2 = \frac{1}{\rho} \left(f_1 + f_2 \, e^{i\omega t} \right) \frac{l}{U} \ . \tag{2.4.4}$$

Once the contours have passed the disk, there is no change in their circulation anymore (position (iii)). This means that per unit of length in the x-direction, the free vorticity strength γ_a, behind the disk boundary becomes

$$\gamma_a(x,t) = \frac{\Gamma_2}{l} = \frac{1}{U\rho}\left(f_1 + f_2\, e^{i(\omega t+\mu x)}\right) , \quad \mu = \frac{\omega}{U} , \quad r = a , \qquad (2.4.5)$$

where γ_a is reckoned positive with a right-hand screw in the negative φ-direction (Figure 2.4.1).

The configuration of ring vorticity on the semi-infinite cylinder surface $r = a$, $x < 0$, describes entirely the velocity field induced by the external force field $\vec{F}$. It induces a disturbance velocity with a non-zero component normal to the duct. We will assume a circular vorticity distribution on the duct, to compensate this normal velocity. As a consequence of the time dependence of the disturbance velocity, this vorticity distribution will also be a function of time, hence in general free vorticity will be shed from the trailing edge of the duct. The interaction of free vorticity behind duct and disk will influence the efficiency of the system.

In the next section we calculate the kinetic energy per length period $2\pi/\mu$ in the x-direction of the fluid far downstream of the disk, when the duct is absent. From this follows the efficiency of the time-dependent actuator disk alone.

2.5. Efficiency of Unsteady Actuator Disk, without Duct

We will calculate in this section the kinetic energy left behind by the actuator disk, when it moves with a velocity U in the positive x-direction, through a fluid at rest. This can be carried out, in terms of the formulation given in the previous section, by considering the kinetic energy E of for instance the real part of the disturbance velocities over one period of length $(2\pi/\mu)$ in the x-direction for some chosen fixed time. For this time we choose $t = 0$, then we have to consider the vorticity (2.4.5)

$$\gamma_a(x,0) = \frac{1}{U\rho}\left(f_1 + f_2\, e^{i\mu x}\right)$$

$$= \gamma_1 + \gamma_2(x) , \qquad x < 0 , \quad r = a , \qquad (2.5.1)$$

where γ_1 represents the steady part of the vorticity and $\gamma_2(x)$ the unsteady part.

We denote the disturbance velocity field infinitely far behind the disk, say at $x = -\infty$, induced by the vorticity (2.5.1), by $\vec{v}_a(x,r)$. Then we split also $\vec{v}_a(x,r)$ into two parts

$$\vec{v}_a(x,r) = \vec{v}_1(r) + \vec{v}_2(x,r) , \qquad (2.5.2)$$

where $\vec{v}_1$ is induced by γ_1 and $\vec{v}_2$ by γ_2.

The velocity field $\vec{v}_1(r)$ follows from (2.1.13)

$$\vec{v}_1(r) = \frac{-f_1}{\rho U}\vec{e}_x , \qquad r < a ;$$

$$\vec{v}_1(r) = (0,0,0) , \qquad r > a . \qquad (2.5.3)$$

For $\vec{v}_2$ we find by a symmetry consideration with respect to γ_2, that

$$\int_x^{x+2\pi/\mu} v_{2x}(\xi, r)\, d\xi = 0 \ , \qquad \vec{v}_2(x, r) = \vec{v}_2(x + 2\pi/\mu, r) \ . \tag{2.5.4}$$

By the first part of (2.5.4) we can write for the kinetic energy E of the real part of the velocity field $\vec{v}_a$ in a slab G of space, of width $2\pi/\mu$ in the x-direction and far behind the disk

$$E = \tfrac{1}{2}\,\rho \int_G \{\mathrm{Re}\,\vec{v}_a\}^2\, d\mathrm{Vol}$$

$$= \tfrac{1}{2}\,\rho \int_G \{\mathrm{Re}\,\vec{v}_1\}^2\, d\mathrm{Vol} + \tfrac{1}{2}\,\rho \int_G \{\mathrm{Re}\,\vec{v}_2\}^2\, d\mathrm{Vol} = E_1 + E_2 \ . \tag{2.5.5}$$

Hence we can simply add, far downstream of the disk, the kinetic energies E_1 and E_2 of the steady part of the slip stream and of the unsteady part respectively.

By (2.5.3) we find for E_1

$$E_1 = \tfrac{1}{2}\,\rho \cdot \pi a^2 \cdot \frac{2\pi}{\mu} \cdot \left(\frac{f_1}{\rho U}\right)^2 = \frac{\pi^2 a^2}{\mu}\,\frac{f_1^2}{\rho U^2} \ . \tag{2.5.6}$$

For the calculation of E_2 we consider two parts G_1 and G_2 of the slab G, namely $0 < r < a$ and $a < r < \tilde{r}$ respectively and take the limit $\tilde{r} \to \infty$. When Φ_2 is the potential of $\vec{v}_2$, we have that at $r = a$ the normal derivative $\partial\Phi_2/\partial n$ is continuous (preservation of fluid) while Φ_2 itself is discontinuous

$$[\Phi_2]_-^+ = \Phi_2^+ - \Phi_2^- = \lim_{r \downarrow a} \Phi_2 - \lim_{r \uparrow a} \Phi_2 \ . \tag{2.5.7}$$

Then we find

$$E_2 = \tfrac{1}{2}\,\rho \int_{G_1+G_2} (\mathrm{grad}\,\mathrm{Re}\,\Phi_2)^2\, d\mathrm{Vol}$$

$$= \tfrac{1}{2}\,\rho \int_{G_1+G_2} \mathrm{div}\,\{\mathrm{Re}\,\Phi_2 \cdot \mathrm{grad}\,\mathrm{Re}\,\Phi_2\}\, d\mathrm{Vol}$$

$$= -\tfrac{1}{2}\,\rho\, a \int_0^{2\pi/\mu} \int_0^{2\pi} \left\{ \mathrm{Re}\,[\Phi_2]_-^+ \cdot \mathrm{Re}\,\frac{\partial\Phi_2}{\partial n} \right\} d\varphi\, d\xi \ . \tag{2.5.8}$$

In (2.5.8) remains only the integral over the cylindrical surface $(0 \le \xi \le 2\pi/\mu, r = a)$, because by periodicity the contributions of the planes $\xi = 0$ and $\xi = 2\pi/\mu$ cancel each other, while the contribution of the cylindrical part $(0 \le \xi \le 2\pi/\mu, r = \tilde{r})$ tends to zero for $\tilde{r} \to \infty$.

For $[\Phi_2]_-^+$ we find

$$[\Phi_2]_-^+ = -\int\limits_x^0 \gamma_2(\xi)\, d\xi = \frac{-i}{\mu}\,\gamma_2(x) + \frac{i f_2}{U\rho\,\mu} \ . \qquad (2.5.9)$$

The normal derivative in (2.5.8) is independent of φ, therefore we can take $\varphi = 0$. By the law of Biot and Savart (1.1.15) we obtain

$$\frac{\partial \Phi_2}{\partial n}(x,a,0) = \frac{a}{4\pi}\,\vec{e}_y \cdot \oint\limits_{-\infty}^{\infty}\int\limits_0^{2\pi} \gamma_2(\xi)\,\frac{(\vec{e}_\theta \times \vec{R})}{R^3}\,d\theta\,d\xi \ , \qquad (2.5.10)$$

where the integral over ξ has to be taken as a Cauchy principle value and $\vec{e}_y$ and $\vec{e}_\theta$ are unit vectors in the y- and the $\theta(=\varphi)$-direction respectively. The components of $\vec{e}_\theta$ and $\vec{R}$ are in the Cartesian system

$$\vec{e}_\varphi = (0, -\sin\theta, \cos\theta) \ ;$$

$$\vec{R} = (x-\xi, a(1-\cos\theta), -a\sin\theta) \ , \qquad R = |\vec{R}| \ . \qquad (2.5.11)$$

Substitution of (2.5.1) and (2.5.11) into (2.5.10) yields

$$\frac{\partial \Phi_2}{\partial n}(x,a,0) = i\nu K(\nu)\,\gamma_2(x) \ , \qquad \nu = a\mu = \frac{a\omega}{U} \ , \qquad (2.5.12)$$

$$K(\nu) = -\frac{1}{\pi\nu}\int\limits_0^\infty\int\limits_0^\pi \frac{\eta\,\sin\nu\eta\,\cos\theta\,d\theta\,d\eta}{\{\eta^2 + 2(1+\cos\theta)\}^{3/2}} \ . \qquad (2.5.13)$$

Using (2.5.9) and (2.5.12), we find by (2.5.8)

$$E_2 = \frac{\pi^2 a^2}{\mu}\,\frac{f_2^2}{\rho U^2}\,K(\nu) = \left(\frac{f_2}{f_1}\right)^2 K(\nu)\,E_1 \ . \qquad (2.5.14)$$

The useful work W delivered by the disk over one period of length $2\pi/\mu$ in the x-direction is

$$W = \pi a^2 f_1 \cdot \frac{2\pi}{\mu} = 2\pi^2\,\frac{a^2 f_1}{\mu} \ . \qquad (2.5.15)$$

Hence the efficiency η_a of the actuator disk alone becomes

$$\eta_a = \frac{W}{W + E_1 + E_2} = \left[1 + \frac{f_1}{2\rho U^2}\left\{1 + \left(\frac{f_2}{f_1}\right)^2 K(\nu)\right\}\right]^{-1} \ . \qquad (2.5.16)$$

In the following we keep all quantities, which determine the situation, fixed except the frequency ω of the superimposed unsteady force field. When $\omega \to 0$ and

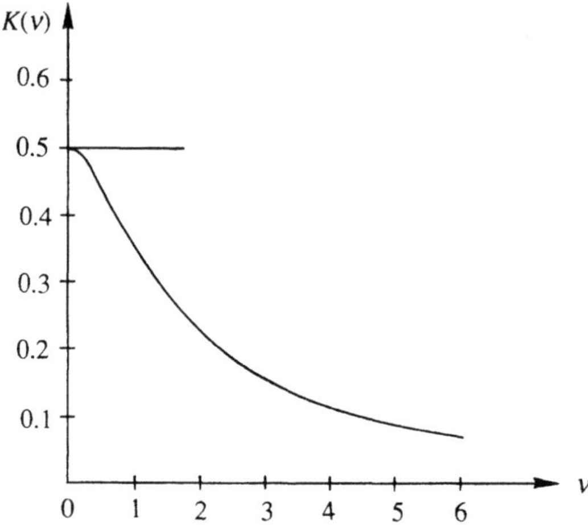

Fig. 2.5.1. $K(\nu)$ as a function of $\nu = a\omega/U$.

hence $\nu = a\omega/U \to 0$, the time-dependent part of the slip stream remains periodic although its length period $2\pi/\mu = 2\pi U/\omega$ in the x-direction becomes very large. Then it follows from (2.5.14) and Figure 2.5.1 that

$$E_2 \to 0.5 \left(\frac{f_2}{f_1}\right)^2 E_1 \ , \qquad \omega \to 0 \ . \tag{2.5.17}$$

This result is different from the case when $\omega = 0$, because then we have no unsteady force field (2.4.1) and it is found easily that

$$E_2 = \left(\frac{f_2}{f_1}\right)^2 E_1 \ , \qquad \omega = 0 \ . \tag{2.5.18}$$

In case $\omega \to \infty$, hence $\nu \to \infty$, it follows from (2.5.13) that $K(\nu) \to 0$, hence

$$E_2/E_1 \to 0 \ , \qquad \omega \to \infty \ , \tag{2.5.19}$$

and the actuator disk resembles more and more an ideal propeller (below (2.2.7)). The reason is that then the period in the x-direction of $\gamma_2(x)$ tends to zero and hence the velocity induced by γ_2 at any point which is at a finite distance of it tends to zero.

From (2.5.16) we find that η_a increases when ω increases, this is because then $K(\nu)$ (Figure 2.5.1) decreases.

2.6. Efficiency of Unsteady Actuator Disk, with Duct

We now return to the configuration of Figure 2.4.1. In the first instance we only need to consider the unsteady part of the load. The reason is that the steady part f_1 of the load (2.6.1) causes steady vorticity at the duct, by which no free vorticity is shed from its trailing edge. So in order to calculate the change in the free vorticity behind the disk which is caused by the duct and from which follows the change in efficiency, we have to consider only the unsteady part f_2 of the load.

At the duct we assume vorticity of strength

$$\gamma_3(x)\, e^{i\omega t} \ , \quad 0 < x < L \ , \quad r = a \ , \tag{2.6.1}$$

where $\gamma_3(x)$ is an unknown function. Because the circulation of the duct is time dependent, free vorticity is shed of which the intensity is denoted by

$$\alpha\, e^{i(\omega t+\mu x)} \ , \quad x < 0 \ , \quad r = a \ , \tag{2.6.2}$$

where α is an unknown constant. This constant is directly coupled to $\gamma_3(x)$ because the free vorticity shed by the duct equals minus the rate of change of its circulation. Hence at $x = 0$

$$U\alpha\, e^{i\omega t} = -\frac{\partial}{\partial t}\int\limits_0^L \gamma_3(x)\, e^{i\omega t}\, dx \ , \quad \alpha = -\frac{i\omega}{U}\int\limits_0^L \gamma_3(x)\, dx \ . \tag{2.6.3}$$

At the cylindrical duct we have the boundary condition that the normal component of the induced velocity has to be zero. Because our problem is axisymmetric we can consider this boundary condition for $\varphi = 0$,

$$v_y(x,t) = 0 \ , \quad 0 < x < L \ , \quad r = r_a \ , \quad \varphi = 0 \ . \tag{2.6.4}$$

By the Kutta condition we have that the resulting vorticity at the duct and behind the duct has to be continuous at the trailing edge. Hence we have also the relation

$$\frac{f_2}{U\rho} + \alpha = \frac{f_2}{U\rho} - \frac{i\omega}{U}\int\limits_0^L \gamma_3(x)\, dx = \gamma_3(0) \ , \tag{2.6.5}$$

where the left-hand side represents the sum of the vorticities caused by f_2 (2.5.1) and by the duct (2.6.2) at $x = 0$.

Using the law of Biot and Savart (1.1.15), analogously as we did in (2.5.10), we can write the boundary condition (2.6.4) as

$$\begin{aligned} \vec{e}_y \cdot \Bigg[& \frac{1}{4\pi} \oint\limits_0^L \int\limits_0^{2\pi} \gamma_3(\xi)\, \frac{(\vec{e}_\theta \times \vec{R})}{R^3}\, d\theta\, d\xi \\ & + \frac{1}{4\pi}\int\limits_{-\infty}^0 \int\limits_0^{2\pi} \left(\frac{f_2}{U\rho} + \alpha\right) e^{i\mu\xi}\, \frac{(\vec{e}_\theta \times \vec{R})}{R^3}\, d\theta\, d\xi \Bigg] = 0 \ , \\ & 0 < x < L \ , \qquad r = a \ , \qquad \varphi = 0 \ , \end{aligned} \tag{2.6.6}$$

where the Cartesian components of $\vec{e}_\theta$ and $\vec{R}$ are given in (2.5.11). In this singular integral equation we still have the unknown constant α. However, by the linearity of the equation we can write its solution as

$$\gamma_3(x) = \frac{f_2}{U\rho}\,\gamma_{31}(x) + \alpha\,\gamma_{32}(x)\ , \tag{2.6.7}$$

where $\gamma_{31}(x)$ and $\gamma_{32}(x)$ have to be determined numerically. Then substitution of (2.6.7) into (2.6.5) yields an equation for α, by which α can be calculated. Hence $\gamma_3(x)$ is known.

In order to determine the efficiency of the ducted actuator disk, we have to take into account again the steady load f_1 (2.5.1). The lost kinetic energy E per period of length in the x-direction can also in this case be written as

$$E = E_1 + E_3\ , \tag{2.6.8}$$

where E_1 is given in (2.5.6) and where now E_3 belongs to the kinetic energy of the real part of the sum of the two vortex layers $\gamma_2(x)$ of (2.5.1) and the vortex layer (2.6.2). Then by (2.6.5) we find analogously to (2.5.14)

$$E_3 = \frac{\rho\,\pi^2 a^2}{\mu}\,\gamma_3^2(0)\,K(\nu)\ . \tag{2.6.9}$$

For the efficiency η_d of the actuator disk in combination with the duct, we find analogously to (2.6.15) and after some algebraic reductions

$$\eta_d = \frac{W}{W + E_1 + E_3} = \left[1 + \frac{f_1}{2\rho\,U^2}\left\{1 + \left(\frac{f_2}{f_1}\right)^2 \frac{E_3}{E_2}\,K(\nu)\right\}\right]^{-1}\ . \tag{2.6.10}$$

We now consider the dimensionless quantity E_3/E_2 which occurs in (2.6.10). This quantity does not depend on f_2, because both E_2 and E_3 are proportional to f_2^2. Hence it depends only on a, ω, $U\rho$ and L. Applying the shorthand method of dimension analysis (Appendix E), we have to consider

$$\begin{aligned} 0 = [E_3/E_2] &= \left[a^{\beta_1}\omega^{\beta_2}U^{\beta_3}\rho^{\beta_4}L^{\beta_5}\right] \\ &= [l]^{\beta_1}\cdot[t]^{-\beta_2}\cdot([l][t]^{-1})^{\beta_3}\cdot([m][l]^{-3})^{\beta_4}\cdot[l]^{\beta_5}\ . \end{aligned} \tag{2.6.11}$$

Equating to zero the powers of $[l]$, $[m]$ and $[t]$ we can choose $\beta_1 = -\beta_3 - \beta_5$, $\beta_2 = -\beta_3$, $\beta_4 = 0$. Hence

$$\frac{E_3}{E_2} = \left(\frac{U}{a\omega}\right)^{\beta_3}\cdot\left(\frac{L}{a}\right)^{\beta_5} \rightarrow \frac{E_3}{E_2} = g(\nu, L/a)\ ,\quad \nu = \frac{a\omega}{U}\ , \tag{2.6.12}$$

where g is an unknown function which can be calculated by (2.5.14) and (2.6.9) when $\gamma_3(x)$ is known. Then the efficiency η_d (2.6.10) can be written as

$$\eta_d = \left[1 + \frac{f_1}{2\rho\,U^2}\left\{1 + \left(\frac{f_2}{f_1}\right)^2 g(\nu, L/a)K(\nu)\right\}\right]^{-1}\ . \tag{2.6.13}$$

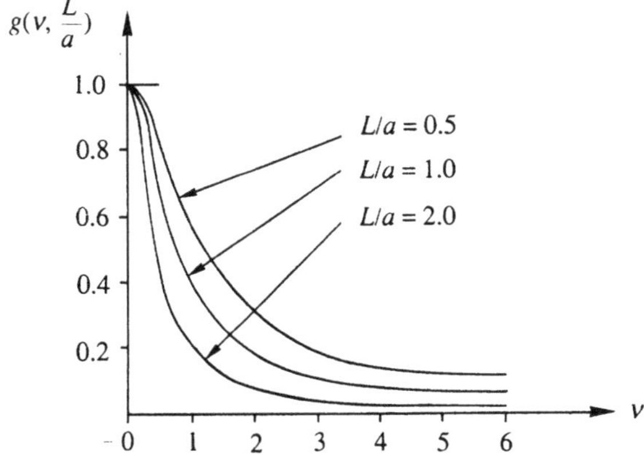

Fig. 2.6.1. g as a function of $\nu = a\omega/U$, for different relative lengths L/a of the duct.

The function $g(\nu, L/a)$ is drawn in Figure 2.6.1.

When $\nu \to 0$ the length of the period of the vorticity of the disk tends to infinity and at each moment behaves more or less as a vorticity sheet of constant strength. Hence the duct will practically shed no vorticity and hence cannot regain a substantial part of the energy E_2, so $g(\nu, L/a) \to 1$ for $\nu \to 0$ (see (2.5.16)).

The longer the shroud or the larger the value of L/a, the better it is able to shed free vorticity needed to compensate the vorticity γ_2 behind the periodic disk. This is in agreement with Figure 2.6.1 where for fixed ν the values of $g(\nu, L/a)$ become smaller for larger values of L/a. It is to be expected that for "$L/a = \infty$", $g(\nu, \infty) = 0$, because then the periodic part of the load of the disk cannot accelerate anymore the infinite mass of the fluid inside the half-infinite tube. Hence the jet becomes stationary and the efficiency has to become equal to the efficiency given in (2.2.7) with $T/S = f_1$.

2.7. Steady Axisymmetric Force Field in a Homogeneous Flow

We will derive some equations which have a bearing on non-linear actuator disk theory. This section and the next one are partly generalisations of Wu's [77]. We use cylindrical coordinates (x, r, φ) (Figure 2.7.1). The force field $\vec{F}$ is axisymmetric and defined in some region B of space,

$$\vec{F}(x,r) = (F_x(x,r), F_r(x,r), F_\varphi(x,r)) \;, \quad (x,r,\varphi) \in B \;;$$
$$\vec{F} = 0 \;, \qquad (x,r,\varphi) \notin B \;, \tag{2.7.1}$$

where we use the physical components of $\vec{F}$ (below (B.1.10)). The velocity field with its physical components are denoted by $\vec{v}(x,r) = (v_x, v_r, v_\varphi)$. The force field is placed in a homogeneous flow with velocity U in the positive x-direction. Because

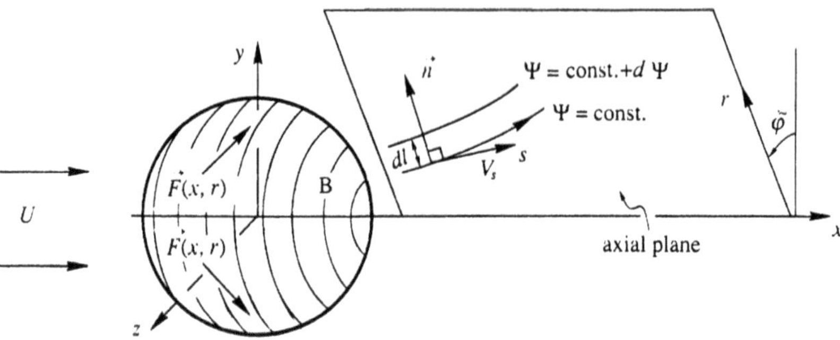

Fig. 2.7.1. External force field in parallel flow U.

the problem is axisymmetric, its quantities depend only on the two coordinates x and r. Then we can introduce the stream function $\Psi = \Psi(x, r)$ (1.20.23) with

$$v_x = \frac{1}{r}\frac{\partial \Psi}{\partial r}\ , \qquad v_r = -\frac{1}{r}\frac{\partial \Psi}{\partial x}\ , \tag{2.7.2}$$

hence the flow field is determined by Ψ and the velocity component $v_\varphi(x, r)$.

We write the equation of motion for this stationary case as (1.2.6)

$$\vec{v} \times \vec{\omega} = \text{grad}\left(\frac{p}{\rho} + \tfrac{1}{2}\,\vec{v}^{\,2}\right) - \frac{1}{\rho}\,\vec{F} = \text{grad}\,H - \frac{1}{\rho}\,\vec{F}\ , \tag{2.7.3}$$

where $\vec{\omega} = \text{rot}\,\vec{v}$ and we introduced the head $H(x, r)$ of the flow. The physical components of grad H and of $\vec{\omega}$ are given by (B.2.7) and (B.2.10)

$$\text{grad}\,H(x, r) = \left(\frac{\partial H}{\partial x}, \frac{\partial H}{\partial r}, 0\right)\ , \tag{2.7.4}$$

$$\vec{\omega} = (\omega_x, \omega_r, \omega_\varphi) = \left(\frac{1}{r}\frac{\partial}{\partial r}\,r v_\varphi\ , -\frac{\partial}{\partial x}\,v_\varphi\ , \frac{\partial}{\partial x}\,v_r - \frac{\partial}{\partial r}\,v_x\right)\ . \tag{2.7.5}$$

Hence in cylindrical coordinates the equations of motion, which are the three components of (2.7.3), become

$$v_x\frac{\partial}{\partial x}\,v_x + v_r\frac{\partial}{\partial r}\,v_x = -\frac{1}{\rho}\frac{\partial p}{\partial x} + \frac{1}{\rho}\,F_x\ , \tag{2.7.6}$$

$$v_x\frac{\partial}{\partial x}\,v_r + v_r\frac{\partial}{\partial r}\,v_r - \frac{1}{r}\,v_\varphi^2 = -\frac{1}{\rho}\frac{\partial p}{\partial r} + \frac{1}{\rho}\,F_r\ , \tag{2.7.7}$$

$$v_x\frac{\partial}{\partial x}\,v_\varphi + v_r\frac{\partial}{\partial r}\,v_\varphi + \frac{1}{r}\,v_r v_\varphi = \frac{1}{\rho}\,F_\varphi\ . \tag{2.7.8}$$

By taking the scalar product of $\vec{v}$ and $\vec{\omega}$ with (2.7.3) we find

$$\vec{v} \cdot \operatorname{grad} H = \frac{1}{\rho}\, \vec{v} \cdot \vec{F} \ , \qquad \vec{\omega} \cdot \operatorname{grad} H = \frac{1}{\rho}\, \vec{\omega} \cdot \vec{F} \ , \tag{2.7.9}$$

respectively. Outside of the force region B, hence when $\vec{F} = 0$, we obtain

$$\vec{v} \cdot \operatorname{grad} H = 0 \ , \qquad \vec{\omega} \cdot \operatorname{grad} H = 0 \ , \tag{2.7.10}$$

which means that outside B the axisymmetric stream tubes and vortex tubes coincide with surfaces H = const.

We now restrict our attention to an axial plane $\varphi = \tilde{\varphi}$ (Figure 2.7.1), in which we have the coordinates x and r. In this plane we consider differentiation along the lines Ψ = const. and differentiation with respect to Ψ. The latter differentiation can be applied only to functions which depend on Ψ alone. Along the lines Ψ = const. we introduce a parameter s which measures length, of which the positive direction is the same as the direction of the velocity component along these lines.

When some function $G(x, r)$ is sufficiently smooth we find

$$\frac{\partial G}{\partial s} = \left(\frac{\partial G}{\partial x}, \frac{\partial G}{\partial r}\right) \cdot \left(\frac{\partial x}{\partial s}, \frac{\partial r}{\partial s}\right) = \left(\frac{\partial G}{\partial x}, \frac{\partial G}{\partial r}\right) \cdot \left(\frac{v_x}{V_s}, \frac{v_r}{V_s}\right) \ , \tag{2.7.11}$$

where $V_s = \left(v_x^2 + v_r^2\right)^{1/2}$ and the second equality is correct because s measures length. By this we find

$$\frac{\partial}{\partial s} = \frac{1}{V_s}\left(v_x \frac{\partial}{\partial x} + v_r \frac{\partial}{\partial r}\right) \ , \qquad V_s = (v_x^2 + v_r^2)^{1/2} \ . \tag{2.7.12}$$

Now we consider first differentiation in the direction normal to the lines Ψ = const. A unit normal $\vec{n}$ to these lines is given by

$$\vec{n} = \left(-\frac{v_r}{V_s}, \frac{v_x}{V_s}\right) \ . \tag{2.7.13}$$

Hence we have

$$\frac{\partial G}{\partial n} = \left(\frac{\partial G}{\partial x}, \frac{\partial G}{\partial r}\right) \cdot \left(-\frac{v_r}{V_s}, \frac{v_x}{V_s}\right) \ , \tag{2.7.14}$$

or

$$\frac{\partial}{\partial n} = \frac{1}{V_s}\left(-v_r \frac{\partial}{\partial x} + v_x \frac{\partial}{\partial r}\right) \ . \tag{2.7.15}$$

By the definition of a stream function, the difference $d\Psi$ of two lines Ψ = const. and Ψ = const. + $d\Psi$ which are at some place dl apart (Figure 2.7.1) is

$$d\Psi = V_s r \ dl \ . \tag{2.7.16}$$

From this it follows that $\vec{n}$ (2.7.13) points in the direction of increasing values of Ψ.
When some function $\tilde{G}(\Psi)$ depends *only* on Ψ, we find by (2.7.16)

$$\frac{d\tilde{G}}{d\Psi} = \frac{dl}{d\Psi}\frac{\partial \tilde{G}}{\partial n} = \frac{1}{rV_s}\frac{\partial \tilde{G}}{\partial n} = -\frac{1}{rV_s^2}\left(v_r\frac{\partial \tilde{G}}{\partial x} - v_x\frac{\partial \tilde{G}}{\partial r}\right) , \tag{2.7.17}$$

hence

$$\frac{d}{d\Psi} = \frac{1}{rV_s^2}\left(-v_r\frac{\partial}{\partial x} + v_x\frac{\partial}{\partial r}\right) . \tag{2.7.18}$$

From (2.7.8) we find by (2.7.12)

$$\frac{\partial}{\partial s}(rv_\varphi) = \frac{r}{\rho V_s}F_\varphi , \tag{2.7.19}$$

which describes the rate of change of the moment of momentum about the x-axis (compare with (1.15.11)). Hence outside the force region B where $\vec{F} = 0$, the quantity rv_φ will depend on Ψ only.

$$rv_\varphi = h(\Psi) , \qquad (x,r,\varphi) \notin B , \tag{2.7.20}$$

where h is some function. It follows from (2.7.19) that for fluid particles, which do not have been in contact with the force field, we have $v_\varphi = 0$.
Consider the first equation of (2.7.9) with $\partial H/\partial\varphi = 0$,

$$v_x\frac{\partial H}{\partial x} + v_r\frac{\partial H}{\partial r} = V_s\frac{\partial H}{\partial s} = \frac{1}{\rho}\,\vec{v}\cdot\vec{F} . \tag{2.7.21}$$

Hence also for the head H we have the (well-known) result that outside the force region $\partial H/\partial s = 0$ and hence

$$H = H(\Psi) , \qquad (x,r,\varphi) \notin B . \tag{2.7.22}$$

We now determine the φ-component of the vector product of $\vec{v}$ with (2.7.3), we find by using (2.7.2) and (2.7.5)

$$V_s^2\left\{\frac{1}{r}\frac{\partial^2\Psi}{\partial x^2} + \frac{\partial}{\partial r}\left(\frac{1}{r}\frac{\partial\Psi}{\partial r}\right)\right\}$$
$$= -\frac{V_s}{r^2}\,(rv_\varphi)\,\frac{\partial}{\partial n}(rv_\varphi) + V_s\frac{\partial H}{\partial n} - \frac{1}{\rho}\,(v_xF_r - v_rF_x) , \tag{2.7.23}$$

where we have to use $\partial/\partial n$ because in the force field both (rv_φ) and H still depend on the two variables x and r and not on Ψ alone. When we are outside of the force region, we know that $(rv_\varphi) = h(\Psi)$ and $H = H(\Psi)$, then we can write (2.7.23) as

$$\frac{\partial^2\Psi}{\partial x^2} + r\frac{\partial}{\partial r}\left(\frac{1}{r}\frac{\partial\Psi}{\partial r}\right) = r^2H'(\Psi) - h(\Psi)h'(\Psi) , \tag{2.7.24}$$

where $h(\Psi)$ and $H(\Psi)$ are still unknown.

The stream function Ψ has to satisfy the limit conditions

$$\frac{1}{r}\frac{\partial\Psi}{\partial r}\to U\ ,\quad \frac{1}{r}\frac{\partial\Psi}{\partial x}\to 0\ ;\quad x\to-\infty \quad \text{or} \quad r\to\infty\ , \tag{2.7.25}$$

$$\frac{1}{r}\frac{\partial\Psi}{\partial x}\to 0\ ,\quad x\to\infty\ . \tag{2.7.26}$$

A modified form of the right-hand side of Equation (2.7.24) is obtained when we make an assumption about the way in which the force field $\vec{F}$ comes into being. We now assume that $\vec{F}$ arises from radially directed axisymmetric bound vorticity which rotates in the force region B with an angular velocity Ω around the x-axis. Because then the force field has to be perpendicular to the relative velocity of the fluid with respect to the bound vorticity, it follows that

$$(\vec{v}-\vec{e}_\varphi\,\Omega\, r)\cdot\vec{F}=0\ , \tag{2.7.27}$$

where $\vec{e}_\varphi$ is the unit vector in the φ-direction. By this it follows from (2.7.21) that

$$V_s\frac{\partial H}{\partial s}=\frac{\Omega\, r}{\rho}\,F_\varphi\ . \tag{2.7.28}$$

Combination of (2.7.19) and (2.7.28) yields

$$\frac{\partial H}{\partial s}=\Omega\frac{\partial}{\partial s}(rv_\varphi)\ \to\ H=\Omega\, rv_\varphi+H_\infty\ ,\quad H_\infty=\frac{p_\infty}{\rho}+\tfrac{1}{2}U^2\ , \tag{2.7.29}$$

where H_∞ is the head of the flow which has not yet been in contact with the force field. Hence in this case, outside the force region B, we find instead of (2.7.24), by using (2.7.29) and (2.7.20),

$$\frac{\partial^2\Psi}{\partial x^2}+r\frac{\partial}{\partial r}\left(\frac{1}{r}\frac{\partial\Psi}{\partial r}\right)=\{r^2\,\Omega-h(\Psi)\}\,h'(\Psi)\ . \tag{2.7.30}$$

The stream function Ψ which results from this equation has also to satisfy the conditions (2.7.25) and (2.7.26).

2.8. Non-Linear Actuator Disk Theory

We now apply the formulas of the previous section to an axisymmetric external force field which is concentrated at the circular region $x=0$, $0\le r\le a$ and hence forms an actuator disk. The force field is written as

$$\vec{F}=(f_x(r),f_r(r),f_\varphi(r))\,\delta(x)\ ,\qquad 0\le r\le a\ ;$$

$$f_x(r)=f_r(r)=f_\varphi(r)=0\ ,\qquad r>a\ , \tag{2.8.1}$$

where $\delta(x)$ is the delta function of Dirac. In order that this disk will deliver a thrust in the situation of Figure 2.7.1, we can assume that mainly $f_x(r) > 0$. Taking $f_\varphi(r) > 0$ means that the disk represents a propeller which rotates in the positive φ-direction, while $f_r(r) \neq 0$ means that we approximate a propeller of which the blades have rake.

When the fluid passes through the disk, it will suddenly change its values of p, v_r and v_φ. First of all it follows from the continuity equation div $\vec{v} = 0$, which reads in this case (B.2.9)

$$\frac{\partial v_x}{\partial x} + \frac{1}{r}\frac{\partial}{\partial r}(r v_r) = 0 \ , \tag{2.8.2}$$

that v_x cannot have a jump discontinuity. Then the first term of (2.8.2) would obtain by the differentiation a $\delta(x)$ function which cannot be compensated by the second term. So we know that v_x is continuous through the disk which is a result we used earlier sometimes without a formal justification. Second we consider the pressure p by using (2.7.6) which reads in connection with (2.8.1)

$$v_x \frac{\partial}{\partial x} v_x + v_r \frac{\partial}{\partial r}\, v_x = -\frac{1}{\rho}\frac{\partial p}{\partial x} + \frac{1}{\rho}\, f_x(r)\, \delta(x) \ . \tag{2.8.3}$$

The left-hand side of this equation cannot compensate, because v_x is continuous, the $\delta(x)$ function of the force component at the right-hand side, hence this has to be accomplished by $\partial p/\partial x$ and we find that the pressure has to have a jump

$$[p]^+_- = p(+0) - p(-0) = f_x(r) \ . \tag{2.8.4}$$

In the same way we find by (2.7.7)

$$v_x \frac{\partial}{\partial x}\, v_r + v_r \frac{\partial}{\partial r}\, v_r - \frac{1}{r}\, v_\varphi^2 = -\frac{1}{\rho}\frac{\partial p}{\partial r} + \frac{1}{\rho}\, f_r(r)\, \delta(x) \ , \tag{2.8.5}$$

that the $\delta(x)$ function at the right-hand side has to be compensated by $\partial v_r/\partial x$ where now v_r has a jump discontinuity, so

$$[v_r]^+_- = \frac{1}{\rho\, v_x}\, f_r(r) \ . \tag{2.8.6}$$

Finally, it follows from (2.7.8)

$$[v_\varphi]^+_- = \frac{1}{\rho\, v_x}\, f_\varphi(r) \ . \tag{2.8.7}$$

Next we want to determine the jump across the disk of $h(\Psi) = r v_\varphi$ and of $H(\Psi)$. From (2.8.7) it follows

$$[h]^+_- = [r v_\varphi]^+_- = \frac{r}{\rho\, v_x}\, f_\varphi(r) \ . \tag{2.8.8}$$

By the definition (2.7.3) of H we find

$$[H]_-^+ = \frac{1}{\rho}[p]_-^+ + \tfrac{1}{2}[v_x^2]_-^+ + \tfrac{1}{2}[v_r^2]_-^+ + \tfrac{1}{2}[v_\varphi^2]_-^+ \ . \tag{2.8.9}$$

The value of $[p]_-^+$ is given in (2.8.4), v_x is continuous at the disk hence $[v_x^2]_-^+ = 0$. So we have still to determine the last two terms of (2.8.9). We assume that it can occur that $v_r(-0) \neq 0$ and $v_\varphi(-0) \neq 0$. This is quite natural with respect to v_r, however we expect $v_\varphi(-0) = 0$. When a fluid particle passes the disk for the first time, this will happen (below (2.7.20)). However, as we will discuss in the next section, it is not unlikely that in the neighbourhood of the edge fluid particles will cross the disk more than once and by this will have $v_\varphi(-0) \neq 0$, before they finally enter into the "slip stream" and are transported to $x = \infty$. Using (2.8.6) and (2.8.7) we have

$$\begin{aligned}[v_r^2]_-^+ &= (v_r^2(+0) - v_r^2(-0)) = (v_r(+0) - v_r(-0))\,(v_r(+0) + v_r(-0)) \\ &= \frac{1}{\rho\, v_x}\, f_r(r)\left(2v_r(-0) + \frac{1}{\rho\, v_x}\, f_r(r)\right) \ ,\end{aligned} \tag{2.8.10}$$

$$[v_\varphi^2]_-^+ = \frac{1}{\rho\, v_x} f_\varphi(r)\left(2v_\varphi(-0) + \frac{1}{\rho\, v_x}\, f_\varphi(r)\right) \ . \tag{2.8.11}$$

Hence

$$\begin{aligned}[H]_-^+ = \frac{1}{\rho}\, f_x(r) + \frac{1}{2\rho\, v_x}\Big\{ & f_r(r)\left(2v_r(-0) + \frac{1}{\rho\, v_x}\, f_r(r)\right) \\ & + f_\varphi(r)\left(2v_\varphi(-0) + \frac{1}{\rho\, v_x}\, f_\varphi(r)\right)\Big\} \ .\end{aligned} \tag{2.8.12}$$

In the remaining part of this section, we assume that there exists through the edge $r = a$ of the disk a continuous boundary stream tube $\Psi = \tilde{C}$, which starts at $x = -\infty$ and ends at $x = \infty$, and that the fluid particles inside this stream tube cross the disk only once. As remarked, this will be questioned in the next section. Also we assume that inside the boundary stream tube the other stream tubes Ψ = const. cross the disk continuously, be it that their slope at the crossing will have a jump by the possible discontinuity of v_r (2.8.6). By the first assumption we have

$$rv_\varphi(x,r) = 0 \ , \qquad H(x,r) = \frac{p_\infty}{\rho} + \tfrac{1}{2}\, U^2 \ , \qquad x < 0 \ . \tag{2.8.13}$$

Downstream of the disk, hence again outside of the force region, we have by (2.8.6) and (2.8.12)

$$h(\Psi) = rv_\varphi(+0,r) = \frac{r}{\rho\, v_x}\, f_r(r) \ , \quad x > 0 \ , \tag{2.8.14}$$

$$H(\Psi) = \frac{p_\infty}{\rho} + \tfrac{1}{2}\,U^2 + \frac{1}{\rho}\,f_x(r)$$
$$+\frac{1}{2\rho\,v_x}\left\{f_r(r)\left(2v_r(-0)+\frac{1}{\rho\,v_x}\,f_r(r)\right)+\frac{1}{\rho\,v_x}\,f_\varphi^2(r)\right\}\ ,$$
$$x > 0\ . \tag{2.8.15}$$

The problem we have to solve can now be stated as follows. We have to determine under the conditions (2.7.25) and (2.7.26) a function $\Psi = \Psi(x,r)$ which satisfies (2.7.24). Also we have to determine $H(\Psi)$ and $h(\Psi)$. Just behind the disk the values of $h(\Psi)$ and $H(\Psi)$ are denoted in (2.8.14) and (2.8.15), where however $v_x(0,r)$ and $v_r(+0,r)$ are still unknown. Everywhere outside the slip stream region S we have

$$h(\Psi) = 0\ , \qquad H(\Psi) = \frac{p_\infty}{\rho} + \tfrac{1}{2}\,U^2\ , \tag{2.8.16}$$

while the boundary ∂S ($x > 0$) of S is unknown. In case $\Psi(x,r)$ and $h(\Psi)$ are known, v_x and v_r follow from Ψ (2.7.2) and v_φ follows from $h(\Psi)$ (2.7.20).

We now introduce the following dimensionless quantities

$$r^* = r/a\ ,\quad x^* = x/a\ ,\quad \vec{v}^* = \vec{v}/U\ ,\Psi^* = \Psi/Ua^2\ ,H^* = H/U^2\ ,$$
$$h^* = h/aU\ ,\quad (f_x^*, f_r^*, f_\varphi^*) = (f_x, f_r, f_\varphi)/\rho\,U^2\ . \tag{2.8.17}$$

Substitution of (2.8.17) into (2.7.24) and neglecting the * yields the same equation as before. The formulas (2.8.14) and (2.8.15), valid behind the disk become after neglecting *

$$h(\Psi) = \frac{r}{v_x}\,f_r(r)\ , \qquad x > 0\ , \tag{2.8.18}$$

$$H(\Psi) = \frac{p_\infty}{\rho\,U^2} + \tfrac{1}{2} + f_x(r)$$
$$+\frac{1}{2v_x}\left\{f_r(r)\left(2v_r(-0)+\frac{1}{v_x}\,f_r(r)\right)+\frac{1}{v_x}\,f_\varphi^2(r)\right\}\ ,$$
$$x > 0\ . \tag{2.8.19}$$

Introducing the disturbance stream function ψ

$$\Psi(x,r) = \tfrac{1}{2}\,r^2 + \psi(x,r)\ , \tag{2.8.20}$$

we find for the differential equation (2.7.24) which, as has been said, has the same form as the dimensionless one

$$L\left(\frac{\psi}{r}\right) = r\left\{\frac{\partial^2}{\partial x^2}+\frac{\partial^2}{\partial r^2}+\frac{1}{r}\,\frac{\partial}{\partial r}-\frac{1}{r^2}\right\}\left(\frac{\psi}{r}\right)$$
$$= r^2 H'(\Psi) - h(\Psi)h'(\Psi)\ . \tag{2.8.21}$$

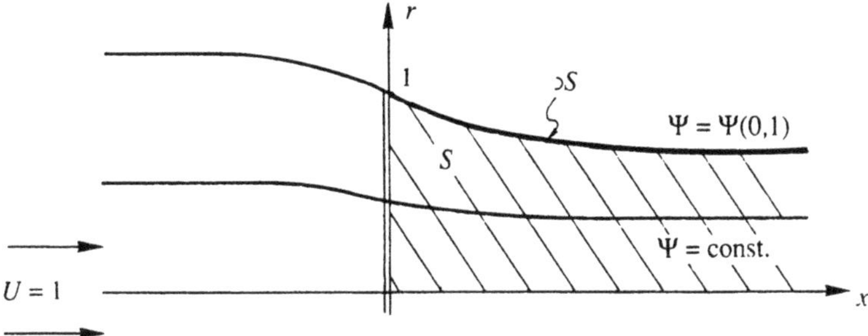

Fig. 2.8.1. Dimensionless configuration of slip stream region S (hatched).

The limit conditions (2.7.25) and (2.7.26) for ψ become homogeneous and read

$$\frac{1}{r}\frac{\partial\psi}{\partial r} \to 0\ , \quad \frac{1}{r}\frac{\partial\psi}{\partial x} \to 0\ ; \quad x \to -\infty \quad \text{or} \quad r \to \infty \tag{2.8.22}$$

$$\frac{1}{r}\frac{\partial\psi}{\partial x} \to 0\ ; \quad x \to \infty\ . \tag{2.8.23}$$

We now follow [77] and bring this problem in a form which is more amenable to numerical calculations by converting (2.8.21) into an integral equation. The Green function $K(x-\tilde{x}, r, \tilde{r})$ belonging to the differential operator L at the left-hand side of (2.8.21), is the solution of

$$L\left(K(x-\tilde{x}, r, \tilde{r})\right) = -\delta(x-\tilde{x})\delta(r-\tilde{r}), \tag{2.8.24}$$

where δ is the delta function of Dirac and L differentiates with respect to x and r. In connection with (2.8.22) and (2.8.23) K has to vanish for $x \to \infty$ and $r \to \infty$. Using the method of Hankel transforms [62] the following solution can be found

$$\begin{aligned} K(x-\tilde{x}, r, \tilde{r}) &= \tfrac{1}{2}\int_0^\infty e^{-|x-\tilde{x}|t} J_1(rt)J_1(\tilde{r}t)\, dt \\ &= \frac{1}{2\pi(r\tilde{r})^{1/2}} Q_{1/2}\left(\frac{(x-\tilde{x})^2 + r^2 + \tilde{r}^2}{2r\tilde{r}}\right)\ , \end{aligned} \tag{2.8.25}$$

where $Q_{1/2}$ is a Legendre function of the second kind ([23] II, p. 203).

Using (2.8.25) we can write (2.8.21) as

$$\begin{aligned} \psi(x,r) = -r \iint_S & K(x-\tilde{x}, r, \tilde{r}) \left\{\tilde{r}^2 H'(\Psi(\tilde{x},\tilde{r}))\right. \\ & \left. - h(\Psi(\tilde{x},\tilde{r}))h'(\Psi(\tilde{x},\tilde{r}))\right\}\, d\tilde{x}\, d\tilde{r}\ , \end{aligned} \tag{2.8.26}$$

where the region of integration is the slip stream region S (Figure 2.8.1) because outside S the tangential velocity $v_\varphi = 0$, hence $h(\Psi) = 0$ and $H(\Psi) = (p_\infty/\rho) + \frac{1}{2}U^2 = \text{const.}$, hence $H'(\Psi) = 0$.

The boundary of the slip stream region S or of the stream tube which passes through the edge $x = 0, r = 1$ of the disk (dimensionless case) is defined by

$$\Psi(x,r) = \Psi(0,1) \quad \text{or} \quad \psi(x,r) = \psi(0,1) + \tfrac{1}{2}(1-r^2) \ . \tag{2.8.27}$$

We describe an iteration procedure to solve the integral equation (2.8.26) when the force field $\big(f_x(r), f_r(r), f_\varphi(r)\big)\,\delta(x)$ is given. First we choose some estimate $\Psi_0(x,r)$ of $\Psi(x,r)$, for instance by solving the linearized problem. Then also the approximation $\psi_0(x,r)$ of $\psi(x,r)$ is known. By using (2.8.4), the continuity of v_x at the disk, (2.8.10) and (2.8.11) we find $p_0, v_{x0}, v_{r0}, v_{\varphi 0}$ just behind the disk ($x = +0$) as functions of r. Hence $h(\Psi_0)$, $h'(\Psi_0)$ and $H'(\Psi_0)$ can be calculated as functions of r, just behind the disk. Because these quantities are constant downstream of the disk approximately along lines $\Psi_0(x,r) = \text{const.}$ we know the approximations $h(\Psi_0)$, $h'(\Psi_0)$ and $H'(\Psi_0)$ in the whole area S_0 which is bounded by the line $\Psi_0(x,r) = \Psi_0(0,1)$. Then (2.8.26) gives again an approximation $\psi_1(x,r)$ of $\psi(x,r)$. Repeating this procedure by replacing ψ_0 by ψ_1 yields again an approximation $\psi_2(x,r)$, etc., and it is hoped that this iteration will converge and will give the desired solution of the problem.

It is clear that the same iteration procedure can be applied to the integral equation which arises from the more special differential equation (2.7.30). Then we have to replace the expression between braces in the integrand of (2.8.26) by the right-hand side of (2.7.30), in which for the corresponding dimensionless case we replace Ω by the dimensionless number $\lambda^{-1} = \Omega a/U$.

We conclude this section by showing some of the numerical results obtained in [21] by using a method based on the preceding considerations. The results are for the special form (2.7.30) of the partial differential equation for Ψ, hence when we have at the disk only radial bound vorticity which rotates with the angular velocity Ω. The bound vorticity $\Gamma(r)$ is characterized by

$$rv_\varphi(+0,r) = C = \text{const.} \ , \qquad 0 \le r \le 1 \ , \tag{2.8.28}$$

because $v_\varphi(-0,r) = 0$ by the assumption just below (2.8.12). This means that $\Gamma(r)$ per unit of length in the φ-direction, has the strength C/r. This 2-dimensional vorticity field is without divergence for $r \neq 0$ and for $r \neq 1$, hence no free vorticity will be shed in the slip stream with the exception of the positive x-axis, where a concentrated "hub" vortex of strength $2\pi C$ is present. At the boundary ∂S of the slip stream we have a layer of shed free vorticity.

When we approximate the disk by a finite number of N concentrated radial bound vortices which rotate with the angular velocity Ω in the plane of the disk, the strength of each of the vortices has to be independent of r, namely $2\pi C/N$, $(0 \le r \le 1)$. Hence the disk under consideration is to some extent a model of a screw propeller of which the circulation around its blades is independent of r.

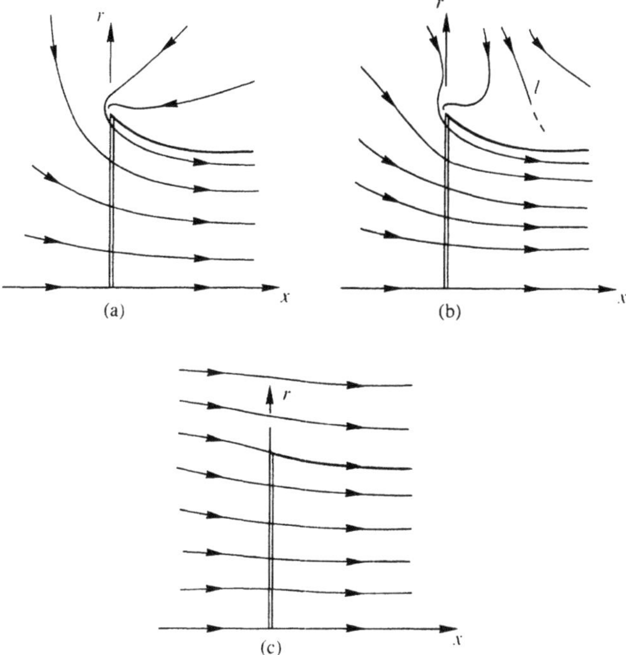

Fig. 2.8.2. Streamlines induced by actuator disks; $rv_\varphi(+0, r) = C$; $\lambda = U/\Omega a$, (a) $\lambda = 0$, (b) $\lambda = 0.01$, (c) $\lambda = 0.1$.

We remark that assumption (2.8.28) does not determine the force component $f_\varphi(r)$. The jump $[v_\varphi]_-^+ = C/r$, which follows from (2.8.28) because we assumed (below (2.8.12)) that $v_\varphi(-0, r) = 0$, is related to $f_\varphi(r)$ by (2.8.7). In this relation $v_x(0, r)$ is still unknown, hence $f_\varphi(r)$ can be determined after the problem is solved. Analogously $f_x(r)$ can be obtained only after the solution of the problem. We do know however that $f_r(r) = 0$.

Because $v_\varphi = 0$ outside the slip stream, the function $(rv_\varphi) = h(\Psi)$, has a jump across the slip stream boundary ∂S, hence $h'(\Psi)$ has a delta function of Dirac character at this boundary. This makes it possible to carry out one of the integrations in (2.8.26), where the form between braces has to be replaced by the right-hand side of (2.7.30). Then a one-dimensional integral equation is obtained. For more details of the calculations we refer to [21], where also the vortex representation of the disk is discussed.

In Figure 2.8.2 the results for $rv_\varphi(+0, r) = C$ and for three values of $\lambda = U/\Omega a$, namely $\lambda = 0$, $\lambda = 0.01$ and $\lambda = 0.1$ are given. It turns out that the shape of the stream lines is nearly independent of the constant C.

It is seen that in the case of $\lambda = 0.01$ (Figure 2.8.2 (b)) a dividing streamline l seems to exist. Fluid particles to the left of l will pass through the disk while fluid particles to the right of l do not. From the calculations it is not clear whether l ends at the boundary of the slip stream or whether l turns downstream to $x = \infty$. In Section 2.10 we make it plausible that l ends at the boundary for a finite value of $x > 0$.

For small values of λ, which correspond to small values of the incoming velocity or to heavy loading of the disk, the boundary of the slip stream is very steep in the neighbourhood of the edge of the disk. It has been assumed in the calculations that the vorticity at the boundary of the slip stream had a singular behaviour at the edge, which as a function of x had the character $O(x^{-1/2})$ for $x \to 0$. In the next section we return to the behaviour of the flow at the rim of the disk. In [21] still other radial bound vortex distributions have been considered so that a better approximation to the flow induced by a real propeller is obtained.

2.9. About the Singularity at the Edge of a Disk, Non-Linear Theory

We will now try to obtain insight in the singular behaviour of the flow at the edge of an actuator disk. It has to be remarked that the theory given in this section does not give a rigorous description of this singular behaviour, but it does strongly indicate that extraordinary phenomena can happen at the edge. We consider the special case of a constant normal load at a circular disk and no incoming flow, hence $U = 0$.

As in Section 2.4 we can also here, because the external force field at the disk is assumed to be perpendicular to it and of constant strength, deform the disk provided that its edge remains at the same place. Then the velocity field induced by the deformed disk is the same as the velocity field of the original one.

We shortly repeat the argumentation. In Figure 2.9.1 we have the cross section of two "different" actuator surfaces B_1 and B_2 with the same edge. The region in between these surfaces is denoted by V_{12}. The external force field $\vec{F}_i$ at B_i $(i = 1, 2)$ can be written as

$$\vec{F}_i = f\delta(n)\,\vec{n}_i \ , \qquad (i = 1, 2) \ , \tag{2.9.1}$$

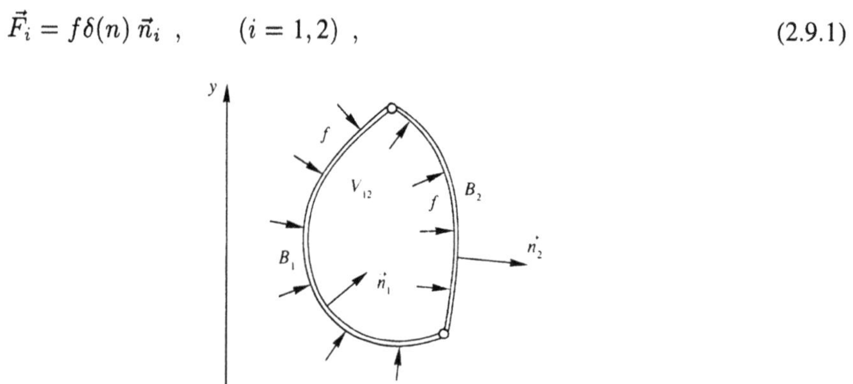

Fig. 2.9.1. Cross-section of two different actuator surfaces B_1 and B_2, with the same edge and the same normal loading.

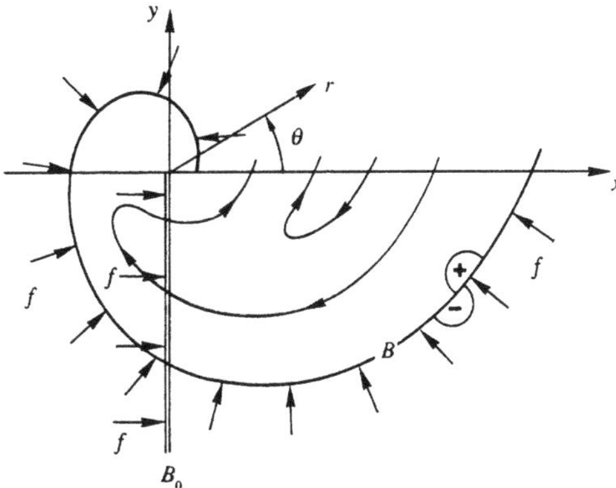

Fig. 2.9.2. The half-infinite actuator surface B_0, the deformed actuator surface B and stream lines.

where f is a constant, $\vec{n}_i$ is the unit normal at B_i and $\delta(n)$ is the delta function of Dirac with respect to a length coordinate normal to B_i in the neighbourhood of B_i. Then we can write

$$\vec{F}_1 - \vec{F}_2 = \text{grad}\, K(x, y, z) \ , \tag{2.9.2}$$

where $K(x, y, z) = \text{const.} = f$ for $(x, y, z) \in V_{12}$ and is zero elsewhere. From this it follows by (1.3.5) that B_1 and B_2 induce the same velocity field.

In our "simple" problem, as stated in the first paragraph of this section, it seems by the foregoing considerations that there is no local direction at the edge of the actuator disk which is specific for the problem. The equivalent actuator surfaces can meet the edge under any direction while their fields of flow are the same. Also, because $U = 0$, there is no local direction of the ambient flow. Then the idea arises that there will not be a specific direction of the vortex sheet by which it leaves the edge. This happens for instance when this sheet has a spiral behaviour around the edge.

To check if this idea can be true we consider the still more "simple" case of the actuator half plane, then the problem is 2-dimensional. Schmidt and Sparenberg [57] have given the following discussion of this case. We consider (Figure 2.9.2) the half plane B_0 ($x = 0, y \leq 0, -\infty < z < \infty$), on which the force field f per unit of area acts in the positive x-direction. We now deform B_0 such that it lies along its own, still unknown free vortex sheet, which is assumed to spiral from the origin (or in fact from the z-axis) towards infinity. This spiral is the new actuator surface B, which replaces the original half plane B_0 and on which we assume the same normal load f. By definition of a free vortex sheet the normal component of the velocity at this sheet is zero, or $v_n = 0$ and because of the applied normal load the pressure at this sheet exhibits a jump $[p]_-^+ = f$.

The problem being 2-dimensional, we use a complex representation in the (x, y) plane. In the remaining part of this section z will no longer be the third coordinate,

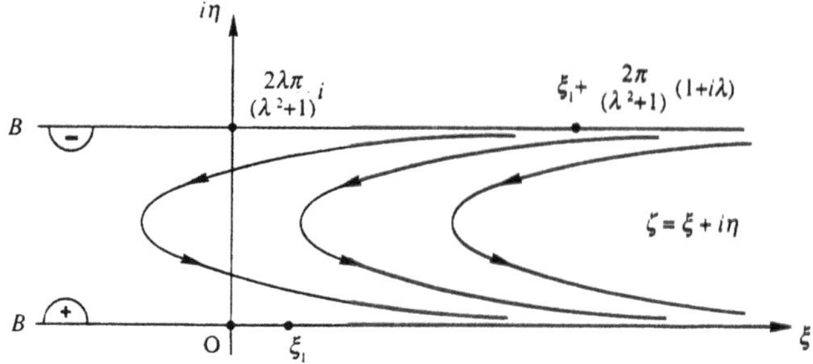

Fig. 2.9.3. Flow region in ζ plane with stream lines.

but

$$z = (x + iy) = r\, e^{i\theta} \ . \tag{2.9.3}$$

We assume the vortex sheet B to have the representation

$$z = e^{\lambda\theta}\, e^{i\theta} \ , \quad \lambda > 0 \ , \quad -\infty < \theta < \infty \ . \tag{2.9.4}$$

In order to formulate the conditions for the normal velocity v_n and pressure p at B, we introduce the complex velocity potential

$$h(z) = \varphi(x, y) + i\psi(x, y) \ , \quad \frac{dh}{dz} = v_x - iv_y \ , \tag{2.9.5}$$

where φ is the real velocity potential and ψ the stream function. Because at B we have $v_n = 0$, B is a streamline and hence ψ has to be constant along B. This constant we take zero, hence

$$\lim_{z \to B} \operatorname{Im} h(z) = 0 \ . \tag{2.9.6}$$

In the whole field of flow we have $H = p/\rho + \frac{1}{2}\, \vec{v}^2 = \text{const.}$, because the fluid particles do not cross the new actuator surface anymore. Then the condition at B for the pressure becomes

$$[p]_-^+ = -\frac{\rho}{2} \left[\left| \frac{dh}{dz} \right|^2 \right]_-^+ = f \ . \tag{2.9.7}$$

Consider the conformal mapping from the $z = x + iy$ plane to the $\zeta = \xi + i\eta$ plane (Figure 2.9.3) by

$$z = e^{(\lambda+i)\zeta} = e^{(\lambda\xi - \eta)}\, e^{i(\xi + \lambda\eta)} \ . \tag{2.9.8}$$

This transformation maps the strip

$$0 < \operatorname{Im} \zeta < \frac{2\lambda\pi}{(\lambda^2 + 1)} \ , \tag{2.9.9}$$

of the ζ plane uniquely onto the region (z plane $- B$). In the latter region we have for $r \to \infty$ that $\theta \to \infty$ and for $r \to 0$ that $\theta \to -\infty$. On the region (2.9.9) we consider the transformed complex potential

$$g(\zeta) = h(z(\zeta)) = h\left(e^{(\lambda+i)\zeta}\right) . \tag{2.9.10}$$

Condition (2.9.6) then becomes

$$\operatorname{Im} g(\zeta) = 0 \ ; \qquad \eta = 0 \ , \quad \eta = \frac{2\lambda\pi}{\lambda^2+1} . \tag{2.9.11}$$

In order to be able to apply condition (2.9.7), we first determine in the ζ plane which points ζ_1 and ζ_2 on the lines $\eta = 0$ and $\eta = 2\lambda\pi/(\lambda^2+1)$ respectively, correspond to one point $\tilde{z}$ at B in the z plane, where ζ_1 corresponds to $\tilde{z}$ situated at the + side of B and ζ_2 corresponds to the same $\tilde{z}$ but now situated at the − side. A substitution shows that this happens to be the case for the points

$$\zeta_1 = \xi \ , \quad \zeta_2 = \xi + \frac{2\pi}{(\lambda^2+1)} + i\frac{2\lambda\pi}{(\lambda^2+1)} \ , \qquad -\infty < \xi < \infty . \tag{2.9.12}$$

Using the relation

$$\frac{dh}{dz} = \frac{e^{-(\lambda+i)\zeta}}{(\lambda+i)}\frac{dg}{d\zeta} \ , \tag{2.9.13}$$

we can write (2.9.7) as

$$\frac{e^{-2\lambda\xi_1}}{(\lambda^2+1)}\left\{\left|\frac{dg}{d\zeta}\right|^2\Bigg|_{\zeta=\zeta_1} - \left|\frac{dg}{d\zeta}\right|^2\Bigg|_{\zeta=\zeta_2}\right\} = -\frac{2}{\rho} f . \tag{2.9.14}$$

As a function $g(\zeta)$ which satisfies (2.9.11) and (2.9.14), we try

$$g(\zeta) = \alpha\, e^{\beta\zeta} \ , \tag{2.9.15}$$

with α and β real. Then condition (2.9.11) becomes

$$\alpha\, e^{\beta\xi} \sin\beta\eta = 0 \ ; \quad -\infty < \xi < \infty \ , \quad \eta = 0 \quad \text{or} \quad \eta = \frac{2\lambda\pi}{(\lambda^2+1)} . \tag{2.9.16}$$

Hence

$$\beta = \frac{(\lambda^2+1)}{2\lambda} k \ , \qquad k = \pm 1, \pm 2, \ldots \tag{2.9.17}$$

Condition (2.9.14) yields

$$\alpha^2\beta^2 \frac{e^{2(\beta-\lambda)\xi}}{(\lambda^2+1)}(1 - e^{4\pi\beta/(\lambda^2+1)}) = -\frac{2}{\rho} f \ , \quad -\infty < \xi < \infty \ , \tag{2.9.18}$$

from which it follows that

$$\lambda = \beta \ , \quad \frac{\alpha^2\beta^2}{(\lambda^2+1)}(1 - e^{4\pi\beta/(\lambda^2+1)}) = -\frac{2}{\rho} f \ . \tag{2.9.19}$$

Equations (2.9.17) and (2.9.19) are the equations for the determination of the unknowns α, β and λ. Substituting $\lambda = \beta$ into (2.9.17), we find

$$(2-k)\lambda^2 = k \ , \qquad k = \pm 1, \pm 2, \dots \tag{2.9.20}$$

The only possibility to satisfy (2.9.20) is $\lambda = k = 1$, hence

$$\lambda = k = \beta = 1 \ . \tag{2.9.21}$$

From (2.9.19) we obtain by taking the positive root

$$\alpha = 2\left\{\frac{f}{\rho\,(e^{2\pi}-1)}\right\}^{1/2} \ . \tag{2.9.22}$$

After some elementary calculations we find

$$\left[\frac{dh}{dz}\right]_-^+ = \frac{\alpha}{(1+i)}\, e^{-i\xi}(1+e^{\pi}) \ . \tag{2.9.23}$$

Because the velocities at both sides of B (Figure 2.9.2) are tangential to B, the vorticity at B follows from (2.9.23) and (2.9.22). We find the value

$$\left\{\frac{2f}{\rho}\frac{(e^{\pi}+1)}{(e^{\pi}-1)}\right\}^{1/2} , \tag{2.9.24}$$

which is a constant, the vorticity is reckoned positive with a right-hand screw in the anti-clockwise direction.

The absolute value $|dh/dz|$ of the velocity is uniformly bounded in the whole plane,

$$\frac{\alpha}{\sqrt{2}} \le \left|\frac{dh}{dz}\right| \le \frac{\alpha}{\sqrt{2}}\, e^{\pi} \ . \tag{2.9.25}$$

Since H is constant in the whole field of flow when we consider the deformed actuator surface B, the pressure p is also uniformly bounded. This is, however, not true for the original half-infinite flat actuator plane B_0 (Figure 2.9.2) as we will discuss now.

The stream lines in the ζ plane are drawn in Figure 2.9.3, which are the lines of constant stream function values in that plane

$$\operatorname{Im} g(\zeta) = \alpha\, e^{\xi} \sin\eta = \text{const.} \tag{2.9.26}$$

These lines transformed back to the original (x, y) plane, are drawn schematically in Figure 2.9.2. They enter the spiral core deeper and deeper and come out again. This means that for each integer $N > 0$ we can find streamlines which cut B_0, N

consecutive times in one direction and then N consecutive times in the opposite direction. However when a fluid particle crosses B_0 in a direction opposite to the force direction its pressure suddenly drops by the amount f. This means that very close to the edge O of B_0 there are regions with arbitrarily low pressures. It is not difficult to show that the pressure in case of B_0, tends logarithmically to minus infinity when we approach the edge of the disk.

By (2.9.4) and (2.9.21) we find for the spiral, on which the constant vorticity of strength (2.9.24) is located,

$$z = r(\theta)\, e^{i\theta} = e^{\theta}(\cos\theta + i\sin\theta) \ , \quad -\infty < \theta < \infty. \tag{2.9.27}$$

Hence for half a turn around the origin O, the radius $r(\theta)$ is multiplied by

$$e^{\pi} \approx 23.2 \ . \tag{2.9.28}$$

We now check if the spiral B (2.9.27) with its constant vorticity (2.9.24) yields the velocity field induced by the half infinite flat actuator disk B_0. This we will do because we are anxious about the influence of the part of B which is "at infinity". First we define for some point where the spiral B cuts the positive x-axis that $\theta = 0$, by which each point at B has an unambiguous value of θ. Now choose an integer $N > 0$ and consider the full turn $B(N)$ of B around the origin O, defined by $2\pi N - \pi/2 \leq \theta \leq 2\pi(N+1) - \pi/2$. This full turn represents with respect to the induced velocities, exactly the interval $0 > y(N) > y > y(N+1)$ of B_0, where $(0, y(N))$ is the point of B with $\theta = 2\pi N - \pi/2$. Now it is easy to calculate the velocities induced by $B(N)$ in any finite fixed region containing O. It is found that these velocities tend exponentially to zero for $N \to \infty$. Further it is clear that any part of B_0 with $y < \tilde{y}$ for some $\tilde{y} < 0$, is covered in this way by a denumerable number of intervals $y(N) > y > y(N+1)$. Hence we can find the influence of B_0 in any finite fixed region around O, by considering the limit $N \to \infty$ of the influence of parts of B with $-\infty < \theta \leq 2\pi N - \pi/2$.

The above-mentioned results have been used in [57] for the numerical calculation of the flow pattern of a circular actuator disk of radius R with a constant normal load placed in a fluid without incoming velocity. The position of the vortex sheet behind the disk is determined, when the incoming flow is zero, by ρ, R and f. Its shape can be described by the quotient of the radius $r(x)$ at the place x, and x itself, hence by $r(x)/x$ which is dimensionless. We now show by a simple dimension analysis, in the abridged form as described in the last paragraph of Appendix E, that this shape is independent of ρ, R and f.

We have

$$\begin{aligned} 0 &= [r(x)/x] = [\rho]^{\beta_1} \cdot [R]^{\beta_2} \cdot [f]^{\beta_3} \\ &= ([m]\,[l]^{-3})^{\beta_1} \cdot [l]^{\beta_2} \cdot ([m]\,[l]^{-1}[t]^{-2})^{\beta_3} \ . \end{aligned} \tag{2.9.29}$$

So we find that $\beta_1 = \beta_2 = \beta_3 = 0$ and hence ρ, R and f do not enter the dimensionless shape of the slip stream.

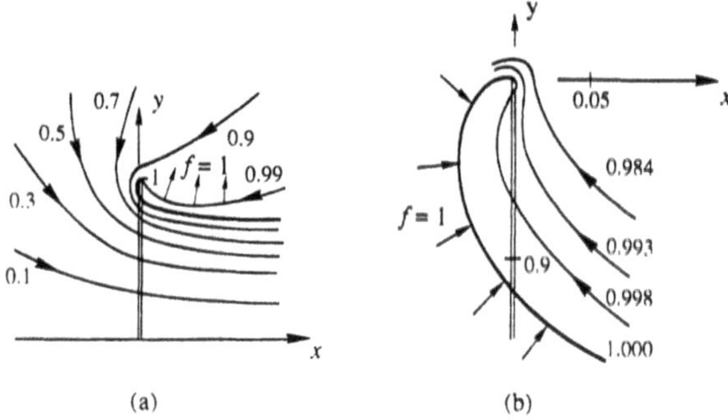

Fig. 2.9.4. Stream lines induced by actuator disk of radius 1, constant normal load $f = 1$, no incoming flow, (a) survey, (b) edge region.

From this it follows that we can take ρ, R and f all equal to 1. It has been assumed in the numerical calculations that sufficiently close to the edge of the disk, the afore-mentioned spiralling behaviour (2.9.27) of the sheet dominates, while its vorticity is given by (2.9.24). The original disk is replaced by an actuator surface which is axisymmetric and situated along the unknown boundary of the slip stream (Figure 2.9.4 (a)). This means in essence that we have calculated the shape of a rigid tube in which we have at $x = \infty$ a parallel flow in the positive x-direction and across which we have a constant pressure jump. Approaching the orifice from the inside of this tube, the tube widens, curves backwards and tends spiralling to the edge of the disk.

In Figure 2.9.4 we have also drawn the computed stream lines of the flow pattern induced by the disk. At large distances from the disk and sufficiently away from the positive x-axis the pattern resembles the flow induced by a sink. The number at a stream line denotes the relative value of the stream function, it is the value of the stream function at that line divided by the value of the stream function at the boundary of the slip stream formed by the vortex sheet.

It is noted that in the neighbourhood of the edge we have a region where the fluid particles cross the disk more than once. This region occupies about 23% of the total area of the disk. The amount of fluid passing this region however is relatively very small, as follows from the number on each stream line. In other words the fluid is nearly stagnant at that region of the disk. This result gives perhaps some information about the flow induced by a propeller in the static condition at bollard.

2.10. Miscellaneous Remarks about Non-Linear Actuator Disk Theory

We reconsider the general method of Section 2.8 from the point of view of the more special theory of Section 2.9, which however could be carried out in greater detail. We start with the remark that the circular actuator disk of finite dimensions

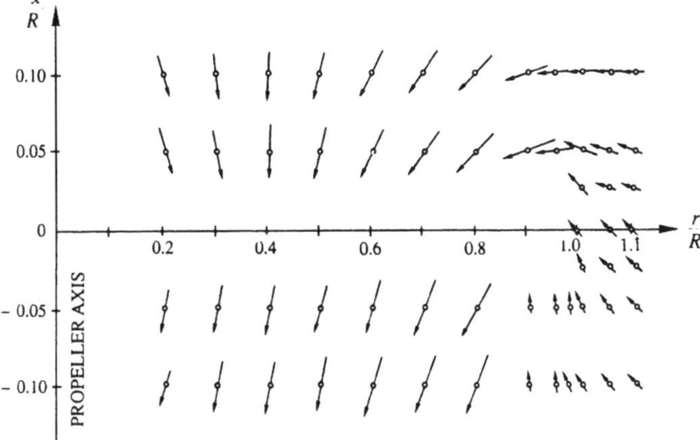

Fig. 2.10.1. Experimental investigation of velocity field.

of Section 2.9 is a limit case of the actuator disk considered in Section 2.8 for the case of the rotating bound vorticity. First, the incoming flow U has to tend to zero. Second, the angular velocity Ω, of the bound vorticity $\Gamma = C/r$ (below (2.8.28)), has to increase unboundedly while we keep $C\Omega = \text{const}$. By the latter limit the force field becomes perpendicular to the disk and becomes of constant strength. This means that it is not improbable that the spiralling flow we found in Section 2.9 will also occur to some extent in the more realistic case of Section 2.8, under appropriate conditions.

In [21] the shape of the boundary of the slip stream is determined by solving a one-dimensional integral equation (not explicitly discussed in Section 2.8, formula (33) in [21]), which has as its domain $0 \leq x \leq \infty$. Hence it is assumed that the shed free vorticity does not come in front of the disk. However, it is seen, as is stated there, that the free vortex sheet in some cases "tries to linger in the disk plane". By this it seems not astonishing that when more freedom is given to the shed vortex sheet it will come in front of the disk, as in Figure 2.9.4.

When a spiralling behaviour of the fluid flow occurs around the edge of the disk, in the more general case of Section 2.8, some difficulty can arise by force fields $f_r(r)\delta(x)$ and $f_\varphi(r)\delta(x)$. According to (2.8.6) and (2.8.7) the jumps in v_r as well as those in v_φ of particles which cross the disk several times, will be added because v_x changes sign when a particle passes the disk in the inverse direction from its $(+0)$ side to its (-0) side. It follows, however, from Figure 2.9.4 (b) that there are particles which cross the disk with a very small value of $|v_x|$, hence these particles can have large jumps in v_r and v_φ. Therefore, it is possible that a complicated flow pattern occurs in the neighbourhood of the edge of the disk.

In Figure 2.10.1 we have drawn the measured values of the velocity field induced by a four-bladed aircraft propeller working in the static condition (incoming flow is zero) as given by Adams [1]. The axial direction is drawn vertically and has an expanded scale with respect to the scale of the radial direction. The velocity vectors,

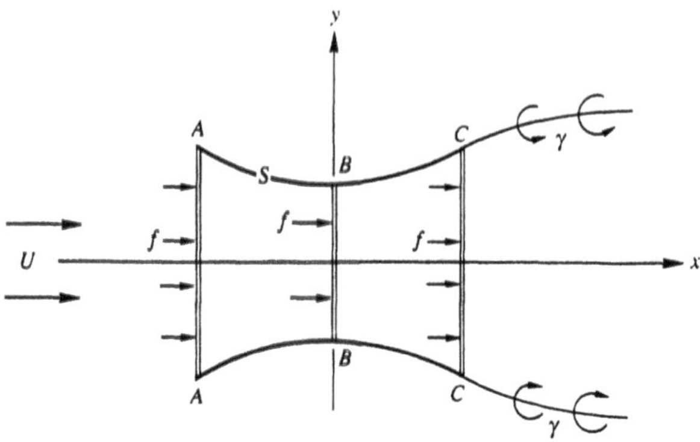

Fig. 2.10.2. Shroud S with three different positions of actuator disk with constant normal load.

however, have been drawn at their true angle and show the relative magnitude of the velocity. The curvature of the flow for small values of r/R, where R is the radius of the propeller, is caused by the hub. It is seen that in the neighbourhood of the edge of the propeller disk there exists a reverse flow which is not unlikely to cross the disk from behind.

From the calculations carried out in [21], as shown in Figure 2.8.2 (b) for $\lambda = 0.01$, it is likely that a dividing stream line can occur. This phenomenon seems to be compatible with the theory of Section 2.9. There the circular disk was replaced by the actuator surface along the free vorticity tube, across which now a pressure jump of strength f has to occur. This means that a cross section of this tube with a plane through the x-axis, has become a "profile" with constant pressure jump. For profiles however dividing stream lines are common and they end at the stagnation point of the flow at the profile, hence in this case at the free vortex sheet.

We remark that it is suggested by the discussed results that a profile of zero thickness with a constant pressure jump, as sometimes used in hydrodynamics, should have theoretically in the non-linear theory a spiralling behaviour in the neighbourhood of the leading and of the trailing edge.

We mention still another application of the deformation of an actuator disk with a constant load. Consider such a disk situated in a shroud S for instance of zero thickness and of the shape as drawn in Figure 2.10.2. Then we can ask for the place of the disk inside the shroud which causes the largest thrust contribution of the shroud. Our considerations will be non-linear hence the slopes of S and the load f of the disk need not to be small.

First we start with the position AA of the disk at the leading edge of S. Suppose that the load f of the disk is such that the demanded thrust T occurs for this combination of disk and shroud. Now we can consider another position BB of the disk at the smallest cross-section (the throat) of the shroud. Then by the same reasoning as in Section 2.4 where we used (1.3.5) and (1.3.7) it is clear that the velocity field in both cases (disk AA or disk BB) is the same. Hence the kinetic energy losses and the thrust of the disk-shroud combinations are the same. However, the pressure field has changed by a constant negative disturbance pressure $\Delta p = -f$ in between AA and BB inside S. Hence a thrust contribution is created by the part of S in between AA and BB. When we consider the disk situated at CC, this thrust contribution is destroyed by the opposite slope of the part BC of the shroud S. Hence the largest contribution of the shroud to the thrust occurs when the disk is placed at position BB.

An analogous reasoning can be given for a shroud of finite thickness and different shape by considering the slopes at the inner side of the shroud.

Finally we discuss a family of exact solutions of the non-linear actuator disk equations. These solutions seem to be of no practical importance, however they elucidate the use of the external force field at the disk. The disk is the circular region $x = 0, 0 \le r \le a$, as in Section 2.8. The external force field is assumed to be of the form (2.8.1). We consider the following family of divergenceless velocity fields.

$$v_x = U \ , v_r = v_\varphi = 0 \ ; \quad x < 0 \ ; \quad 0 < x \ , \quad a < r \ , \tag{2.10.1}$$

$$v_x = U \ , v_r = 0 \ , v_\varphi = \tilde{v}_\varphi(r) \ ; \quad 0 < x \ , \ 0 < r < a \ , \tag{2.10.2}$$

where U is the incoming velocity and $\tilde{v}_\varphi(r)$ is some given function. Without a restriction of generality we take the pressure $p = 0$ in the undisturbed flow regions $x < 0$ and $0 < x, a < r$.

It follows from (2.8.3), for $x > 0$

$$\frac{\partial p}{\partial x} = 0 \ \rightarrow \ p = p(r) \ , \tag{2.10.3}$$

and from (2.8.5) also for $x > 0$.

$$\frac{1}{r}\,\tilde{v}_\varphi^2(r) = \frac{1}{\rho}\,\frac{\partial p}{\partial r} \rightarrow p(r) = -\rho \int_r^a \frac{\tilde{v}_\varphi^2(\tilde{r})}{\tilde{r}}\, d\tilde{r} \ . \tag{2.10.4}$$

Then by (2.8.4), (2.8.6) and (2.8.7) we find

$$f_x(r) = -\rho \int_r^a \frac{\tilde{v}_\varphi^2(\tilde{r})}{\tilde{r}}\, d\tilde{r} \ , \quad f_r = 0 \ , \quad f_\varphi(r) = \rho\, U \tilde{v}_\varphi(r) \ , \tag{2.10.5}$$

respectively.

Herewith we have found the external force field at the disk. It is clear that $f_x(r)$ shields the region of zero pressure at the $-$ side of the disk from the negative pressures at the + side of the disk. The negative pressures are needed to let the fluid particles follow their circular paths. The special case $\tilde{v}_\varphi(r)$ = const. r is discussed by van Kuik in [44].

Chapter 3

The Ship Screw

Our next subject will be the screw propeller which is up to now the most important device for the propulsion of ships. It consists of a number of helicoidally shaped blades connected to a hub. The number of blades can vary from two upto about six. The hub is mounted on a shaft (Figure 3.1) which is rotated by the engine. The blades have to be designed so that at a given rotational velocity, a prescribed thrust is produced which moves the ship with the desired speed. This yields a difficult problem because the screw propeller behind a ship is working in a complicated field of flow, for several reasons. We indicate some of these.

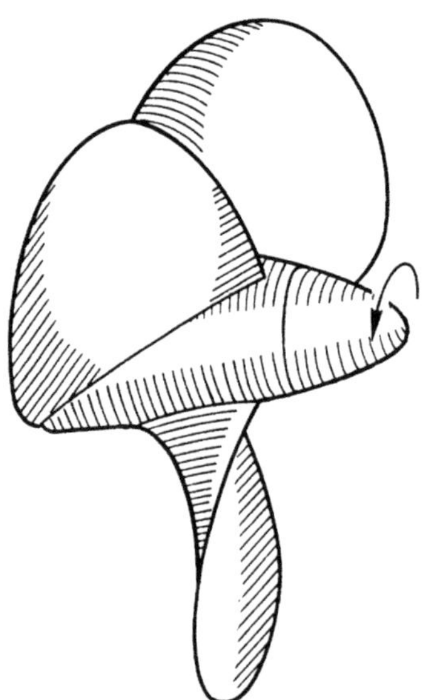

Fig. 3.1. A three-bladed screw propeller.

First, the water has to follow the ship's form, hence it has to converge at the stern. Second, the water flowing closely along the ship is dragged with it by viscosity and becomes turbulent. Third, the wave pattern induced by the hull at the free surface causes a velocity field which varies with depth. When at the stern a crest is formed, this velocity field is in the direction of the motion of the ship. When a trough is formed, the inverse happens. When the ship is towed, the resulting disturbances in the absence of the propeller are called the nominal wake of the hull ([45], p. 388). It has to be observed that when the propeller is present and develops thrust, the wake is changed.

The hull of the ship has influence still in another way. As a rigid boundary of the flow domain it will hamper the water to be set into motion by the propeller. The same holds for the rudder. Also the free surface acts as a boundary of the region in which the propeller operates.

As a result of the mentioned inhomogeneities of the inflow and because of the motion of the propeller blades with respect to the boundaries of the region of operation unsteady loading of the blades will occur and possibly also unsteady cavitation. The latter phenomenon is undesirable from the point of view of erosion caused by the imploding cavitation bubbles.

Another essential difficulty in propeller theory is that for realistic ship screws non-linear effects are important. For instance, the axially and radially induced velocities just behind the propeller which deform the free vortex sheets shed by the blades. Also the roll up of the vortex sheets themselves (Section 1.22) is an aspect of the non-linearities which occur. In relation to the optimization of a propeller we mention in Section 5.13 (see (5.13.4)) a method to cope with the first mentioned effect in an approximate way.

Considering such difficulties it seems sensible in a first approach to the problem to make simplifications. Our theory will be linear, we neglect viscosity and we assume no influence of the ship's hull and the free surface. The incoming flow will be homogeneous. We neglect the influence of the hub because the thrust is mostly delivered by parts of the blades which are not too close to the axis of rotation. Hence the blades are moving "freely" through the fluid along prescribed helicoidal paths. We also assume that the fluid keeps contact with the blades so that no cavitation as a result of low pressures will occur.

In the first instance we consider (Section 3.3) the lifting surface theory for a simple helicoidal propeller as an application of the general lifting surface theory of Sections 1.10–1.14. The advantage of using the general theory is that there the Hadamard principle value method is proved to be valid for arbitrarily curved reference surfaces. After this we discuss (Section 3.4) the lifting surface theory by means of vorticity distributions. The theoretical validity of this latter approach follows from the preceding general Hadamard principle value method.

We remark that although the following discussion is with regard to screw propellers with simple straight generator lines, perpendicular to the axis of rotation, the theory is easily extended to more general curved generator lines or to blades with end plates (Section 6.1). For general works about screw propellers we refer to [8] and [28].

3.1. The Geometry of the Screw Propeller

We will give a description of the geometry of a screw propeller, adapted to our intention to derive a linear hydrodynamical theory. For a more technical description we refer to [45]. We use a cylindrical coordinate system (x, r, φ) (Figure 3.1.1) of which the x-axis is along the axis of rotation of the screw. With respect to this coordinate system we have an incoming homogeneous parallel flow with velocity U in the positive x-direction.

We consider a screw propeller with N blades and assume as its helicoidal reference surfaces H_ν,

$$H_\nu : \varphi + ax - \omega t = \frac{2\pi\nu}{N} , \quad a = \frac{\omega}{U} , \quad \omega > 0 , \quad \nu = 0, 1, \ldots, N-1 , \tag{3.1.1}$$

hence we take $g(r) = 0$ in the more general formula (1.9.15). On each H_ν we choose (x, r) as a 2-dimensional coordinate system. This means that a point (x, r) at H_ν is the point $(x, r, -ax + \omega t + 2\pi\nu/N)$ in 3-dimensional space. Then we define on each H_ν a planform W_ν of the propeller blade $\tilde{W}_\nu$ by

$$W_\nu : x_l(r) \leq x \leq x_t(r) , \quad r_1 \leq r \leq r_2 , \quad (x, r, \varphi) \in H_\nu , \tag{3.1.2}$$

where $x_l(r)$ and $x_t(r)$ are given functions which define the projections of the leading edge and the trailing edge of the blade $\tilde{W}_\nu$ on the corresponding reference surface H_ν and r_1 and r_2 are the inner and outer radius, respectively.

Each planform W_ν (3.1.2), as an impermeable rigid surface of finite extent, does not disturb the incoming parallel flow. A screw propeller, however, has to produce thrust and, being a periodically moving body of finite extent, has to shed vorticity (Section 1.16) and hence will disturb the parallel flow. Because we want to develop a linear theory, the assumption is made that the disturbance velocities (v_x, v_r, v_φ) are sufficiently small. This happens to be the case when the blades $\tilde{W}_\nu$ are in the neighbourhood of the planforms W_ν. We represent the $\tilde{W}_\nu$ by

$$\varphi - \omega t + ax + h^j(x, r) = \frac{2\pi\nu}{N} , \quad (x, r) \in W_\nu , \quad j = (-), (+) , \tag{3.1.3}$$

$$h^{(-)}(x, r) \geq h^{(+)}(x, r) , \tag{3.1.4}$$

$$h^{(-)}(x_l(r), r) = h^{(+)}(x_l(r), r) , \quad h^{(-)}\left(x_t(r), r\right) = h^{(+)}\left(x_t(r), r\right) , \tag{3.1.5}$$

where $h^{(-)}(x, r)$ and $h^{(+)}(x, r)$ determine the "back" and the "face" of the blade respectively. The values of $|h^{(-)}|$ and $|h^{(+)}|$ are small, say $O(\varepsilon)$, so that a linear theory will be valid. The leading edge of each $\tilde{W}_\nu$ consists of points $\left(x_l(r), r, \varphi\right)$ in space, which satisfy

$$\varphi - \omega t + ax_l(r) + h^{(-)}(x_l(r), r) = \frac{2\pi\nu}{N} , \tag{3.1.6}$$

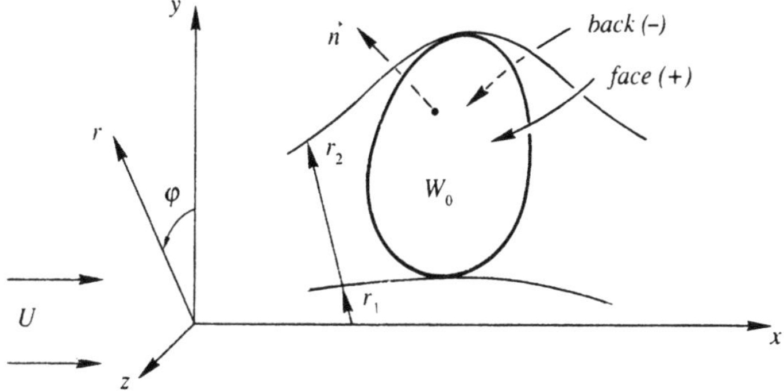

Fig. 3.1.1. Planform W_0 of a screw propeller blade.

the trailing edge is defined analogously with $x_l(r)$ replaced by $x_t(r)$.

We introduced below (3.1.5) the back and the face of a blade. A more physical definition reads, the back ($h^{(-)}$) of a blade is its low pressure side, the face ($h^{(+)}$) of a blade is its high pressure side.

The thickness $D(x,r)$ of a blade for a point P of a planform W_ν is defined as follows. Determine the points of intersection of the normal at W_ν through P, with the back and the face of the blade. The thickness D will be the distance between these two points of intersection. We introduce the unit normal $\vec{n}$ on the planform, pointing in the direction of decreasing values of φ,

$$\vec{n} = (n_x, n_r, n_\varphi) = -\frac{(ar, 0, 1)}{(1 + a^2r^2)^{1/2}} , \tag{3.1.7}$$

where the components are with respect to the cylindrical coordinate system. The normal points from the high pressure side (+) to the low pressure side (–) of the blade. Then D is found to be equal to

$$D(x,r) = \frac{r\{h^{(-)}(x,r) - h^{(+)}(x,r)\}}{(1 + a^2r^2)^{1/2}} , \quad (x,r) \in W_\nu , \tag{3.1.8}$$

hence $D = 0$ at the leading edge and at the trailing edge (3.1.5).

The middle surface W_ν^* of a blade, lying in the middle of the back and the face is given by

$$W_\nu^* : \varphi + ax - \omega t + \tfrac{1}{2}(h^{(-)}(x,r) + h^{(+)}(x,r))$$

$$= \varphi + ax - \omega t + h^*(x,r) = \frac{2\pi\nu}{N} , \tag{3.1.9}$$

where $h^*(x,r)$ is called the camber of the blade.

Next we introduce the local angle of incidence α of the middle surface W_ν^* at some point $(\tilde{x}, \tilde{r}, \tilde{\varphi}) \in W_\nu^*$. This is the angle formed by the helicoidal line at $H_\nu(r = \tilde{r})$ at the point $(\tilde{x}, \tilde{r}) \in W_\nu$, with the tangent plane of W_ν^* at the point $(\tilde{x}, \tilde{r}, \tilde{\varphi})$. The angle α follows, within the accuracy of the linear theory, from the scalar product of the unit vector tangent to the helicoidal line at $(\tilde{x}, \tilde{r}) \in W_\nu$, with the unit vector perpendicular to W_ν^* at $(\tilde{x}, \tilde{r}, \tilde{\varphi})$. Leaving aside the tilde notation we obtain

$$\alpha(x,r) \approx -\frac{(1,0,-ar)}{(1+a^2r^2)^{1/2}} \cdot \frac{\{r(a+\partial h^*/\partial x), r\,\partial h^*/\partial r, 1\}}{\{r^2(a+\partial h^*/\partial x)^2 + (r\,\partial h^*/\partial r)^2 + 1\}^{1/2}}$$

$$\approx \frac{-r}{(1+a^2r^2)} \frac{\partial}{\partial x} h^*(x,r) \ , \qquad (3.1.10)$$

where the equality sign $\approx$ means accuracy up to and including the first order of $h^*(x,r) = O(\varepsilon)$.

A profile of a blade is defined as the cross section of the blade with a cylinder r = const. The skeleton line or the camber line of a profile is the intersection of the middle surface W_ν^* (3.1.9) with this cylinder.

At the back (–) and the face (+) of a blade we have to satisfy the boundary conditions for the flow, stating that the fluid velocities are tangent to these surfaces. Because our theory is linear, these conditions will not be demanded at points (x, r, φ) of the back or the face themselves, but at the points (x, r) of the planform W_ν.

In the following we shall often identify the blade $\tilde{W}_\nu$ with its planform W_ν. A chord of the blade is then by definition a line r = const. at W_ν. Chord lines can be called reference lines for the profiles because the profiles lie in their neighbourhood and because they are in the direction of the undisturbed relative fluid velocity.

We now state the problem we want to solve by lifting surface theory. Assume that the load $Q(x,r)$, which is defined as the difference in pressure between the face and the back of a blade

$$Q(x,r) = p(\text{face } (+)) - p(\text{back } (-)) \ , \qquad (x,r) \in W_\nu \ , \qquad (3.1.11)$$

is given as a function of position. Besides this we assume that, for instance by demands of strength and stiffness of the blade, the thickness $D(x,r)$ is known. Then the question is, how do we have to choose the middle surface W_ν^* of the blades, so that we obtain the desired load (3.1.11). When W_ν^* is found, we can construct the back and the face of each blade, because the thickness is prescribed.

Our theory being linear, this problem can be split into two separate parts. First, what is the camber $h_D^*(x,r)$ (3.1.9) of each blade which yields the load $Q(x,r) = 0$, while the blades have the prescribed thickness $D(x,r)$. Second, what is the camber $h_Q^*(x,r)$ which yields the prescribed load $Q(x,r)$, while $D(x,r) = 0$. Then the total camber needed to satisfy (3.1.11) follows from

$$h^*(x,r) = h_D^*(x,r) + h_Q^*(x,r). \qquad (3.1.12)$$

The splitting of the problem into two independent parts is interesting from the following point of view. By changing either the thickness $D(x,r)$ or the load

$Q(x,r)$ separately, we need only to take into account the changing quantity. For instance, a multiplication of $D(x,r)$ or $Q(x,r)$ by a constant, results in a simple multiplication of $h_D^*(x,r)$ or $h_Q^*(x,r)$ by the same constant.

When instead of a helicoidal screw propeller blade we consider a "flat" wing, we clearly have $h_D^*(x,r) = 0$.

3.2. Screw Blades with Thickness and without Load

We now discuss the first part of the problem, described at the end of the previous section: the design of a screw blade with a prescribed thickness distribution $D(x,r) \neq 0$, without pressure difference between back and face. Hence we have to calculate for $Q(x,r) = 0$ the camber of the blade $h_D^*(x,r)$. This can be carried out for any fixed time, say $t = 0$ (3.1.1).

First we consider separately $h_D^*(x,r)$ of the blade $\tilde{W}_0$ alone. The planform W_0 of this blade is part of the reference surface H_0 (3.1.1). Afterwards we discuss the influence of the other blades of the propeller.

Consider a layer of sources placed at W_0. We will show that such a layer is sufficient to represent the induced flow in this case. The layer induces a disturbance potential Φ (1.1.8)

$$\Phi(x,r,\varphi) = -\frac{1}{4\pi} \iint\limits_{W_0} \frac{m}{R}\, dS \ , \tag{3.2.1}$$

where m is the local strength per unit of area of the source layer and $R = |\vec{R}\,|$ is the length of vector $\vec{R}$ from the point $(\tilde{x},\tilde{r},\tilde{\varphi})$, where the element of area dS is situated, to the point (x,r,φ) where we calculate the potential. These quantities are in formula

$$\vec{R} = (x-\tilde{x}, r-\tilde{r}\cos(\varphi-\tilde{\varphi}), \tilde{r}\sin(\varphi-\tilde{\varphi})) \ , \tag{3.2.2}$$

$$R = |\vec{R}\,| = \left\{(x-\tilde{x})^2 + r^2 + \tilde{r}^2 - 2r\tilde{r}\cos(\varphi-\tilde{\varphi})\right\}^{1/2} \ , \tag{3.2.3}$$

$$dS = d\sigma\, d\tilde{r} = (1+a^2\tilde{r}^2)^{1/2}\, d\tilde{x}\, d\tilde{r} \ , \tag{3.2.4}$$

where the components of $\vec{R}$ are with respect to the cylindrical coordinate directions at the point (x,r,φ) and $d\sigma$ is an element of length along a helicoidal line at W_0.

The limiting values of the normal derivative of Φ at the back (–) and the face (+) become ([36], p. 164)

$$\frac{\partial\Phi^{(-)}}{\partial n} = \frac{m}{2} - \frac{1}{4\pi} \iint\limits_{W_0} m\, \frac{\partial}{\partial n}\, \frac{1}{R}\, dS \ , \tag{3.2.5}$$

$$\frac{\partial\Phi^{(+)}}{\partial n} = -\frac{m}{2} - \frac{1}{4\pi} \iint\limits_{W_0} m\, \frac{\partial}{\partial n}\, \frac{1}{R}\, dS \ , \tag{3.2.6}$$

respectively, where $\partial/\partial n$ means differentiation in the direction of the normal $\vec{n}$ given in (3.1.7). The point at the planform W_0, where the normal derivative of Φ is considered, will be denoted by $(x_s, r_s, \varphi_s) = (x_s, r_s, -ax_s)$. Then by (3.1.7), (3.2.3) and (B.2.7) we find

$$\begin{aligned}\frac{\partial}{\partial n}\frac{1}{R} &= \vec{n}_s \cdot \underset{s}{\text{grad}}\,\frac{1}{R}\\ &= -\frac{(ar_s, 0, 1)}{(1+a^2r_s^2)^{1/2}} \cdot -\frac{(x_s-\tilde{x}, r_s - \tilde{r}\cos(\varphi_s - \tilde{\varphi}), \tilde{r}\sin(\varphi_s - \tilde{\varphi}))}{R^3}\\ &= \frac{1}{(1+a^2r_s^2)^{1/2}}\,\frac{\{ar_s(x_s-\tilde{x}) + \tilde{r}\sin(\varphi_s-\tilde{\varphi})\}}{R^3}\ ,\end{aligned} \tag{3.2.7}$$

where the index s at $\vec{n}$ means $\vec{n}$ at the place (x_s, r_s, φ_s) and the index s at grad means differentiation with respect to x_s, r_s and φ_s. The double integrals in (3.2.5) and (3.2.6) exist because the singular behaviour of the integrands is only of $O(R^{-1})$. For the difference of the normal components of the disturbance velocity at both sides of the blade, we find from (3.2.5) and (3.2.6) the well-known relation

$$\frac{\partial \Phi^{(-)}}{\partial n} - \frac{\partial \Phi^{(+)}}{\partial n} = m\ . \tag{3.2.8}$$

The condition that the fluid flows along the back (–) and along the face (+) of the blade, follows from the substitution of (3.1.3) into (1.2.18). This yields, when we neglect second order terms,

$$U\frac{\partial h^j}{\partial x} = -\left(av_{xj} + \frac{1}{r}\,v_{\varphi j}\right)\ , \qquad j = (-), (+)\ , \tag{3.2.9}$$

where here v_{xj} and $v_{\varphi j}$ denote disturbance velocity components at the back $j = (-)$ or at the face $j = (+)$. The difference of the normal components of the disturbance velocity at both sides has the value

$$\begin{aligned}&\frac{\partial \Phi^{(-)}}{\partial n} - \frac{\partial \Phi^{(+)}}{\partial n}\\ &\quad = \vec{n}\cdot(v_x(-) - v_x(+), v_r(-) - v_r(+), v_\varphi(-) - v_\varphi(+))\ .\end{aligned} \tag{3.2.10}$$

Using (3.2.8), (3.1.7) and (3.2.9) we find from (3.2.10)

$$m(x,r) = \frac{Ur}{(1+a^2r^2)^{1/2}}\left(\frac{\partial}{\partial x}\,h^{(-)}(x,r) - \frac{\partial}{\partial x}\,h^{(+)}(x,r)\right)\ . \tag{3.2.11}$$

By the definition of the thickness $D(x,r)$ (3.1.8) of the blade, this can be written as

$$m(x,r) = U\frac{\partial}{\partial x}\,D(x,r) = U(1+a^2r^2)^{1/2}\frac{\partial}{\partial \sigma}\,D(x,r)\ , \tag{3.2.12}$$

where again σ is a length parameter along a helicoidal line in the planform W_0, hence for $r = \text{const}$.

This formula could have been derived more directly. The quantity

$$U(1+a^2r^2)^{1/2} = (U^2+\omega^2r^2)^{1/2} \ , \tag{3.2.13}$$

is the relative undisturbed velocity of the fluid with respect to W_0. Hence the right-hand side of (3.2.12) can be interpreted as the difference in normal velocity of the fluid at both sides of the blade, because $\partial D/\partial\sigma$ is the difference in slope of the sides of the blade. Then by (3.2.8) we obtain (3.2.12).

Now we derive a relation between the camber h_D^* (3.1.12) of the blade and the disturbance velocities. Consider the sum of the normal components of the disturbance velocities

$$\frac{\partial\Phi^{(-)}}{\partial n} + \frac{\partial\Phi^{(+)}}{\partial n}$$

$$= \vec{n}\cdot(v_x(-)+v_x(+), v_r(-)+v_r(+), v_\varphi(-)+v_\varphi(+)) \ . \tag{3.2.14}$$

Using (3.1.7), (3.1.9), (3.2.5), (3.2.6) and (3.2.9), we find

$$\frac{rU}{\left(1+a^2r^2\right)^{1/2}}\frac{\partial}{\partial x}\,h_D^* = rU\frac{\partial}{\partial\sigma}\,h_D^* = -\frac{1}{4\pi}\iint\limits_{W_0} m\,\frac{\partial}{\partial n}\,\frac{1}{R}\,dS \ . \tag{3.2.15}$$

Then by (3.1.1) (for $t = 0$ and $\nu = 0$), (3.2.4), (3.2.7) and (3.2.12) this relation can be written more explicitly as

$$r_s\frac{\partial}{\partial x_s}\,h_D^*(x_s,r_s) = -\frac{1}{4\pi}\iint\limits_{W_0}\frac{\partial D(\tilde{x},\tilde{r})}{\partial\tilde{x}}$$

$$\cdot\frac{\{ar_s(x_s-\tilde{x})-\tilde{r}\sin a(x_s-\tilde{x})\}(1+a^2\tilde{r}^2)^{1/2}}{\{(x_s-\tilde{x})^2+r_s^2+\tilde{r}^2-2r_s\tilde{r}\cos a(x_s-\tilde{x})\}^{3/2}}\,d\tilde{x}\,d\tilde{r} \ , \tag{3.2.16}$$

where (x_s, r_s) is the point of W_0 under consideration and $\tilde{x}$ and $\tilde{r}$ are the variables of integration.

This formula is valid when there was only one blade $\tilde{W}_0$. It is, however, not difficult to write down the influence of the other blades $\tilde{W}_1, \ldots, \tilde{W}_{N-1}$, of which the disturbance velocities have no singular behaviour at $\tilde{W}_0$. Then we have to replace in (3.2.2) $\tilde{\varphi}$ not by $-a\tilde{x}$, as we did in (3.2.16), but by $-a\tilde{x}+2\pi\nu/N$. We find

$$r_s\frac{\partial}{\partial x_s}\,h_D^*(x_s,r_s) = r_s(1+a^2r_s^2)^{1/2}\frac{\partial}{\partial\sigma}\,h_D^*(x_s,r_s) = -\frac{1}{4\pi}\sum_{\nu=0}^{N-1}\iint\limits_{W_0}\frac{\partial D(\tilde{x},\tilde{r})}{\partial\tilde{x}}$$

$$\cdot\frac{\{ar_s(x_s-\tilde{x})-r\sin(a(x_s-\tilde{x})+2\pi\nu/N)\}\left(1+a^2\tilde{r}^2\right)^{1/2}}{\{(x_s-\tilde{x})^2+r_s^2-\tilde{r}^2-2r_s\tilde{r}\cos(a(x_s-\tilde{x})+2\pi\nu/N)\}^{3/2}}\,d\tilde{x}\,d\tilde{r} \ . \tag{3.2.17}$$

The meaning of the results up to now is the following. The right-hand side of (3.2.17) represents the normal velocity component induced by the source density $m(x, r)$ at "a distance" from the point under consideration (x_s, r_s) on W_0. It is easily seen that the local angle of incidence α (3.1.10) of the blade has by (3.2.17) just the value that the disturbed flow is tangent to the middle surface W_0^* (3.1.9). The density $m(x_s, r_s)$, hence at the point under consideration, takes care of the difference in slope at both sides of the blade (3.2.12).

The camber $h_D^*(x, r)$ is not uniquely defined by (3.2.17). We still have the freedom to choose at the planform W_0 some suitable line, connecting the root and the tip of the blade, along which $h_D^*(x, r)$ is supposed to be zero. Then by integration of (3.2.17) we can determine $h_D^*(x, r)$.

One question is left: are the pressures at the back and the face for each point (x, r) of W_0 equal to each other so that $Q(x, r) = 0$? The only thing we know a priori is that no resultant non-zero mean force or moment can be exerted on the fluid by the propeller. This follows from Section 1.16, because the flow here is represented by sources and sinks only, so no vorticity will be shed into the fluid.

To answer this question we consider the linearized version of Bernoulli's law for the unsteady case (1.2.12), where $K = 0$ because we have no external force field and where we can take $h(t) = 0$, because we are looking for pressure differences, hence

$$p = -\rho\left(U\frac{\partial\Phi}{\partial x} + \frac{\partial\Phi}{\partial t}\right) . \tag{3.2.18}$$

We apply this equation in the neighbourhood of the back and the face of one of the blades. The induced velocity field rotates with the blade hence we can write $\Phi = \Phi(x, r, \varphi - \omega t)$ and we can replace the derivative with respect to t in (3.2.18) by a derivative with respect to φ,

$$p = -\rho\left(U\frac{\partial\Phi}{\partial x} - \frac{1}{r}\frac{\partial\Phi}{\partial\varphi}\,\omega r\right) = (U, 0, -\omega r)\cdot \operatorname{grad}\Phi . \tag{3.2.19}$$

The vector $(U, 0, -\omega r)$ is the relative velocity of a particle of the undisturbed flow, with respect to the planform of the blade. This means that (3.2.19) is the rate of change of Φ, observed by a particle moving with respect to the planform along a helicoidal line $r = \text{const.}$ with a velocity $(U^2 + \omega^2 r^2)^{1/2}$. Hence we can write

$$p = -\rho\,(U^2 + \omega^2 r^2)^{1/2}\,\frac{\partial\Phi}{\partial\sigma} , \tag{3.2.20}$$

where σ is a length parameter along the helicoidal line. It is well known, however, that a tangential derivative of a potential of a layer of sources is continuous across the layer. Hence p has at both sides of the blade the same value and hence $Q = 0$.

The influence of the thickness of a screw blade on its camber is discussed for instance by Jacobs and Tsakonas [32] and by Kerwin and Leopold [38]. It turns out that for conventional screw propellers this influence on the total camber of the blade is small.

3.3. Screw Blades of Zero Thickness, Prescribed Load, 1

Next we discuss the screw propeller with blades of zero thickness ($D(x,r)=0$) and prescribed load $Q(x,r)$, by means of the result of the lifting surface theory as given in Section 1.14. The propeller has N blades $\tilde{W}_\nu$ which are denoted in agreement with (3.1.1) and (3.1.2) as

$$\tilde{W}_\nu : \varphi - \omega t + ax + h_Q^*(x,r) = \frac{2\pi\nu}{N} , \quad \nu = 0,1,\ldots,N-1 , \tag{3.3.1}$$

$$x_l(r) \le x \le x_t(r) , \qquad r_1 \le r \le r_2 , \tag{3.3.2}$$

where we have to determine $h_Q^*(x,r)$ so that the desired load $Q(x,r)$ occurs. The condition (1.2.18) for the fluid flow to be tangent to the blade becomes

$$\begin{aligned} U\frac{\partial h_\varphi^*}{\partial x} &= -\left(av_x + \frac{v_\varphi}{r}\right) \\ &= \frac{(1+a^2r^2)^{1/2}}{r}\,\vec{n}\cdot\vec{v} = \frac{(1+a^2r^2)^{1/2}}{r}\,v_n , \end{aligned} \tag{3.3.3}$$

where v_n is the component of the disturbance velocity $\vec{v}$ normal to the reference surface.

First we consider again, as in the previous section, the blade $\tilde{W}_0$ as if it was rotating alone in the fluid. Afterwards we discuss the influence of the other blades on the camber $h_Q^*(x,r)$. It is again sufficient to consider the screw in its position at $t=0$.

In order to apply the result of Section 1.4, we have to choose $\vec{q}(\lambda,\mu)$ (1.10.1), so that it describes the reference surface (3.1.1) for $t=0$ and $\nu=0$. This can be done as follows

$$\vec{q}(\lambda,\mu) = (-\lambda, \mu\cos a\lambda, \mu\sin a\lambda) = (x,y,z) , \tag{3.3.4}$$

where the components of $\vec{q}$ are with respect to the Cartesian coordinate system. It follows from (3.3.4) that the cylindrical coordinates of a point defined by (λ,μ) become

$$x = -\lambda , \quad r = \left(y^2+z^2\right)^{1/2} = \mu , \quad \varphi = a\lambda = -ax . \tag{3.3.5}$$

Differentiation of (3.3.4) yields

$$\frac{\partial\vec{q}}{\partial\lambda} = (-1, -a\mu\sin a\lambda, a\mu\cos a\lambda) , \tag{3.3.6}$$

$$\frac{\partial\vec{q}}{\partial\mu} = (0, \cos a\lambda, \sin a\lambda) . \tag{3.3.7}$$

Then the unit normal (1.10.4) to the reference surface becomes

$$\begin{aligned} \vec{n} &= \left(\frac{\partial\vec{q}}{\partial\lambda}\times\frac{\partial\vec{q}}{\partial\mu}\right)\cdot\left|\frac{\partial\vec{q}}{\partial\lambda}\times\frac{\partial\vec{q}}{\partial\mu}\right|^{-1} \\ &= \frac{(-a\mu, \sin a\lambda, -\cos a\lambda)}{\left(1+a^2\mu^2\right)^{1/2}} = \frac{-(ar, \sin ax, \cos ax)}{\left(1+a^2r^2\right)^{1/2}} , \end{aligned} \tag{3.3.8}$$

which is the same vector (Cartesian components) as $\vec{n}$ (3.1.7) (cylindrical components). Hence both normals, which were defined independently, have the same sense. Then by (3.3.3), (1.14.17) and (1.14.18) it follows that

$$4\pi\rho\, v_n(\vec{q}_s) = \frac{4\pi\rho\, r_s U}{(1+a^2r_s^2)^{1/2}} \frac{\partial h_Q^*(\vec{q}_s)}{\partial x} = 4\pi\rho\, r_s U \frac{\partial h_Q^*(\vec{q}_s)}{\partial \sigma}$$

$$= \lim_{\beta\to 0} -\left\{ \iint\limits_{\tilde{\Omega}\setminus B_\beta} Q(\tilde{\lambda},\tilde{\mu})\, \tilde{K}_0(\tilde{\lambda}_s,\tilde{\mu}_s,\tilde{\lambda},\tilde{\mu})\, d\tilde{\lambda}\, d\tilde{\mu} \right.$$

$$\left. + \frac{c_0}{U\beta} \int\limits_{\tilde{\lambda}_s}^{\mathcal{L}(\tilde{\mu}_s)} Q(\tilde{\lambda},\tilde{\mu}_s)\, d\tilde{\lambda} \right\}, \qquad (3.3.9)$$

where we have replaced the external force field $\tilde{f}$ by $-Q$ (action equals reaction), $(\tilde{\lambda},\tilde{\mu})$ are the local coordinates (1.14.8) at the planform, $\vec{q}_s = \vec{q}(\tilde{\lambda}_s,\tilde{\mu}_s)$ is the point under consideration and σ is again a length parameter along a helicoidal line. The coordinates $\tilde{\lambda}$ and $\tilde{\mu}$ are for $t = 0$ equal to λ and μ, respectively.

We repeat the definitions (1.14.19), (1.14.13), (1.10.19) and (1.10.20)

$$\tilde{K}_0(\tilde{\lambda}_s,\tilde{\mu}_s,\tilde{\lambda},\tilde{\mu}) = \int\limits_{-\infty}^{0} K(\tilde{\lambda}_s,\tilde{\mu}_s,\tilde{\lambda}+U\tau,\tilde{\mu})\, d\tau\ , \qquad (3.3.10)$$

$$K(\lambda_s,\mu_s,\lambda,\mu) = \left\{ \frac{\partial H_1}{\partial\lambda}(\vec{q}_s,\vec{q}) + \frac{\partial H_2}{\partial\mu}(\vec{q}_s,\vec{q}) \right\}, \qquad (3.3.11)$$

$$H_1(\vec{q}_s,\vec{q}) = -\frac{(\vec{q}_s-\vec{q})}{R^3(\vec{q}_s-\vec{q})} \cdot \left(\frac{\partial\vec{q}}{\partial\mu} \times \vec{n}_s \right), \qquad (3.3.12)$$

$$H_2(\vec{q}_s,\vec{q}) = \frac{(\vec{q}_s-\vec{q})}{R^3(\vec{q}_s-\vec{q})} \cdot \left(\frac{\partial\vec{q}}{\partial\lambda} \times \vec{n}_s \right), \qquad (3.3.13)$$

where by (3.3.4)

$$R(\vec{q}_s-\vec{q}) = |\vec{q}_s-\vec{q}|$$
$$= \{(\lambda_s-\lambda)^2 + \mu_s^2 + \mu^2 - 2\mu_s\mu\cos a(\lambda_s-\lambda)\}^{1/2}\ . \qquad (3.3.14)$$

By straight-forward calculations we obtain

$$H_1(\vec{q}_s,\vec{q}) = -\frac{\{(\lambda_s-\lambda)\cos a(\lambda_s-\lambda) + a\mu_s^2\sin a(\lambda_s-\lambda)\}}{(1+a^2\mu_s^2)^{1/2}\, |\vec{q}_s-\vec{q}|^3}, \qquad (3.3.15)$$

$$H_2(\vec{q}_s,\vec{q}) = \tag{3.3.16}$$

$$\frac{\{a\mu(\lambda_s-\lambda)\sin a(\lambda_s-\lambda)-a^2\mu_s^2\mu\cos a(\lambda_s-\lambda)+a^2\mu_s\mu^2-\mu_s+\mu\cos a(\lambda_s-\lambda)\}}{\left(1+a^2\mu_s^2\right)^{1/2}\ |\vec{q}_s-\vec{q}|^3}\ .$$

Then we find by (3.3.11)

$$K(\lambda_s,\mu_s,\lambda,\mu) = -\frac{\{a^2\mu_s\mu+\cos a(\lambda_s-\lambda)\}}{(1+a^2\mu_s^2)^{1/2}\ |\vec{q}_s-\vec{q}|^3}$$

$$+\frac{3\{a\mu_s(\lambda_s-\lambda)-\mu\sin a(\lambda_s-\lambda)\}\{a\mu(\lambda_s-\lambda)-\mu_s\sin a(\lambda_s-\lambda)\}}{(1+a^2\mu_s^2)^{1/2}|\vec{q}_s-\vec{q}|^5}\ . \tag{3.3.17}$$

We remark that in order to arrive at (3.3.17), we have to reduce a part of that part of $\partial H_2/\partial\mu$ in (3.3.11) which has $|\vec{q}_s-\vec{q}|^5$ in its denominator, to a form which has $|\vec{q}_s-\vec{q}|^3$ in its denominator.

The constant c_0 which occurs in the last term of (3.3.9) is defined in (1.13.19),

$$c_0 = \frac{4\ |\partial\vec{q}_s/\partial\lambda|}{|\partial\vec{q}_s/\partial\lambda\times\partial\vec{q}_s/\partial\mu|} = 4\ . \tag{3.3.18}$$

Changing from (λ,μ) to (x,r) (3.3.5) and to the other notations which are used in this chapter, we can write (3.3.9) as

$$4\pi\rho\, v_n(x_s,r_s) = \frac{4\ \pi\rho\ r_s U}{\left(1+a^2r_s^2\right)^{1/2}}\frac{\partial h_Q^*}{\partial x}(x_s,r_s) = 4\pi\rho\ r_s U\frac{\partial h_Q^*}{\partial\sigma}(x_s,r_s)$$

$$= \lim_{\beta\to 0}\Bigg[\left(\int\limits_{r_1}^{r-\beta}+\int\limits_{r+\beta}^{r_2}\right)\int\limits_{x_l(r)}^{x_t(r)} Q(x,r)M_0(x_s,r_s,x,r)\ dx\ dr$$

$$-\frac{4}{U\beta}\int\limits_{x_l(r_s)}^{x_s} Q(x,r_s)\ dx\Bigg]\ , \tag{3.3.19}$$

$$M_0(x_s,r_s,x,r) = \frac{1}{U\left(1+a^2r_s^2\right)^{1/2}}\int\limits_{-\infty}^{(x_s-x)}\Bigg[\frac{\{a^2r_sr+\cos a\tau\}}{\{\tau^2+r_s^2+r^2-2r_sr\cos a\tau\}^{3/2}}$$

$$-\frac{3\ \{ar_s\tau-r\sin a\tau\}\{ar\tau-r_s\sin a\tau\}}{\{\tau^2+r_s^2+r^2-2r_sr\cos a\tau\}^{5/2}}\Bigg]\ d\tau\ . \tag{3.3.20}$$

Estimating the singular behaviour of $M_0(x_s, r_s, x, r)$, we find.

$$M_0(x_s, r_s, x, r) \approx \frac{2(1+a^2r_s^2)^{1/2}}{(r_s-r)^2} - \frac{a^2 r_s}{(1+a^2r_s^2)^{1/2}(r_s-r)}$$
$$+ O(\ln|r_s - r|) \ ; \quad x < x_s \ , \quad r \to r_s \ . \tag{3.3.21}$$

This singularity of $O((r_s - r)^2)$ is not integrable, which is the reason that the Hadamard principle value in (3.3.19) occurs.

We now shortly discuss the influence of the other blades with $\nu = 1, 2, \ldots, N-1$ in (3.3.1). At these blades we have instead of (3.3.4) and (3.3.5)

$$\vec{q}(\lambda, \mu) = \left(-\lambda, \mu\cos\left(a\lambda + \frac{2\pi\nu}{N}\right), \mu\sin\left(a\lambda + \frac{2\pi\nu}{N}\right)\right) = (x, y, z) \ , \tag{3.3.22}$$

$$x = -\lambda, \quad r = (y^2 + z^2)^{1/2} = \mu, \quad \varphi = a\lambda + \frac{2\pi\nu}{N} = -ax + \frac{2\pi\nu}{N}. \tag{3.3.23}$$

The load $Q(x, r)$ of these blades induces no singular velocity at the blade $\nu = 0$, because they have a finite distance to this blade. Then it is easily found that the influence of these blades follows from (3.3.19) when we neglect the last singular term and replace $M_0(x_s, r_s, x, r)$ by

$$M_\nu(x_s, r_s, x, r)$$
$$= \frac{1}{U(1+a^2r_s^2)^{1/2}} \int\limits_{-\infty}^{(x_s - x)} \left[\frac{\{a^2 r_s r + \cos(a\tau + 2\pi\nu/N)\}}{\{\tau^2 + r_s^2 + r^2 - 2r_s r\cos(a\tau + 2\pi\nu/N)\}^{3/2}} \right.$$
$$\left. - \frac{3\{a r_s \tau - r\sin(a\tau + 2\pi\nu/N)\}\{a r\tau - r_s\sin(a\tau + 2\pi\nu/N)\}}{\{\tau^2 + r_s^2 + r^2 - 2r_s r\cos(a\tau + 2\pi\nu/N)\}^{5/2}} \right] d\tau \ . \tag{3.3.24}$$

Then for the propeller with N blades (3.3.19) becomes

$$4\pi\rho\, v_n(x_s, r_s) = \frac{4\,\pi\rho\, r_s U}{\left(1+a^2r_s^2\right)^{1/2}} \frac{\partial h_Q^*}{\partial x}(x_s, r_s) = 4\pi\rho\, r_s U \frac{\partial h_Q^*}{\partial\sigma}(x_s, r_s)$$
$$= \lim_{\beta\to 0} \left\{ \left(\int\limits_{r_1}^{r-\beta} + \int\limits_{r+\beta}^{r_2} \right) \int\limits_{x_l(r)}^{x_t(r)} Q(x, r) M_0(x_s, r_s, x, r)\, dx\, dr - \frac{4}{U\beta} \int\limits_{x_l(r)}^{x_s} Q(x, r_s)\, dx \right\}$$
$$+ \sum_{\nu=1}^{N-1} \int\limits_{r_1}^{r_2} \int\limits_{x_l(r)}^{x_t(r)} Q(x, r) M_\nu(x_s, r_s, x, r)\, dx\, dr \ . \tag{3.3.25}$$

As we mentioned already for the camber $h_D^*(x, r)$ in the previous section, also the camber $h_Q^*(x, r)$ is not defined uniquely by (3.3.25). Again we can choose at the planform of each blade a line, connecting the root and the tip of the blade, along which h_Q^* is supposed to be zero. Then by integrating with respect to the length parameter σ, along each helicoidal line r = const., we can determine h_Q^* from (3.3.25).

3.4. The Meaning of the Hadamard Principle Value

In this section we discuss the meaning of the Hadamard principle value. This is important because it leads us to a conclusion about the validity of the Cauchy principal value of an integral which will be used in the next section for the representation of the blades of zero thickness by means of a vortex layer. We emphasize that the validity of the Cauchy principal value, does not follow alone from the singularity of the integrand of the integral over a vortex layer, because it has to be proved that no other local singular contributions occur.

The singular kernel $M_0(x_s, r_s, x, r)$ which occurs in (3.3.25) is seemingly related (see (1.10.9)) to the induced normal velocity component at the point (x_s, r_s) by a singular unit force acting at the fluid at the point (x, r), where normal means in the direction of $\vec{n}(x_s, r_s)$ (3.1.7). There are however two differences. First, we have a factor $4\pi\rho$ at the left-hand side of (3.3.25) and second, $(dx\ dr)$ is not an element of area dS, but

$$dS = (1 + a^2 r^2)^{1/2}\, dx\, dr\ . \tag{3.4.1}$$

These two differences cause that the values of $M_0(x_s, r_s, x, r)$ can be calculated by applying the law of Biot and Savart to the vorticity of a singular force $\vec{F}$ normal to the blade of strength

$$|\vec{F}\,| = 4\pi\rho\,(1 + a^2 r^2)^{1/2}\ . \tag{3.4.2}$$

The vorticity which belongs to such a force follows from Figure 1.8.1 (b) and is drawn in Figure 3.4.1. It consists approximately of a short bound vortex of length 2ε and of two trailing vortices, a distance 2ε apart. The strength Γ^* of the bound vortex has by (1.8.3) and (3.4.2) the value

$$\Gamma^* = \frac{1}{2\rho\,\varepsilon\,\left(U^2 + \omega^2 r^2\right)^{1/2}}\, 4\pi\rho\,(1 + a^2 r^2)^{1/2} = \frac{2\pi}{\varepsilon U}\ . \tag{3.4.3}$$

The free tip vortices are of equal strength and lie along the two helicoidal lines at $\varphi + ax = 0$, stretching from the tips of the short bound vortex to $x = \infty$. These three vortices form an elementary horse-shoe vortex.

Now we consider in the Cartesian (x, r) plane of integration (Figure 3.4.2) an elementary horse-shoe vortex of which one tip vortex lies along the line $r = r_s - \beta$. This tip vortex is not largely compensated by a neighbouring one, as happens to be the case with the tip vortex along $r = r_s - \beta - 2\varepsilon$. It means that along $r = r_s - \beta$

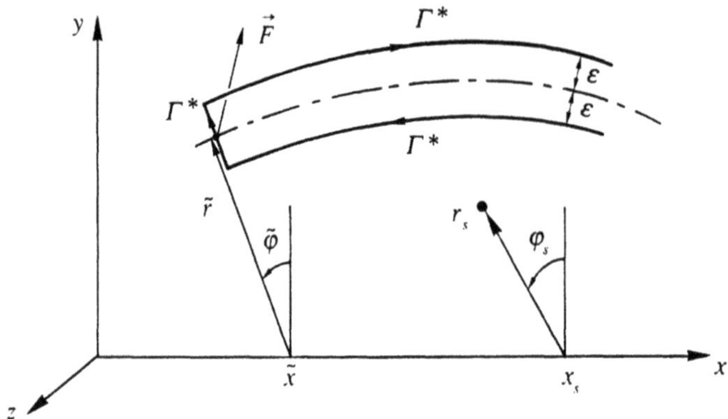

Fig. 3.4.1. Vortex "representation" in 3-dimensional space of $M_0(x_s, r_s, \tilde{x}, \tilde{r})$.

we have at x, a concentrated vortex of finite strength

$$\Gamma(x) = \int_{x_l(r_s)}^{x} 2\varepsilon\, Q(\tilde{x}, r_s)\, d\tilde{x} \cdot \frac{2\pi}{\varepsilon U} = \frac{4\pi}{U} \int_{x_l(r_s)}^{x} Q(\tilde{x}, r_s)\, d\tilde{x} \ , \tag{3.4.4}$$

because the strength at x is also caused by the elementary horse-shoe vortices starting between $x_l(r_s)$ and x. An analogous reasoning can be given for the upper-side of the excluded strip (Figure 3.4.2), hence for $r = r_s + \beta$.

The result is that along $r = r_s - \beta$ and along $r = r_s + \beta$, there are two concentrated vortices of opposite rotation, which at $x = x_s$ have the strength

$$\Gamma(x_s) = \frac{4\pi}{U} \int_{x_l(r_s)}^{x_s} Q(\tilde{x}, r_s)\, d\tilde{x} \ . \tag{3.4.5}$$

Because β is very small, these concentrated vortices can in the neighbourhood of (x_s, r_s) be considered as infinitely long straight vortices of strength (3.4.5). Then they induce at (x_s, r_s) a velocity normal to the reference surface of magnitude

$$2 \cdot \frac{\Gamma(x_s)}{2\pi\beta} = \frac{4}{\beta U} \int_{x_s(r_s)}^{x_s} Q(\tilde{x}, r_s)\, d\tilde{x} \ . \tag{3.4.6}$$

The meaning of the Hadamard principal value is now that this singular contribution (3.4.6) is annihilated by the singular last term in (3.3.25).

When we want to give a more accurate discussion, we have to take into account the small deviations of the values of $Q(x_s, r_s - \beta)$ and $Q(x_s, r_s + \beta)$ from the value of $Q(x_s, r_s)$. However, when we assume $Q(x, r)$ to be sufficiently smooth, these

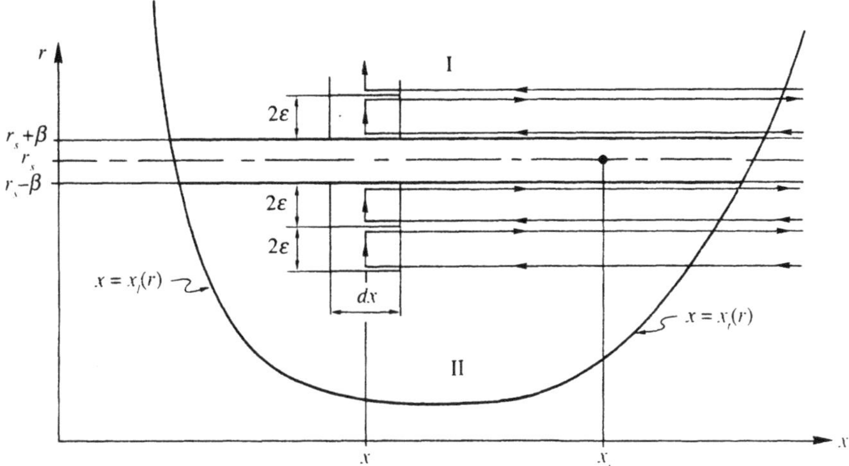

Fig. 3.4.2. Elementary horse-shoe vortices at the planform W_0.

deviations are of $O(\beta)$ and opposite at $r = r_s - \beta$ and $r = r_s + \beta$, hence their influence cancels at $r = r_s$.

So we come to the conclusion that the normal component v_n of the induced velocity at the point (x_s, r_s) (Figure 3.4.2), as given in (3.2.25) consists of three parts. First, a part induced by the continuous bound vorticity at the interior of each of the two regions of the planform, namely region I above the excluded strip and region II below this strip, as well as by the free vorticity shed by this bound vorticity. Second, a part induced by two concentrated vortices of finite strength (3.4.4) and of opposite rotation along the boundaries $r = r_s - \beta$ and $r = r_s + \beta$ of the excluded strip. Third, a part with factor β^{-1} in (3.3.25) which annihilates the artificial singular velocity component of the concentrated vortices of the second part. This singular part is, because the concentrated vortices have finite strength, the only contribution by them to v_n, in the limit $\beta \to 0$.

We now consider the continuous vorticity alone, which is not free of divergence anymore, because therefore the concentrated vortices are needed. Then we can in principle calculate by means of the law of Biot and Savart the induced velocity at the point (x_s, r_s), which still is in the void excluded strip between $r = r_s - \beta$ and $r = r_s + \beta$. By taking the limit $\beta \to 0$, it follows from the above that we find the correct induced normal velocity component at (x_s, r_s). This is the *basis* for the considerations in the next section.

3.5. Screw Blades of Zero Thickness, Prescribed Load, 2

We start in this section from the vorticity distribution on the blades. Because one of the important applications of lifting surface theory is the design of the screw blades,

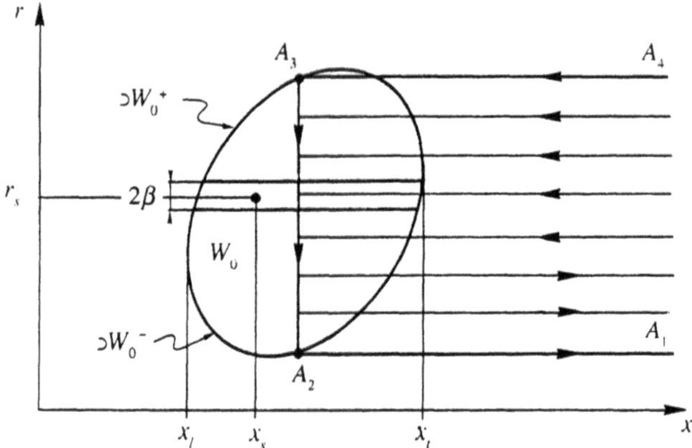

Fig. 3.5.1. The planform W_0 with a basic vortex system.

we assume that the load $Q(x, r)$ of the blades is finite and bounded everywhere. This is in agreement with the desire to avoid cavitation in case of a conventional propeller. We could even have assumed that the load tends to zero at the whole edge of the blade, but in practice sometimes a finite load is preferred at the leading edge.

We consider in this section again a propeller with N blades as in Section 3.3 and start also here by considering in first instance only one blade, namely $\tilde{W}_0$ (3.3.1). In Figure 3.5.1 we have drawn the planform W_0 in the Cartesian (x, r) plane. We will consider as a basic vortex system, the bound vortex A_2A_3, together with its shed free vorticity. This free vorticity consists of two parts. First, the vorticity shed by the interior of the bound vortex A_2A_3 and second, the vortices A_2A_1 and A_3A_4. These latter vortices do not occur when we assume that the load $Q(x, r)$ tends continuously to zero at the edge ∂W_0 of W_0. Although we talk about vortices, it is clear that we have vortex densities and also it is clear that in space the vorticity is situated at the helicoidal reference surface (3.1.1) ($\nu = 0$).

We discuss first the influence numbers of elements of the mentioned basic vortex system and start with the induced velocity at the point (x_s, r_s) at W_0, by a vortex element of length dr of the bound vortex A_2A_3. Using the cylindrical coordinate system (x, r, φ), we have drawn in Figure 3.5.2 the relevant quantities. For convenience we restate the formulas for the vector $\vec{R}$ (3.2.2), its length R (3.2.3) and the unit normal $\vec{n}$ (3.1.7), we add the unit vector $\vec{r}$ in the r-direction. All the mentioned vectors are given by their components in the coordinate directions at the point (x_s, r_s, φ_s), this is done in order to be able to determine the scalar or vector product of two vectors and to carry out integrations.

$$\vec{R} = (x_s - \tilde{x}, r_s - \tilde{r}\cos(\varphi_s - \tilde{\varphi}), \tilde{r}\sin(\varphi_s - \tilde{\varphi})) \ , \tag{3.5.1}$$

$$R = |\vec{R}| = \{(x_s - \tilde{x})^2 + r_s^2 + r^2 - 2r_s\tilde{r}\cos(\varphi_s - \tilde{\varphi})\}^{1/2} \ , \tag{3.5.2}$$

$$\vec{n}_s = -(ar_s, 0, 1)(1 + a^2r_s^2)^{-1/2} \ , \tag{3.5.3}$$

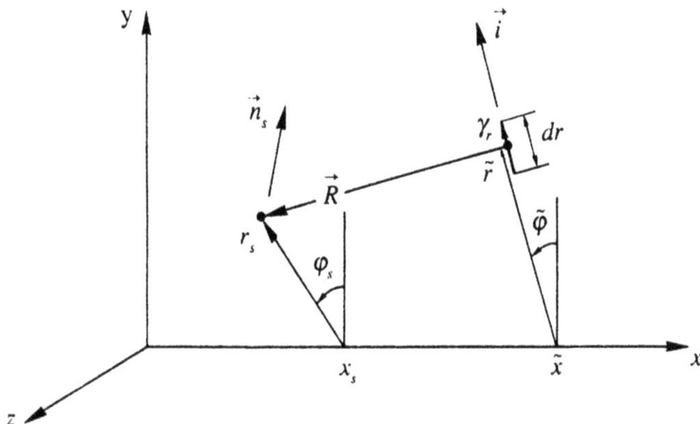

Fig. 3.5.2. Element $\gamma_r\ dr$ of bound vortex A_2A_3.

$$\vec{\imath} = (0, \cos(\varphi_s - \tilde{\varphi}), -\sin(\varphi_s - \tilde{\varphi})) \ . \tag{3.5.4}$$

The strength of the vortex element $\gamma_r\ dr$ is coupled with a right-hand screw to the positive r-direction. By the law of Biot and Savart (1.1.15), we find for the induced velocity component dv_n normal to the planform and at the point (x_s, r_s)

$$\begin{aligned} dv_n(x_s, r_s) &= \frac{\gamma_r \vec{n} \cdot (\vec{\imath} \times \vec{R})\ dr}{4\pi R^3} \\ &= \frac{-\gamma_r\ dr}{4\pi\left(1 + a^2 r_s^2\right)^{1/2}} \frac{\{a r_s^2 \sin(\varphi_s - \tilde{\varphi}) - (x_s - \tilde{x})\cos(\varphi_s - \tilde{\varphi})\}}{\{(x_s - \tilde{x})^2 + r_s^2 + \tilde{r}^2 - 2 r_s \tilde{r} \cos(\varphi_s - \tilde{\varphi})\}^{3/2}} \ . \end{aligned} \tag{3.5.5}$$

Because both (x_s, r_s, φ_s) and $(\tilde{x}, \tilde{r}, \tilde{\varphi})$ are situated at the helicoidal reference surface (3.1.1) ($\nu = 0$), we have $\varphi_s = -a x_s$ and $\tilde{\varphi} = -a\tilde{x}$. Then we can write (3.5.5) as

$$dv_n(x_s, r_s) = \frac{\gamma_r\ dr}{4\pi(1 + a^2 r_s^2)^{1/2}} N_{01}(x_s, r_s, \tilde{x}, \tilde{r}) \ , \tag{3.5.6}$$

$$N_{01}(x_s, r_s, \tilde{x}, \tilde{r}) = \frac{\{a r_s^2 \sin a(x_s - \tilde{x}) + (x_s - \tilde{x}) \cos a(x_s - \tilde{x})\}}{\{(x_s - \tilde{x})^2 + r_s^2 + \tilde{r}^2 - 2 r_s \tilde{r} \cos a(x_s - \tilde{x})\}^{3/2}} \ . \tag{3.5.7}$$

For the singularity of N_{01} we find

$$\lim_{\tilde{x} \to x_s, \tilde{r} \to r_s} N_{01}(x_s, r_s, \tilde{x}, \tilde{r}) = \frac{(1 + a^2 r_s^2)(x_s - \tilde{x})}{\{(1 + a^2 r_s^2)(x_s - \tilde{x})^2 + (r_s - \tilde{r})^2\}^{3/2}} . \tag{3.5.8}$$

Next we determine the influence number of a half-infinite helicoidal vortex line L stretching from $x = \tilde{x}$ towards $x = +\infty$ (Figure 3.5.3). The helicoidal line L is given by

$$L : a\xi + \theta = 0 \ , \quad r = \tilde{r} \ , \quad \xi \geq \tilde{x} \ , \tag{3.5.9}$$

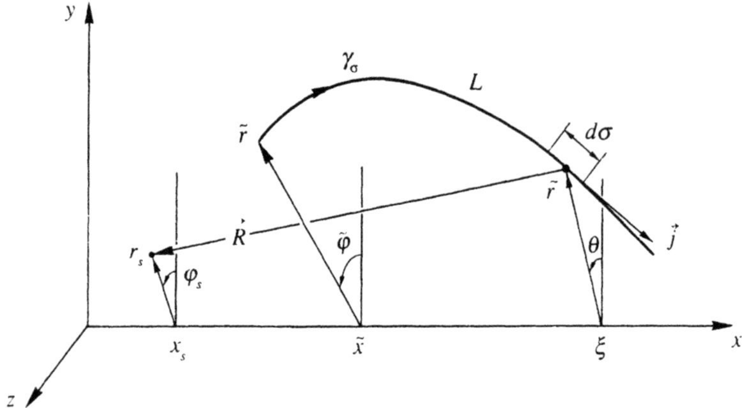

Fig. 3.5.3. Half-infinite helicoidal vortex line L of strength γ_σ.

Along L we have the length parameter σ with $\sigma = 0$ for $x = \tilde{x}$. An element of length $d\sigma$ of L has the value

$$d\sigma = (1 + a^2\tilde{r}^2)^{1/2}\, d\xi \ . \tag{3.5.10}$$

The strength of the vortex is γ_σ, coupled with a right-hand screw to the positive σ-direction. Here we need the unit vector $\vec{j}$, tangent to L at the point $(\xi, \tilde{r}, \theta)$. Its components with respect to the coordinate directions at (x_s, r_s, φ_s) are

$$\vec{j} = (1, -a\tilde{r}\sin(\varphi_s - \theta), -a\tilde{r}\cos(\varphi_s - \theta)) \cdot (1 + a^2\tilde{r}^2)^{-1/2} \ . \tag{3.5.11}$$

Then again by the law of Biot and Savart (1.1.15), we find for the induced velocity component normal to the planform W_0 at the point (x_s, r_s)

$$v_n(x_s, r_s) = \frac{\gamma_\sigma}{4\pi} \int\limits_{\tilde{x}}^{\infty} \frac{\vec{n}_s \cdot (\vec{j} \times \vec{R})}{R^3}\, d\sigma$$

$$= \frac{\gamma_\sigma}{4\pi(1 + a^2 r_s^2)^{1/2}} \int\limits_{\tilde{x}}^{\infty}$$

$$\frac{\{-r_s(1 - a^2\tilde{r}^2) + \tilde{r}(1 - a^2 r_s^2)\cos a(x_s - \xi) + (x_s - \xi)a\tilde{r}\sin a(x_s - \xi)\}}{\{(x_s - \xi)^2 + r_s^2 + \tilde{r}^2 - 2r_s\tilde{r}\cos a(x_s - \xi)\}^{3/2}}\, d\xi$$

$$= \frac{\gamma_\sigma}{4\pi(1 + a^2 r_s^2)^{1/2}} N_{02}(x_s, r_s, x, r) \ , \tag{3.5.12}$$

$$N_{02}(x_s, r_s, \tilde{x}, \tilde{r})$$

$$= \int\limits_{-\infty}^{(x_s - \tilde{x})} \frac{\{-r_s(1 - a^2\tilde{r}^2) + \tilde{r}(1 - a^2 r_s^2)\cos a\tau + \tau a\tilde{r}\sin a\tau\}}{\{\tau^2 + r_s^2 + \tilde{r}^2 - 2r_s\tilde{r}\cos a\tau\}^{3/2}}\, d\tau \ . \tag{3.5.13}$$

For the singularity of N_{02} we find

$$\lim_{\tilde{r}\to r_s} N_{02}(x_s, r_s, \tilde{x}, \tilde{r}) \approx \frac{2(1+a^2r_s^2)}{(r_s-\tilde{r})} \ , \qquad x_s > \tilde{x} \ . \tag{3.5.14}$$

We now return to the basic vortex system as drawn in Figure 3.5.1. First, we consider the bound vortex A_2A_3 as a strip on the blade of constant width $d\sigma$, where σ is for each r a length parameter along the helicoidal line through r. Then we find from (1.8.3) the following relation between the load $Q(x,r)$ and the bound vorticity $\gamma_r(x,r)$, which is now per unit of length in the σ-direction,

$$-\rho\, V(r)\gamma_r(x,r)\, d\sigma\, dr = Q(x,r)\, d\sigma\, dr \ , \quad V(r) = U(1+a^2r^2)^{1/2} \ , \tag{3.5.15}$$

hence

$$\gamma_r(x,r) = \frac{-Q(x,r)}{\rho\, U(1+a^2r^2)^{1/2}} \ . \tag{3.5.16}$$

So the induced normal velocity component v_{n1}, caused by the bound vorticity at the blade becomes

$$v_{n1}(x_s,r_s) = \frac{-1}{4\pi(1+a^2r_s^2)^{1/2}} \int\int_{W_0-B_\beta} Q(x,r)N_{01}(x_s,r_s,x,r)\, dx\, dr \ , \tag{3.5.17}$$

where we excluded, as in the previous two sections, the strip B_β of width 2β around r_s.

Second, the normal velocity component v_{n2}, induced by the free vorticity γ_σ^* shed from the interior of A_2A_3. Because a vorticity field is divergenceless it follows that for fixed x

$$\gamma_\sigma^*(x,r)\, dr = -\frac{\partial}{\partial r}(\gamma_r(x,r)\, d\sigma)\, dr = \frac{\partial}{\partial r}\left(\frac{Q(x,r)\, dx}{\rho\, U}\right)\, dr \ , \tag{3.5.18}$$

where we used (3.5.16). Note that $\gamma_\sigma^*(x,r)$ is not the total free vorticity at some point of the blade, it belongs only to one basic vortex system. We find by (3.5.12)

$$v_{n2}(x_s,r_s) = \frac{1}{4\pi\rho\, U(1+a^2r_s^2)^{1/2}} \int\int_{W_0-B_\beta} \frac{\partial Q(x,r)}{\partial r} N_{02}(x_s,r_s,x,r)\, dx\, dr \ . \tag{3.5.19}$$

Third, the normal velocity component v_{n3}, induced by the tip vorticity A_2A_1 and A_3A_4 of the bound vorticity A_2A_3. We divide the boundary ∂W_0 of W_0 into two parts, ∂W_0^+ from $x = x_l$ to $x = x_t$ (Figure 3.5.1) at the "upper" side of W_0, and

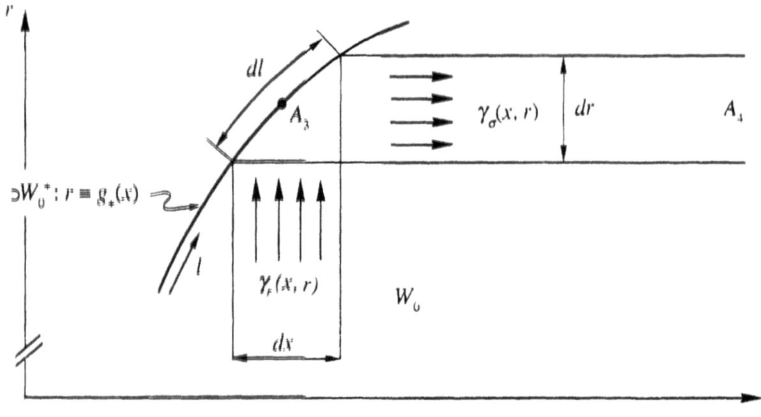

Fig. 3.5.4. The upper part ∂W_0^+ of the boundary of W_0.

∂W_0^- at the "lower" side of W_0. Here x_l and x_t are the extreme values of W_0 in the x-direction. The parts ∂W_0^+ and ∂W_0^- of the boundary are given by

$$\partial W_0^+ : r = g_+(x) \; , \qquad \partial W^- : r = g_-(x) \; , \qquad x_l \leq x \leq x_t \; . \tag{3.5.20}$$

Along each of the parts ∂W^+ and ∂W^- we have a length parameter l. We start with $A_3 A_4$, hence at ∂W_0^+ (Figure 3.5.4). From the conservation of vorticity we find

$$\gamma_\sigma(x,r)dr = \gamma_r(x,r)\, d\sigma = -\frac{Q(x,r)}{\rho\, U}\, dx = \frac{-Q(x,r)\, dl}{\rho\, U(1+g_+'(x)^2)^{1/2}} \; , \tag{3.5.21}$$

With respect to $A_2 A_1$ at the lower part ∂W_0^-, we find analogously

$$\gamma_\sigma(x,r)\, dr = \frac{Q(x,r)\, dl}{\rho\, U(1+g_-'(x)^2)^{1/2}} \; . \tag{3.5.22}$$

Hence by (3.5.12) we find for the contribution of the tip vortices of the bound vorticity $A_2 A_3$

$$v_{n3}(x_s, r_s) = \frac{1}{4\pi\rho\, U(1+a^2 r_s^2)^{1/2}} \cdot \Bigg\{ \int_{\partial W_0^+} -Q(x,r) N_{02}(x_s, r_s, x, r) \frac{dl}{(1+g_+'(x)^2)^{1/2}} + \int_{\partial W_0^-} Q(x,r) N_{02}(x_s, r_s, x, r) \frac{dl}{(1+g_-'(x)^2)^{1/2}} \Bigg\} \; , \tag{3.5.23}$$

where we have to exclude from ∂W_0^+ and ∂W_0^- those parts which belong to the boundary of B_β.

It follows from the consideration at the end of the previous section, that when we take the limit $\beta \to 0$ in (3.5.17), (3.5.19) and (3.5.23), we find the *correct value* of

$$v_{0n}(x_s, r_s) = \lim_{\beta \to 0} \{v_{n1}(x_s, r_s) + v_{n2}(x_s, r_s) + v_{n3}(x_s, r_s)\} \ . \tag{3.5.24}$$

By (3.5.14) it is clear that in the limit $\beta \to 0$ the integrals (3.5.19) and (3.5.23) for $v_{n,2}$ and $v_{n,3}$ respectively are calculated by means of the Cauchy principal value. Hence the Hadamard principal value of Section 3.3 is reduced to the less singular Cauchy principal value. Such a reduction can also be obtained by partial integration in the r-direction of the integral over M_0 in (3.3.25).

Next we have to consider the influence of the other blades. For this we use instead of the kernels N_{01} (3.5.7) and N_{02} (3.5.13), the kernels $N_{\nu 1}$ and $N_{\nu 2}$ which are determined as follows. The kernel $N_{\nu 1}$ is obtained by the substitution $\tilde{\varphi} = -a\tilde{x} + 2\pi\nu/N$ into (3.5.5) (see the text below (3.5.5)), we find

$$N_{\nu 1}(x_s, r_s, \tilde{x}, \tilde{r})$$

$$= \frac{\{ar_s^2 \sin(a(x_s - \tilde{x}) + 2\pi\nu/N) + (x_s - \tilde{x})\cos(a(x_s - \tilde{x}) + 2\pi\nu/N)\}}{\{(x_s - \tilde{x})^2 + r_s^2 + \tilde{r}^2 - 2r_s\tilde{r}\cos\left(a(x_s - \tilde{x}) + 2\pi\nu/N\right)\}^{3/2}} \ . \tag{3.5.25}$$

The kernel $N_{\nu 2}$ becomes analogously

$$N_{\nu 2}(x_s, r_s, \tilde{x}, \tilde{r}) = \int_{-\infty}^{(x_s - \tilde{x})} \tag{3.5.26}$$

$$\frac{\{-r_s(1 - a^2\tilde{r}^2) + \tilde{r}(1 - a^2 r_s^2)\cos(a\tau + 2\pi\nu/N) + \tau a\tilde{r}\sin(a\tau + 2\pi\nu/N)\}}{\{\tau^2 + r_s^2 + \tilde{r}^2 - 2r_s\tilde{r}\cos(a\tau + 2\pi\nu/N)\}^{3/2}} \, d\tau \ .$$

Then we have to calculate the expressions (3.5.17), (3.5.19) and (3.5.23) but now with N_{01} and N_{02} replaced by $N_{\nu 1}$ and $N_{\nu 2}$ respectively. After this we sum the results over $\nu = 1, \ldots, N-1$ and add the results to $v_{0n}(x_s, r_s)$ of (3.5.24), by which we obtain the total induced normal velocity component v_n at the point (x_s, r_s) of W_0.

The camber h_Q^* (3.1.13) follows for instance from (3.3.9)

$$\frac{\partial h_Q^*}{\partial \sigma}(x, r) = \frac{v_n(x, r)}{Ur} \ , \tag{3.5.27}$$

again by integration along a helicoidal line, when we have chosen at W_0 some line from the root of the blade to its tip, where $h_Q^*(x, r)$ is supposed to be zero.

We now remark on the case that the load $Q(x, r)$ becomes infinite at the leading edge. This happens in general when the problem is considered that the geometry

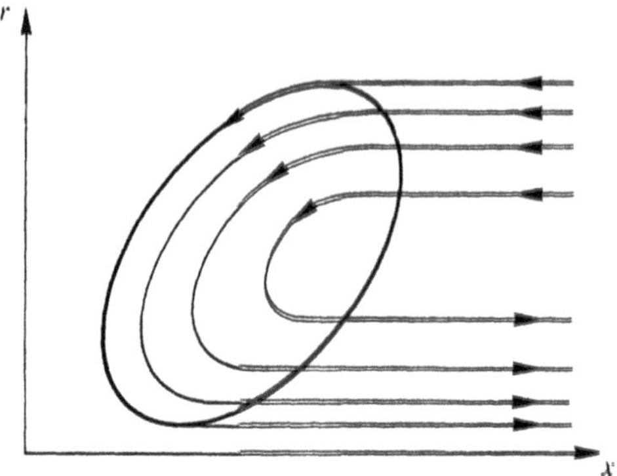

Fig. 3.5.5. Another picture of the vorticity on the blade.

of the blade and its profiles are prescribed and $Q(x,r)$ has to be calculated. Then the integrals (3.5.19) and (3.5.23) for v_{n2} and v_{n3} have become meaningless, both integrals have integrands which yield an infinite value. Then the interplay between v_{n2} and v_{n3} has to be considered in more detail. They have to be combined in one integral over the shed vorticity. However, being interested in the design of a blade we will not discuss this any further.

Finally, we say something about the shape of the vortex lines at a blade. Because the vorticity is free of divergence, the vortex lines will represent the flow of a 2-dimensional incompressible fluid on the curved surface consisting of the blade and the helicoidal surface behind it. Then the vorticity in the neighbourhood of the leading edge has to be tangent to it and we obtain a picture of the vorticity lines as given in Figure 3.5.5. In this picture it is assumed that $Q(x,r) = 0$ at the trailing edge, otherwise the vorticity lines would have a kink at the edge. See the text belonging to (1.18.15) and also Figure 6.5.4 for an application of the vorticity lines at a blade with end plates. It is also clear that with this representation of the vorticity, the problem of the non-existence of the integrals (3.5.19) and (3.5.23), as we mentioned in the previous paragraph, does not exist anymore.

3.6. Some Additional Remarks

In reality the ship screw has to perform its task behind a ship. This means, as we discussed in the beginning of this chapter, that the inflow is not homogeneous as we assumed in our theory. The perturbations caused by the hull, depend besides on x and r also on the angular coordinate φ of our cylindrical coordinate system. Then the load will have a periodic character. In order that the phenomena can be described by a linear theory, the disturbance velocities at the plane of the screw must remain "sufficiently" small. Hence the assumption has to be made that the total velocity $\vec{U}_0$

behind the ship when the propeller is absent, hence the nominal wake, can be written as

$$\vec{U}_0 = (U + v_{0x}, v_{0r}, v_{0\varphi}) \ , \tag{3.6.1}$$

where the disturbance velocity components of the nominal wake v_{0x}, v_{0r} and $v_{0\varphi}$ have to satisfy the relations

$$\frac{v_{0x}}{U} \ll 1 \ , \qquad \frac{v_{0r}}{U} \ll 1 \ , \qquad \frac{v_{0\varphi}}{U} \ll 1 \ . \tag{3.6.2}$$

Possibly this demand can be relieved to some extent by comparing v_{0x}, v_{0r} and $v_{0\varphi}$ at a radius r not with the incoming velocity U but with the velocity $U(1 + a^2r^2)^{1/2}$ of a blade profile relative to the water.

In order to calculate by means of a linear theory the fluctuations of the load during each revolution, we can consider the N planforms of the blades to be impermeable and to have no camber, hence to be parts of the helicoidal surfaces

$$\varphi + ax - \omega t = \frac{2\pi\nu}{N} \ , \qquad \nu = 0, 1, \ldots, N - 1 \ . \tag{3.6.3}$$

Having found in one way or another the fluctuating pressures on these planforms moving in the incoming flow (3.6.1), we can add them to the pressures of the screw with thickness and camber, working in the undisturbed parallel flow. This is allowed because our theory is linear.

A consequence of the disturbed incoming velocity field is that cavitation is likely to occur. This is undesirable for reasons of erosion and deformation of the blades, vibration and noise. For supercavitating propellers we refer to [29] and Kruppa [43].

A more direct approach to lifting surface theory is possible by replacing the continuous vorticity at the blades and the continuous trailing vorticity by concentrated vortex lines. At the blade these vortex lines consist of two different types, bound vortex lines perpendicular to the local relative velocity $\vec{V}$ of the fluid and free vortices aligned with $\vec{V}$. Suppose that the loading Q at the blade is prescribed as a function of position. Then we can assign to each elementary bound vortex $(A-D)$ (Figure 3.6.1) a small region ΔS of the blade and demand that the vortex strength Γ of $(A - D)$ is such that

$$\rho\,\Gamma|\vec{V}\,| \cdot |(A - D)| = Q\Delta S \ . \tag{3.6.4}$$

From the elementary bound vortices follow at the blade the elementary free vortices, for instance $(A - B)$ and $(D - C)$ by the property that a vortex field is free of divergence. Then at the "midpoints" of the vortex "rectangles" such as the point P, the component of the velocity normal to the blade is calculated by means of the law of Biot and Savart. From this normal component follows again the camber of the profiles of the blades. Care has to be taken in the neighbourhood of the edges of the blade. A lot of experience has been obtained with this powerful method which has the advantage of being simple and of which the results seem to be reliable. A

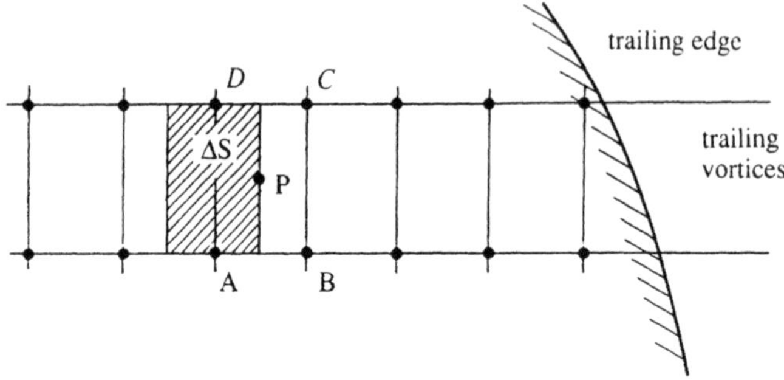

Fig. 3.6.1. Vortex lattice.

theoretical drawback is that the convergence of the approximate solution to the exact values, when the size of the grid elements decreases, is probably not yet proved for generally curved reference surfaces.

An analogous discrete treatment can be given to the influence of the thickness of the blades, by replacing the continuous layer of sources and sinks of Section 3.2, by concentrated source or sink lines along the bound vortex lines.

Non-linear effects can be treated also by this method. For instance the vortex grid need not to be placed at the helicoidal reference surface but can be placed at the profile itself. Also in the neighbourhood of the relatively thick roots of the blades a grid can be placed at both sides of the blade or a surface panel method can be applied.

We conclude this chapter by mentioning some more complicated propulsion devices based on the screw propeller. The most simple extension of the conventional propeller is the propeller with winglets or end plates at the tips of its blades. These winglets can spread out the tip vorticity of a blade and by this reduce the kinetic energy left behind in the water. A draw back of the end plates is that they have a large velocity with respect to the water and by the viscosity of the water they need extra power from the engine. So it has to be judged carefully if their application will raise the efficiency in given circumstances. We return to this subject in Section 6.6.

Next there is the very important combination of a screw propeller with an annular "hydrofoil", called duct, shroud or nozzle. We distinguish between the accelerating duct in which the fluid is accelerated and the decelerating duct in which the fluid is decelerated (Figure 3.6.2). The first one can be used in the case of a heavily loaded screw, in order to improve the efficiency of the propeller. The second one is used to increase the pressure in the neighbourhood of the screw, hence it can be used to delay cavitation. Just as the end plates, also a duct has the ability to spread the tip vorticity of the blades when the slit between blade tips and duct is sufficiently narrow. We do not enter here into the complicated hydrodynamic problems connected with this configuration but refer for instance to Kinnas and Coney [39].

We also mention the combination of two screw propellers, one working behind the other. Tandem propellers consist of two propellers mounted on the same shaft

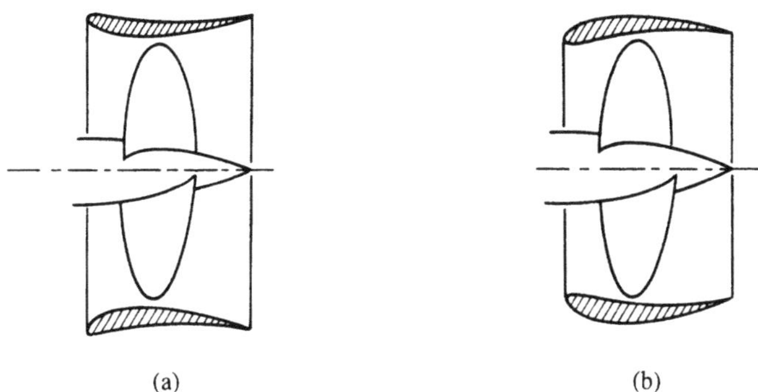

Fig. 3.6.2. Ducted propeller, (a) accelerating type, (b) decelerating type.

hence rotating in the same direction with the same angular speed. Contra-rotating propellers are mounted on two coaxial contrary-turning shafts. They are used to reduce the kinetic energy connected to the rotation of the fluid behind a propeller. The theory of this configuration is complicated because it is essentially unsteady, the blades of the hindmost propeller cut periodically the free vortex sheets of the foremost one.

Another way of reducing the kinetic energy belonging to the rotation of the fluid behind a screw propeller, is to place a (preswirl) stator in front of it. This stator has to induce an overal rotation in the fluid contrary to the rotation induced by the screw. Also it is possible to place a (postswirl) stator behind a propeller, which then extracts rotational energy out of the fluid.

Finally we mention the screw propeller with vane wheel. The vane wheel consists of a freely rotating hub installed on the propeller shaft behind the propeller. On its hub are mounted a number of blades or vanes. The diameter of the vane wheel is larger than the diameter of the screw propeller. The inner parts of the vanes are driven by the slip stream of the screw propeller, while their tips contribute to the propulsion. For an elaborate discussion of this device we refer to Grimm [22]. The question arises if a good designed propeller with a diameter equal to the diameter of the vane wheel, cannot function equally well.

There is now a large amount of literature about the subjects mentioned in this section. Many of the recent investigations attack these problems numerically by means of lifting surface or surface panel methods. It seems appropriate to refer for an up-to-date discussion of these articles to a general survey such as is given in the Proceedings of the International Towing Tank Conferences, which are held every three years. The last one before the publication of this work was held in 1993 [31].

Chapter 4

Unsteady Propulsion

In this chapter we discuss some aspects of unsteady propulsion. First we have to give a meaning to the expression "unsteady propeller". The most simple one seems to be: a propeller is unsteady when no inertial reference frame exists with respect to which the induced flow is time independent. This, however, is not appropriate because then nearly all conceivable propulsion systems are unsteady, with the only exception probably of the sails of a yacht in steady motion. Even such a reference frame does not exist for the screw propeller working in a homogeneous flow. A better description is: a propeller is unsteady when the fluid flow relative to its lifting surface is time dependent while this time dependency is essential for its functioning. The second part of this definition is vague to some degree, it is intended to exclude for instance the screw propeller in the wake of a ship. Essential unsteady propulsion occurs in the case of the Voith-Schneider propeller (Section 4.9), contra-rotating propellers, the propulsion wheels of a paddle boat and the sculling of a small boat by means of an oar at its stern.

In nature almost all animal propulsion in water or in air is unsteady, the flapping motion of the tail of a fish or of the wings of a bird. An exception is the "rotating" motion of the helicoidal tail of some micro-organisms which screw themselves through a fluid. In that case, however, the inertia forces caused by the fluid can be neglected with respect to the viscous forces (Stokes flow, Reynolds number of $O(10^{-6})$, see for instance Chwang and Wu [12].

The unsteady propulsion we will consider here is assumed to happen in an incompressible and inviscid fluid while the propulsion occurs by lift of profiles and by suction forces at their leading edge. This excludes the propulsion wheels of a paddle boat, which produce thrust by resistance. What is left are propulsion systems consisting of possibly flexible lifting surfaces making flapping motions. These motions are assumed to be periodic and are allowed to have a small $O(\varepsilon)$ or a large $O(\varepsilon^0)$ amplitude.

In the Sections 4.1 and 4.2 we give a classification of a number of unsteady propulsion regimes. Some of these are discussed more in detail in the remaining part of this chapter.

An interesting subject is the design of unsteady propulsion by means of two or more wings oscillating with large amplitude motions behind a boat or ship, which can compete with the screw propeller. We will discuss this subject in Sections 6.8–6.12.

4.1. Concepts of Unsteady Propulsion, Linear Theory

Up to now only unsteady propulsion systems which shed a small amount ($O(\varepsilon)$) of free vorticity per unit of time can be treated by analytic means. This is the reason that here we will restrict ourselves to such systems which, as we mentioned before, are working in an unbounded region filled with an incompressible and inviscid fluid. We assume that the propulsion systems are periodic in their action which means that after a period of time τ the field of flow, with respect to the propeller, will appear again.

In this section we consider a number of different regimes of working of propulsion systems. They have in common that they can be described by a linear theory. In the next section we discuss systems which, although they shed a small amount of vorticity per unit of time, need partly a non-linear theory.

We will use the denomination "flexible lifting surface", by this we will understand a surface of which the shape is prescribed as a function of time. Hence we do not have in view an elastic plate of which the shape depends on a priori unknown induced fluid pressures. For such wings we refer for instance to Szeless [68].

Regime (1.a): Small-amplitude, one flexible wing.

In the following we repeat for convenience shortly some concepts discussed in Section 1.9. Consider a lifting surface $\tilde{W}$ without thickness moving along a flat reference strip H (Figure 4.1.1) which is part of the plane $y = 0$ and which stretches along the x-axis. This flexible lifting surface moves in an ε neighbourhood of H, which means that its amplitude, its slopes and its curvatures are $O(\varepsilon)$

$$\tilde{W} : y = f(x, z, t) \ , \qquad (x, z) \in W \ , \tag{4.1.1}$$

where W is the planform of $\tilde{W}$ and $f(x + U\tau, z, t + \tau) = f(x, z, t) = O(\varepsilon)$.

If $\tilde{W}$ moves exactly along the strip H it does not cause any fluid flow and hence does not shed free vorticity. This motion will be called the "base motion". Small periodic deviations of $O(\varepsilon)$ of the base motion will be called the "added motion". Because our theory will be linear, the boundary conditions related to the lifting surface $\tilde{W}$ will be satisfied at the planform W. The bound and free vorticity of $\tilde{W}$ are assumed to be situated at H, the same holds for the free vorticity behind $\tilde{W}$. See Section 1.4, the paragraph in italics.

The splitting of the motion of lifting surfaces into two parts, namely the base motion and the added motion, is important for the way in which we will treat the linearized theory of propulsion systems. The base motion is always assumed to be such that it sheds no vorticity into the fluid, while the added motion sheds vorticity of

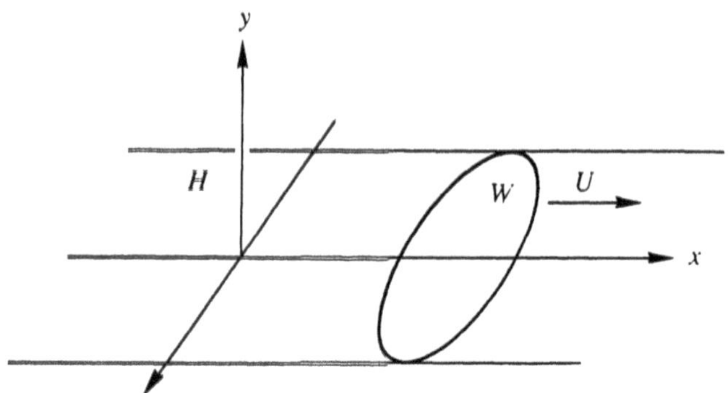

Fig. 4.1.1. Planform W of small-amplitude flexible wing $\tilde{W}$.

$O(\varepsilon)$ by which thrust is generated. These two parts of the motion of a lifting surface are essential in connection with the optimization theory discussed in the next chapter.

The pressure differences between the two sides of $\tilde{W}$ are $O(\varepsilon)$ and also the slopes of $\tilde{W}$ in the x-direction are $O(\varepsilon)$, hence the component in the x-direction of the force exerted at $\tilde{W}$ is $O(\varepsilon^2)$. The leading edge suction K (1.21.13) is also $O(\varepsilon^2)$, because $Q(t)$ (1.21.1) is $O(\varepsilon)$. Hence the time-dependent total thrust $T = T(t)$ is $O(\varepsilon^2)$.

We assume in the following that the propelled body, which in reality moves through a slightly viscous fluid and experiences some wave resistance, has a sufficiently large mass. Then its velocity U is, in spite of the fluctuating thrust $T(t)$, nearly constant. The useful power of the propeller then equals $U\bar{T}$, where $\bar{T}(\varepsilon^2)$ is the mean value with respect to time of $T(t)$, which both are of $O(\varepsilon^2)$.

The shed free vorticity is $O(\varepsilon)$, hence the induced velocities behind the propeller are $O(\varepsilon)$. From this it follows that the mean value of the kinetic energy $E(\varepsilon^2)$ added to the fluid per unit of time is $O(\varepsilon^2)$. Hence (Section 1.17) the efficiency η of such a propulsion system becomes

$$\eta = \frac{U\bar{T}(\varepsilon^2)}{U\bar{T}(\varepsilon^2) + E(\varepsilon^2)} = 1 - O(\varepsilon^0) \ . \tag{4.1.2}$$

It is seen that this efficiency is independent of $\bar{T}$ because both $\bar{T}$ and E are proportional to ε^2.

Regime (1.b): Small-amplitude, two flexible wings, one behind the other.

Suppose we have another wing $\tilde{W}_2$, which has the same span as $\tilde{W}$ and which moves behind $\tilde{W}$. Then $\tilde{W}_2$ can deform periodically in such a way that its shed free vorticity $\vec{\gamma}_2(x, z)$ is opposite to the free vorticity $\vec{\gamma} = \vec{\gamma}(x, z)$ shed by $\tilde{W}$. Hence in the linearized theory we do not have shed vorticity at H behind the two wings. By Section 1.16 it seems that then the mean value of the sum of the thrusts $T(t)$ of $\tilde{W}$ and $T_2(t)$ of $\tilde{W}_2$ has to be zero. This, however, is not true, the reason is that $\vec{\gamma}(x, z)$

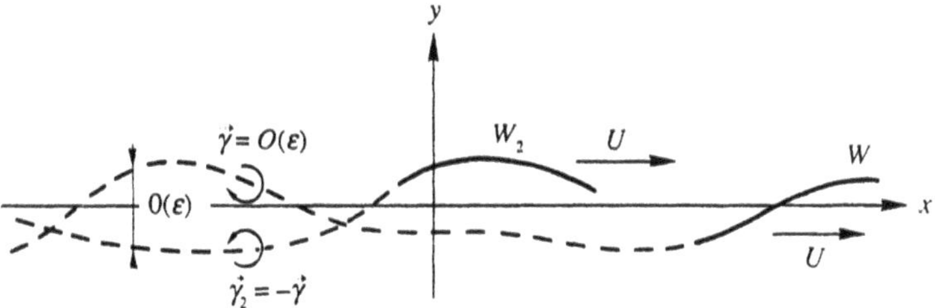

Fig. 4.1.2. Two free vortex layers, a distance of $O(\varepsilon)$ apart.

and $\vec{\gamma}_2(x,z)$ do not coincide exactly but are a distance of $O(\varepsilon)$ apart, which is caused by the fact that $\tilde{W}_2$ passes periodically above and below $\vec{\gamma}(x,z)$. The linearized theory however does not take into account separately these two free vortex layers, which lie close to each other.

We remark that the base motion of this propeller is clearly the one that both wings $\tilde{W}$ and $\tilde{W}_2$ move exactly along the strip H, because then the fluid is not disturbed. The added motion is the motion of $\tilde{W}$ and $\tilde{W}_2$ as described in the preceding paragraph.

In Figure 4.1.2 we have made a cross section of the two free vorticity layers $\vec{\gamma}(x,z)$ and $\vec{\gamma}_2(x,z) = -\vec{\gamma}(x,z)$. When we look far behind the two wings, we find that in the narrow region between the two vorticity layers the induced velocity is $O(\varepsilon)$. Outside this narrow region the induced velocity is only $O(\varepsilon^2)$ because at each finite distance of $O(\varepsilon^0)$ from the two layers they nearly compensate each other. The area of the narrow region, per period of length in the x-direction is $O(\varepsilon)$. So the mean value of the lost kinetic energy $E(\varepsilon^3)$ per unit of time is $O(\varepsilon^3)$. This means that we have to replace (4.1.2) by

$$\eta = \frac{U\{\bar{T}(\varepsilon^2) + \bar{T}_2(\varepsilon^2)\}}{U\{\bar{T}(\varepsilon^2) + \bar{T}_2(\varepsilon^2)\} + E(\varepsilon^3)} = 1 - O(\varepsilon) \ , \tag{4.1.3}$$

where $\bar{T}_2(\varepsilon^2)$ of $O(\varepsilon^2)$ is the mean value of the thrust of $\tilde{W}_2$. It follows that now the deviation of the efficiency η from 1 is only $O(\varepsilon)$. This means that for sufficiently small values of ε, when the linearized theory is accurate, the efficiency (4.1.2) will be lower than the efficiency (4.1.3). In the latter case η decreases with increasing values of ε or what is the same, with increasing values of the total mean thrust $\bar{T} + \bar{T}_2$.

Regime (2): Large-amplitude, one or more flexible wings.

We consider a periodically curved reference strip H (Figure 4.1.3), which is at rest with respect to the undisturbed fluid and of which the deviations from the (x,z) plane are $O(\varepsilon^0)$. In an ε neighbourhood of H moves a flexible wing $\tilde{W}$ without thickness. This means, as in the previous cases, that the deviations of $\tilde{W}$ from its planform W on H and the deviations of its slopes and curvatures from those of W, are $O(\varepsilon)$. If $\tilde{W}$ moves exactly along H, it does not disturb the fluid and hence does not shed

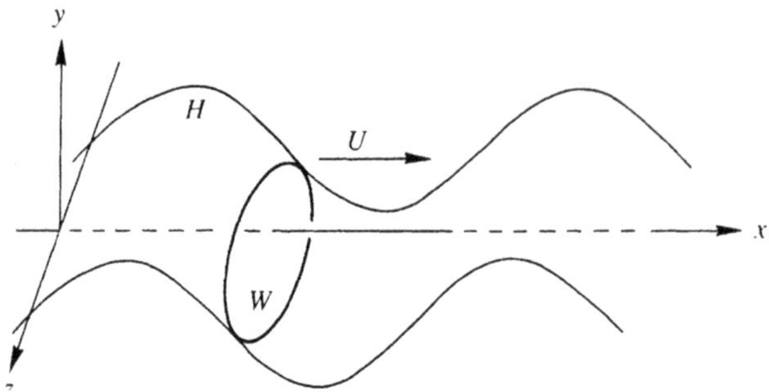

Fig. 4.1.3. Planform W of flexible wing $\tilde{W}$, large-amplitude propulsion.

free vorticity. This motion is the base motion of $\tilde{W}$. Small deviations of $O(\varepsilon)$ from the base motion, which induce velocities and vorticity and by which a mean value of thrust can be generated, is the added motion. The boundary conditions related to the lifting surface $\tilde{W}$ will be satisfied at its planform W on H. The bound and free vorticity of $\tilde{W}$ are, again by the paragraph in italics of Section 1.4, situated at H.

The pressure differences between the two sides of $\tilde{W}$ are $O(\varepsilon)$. With respect to the calculation of the thrust, the angles of $\tilde{W}$ with the x-axis can be replaced within the accuracy of the linear theory, by the angles of W with the x-axis, they are $O(\varepsilon^0)$. From this it follows that the mean value $\bar{T}(\varepsilon)$ of the thrust (in the x-direction) is $O(\varepsilon)$. Because the shed vorticity is $O(\varepsilon)$, it follows that the mean value of the energy losses $E(\varepsilon^2)$ per unit of time are $O(\varepsilon^2)$, hence for the efficiency holds

$$\eta = \frac{U\bar{T}(\varepsilon)}{U\bar{T}(\varepsilon) + E(\varepsilon^2)} = 1 - O(\varepsilon) \ , \tag{4.1.4}$$

where U is the constant velocity with which W moves in the x-direction.

From a comparison of (4.1.4) and (4.1.2) it follows that in the linearized case the large-amplitude propulsion will have a higher efficiency for sufficiently small values of ε than the small-amplitude propulsion of Regime (1.a). It is clear that the proportion between the useful thrust and the inconvenient lateral forces in case of large-amplitude propulsion is $O(\varepsilon^0)$, while for Regime (1.a) this proportion is only $O(\varepsilon)$. This means that also from this point of view the large-amplitude propulsion is more favourable.

It makes no sense to consider a second wing $\tilde{W}_2$ as we did in Regime (1.b), moving behind $\tilde{W}$, which annihilates the free vorticity shed by $\tilde{W}$. Then we would obtain an analogous picture as is drawn in Figure 4.1.2, but now in the neighbourhood of the curved strip H. It is seen that then the momentum of the fluid in the x-direction and far behind the propeller becomes $O(\varepsilon^2)$, hence also the mean value of the thrust of both wings together becomes $O(\varepsilon^2)$. However for the original wing $\tilde{W}$ the thrust was $O(\varepsilon)$, which is "much larger" for small values of ε. So in order to destroy the vorticity of the first wing, the second wing has to have a thrust in the wrong direction

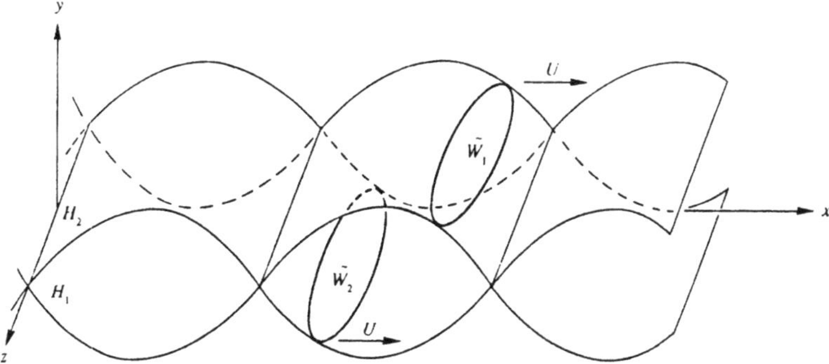

Fig. 4.1.4. Two flexible wings $\tilde{W}_1$ and $\tilde{W}_2$.

and of nearly equal magnitude as the thrust of the first one.

However, it is possible to consider for instance two flexible wings $\tilde{W}$ and $\tilde{W}_2$, as drawn in Figure 4.1.4, which move each in the neighbourhood of its own reference surface H_1 and H_2 respectively. The base motion is then simply that each wing moves exactly along its own reference surface. Then (4.1.4) remains valid although the $O(\varepsilon)$ deficiency of η from 1 can be made smaller in the case of two wings than in the case of one wing. This will be discussed in Chapter 5.

We now make a remark with respect to the formulas (4.1.2), (4.1.3) and (4.1.4) for the efficiency η and direct our attention to (4.1.4), the other two can be treated analogously. The objection can be made that the mean value of the thrust $\bar{T}(\varepsilon)$ of $O(\varepsilon)$ is accurate only up to errors $T^*(\varepsilon^2)$ of $O(\varepsilon^2)$. These errors are of the same order of magnitude as the kinetic energy $E(\varepsilon^2)$ of $O(\varepsilon^2)$ left behind per unit of time. So it seems that the efficiency η cannot be calculated at all. However, we find

$$\eta = \frac{U\{\bar{T}(\varepsilon) + T^*(\varepsilon^2)\}}{U\{\bar{T}(\varepsilon) + T^*(\varepsilon^2)\} + E(\varepsilon^2)} = \frac{U\bar{T}(\varepsilon)}{U\bar{T}(\varepsilon) + E(\varepsilon^2)} + O(\varepsilon^2) \ . \tag{4.1.5}$$

The first term at the right-hand side of (4.1.5) is $(1 - O(\varepsilon))$, hence we can state that the efficiency can be calculated correctly up to and including $O(\varepsilon)$. The error $T^*(\varepsilon^2)$ does not disturb the $O(\varepsilon)$ deviation of η from 1.

4.2. Concepts of Unsteady Propulsion, Semi-Linear Theory

We now want to discuss the possibility of large-amplitude unsteady propulsion by means of rigid lifting surfaces moving in such a way that only a small amount of free vorticity is shed into the fluid. In general a rigid wing moving periodically in one way or another through a fluid will shed free vorticity of $O(\varepsilon^0)$ per unit of time. Hence also kinetic energy is left behind of $O(\varepsilon^0)$ per unit of time. It would be interesting, however, to develop a theory for this type of unsteady propulsion where analogous to the previous sections, there exists a large-amplitude base motion of the rigid wing which sheds no vorticity in the fluid. Then we can superimpose on this base motion of the wing an added motion of $O(\varepsilon)$, by which a thrust of $O(\varepsilon)$ can be developed. The efficiency of the propeller is then also given by (4.1.4).

In such a theory we superimpose on the velocity field of $O(\varepsilon^0)$ caused by the base motion, a velocity field of $O(\varepsilon)$ caused by the thrust producing added motion. The shed free vorticity of $O(\varepsilon)$ is then assumed to move with the velocity field of the base motion. This latter field can be computed rather easily because no free vorticity is present in the fluid.

Regime (3): Large-amplitude, one or more rigid wings of infinite span.

This is a 2-dimensional case in which the phenomena are independent of the z-coordinate (Figure 4.2.1). We describe first the possibility of the base motion for one rigid profile. When no free vorticity is shed into the fluid, the circulation of the profile has to be a constant. We take this constant equal to zero, then we avoid a mean transverse force. In the following we assume that the Kutta condition is satisfied at the sharp trailing edge of the profile. This condition states that the pressure jump over the profile, tends to zero at the trailing edge. Hence the bound vorticity of the profile has to tend to zero at the trailing edge.

We prove the existence of a point $\tilde{Q}$ such that when rotating the profile around $\tilde{Q}$ its circulation is zero. First, consider for the configuration of Figure 4.2.1 a point Q_1, at the y-axis, with a sufficiently large negative value of y. A clockwise rotational velocity of the profile around Q_1 will cause a positive or anti-clockwise circulation around the profile. Next we consider a point Q_2, which we take for instance at a line through O making an angle of $-45°$ with the x-axis, at a sufficiently large distance from O. Then a clockwise rotational velocity of the profile around Q_2 will cause a negative circulation. Hence when we connect Q_1 and Q_2 by a line, there will be by continuity a point $\tilde{Q}$ on this line such that the circulation is zero when the profile rotates clockwise around it.

For each profile we can also find a direction of translation so that the circulation of the profile is zero, this direction is denoted by the line m. When we have found the point $\tilde{Q}$ and a line m we can construct a straight line r of points with the same property as $\tilde{Q}$, namely the line through $\tilde{Q}$ and perpendicular to m. This is correct because a rotational velocity of the profile around any point Q of r can be represented by a rotational velocity around $\tilde{Q}$ and a translational velocity in the direction m. Hence the circulation of the profile is zero. Therefore we can assign to some chosen profile a line r which has a fixed position with respect to it and a direction m, with the above-mentioned properties.

We now take at r some point Q, construct through Q the line m which is perpendicular to r and force the profile to move in the following way. Choose a line L and because we are interested in periodic motions, we assume L to be periodic in the x-direction. The point Q moves along L and we keep the line m tangent to L. Then, irrespective of the shape of L, the circulation around the profile remains zero. This follows from the fact that the motion of the profile at each moment consists of a

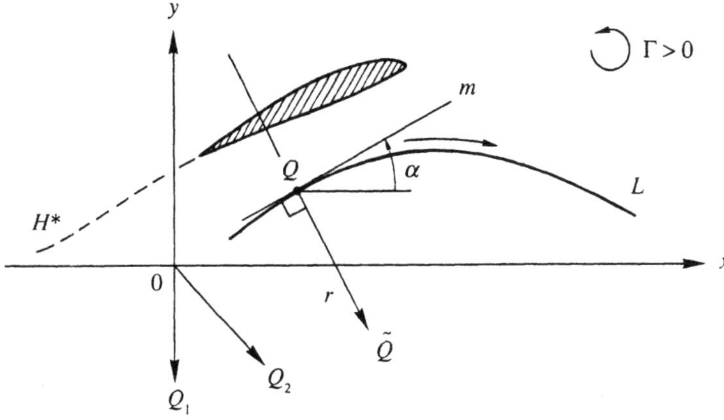

Fig. 4.2.1. Rigid profile moving along L.

rotation around Q and a translation in the direction of m. The motion defined in this way is a base motion of the profile with respect to L. Making another choice of Q on r yields another base motion. The previous result is obtained in a different way in [14].

Next we consider the fluid particles which have passed along the boundary of the profile and have left it at the trailing edge. These particles are lying at a line which we call the wake H^* (Figure 4.2.1). In the neighbourhood of the moving profile the shape of H^* is still changing. When the profile however is a sufficiently large number of periods away, H^* obtains a fixed form which becomes periodic with the same period as L. The wake with its ultimate periodic shape is denoted by H. It is not very difficult, as we mentioned already, to calculate H^* and H because there is no vorticity in the fluid and the time-dependent velocity field depends only on the momentary motion of the profile.

We remark that for the validity of this reasoning, we have to assume that the angle at which both sides of the profile meet each other at the trailing edge, is zero. Otherwise there arises a "stagnation" point at the trailing edge, at which the fluid particles flowing along one side of the boundary of the profile would come temporarily to a relative standstill. In practice, however, we do not bother too much about this.

When the profile has to deliver a non-zero mean value of thrust, an added motion has to be superimposed on the base motion. This motion sheds free vorticity of $O(\varepsilon)$ which is assumed to be situated at H^* and in the progress of time at H. This is the linearization assumption which facilitates calculations because now we know where we have to place the shed free vorticity. This vorticity is transported at H^* with the velocities which follow from the base motion, it comes to rest at H.

From the point of view of propulsion, the profile, which in the previous discussion can have any shape, will in general be chosen to be symmetric because at each side there will be alternately low and high pressures during the motion.

In case a propeller consists of n wings ($n > 1$) parallel to the z-axis and stretching from $z = -\infty$ towards $z = +\infty$, we have to consider n profiles in the (x, y) plane, moving periodically in each others neighbourhood. In this case we cannot explicitly describe a base motion for the profiles, so that they do not shed vorticity. Such a motion will depend besides on the shape of the profiles, also on their relative motion. In order to determine the base motion, we have to solve n coupled integral equations, of which we will give an example in Section 6.9 for the case of two profiles.

Regime (4): Large-amplitude, one rigid wing of finite span.

We consider flat wings of zero thickness and show that only for special wing contours there exist base motions, when we assume that the Kutta condition is satisfied at the trailing edge.

For simplicity we assume the wing to be symmetric with respect to its midspan. We discuss the existence of a wing $\tilde{W}$ that can move tangentially to an arbitrary cylindrical surface G, having a fixed line of tangency $(A\text{–}A)$, without leaving behind free vorticity (Figure 4.2.2).

At each time, such a motion of $\tilde{W}$ can be described as a superposition of a motion of the wing tangent to itself and a rotation around the line of tangency $(A\text{–}A)$. The normal velocities at $\tilde{W}$ are caused only by the rotation around the line $(A\text{–}A)$ and are equal to $\omega\xi$, where ω is the momentary rotational velocity and ξ is the distance of the point under consideration to $(A\text{–}A)$. Therefore we consider the following related problem.

We assume the wing of Figure 4.2.2 lying in the plane $z = 0$, symmetrically with respect to the x-axis. The line of tangency $(A\text{–}A)$ is along the y-axis (Figure 4.2.3 (a)). We now define a wing $\tilde{V}$ without thickness and with its planform V (projection of $\tilde{V}$ on the plane $z = 0$) coinciding with $\tilde{W}$. The wing $\tilde{V}$ has parabolic profiles $(B\text{–}B)$, defined by $z = 1/2\ \varepsilon x^2$. We consider the linearized, stationary situation of placing $\tilde{V}$ in a homogeneous flow U in the negative x-direction. This means that we assume

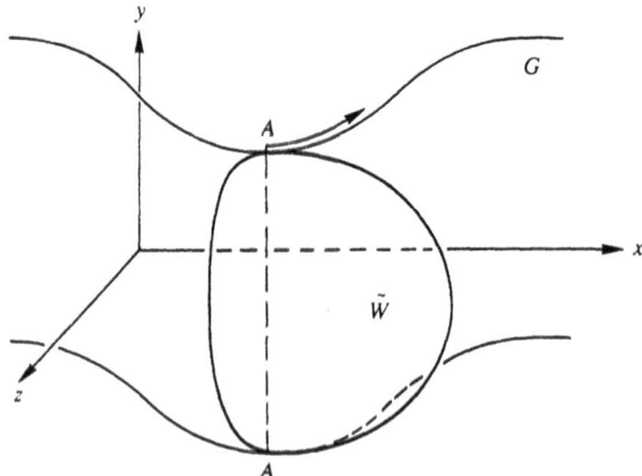

Fig. 4.2.2. Wing $\tilde{W}$ moving tangentially along cylindrical surface G.

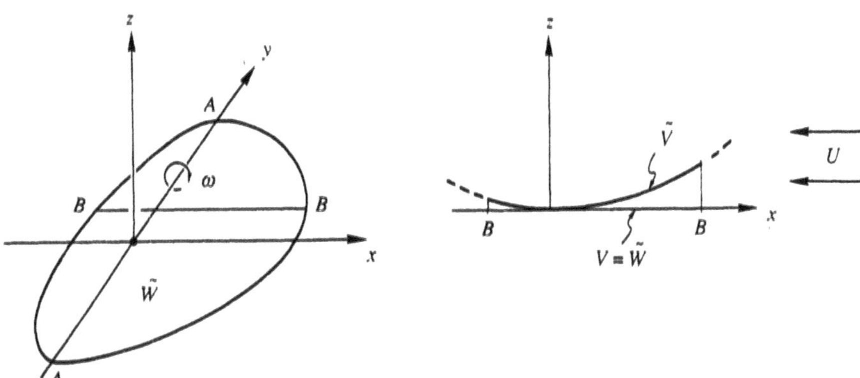

Fig. 4.2.3. Illustration of the stationary problem, (a) wing $\tilde{W}$ rotating around A–A, (b) cross section of wing $\tilde{V}$.

ε to be small and that we prescribe the normal velocity $v_n(x, y) = \varepsilon U x$ on the planform $V = \tilde{W}$ of $\tilde{V}$, which is apart from the constant factor $\varepsilon U/\omega$, the same as in the instationary case. Furthermore, linearization means that the vorticity on $\tilde{V}$ is placed at the planform V and that the free vorticity, which is formed behind $\tilde{V}$ stays in the plane $z = 0$ and does not change in the x-direction.

We now try to construct under a number of constraints, another wing $\tilde{V}^*$ with planform V^*, such that $\tilde{V}^*$ does not shed free vorticity when it is placed in the parallel flow U and is treated by a linear theory. The constraints are as follows. The new planform V^* is also symmetric with respect to the x-axis, it has the same span as V and its chord lengths are the same as those of V. The new wing $\tilde{V}^*$ has parabolic profiles which also are parts of $z = 1/2\ \varepsilon x^2$. Hence in the linearized theory we can also demand the normal velocities $v_n^*\ (x, y) = \varepsilon U x$ at V^*.

Suppose we have found such a wing $\tilde{V}^*$ which sheds no free vorticity, then its planform V^* can be used as a flat rigid wing $\tilde{W}^*(= V^*)$ at which we prescribe the normal velocity ωx, where we replace the small quantity εU of $O(\varepsilon)$ from the previous paragraph by ω which is $O(\varepsilon^0)$. This is allowed because there is no shed vorticity, so the linearized solution for the wing $\tilde{V}^*$ with camber, is an exact non-linear solution for the rotating flat wing $\tilde{W}^*$.

Now we have found a flat rigid wing $\tilde{W}^*$ of zero thickness which does not shed free vorticity, when gliding in the afore-mentioned way (Figure 4.2.2) along the cylindrical surface G. This is a possible base motion for $\tilde{W}^*$, the circulation around each chord of $\tilde{W}^*$ then is equal to zero. In order to create a thrust we have to superimpose on the base motion again an added motion by which free vorticity is shed. This vorticity is situated, in a semi-linear theory, at the wake of $\tilde{W}^*$. The wake is defined in the same way as we did for the wing of infinite span of Regime 3. Hence it consists of those fluid particles which have passed along the surface of $\tilde{W}^*$ when it carries out its base motion and the free vorticity is assumed to be transported by the velocity field of these particles.

Next we discuss shortly that it is probable that we can find a planform V^* and hence a wing $\tilde{W}^*$ with the previous properties. Looking at Figure 4.2.3 (b) we can move each profile (B–B) of the original wing $\tilde{V}$ to the left or to the right, along the dotted parabola $z = 1/2 \ \varepsilon x^2$. Moving B–B to the left means decreasing the mean angle of incidence and moving B–B to the right means increasing this angle. Somewhere in between, there must be a situation in which the lift at this chord equals zero, hence where the circulation around the chord is zero. By moving each profile in an appropriate way, we expect there will be one situation for which the circulation around all profiles vanishes at the same time. Then no vorticity is shed.

We will discuss this problem further in Section 4.8 and we will give there some examples of these wings $\tilde{W}^*$.

It is not clear if it is possible to frame a semi-linear theory for the large-amplitude propulsion by means of two rigid wings of finite span, moving in each others neighbourhood and hence influencing each other. The question is, can we find a base motion of such wings, by which no free vorticity is shed. Of course, there is much freedom in choosing their paths and the shape of their planforms. However, this question seems difficult to answer.

4.3. Small-Amplitude Propulsion, 2-Dimensional

We first discuss the 2-dimensional case of Regime (1.a) of Section 4.1, which is a classical problem. We mention for instance Wu [78], where also accelerated swimming motions are considered. The fluid is inviscid and incompressible. In Figure 4.3.1 we have drawn the swimming profile, without thickness and stretching from $x = -l$ towards $x = l$. The motion of the profile is given by

$$y = h(x,t) \ , \qquad |x| \leq l \ , \tag{4.3.1}$$

where $h(x,t)$ as well as its first and second derivatives with respect to x and t are assumed to exist and to be of $O(\varepsilon)$. The profile is placed in a parallel flow of velocity U which is $O(\varepsilon^0)$.

The thickness of the profile is neglected because in linearized theory it does not change the thrust or lift production. Its influence on the field of flow can be described by a continuous distribution of sources and sinks at $|x| \leq l$, $y = 0$, of which the velocities can be simply added to the flow described here. We remark that this does not hold for screw propellers of which the reference surfaces have a non-zero curvature of $O(\varepsilon^0)$ (Section 3.1) and of which the blades can influence each other sideways. Also it is not true when the profile of Figure 4.3.1 would have a large thickness of $O(\varepsilon^0)$, but then the theory is no longer purely linear. See for instance Uldrick and Siekmann [71].

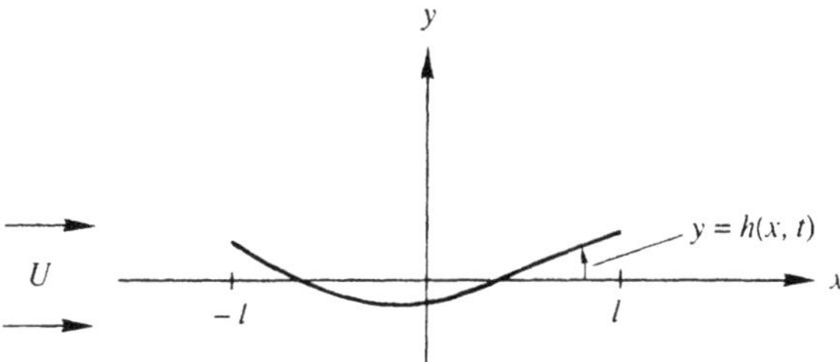

Fig. 4.3.1. 2-dimensional small-amplitude propulsion.

The disturbance velocity components v_x, v_y satisfy the linearized equations of motion (1.2.4) and the equation of conservation of mass

$$\frac{\partial v_x}{\partial t} + U \frac{\partial v_x}{\partial x} = -\frac{1}{\rho} \frac{\partial p}{\partial x} , \tag{4.3.2}$$

$$\frac{\partial v_y}{\partial t} + U \frac{\partial v_y}{\partial x} = -\frac{1}{\rho} \frac{\partial p}{\partial y} , \tag{4.3.3}$$

$$\text{div } (v_x, v_y) = \frac{\partial v_x}{\partial x} + \frac{\partial v_y}{\partial y} = 0 . \tag{4.3.4}$$

For convenience we introduce the function

$$\varphi(x, y, t) = -\frac{1}{\rho} p(x, y, t) , \tag{4.3.5}$$

called acceleration potential because its gradient yields the components of the acceleration of a fluid particle, as given by the left-hand sides of (4.3.2) and (4.3.3). Differentiation of (4.3.2) with respect to x and of (4.3.3) with respect to y and using (4.3.4) yields

$$\frac{\partial^2 \varphi}{\partial x^2} + \frac{\partial^2 \varphi}{\partial y^2} = 0 . \tag{4.3.6}$$

Next we introduce the complex variable $z = x + iy$ and the analytic function

$$f(z, t) = \varphi(x, y, t) + i\psi(x, y, t) , \tag{4.3.7}$$

where ψ is the complex conjugate function of φ. Hence φ and ψ satisfy the Cauchy-Riemann relations

$$\frac{\partial \varphi}{\partial x} - \frac{\partial \psi}{\partial y} = 0 , \qquad \frac{\partial \varphi}{\partial y} + \frac{\partial \psi}{\partial x} = 0 . \tag{4.3.8}$$

The conjugate complex velocity vector is denoted by

$$w(z, t) = v_x(x, y, t) - iv_y(x, y, t) . \tag{4.3.9}$$

Using (4.3.2) and (4.3.3) we find

$$\frac{\partial f}{\partial z} = \frac{\partial f}{\partial x} = \frac{\partial \varphi}{\partial x} + i\,\frac{\partial \psi}{\partial x} = \frac{\partial \varphi}{\partial x} - i\,\frac{\partial \varphi}{\partial y} = \frac{\partial w}{\partial t} + U\,\frac{\partial w}{\partial z} \ . \tag{4.3.10}$$

The boundary condition (1.2.16) for our 2-dimensional problem has the form

$$\left\{\frac{\partial}{\partial t} + (U + v_x)\frac{\partial}{\partial x} + v_y\frac{\partial}{\partial y}\right\}(y - h(x,t)) = 0 \ , \quad |x| < l\,, \ y = 0 \ . \tag{4.3.11}$$

Neglecting terms of $O(\varepsilon^2)$ we find

$$v_y(x,+0,t) = v_y(x,-0,t) = \left(\frac{\partial}{\partial t} + U\,\frac{\partial}{\partial x}\right)h(x,t) = V(x,t) \ , \qquad |x| < l \ , \tag{4.3.12}$$

where we introduced $V(x,t)$, which is a known function when $h(x,t)$ is prescribed and the first equality follows from the preservation of fluid. From the equation of motion (4.3.3) it follows by (4.3.12)

$$-\frac{\partial \psi}{\partial x}(x,0,t) = \frac{\partial \varphi}{\partial y}(x,0,t) = -\frac{1}{\rho}\,\frac{\partial p}{\partial y}(x,0,t) = \left(\frac{\partial}{\partial t} + U\,\frac{\partial}{\partial x}\right)V(x,t) \ , \qquad |x| < l \ , \tag{4.3.13}$$

hence also this expression is known for $-l < x < l$ when $h(x,t)$ is given. We remark that $\partial\varphi/\partial y = -\partial\psi/\partial x$ is continuous for all values of x, because v_y is continuous across any line y = const. We can represent the profile by a distribution of external "forces" parallel to the y-axis. These "forces" are, because our theory is 2-dimensional, in fact forces per unit of length in the x-direction and concentrated on lines perpendicular to the (x,y) plane. Then it follows from (1.4.4) that

$$\varphi(x,0,t) = 0 \ , \qquad l < |x| \ , \tag{4.3.14}$$

$$\varphi(x,-0,t) = -\varphi(x,+0,t) \ , \qquad |x| < l \ , \tag{4.3.15}$$

because the pressure results from a pressure dipole layer (D.2.7).

From (4.3.13) it follows that $\partial\psi/\partial x$ is continuous at the x-axis hence

$$\psi(x,-0,t) = \psi(x,+0,t) \ , \qquad |x| < \infty \ . \tag{4.3.16}$$

For Re $z = x \to -\infty$ we know that $w(z,t) \to 0$, hence by integration of (4.3.10)

$$f(z,t) = Uw(z,t) + \int_{-\infty}^{z} \frac{\partial w}{\partial t}(\zeta,t)\,d\zeta \ . \tag{4.3.17}$$

Comparing in this equation the imaginary parts we find

$$\psi(x,y,t) = -Uv_y(x,y,t) - \int_{-\infty}^{x} \frac{\partial}{\partial t} v_y(\xi,y,t)\, d\xi \ , \tag{4.3.18}$$

where we have chosen a simple path of integration in the complex plane, namely a straight line $y = \text{const}$.

We now consider (4.3.10) as a linear first order partial differential equation for $w(z,t)$. The solution of this equation can be written as

$$w(z,t) = \frac{1}{U} f(z,t) - \frac{1}{U^2} \int_{-\infty}^{z} \frac{\partial f}{\partial t}\left(\zeta, t + \frac{\zeta - z}{U}\right) d\zeta \ , \quad \zeta = \xi + iy \ . \tag{4.3.19}$$

Comparing imaginary parts of this equation, we obtain

$$v_y(x,y,t) = -\frac{1}{U}\psi(x,y,t) + \frac{1}{U^2}\int_{-\infty}^{x} \frac{\partial \psi}{\partial t}\left(\xi, y, t + \frac{\xi - x}{U}\right) d\xi \ . \tag{4.3.20}$$

Substitution of (4.3.12) into (4.3.18) for $y = 0$ and $|x| < l$, hence on the profile, yields

$$\begin{aligned} \psi(x,0,t) &= -\left(U\frac{\partial}{\partial x} + \frac{\partial}{\partial t}\right)\int_{-l}^{x} V(\xi,t)\, d\xi - \int_{-\infty}^{-l} \frac{\partial}{\partial t} v_y(\xi,0,t)\, d\xi \\ &= \psi_1(x,t) + A(t) \ , \qquad |x| < l \ , \end{aligned} \tag{4.3.21}$$

where

$$\psi_1(x,t) = -\left(U\frac{\partial}{\partial x} + \frac{\partial}{\partial t}\right)\int_{-l}^{x} V(\xi,t)\, d\xi \ , \tag{4.3.22}$$

is a known function for $|x| < l$ and the remaining part denoted by $A(t)$ is an unknown real function of time.

We now bring $A(t)$ in a different form. From (4.3.18) we have

$$\psi(-l,0,t) = -Uv_y(-l,0,t) - \int_{-\infty}^{-l} \frac{\partial}{\partial t} v_y(\xi,0,t)\, d\xi \ , \tag{4.3.23}$$

and from (4.3.20)

$$\psi(-l,0,t) = -Uv_y(-l,0,t) + \frac{1}{U}\int_{-\infty}^{-l} \frac{\partial \psi}{\partial t}\left(\xi, 0, t + \frac{\xi + l}{U}\right) d\xi \ . \tag{4.3.24}$$

Combining (4.3.23) and (4.3.24) with the definition of $A(t)$ (4.3.21) it is found that we can write

$$A(t) = \frac{1}{U} \int_{-\infty}^{-l} \frac{\partial \psi}{\partial t} \left(x, 0, t + \frac{x+l}{U} \right) dx \; . \tag{4.3.25}$$

With respect to the unknown function $f(z,t)$ (4.3.7) we have the following data. Its real part φ satisfies (4.3.14) and (4.3.15) and for its imaginary part ψ holds (4.3.21). From this it follows

$$f^+(z,t) + f^-(z,t) = 2i\,(\psi_1(x,t) + A(t)) \; , \qquad |x| < l \; , \tag{4.3.26}$$

$$f^+(z,t) - f^-(z,t) = 0 \; , \qquad |x| > l \; , \tag{4.3.27}$$

where "+" and "−" denote the limit of $f(z,t)$ for $y \to 0$ through positive and negative values, respectively. The type of problem stated in (4.3.26) and (4.3.27) is called a Hilbert problem for the function $f(z,t)$. We remark that $f(z,t)$ will be analytic in the whole complex plane with the exception of the line segment $|x| < l$, $y = 0$, where it exhibits a jump discontinuity. Such a function is called "sectionally holomorphic". In the next section we discuss this Hilbert problem.

4.4. Solution of the Hilbert Problem

The theory of the Hilbert problem is thoroughly discussed in [52]. For direct reference we mention Appendix A, where some results are derived without entering into details.

First we consider the homogeneous part of (4.3.26)

$$X^+(z) + X^-(z) = 0 \; , \qquad |x| < l \; . \tag{4.4.1}$$

The general solution of this equation which has an algebraic behaviour at infinity, a square root singularity at $x = -l$ and which is zero at $x = l$, has by (A.2.5) and (A.3.5) the form

$$X(z) = P(z) \left(\frac{z-l}{z+l} \right)^{1/2} \; , \tag{4.4.2}$$

where $P(z)$ is some polynomial. This choice of $X(z)$ already anticipates the demand that the Kutta condition has to be satisfied at the trailing edge $x = l$ of the profile, in other words that the pressure jump $p^+(x,t) - p^-(x,t) = -\rho\,(\varphi(x,+0,t) - \varphi(x,-0,t))$ vanishes for $x = l$. The definition of the square root in (4.4.2) is as follows. For real values of z hence $z \equiv x$, we assume for $x > l$

$$\left(\frac{z-l}{z+l} \right)^{1/2} = \left(\frac{x-l}{x+l} \right)^{1/2} > 0 \; . \tag{4.4.3}$$

Then $X(z)$ is uniquely defined in the complex z plane by analytic continuation, when the segment $|x| \leq l$, $y = 0$ is a cut in the plane. In the following a square root of a positive number is assumed to be positive.

By (A.2.7) it is seen that a solution (4.3.26) can be written as

$$f(z,t) = \frac{X_1(z)}{2\pi i} \int_{-l}^{l} \frac{2i\{\psi_1(\xi,t) + A(t)\}}{X_1^+(\xi)(\xi - z)} \, d\xi + X_2(z) \; , \tag{4.4.4}$$

where $X_1(z)$ and $X_2(z)$ are solutions of (4.4.1) of the form (4.4.2) with different choices of the polynomial $P(z)$ namely $P_1(z)$ and $P_2(z)$ respectively. It follows from (4.4.3) that

$$X_1^+(\xi) = iP_1(\xi)\left(\frac{l-\xi}{l+\xi}\right)^{1/2} \; , \qquad |\xi| < l \; . \tag{4.4.5}$$

For $X_1(z)$ we take $P_1(z) = (z + l)$ and for $X_2(z)$ we take $P_2(z) = C(t)$, which is a "time-dependent constant". We find for $f(z,t)$

$$f(z,t) = \frac{(z^2 - l^2)^{1/2}}{\pi i} \int_{-l}^{l} \frac{\{\psi_1(\xi,t) + A(t)\}}{(l^2 - \xi^2)^{1/2}(\xi - z)} \, d\xi + C(t)\left(\frac{z-l}{z+l}\right)^{1/2} \; . \tag{4.4.6}$$

These choices of $X_1(z)$ and $X_2(z)$ are not uniquely determined by the mathematical problem, however besides that the Kutta condition has to be satisfied they have to be such that the disturbance pressures tend to zero at infinity, or

$$\lim_{|z|\to\infty} f(z,t) = 0 \; . \tag{4.4.7}$$

Using the integrals

$$\int_{-l}^{l} \frac{d\xi}{(l^2 - \xi^2)^{1/2}} = \pi \; , \qquad \int_{-l}^{l} \frac{d\xi}{(l^2 - \xi^2)^{1/2}(\xi - z)} = -\frac{\pi}{(z^2 - l^2)^{1/2}} \; , \tag{4.4.8}$$

we find from (4.4.6) and (4.4.7)

$$C(t) = -\frac{i}{\pi} \int_{-l}^{l} \frac{\psi_1(\xi,t)}{(l^2 - \xi^2)^{1/2}} \, d\xi - iA(t) \; . \tag{4.4.9}$$

Substitution of (4.4.9) into (4.4.6) yields

$$f(z,t) = iA(t)\left\{1 - \left(\frac{z-l}{z+l}\right)^{1/2}\right\}$$
$$- \frac{i}{\pi} \int_{-l}^{l} \frac{\psi_1(\xi,t)}{(l^2 - \xi^2)^{1/2}} \left\{\frac{(z^2 - l^2)^{1/2}}{(\xi - z)} + \left(\frac{z-l}{z+l}\right)^{1/2}\right\} d\xi \; . \tag{4.4.10}$$

We now have to determine the still unknown function $A(t)$. Taking the imaginary part of (4.4.10), we find for $x < -l$ and $y = 0$,

$$\psi(x,0,t) = A(t)\left\{ 1 - \left(\frac{x-l}{x+l}\right)^{1/2} \right\}$$
$$-\frac{1}{\pi}\int_{-l}^{l} \frac{\psi_1(\xi,t)}{(l^2-\xi^2)^{1/2}}\left\{ -\frac{(x^2-l^2)^{1/2}}{(\xi-x)} + \left(\frac{x-l}{x+l}\right)^{1/2} \right\} d\xi \ . \tag{4.4.11}$$

Substitution of (4.4.11) into (4.3.25) yields

$$0 = \int_{-\infty}^{-l} \left[A'\left(t+\frac{x+l}{U}\right)\left(\frac{x-l}{x+l}\right)^{1/2} + \frac{1}{\pi}\int_{-l}^{l} \frac{\frac{\partial}{\partial t}\psi_1(\xi, t+\frac{x+l}{U})}{(l^2-\xi^2)^{1/2}} \right.$$
$$\left. \cdot\left\{ -\frac{(x^2-l^2)^{1/2}}{(\xi-x)} + \left(\frac{x-l}{x+l}\right)^{1/2} \right\} d\xi \right] dx \ , \tag{4.4.12}$$

where $A(t)$ at the left-hand side of (4.3.25) is cancelled by a part of the integral at the right-hand side after the substitution.

When we assume that for $t < t_0$ the profile is at rest and the homogeneous flow U is undisturbed, we have by (4.3.12), (4.3.22) and (4.3.25),

$$\psi_1(x,t) = 0 \ , \qquad A(t) = 0 \ , \qquad t < t_0 \ . \tag{4.4.13}$$

Furthermore, we can assume that from $t = t_0$ upto $t = t_1 > t_0$ the motion of the profile starts sufficiently smooth and that for $t > t_1$, we have

$$h(x, t+\tau) = h(x,t) \ , \qquad t > t_1 > t_0 \ , \tag{4.4.14}$$

hence the motion has become periodic.

When we replace x in (4.4.12) by $U(\xi - t) - l$ and ξ by x, we obtain

$$\int_{t_0}^{t} A'(\xi)K(t-\xi)\, d\xi \stackrel{\text{def}}{=} \int_{t_0}^{t} A'(\xi)\left\{\frac{U(\xi-t)-2l}{U(\xi-t)}\right\}^{1/2} d\xi$$
$$= \frac{1}{\pi}\int_{t_0}^{t}\int_{-l}^{l} \frac{\frac{\partial}{\partial \xi}\psi_1(x,\xi)}{(l^2-x^2)^{1/2}}$$
$$\cdot\left[\frac{\{(U(\xi-t)-l)^2-l^2\}^{1/2}}{(x-U(\xi-t)+l)} - \left\{\frac{U(\xi-t)-2l}{U(\xi-t)}\right\}^{1/2} \right] dx\, d\xi$$
$$\stackrel{\text{def}}{=} \Psi(t) \ , \tag{4.4.15}$$

where we introduced the kernel $K(t)$ and the known function $\Psi(t)$. This equation is an integral equation of Volterra of the first kind for $A'(\xi)$, see for instance [25].

We do not pursue this general problem but change to a more simple version by assuming $t_0 = -\infty$, $t_1 = -\infty$ and

$$h(x,t) = h(x)\, e^{j\omega t} \;, \quad h(x) = h_1(x) + jh_2(x) \;, \quad -\infty < t < \infty \;, \tag{4.4.16}$$

where j is the imaginary unit in the time domain. Hence we consider a simple harmonic motion. In order to have convergence of a number of integrals which follow and to be able to carry our partial integrations when necessary, we assume that ω has a small negative imaginary (j) part. This means physically that the motion of the profile starts very smoothly "at" $t = -\infty$ and that all disturbances tend exponentially to zero for $t \to -\infty$. We tacitly take the limit $\operatorname{Im}\omega \to 0$ in the results, so that these belong to a simple periodic motion (4.4.16) with a real value of ω.

Also we assume

$$A(t) = A\, e^{j\omega t} \;, \tag{4.4.17}$$

A being an unknown constant.

We first calculate the left-hand side of (4.4.15). Substitution of (4.4.17) and $t_0 = -\infty$ into this left-hand side yields

$$\int_{-\infty}^{t} j\omega A\, e^{j\omega\xi} K(t-\xi)\, d\xi = j\omega A e^{j\omega t} \int_{0}^{\infty} e^{-j\omega\eta} K(\eta)\, d\eta$$

$$= j\alpha A e^{j(\alpha+\omega t)} \int_{1}^{\infty} e^{-j\alpha\xi} \left\{ \frac{1}{(\xi^2-1)^{1/2}} + \frac{\xi}{(\xi^2-1)^{1/2}} \right\} d\xi \;,$$

$$\alpha = \frac{\omega l}{U} \;, \tag{4.4.18}$$

where α is the dimensionless reduced frequency. Making use of the following relations for Hankel functions ([74], pp. 45, 170)

$$H_0^{(2)}(x) = \frac{2j}{\pi} \int_{1}^{\infty} \frac{e^{-jx\xi}}{(\xi^2-1)^{1/2}}\, d\xi \;, \quad (\operatorname{Re} x > 0) \;, \tag{4.4.19}$$

$$\frac{d}{dx} H_0^{(2)}(x) = -H_1^{(2)}(x) \;, \tag{4.4.20}$$

we write (4.4.18) as

$$\tfrac{1}{2}\, \pi\alpha A\, e^{j(\alpha+\omega t)} \{ H_0^{(2)}(\alpha) - jH_1^{(2)}(\alpha) \} \;. \tag{4.4.21}$$

Next we consider the right-hand side of (4.4.15). Introducing $V(x)$ and $\psi_1(x)$ by assuming

$$V(x,t) = V(x)\, e^{j\omega t} \ , \qquad \psi_1(x,t) = \psi_1(x)\, e^{j\omega t} \ , \tag{4.4.22}$$

we find by substituting $\xi = l(-\zeta+1)/(U+t)$ into (4.4.15), after replacing x by ξ

$$\Psi(t) = -\frac{j\alpha}{\pi}\, e^{j(\alpha+\omega t)} \int\limits_{-l}^{l} \left[\int\limits_{1}^{\infty} \frac{\psi_1(\xi)\, e^{-j\alpha\zeta}}{(l^2-\xi^2)^{1/2}} \cdot \left\{ \left(\frac{\zeta+1}{\zeta-1}\right)^{1/2} - \frac{l(\zeta^2-1)^{1/2}}{(\xi+l\zeta)} \right\} d\zeta \right] d\xi \ . \tag{4.4.23}$$

Using (4.4.19) and (4.4.20) we write the first term at the right-hand side of (4.4.23) as

$$-\tfrac{1}{2}\alpha\, e^{j(\alpha+\omega t)} \{H_0^{(2)}(\alpha) - jH_1^{(2)}(\alpha)\} \int\limits_{-l}^{l} \frac{\psi_1(\xi)}{(l^2-\xi^2)^{1/2}}\, d\xi \ . \tag{4.4.24}$$

By (4.3.22), we can write the second term at the right-hand side of (4.4.23) as

$$-\frac{j\alpha}{\pi}\, e^{j(\alpha+\omega t)} l \int\limits_{-l}^{l} \left\{ \frac{UV(\xi) + \omega j \int_{-l}^{\xi} V(\sigma)\, d\sigma}{(l^2-\xi^2)^{1/2}} \right\} \cdot \left\{ \int\limits_{1}^{\infty} \frac{e^{-j\alpha\zeta}(\zeta^2-1)^{1/2}}{(\xi+l\zeta)}\, d\zeta \right\} d\xi \ . \tag{4.4.25}$$

From this expression we consider again the second term. In this term we carry out a partial integration with respect to ζ and use the identity

$$\frac{-l}{(l^2-\xi^2)^{1/2}} \frac{\partial}{\partial\zeta} \frac{(\zeta^2-1)^{1/2}}{(\xi+\zeta l)} = \frac{1}{(\zeta^2-1)^{1/2}} \frac{\partial}{\partial\xi} \frac{(l^2-\xi^2)^{1/2}}{(\xi+\zeta l)} \ , \tag{4.4.26}$$

then, after some arrangements, we can write (4.4.25) as

$$-\frac{j\alpha U}{\pi l}\, e^{j(\alpha+\omega t)} \int\limits_{-l}^{l} \frac{V(\xi)}{(l^2-\xi^2)^{1/2}} \left[\int\limits_{1}^{\infty} e^{-j\alpha\zeta} \left\{ \frac{l\zeta}{(\zeta^2-1)^{1/2}} - \frac{\xi}{(\zeta^2-1)^{1/2}} \right\} d\zeta \right] d\xi$$

$$= \frac{\alpha U}{2l}\, e^{j(\alpha+\omega t)} \int\limits_{-l}^{l} \frac{V(\xi)}{(l^2-\xi^2)^{1/2}} \{jlH_1^{(2)}(\alpha) + \xi H_0^{(2)}(\alpha)\} d\xi \ . \tag{4.4.27}$$

Now (4.4.21) is equal to the sum of (4.4.24) and (4.4.27), from which the former unknown constant A follows as

$$A = -\frac{1}{\pi} \int_{-l}^{l} \frac{\psi_1(\xi)}{(l^2 - \xi^2)^{1/2}} \, d\xi$$

$$+ \frac{U}{\pi l} \int_{-l}^{l} \frac{V(\xi)}{(l^2 - \xi^2)^{1/2}} \frac{\{jlH_1^{(2)}(\alpha) + \xi H_0^{(2)}(\alpha)\}}{\{H_0^{(2)}(\alpha) - jH_1^{(2)}(\alpha)\}} \, d\xi \; . \qquad (4.4.28)$$

Herewith the problem is solved explicitly. The function $f(z)$ follows from (4.4.6), (4.4.9) and (4.4.28).

4.5. Thrust and Efficiency of 2-Dimensional Small-Amplitude Propulsion

In Section 4.3 we introduced the imaginary unit i with $(i)^2) = -1$, in the complex space domain. In Section 4.4 we introduced the complex unit j with $(j)^2 = -1$, in connection with the time dependency of the shape (4.4.16) of the swimming profile. These different imaginary units have no interaction, we leave unaltered the product ij.

The reason that j could be introduced is that the theory is linear and hence both the real part and the imaginary part, with respect to j of the complex solution are each a solution of the problem. When we consider however non-linear quantities such as the thrust and the kinetic energy which are both $O(\varepsilon^2)$ it can, not always, give wrong results to keep the complex j notation. Then quadratic terms are apt to occur in which $j^2 = -1$, hence there can arise a non-physical coupling between the real and the imaginary part of the complex solution. Therefore, we now consider real motions of the profile, given by

$$\tilde{h}(x,t) = \text{Re}_j \; \left(h(x)\, e^{j\omega t}\right) = h_1(x) \cos \omega t - h_2(x) \sin \omega t \; . \qquad (4.5.1)$$

In order to know with respect to which imaginary unit or units the real part of an expression has to be taken, we attach to the symbol Re the index i or j or both indices. In the following, when necessary for clearness, quantities which belong to $\tilde{h}(x,t)$ are provided with a tilde.

The pressures at the upper and the lower side of the profile follow from (4.3.5)

$$\tilde{p}(x, \pm 0, t) = -\rho\, \tilde{\varphi}(x, \pm 0, t) \overset{\text{def}}{=} -\rho\, \text{Re}_{ij}\, f^{\pm}(z,t) \; , \qquad |x| < l \; , \qquad (4.5.2)$$

where $f(z,t)$ is given by (4.4.10), $h(x,t)$ by (4.4.16) and $A(t)$ follows from (4.4.17) and (4.4.28). In connection with (4.4.3) we have

$$\left\{ \left(\frac{z-l}{z+l}\right)^{1/2} \right\}^{\pm} = \pm i \left(\frac{l-x}{l+x}\right)^{1/2} ,$$

$$\{(z^2 - l^2)^{1/2}\}^{\pm} = \pm i \, (l^2 - x^2)^{1/2} \; , \qquad |x| < l \; , \quad y = 0 \; . \qquad (4.5.3)$$

Then because both $\psi_1(x,t)$ and $A(t)$ are real with respect to i, we find

$$\tilde{\varphi}(x,\pm 0,t) = \pm\tilde{A}(t)\left(\frac{l-x}{l+x}\right)^{1/2}$$

$$\pm\frac{1}{\pi}\int_{-l}^{l}\frac{\tilde{\psi}_1(\xi,t)}{(l^2-\xi^2)^{1/2}}\left\{\frac{(l^2-\xi^2)^{1/2}}{(\xi-x)}+\left(\frac{l-x}{l+x}\right)\right\}d\xi\ ,\quad |x|<l\ , \tag{4.5.4}$$

where

$$\tilde{\psi}_1(\xi,t) = \mathrm{Re}_j\ \psi_1(\xi,t)\ ,\qquad \tilde{A}(t) = \mathrm{Re}_j A(t)\ . \tag{4.5.5}$$

The thrust $T(t)$ per unit of span induced by the profile motion (4.5.1), reckoned positive in the negative x-direction, becomes

$$T(t) = -\int_{-l}^{l}[\tilde{p}]_-^+\frac{\partial\tilde{h}}{\partial x}(x,t)\,dx+\frac{1}{4}\rho\,\pi\,\tilde{Q}^{\,2}(t)\ , \tag{4.5.6}$$

$[\tilde{p}]_-^+ = (\tilde{p}(x,+0,t)-\tilde{p}(x,-0,t))$ and the second term at the right-hand side is the suction force (1.21.13), $\tilde{Q}(t)$ being the coefficient of the square root singularity of the vorticity at the leading edge (1.21.1). The coefficient $\tilde{Q}(t)$ can be calculated as follows. First we observe that the pressure jump over the profile equals

$$[\tilde{p}]_-^+ = -\rho\,U\tilde{\Gamma}_b(x,t)\ , \tag{4.5.7}$$

where $\tilde{\Gamma}_b(x,t)$ is by definition the bound vorticity of the profile. In order to calculate the total vorticity $\tilde{\Gamma}_{\mathrm{tot}}(x,t)$ on the profile at the point x and at the time t, we have to add to $\tilde{\Gamma}_b(x,t)$ the floating vorticity shed at appropriate times by $\tilde{\Gamma}_b(\xi,t)$ situated in between the leading edge and x, hence $-l<\xi<x$. We obtain

$$\tilde{\Gamma}_{\mathrm{tot}}(x,t) = \tilde{\Gamma}_b(x,t)-\frac{1}{U}\int_{-l}^{x}\frac{\partial\tilde{\Gamma}_b}{\partial t}\left(\xi,t-\frac{(x-\xi)}{U}\right)d\xi\ . \tag{4.5.8}$$

However,

$$\lim_{x\to -l}\int_{-l}^{x}\frac{\partial\tilde{\Gamma}_b}{\partial t}\left(\xi,t-\frac{(x-\xi)}{U}\right)d\xi = 0\ , \tag{4.5.9}$$

because $\partial\tilde{\Gamma}_b(x,t)/\partial t$ has at most a square root singularity at $x=-l$, which is absolutely integrable. This means that the singular behaviour of $\tilde{\Gamma}_b(x,t)$ and $\tilde{\Gamma}_{\mathrm{tot}}(x,t)$ for $x\to -l$ is the same. Hence using (4.3.15), (4.5.7) and (4.5.2), we find

$$Q(t) = 2\lim_{x\to -l}\frac{(l+x)^{1/2}}{U}\,\tilde{\varphi}(x,+0,t)$$

$$= \frac{2(2l)^{1/2}}{U}\left\{\tilde{A}(t)+\frac{1}{\pi}\int_{-l}^{l}\frac{\tilde{\psi}_1(\xi,t)}{(l^2-\xi^2)^{1/2}}\,d\xi\right\}\ . \tag{4.5.10}$$

When we denote by $k(x,t)$ the external force field parallel to the y-axis, exerted by the profile on the fluid, the power $P(t)$ per unit of span needed to maintain the motion of the profile becomes

$$P(t) = \int_{-l}^{l} k(x,t) \frac{\partial \tilde{h}}{\partial t} (x,t)\, dx = \int_{-l}^{l} [\tilde{p}]_{-}^{+} \frac{\partial \tilde{h}}{\partial t}(x,t)\, dx \ . \tag{4.5.11}$$

We now consider the kinetic energy dE/dt shed per unit of time into the fluid. To this end we assume the fluid to be at rest, while the profile moves with the velocity U in the negative x-direction. Then dE/dt equals the power applied to the fluid (Section 1.17), which is equal to the power $P(t)$ applied to the profile minus the useful work $UT(t)$. In fact this useful work is extracted again out of the system consisting of fluid and profile by some apparatus which keeps the velocity of the profile equal to U. Hence

$$\frac{dE}{dt} = P(t) - UT(t) \ . \tag{4.5.12}$$

From (1.17.9) however, because the action of the profile on the fluid can be represented by the external force field $k(x,t)$ as introduced above, it follows that we also have the representation

$$\frac{dE}{dt} = \int_{-l}^{l} k(x,t)\, v_y(x,0,t)\, dx \ . \tag{4.5.13}$$

The question arises: are (4.5.12) and (4.5.13) equivalent? In order to make this clear, we use a continuity argument based on the last two paragraphs of Section 1.21. When the vorticity on the profile is changed in a very small one-sided $\tilde{\varepsilon}$ neighbourhood of the leading edge (Figure 1.21.3 (a)), so that it is zero at this edge, we have physically practically the same situation as before, but we lost the suction force in (4.5.6). Now this force is distributed over a small neighbourhood of the leading edge and is taken care of by the integral (4.5.13) to any degree of accuracy, by taking $\tilde{\varepsilon}$ sufficiently small. Hence we can compare (4.5.12) and (4.5.13) neglecting the suction force. However, the boundary condition at the profile reads

$$v_y = U\frac{\partial h}{\partial x} + \frac{\partial h}{\partial t} \ , \tag{4.5.14}$$

from which by substitution into (4.5.13) the desired equality follows.

Introducing the mean values of the thrust and the power over one period by

$$\bar{T} = \frac{\omega}{2\pi} \int_{0}^{2\pi/\omega} T(t)\, dt \ , \qquad \bar{P} = \frac{\omega}{2\pi} \int_{0}^{2\pi/\omega} P(t)\, dt \ , \tag{4.5.15}$$

we find for the efficiency of the propulsive profile

$$\eta = U\bar{T}/\bar{P} = 1 - O(\varepsilon^0) \ . \tag{4.5.16}$$

Finally, for the sake of completeness we write the formula for the lift $L(t)$ delivered by the profile

$$L(t) = -\int_{-l}^{l} [p]_-^+ \, dx \ . \tag{4.5.17}$$

4.6. Theoretical and Experimental Results

In order to illustrate the previous theory of small-amplitude propulsion which describes the 2-dimensional case of Regime (1.a), we show some theoretical results which are compared with experimental ones as described by Kelly, Rentz and Siekmann in [37]. The motion of the profile (4.5.1) is given by

$$\tilde{h}(x,t) = \text{Re}_j \ l\{(C_0 + C_1 x + C_2 x^2)\, e^{-j\sigma x}\, e^{j\omega t}\} \ , \qquad |x| \leq l \ . \tag{4.6.1}$$

In Figures 4.6.1–4.6.3 is drawn the dimensionless thrust coefficient, defined by

$$C_T = \frac{\bar{T}}{\pi\rho\, U^2 l W} \ , \tag{4.6.2}$$

versus the dimensionless reduced frequency $\alpha = \omega l/U$, where in (4.6.2) W is the width of the plate. In order to simulate a 2-dimensional flow, the plate is situated between two walls a distance W apart.

First in Figure 4.6.1 the purely heaving motion of a flat rigid metal plate is considered. Then the thrust is caused only by the suction force at the leading edge of the plate. Besides a small systematic deviation caused by the resistance induced by the turbulent boundary layer of the slightly viscous fluid, the trend of the experiments is in good agreement with the theory.

In Figure 4.6.2 the results for a flexible metal profile are given. The deformation is of constant amplitude and its wavelength equals the chord length. For values $\alpha < 3$, there is some discrepancy between theory and measurement. For that region the wave speed of the waves of the plate is less than the free stream velocity U. It was observed in the experiments by means of dye flow patterns, that flow separation occurred in the lee of the peaks of the waves of the plate.

Flow separation did not occur in the case of a wave motion of the plate with quadratic amplitude, at least for values of $\alpha > 1$. This is shown in Figure 4.6.3.

For more information with respect to other profile motions and about the efficiency of the small-amplitude motion we refer to Wu [79] where also many references are given. For the influence of finite thickness of a profile on its propulsive capacity we mention again [71].

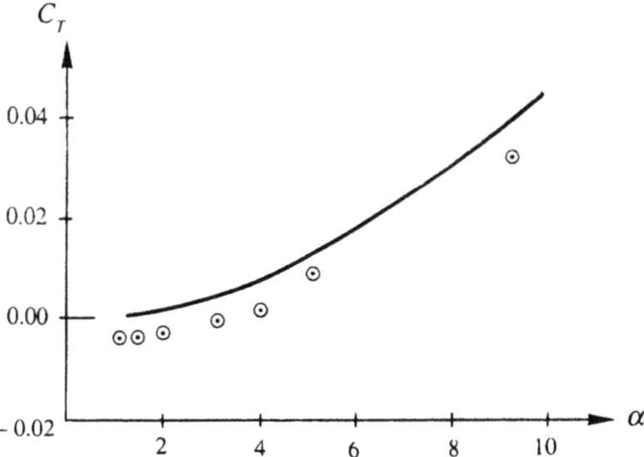

Fig. 4.6.1. Thrust coefficient C_T as a function of reduced frequency α, heaving motion ($C_0 = 1/24$, $C_1 = C_2 = 0$, $\sigma = 0$) of rigid plate, — theory, ⊙ experiment.

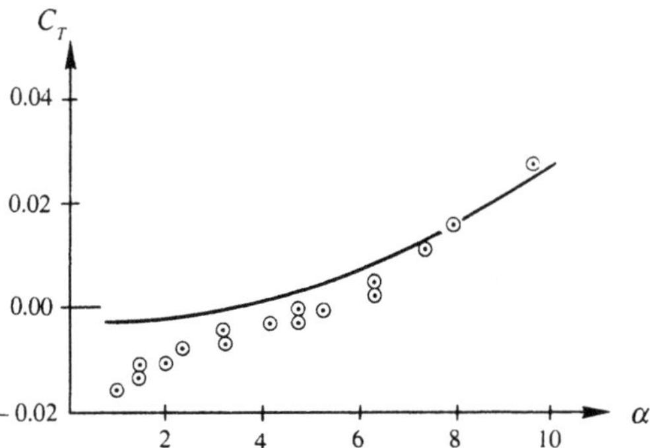

Fig. 4.6.2. Thrust coefficient C_T as a function of reduced frequency α, wave motion with constant amplitude (C_0= 1/12, $C_1 = C_2 = 0$, $\sigma = \pi$), — theory, ⊙ experiment.

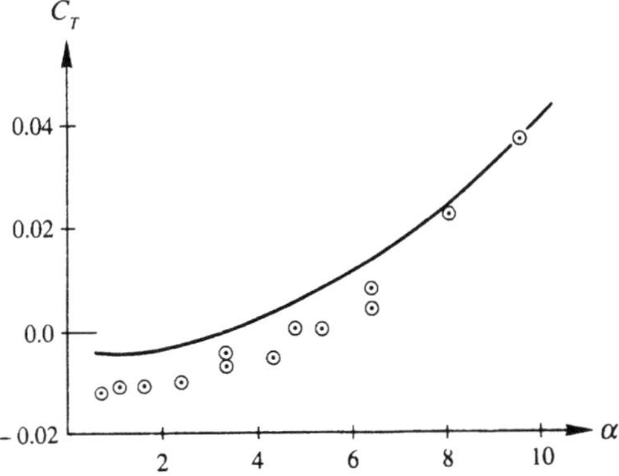

Fig. 4.6.3. Thrust coefficient C_T as a function of reduced frequency α, wave motion with quadratic amplitude ($C_0 = 0.023$, $C_1 = 0.042$, $C_2 = 0.034$, $\sigma = \pi$), — theory, ⊙ experiment.

4.7. Large-Amplitude Unsteady Propulsion, Rigid Profile

We will discuss an example of Regime (3) more closely and consider the 2-dimensional problem of a rigid flat profile without thickness of chord length $2l$. Along the profile we introduce a length parameter s, which is zero at the midpoint M of the chord and which is reckoned positive in the direction of the leading edge. Hence $s = -l$ at the trailing edge A of the profile and $s = l$ at its leading edge B (Figure 4.7.1). At the profile we choose a point R with $s = b$, this point is forced to move along some prescribed periodic line represented by the one-valued periodic function $y = f(x) = O(\varepsilon^0)$ with period $2h$. This point R represents the pivotal axis by which the angular position α_0 of the profile is governed. The velocity of R along $y = f(x)$ can be described arbitrarily in a periodic way.

First we describe the "base motion" of the profile (A–B) which is, as we defined in Section 4.2, the motion for which the circulation around the profile remains zero. This unsteady propeller belongs to Regime (3), so we have to determine the direction m and the line r which is perpendicular to m, belonging to this flat profile. The direction m for which the circulation of the profile is zero, is clearly parallel with the profile. Now we have to find one point Q of the line r. We look for such a point on (A–B) itself and denote its parameter value by $s = a$. When we rotate the profile around Q with rotational velocity ω ($\omega > 0$, anti-clockwise), the velocities of the points of (A–B) are in the direction of the normal $\vec{n}$ and amount to

$$v_n(s) = \omega(s - a) \ . \tag{4.7.1}$$

The vorticity $\gamma_0(s)$ per unit of length, needed on the profile in order that the fluid flows along it, has to satisfy

$$v_n(s) = \frac{1}{2\pi} \oint_{-l}^{l} \frac{\gamma_0(\sigma)}{(s - \sigma)} \, d\sigma \ . \tag{4.7.2}$$

The solution of this singular integral equation, which satisfies the Kutta condition at the trailing edge $s = -l$, hence with $\gamma(-l) = 0$, is (A.3.9)

$$\gamma_0(s) = 2\omega(a + l - s) \left(\frac{l + s}{l - s} \right)^{1/2} \ . \tag{4.7.3}$$

The condition that the total circulation is zero yields

$$\int_{-l}^{l} \gamma_0(s) \, ds = \pi \omega l \, (l + 2a) = 0 \ , \tag{4.7.4}$$

hence

$$a = -\tfrac{1}{2} l \ . \tag{4.7.5}$$

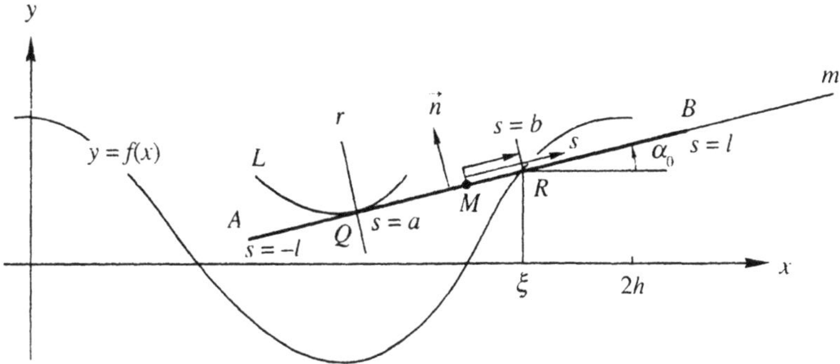

Fig. 4.7.1. Profile $(A\text{–}B)$ moving with point R along $y = f(x)$, base motion.

This means that Q is the well-known three-quarter chord point of the profile. Hence when the flat profile moves with its three-quarter chord point along an arbitrary line L and is tangent to L, its circulation remains identically equal to zero. In our case this can be realized in the following way. The point R $(s = b)$ was assumed to move along the periodic line $y = f(x)$. We start with the profile in some position, for instance the one of Figure 4.7.1, where the angle of the profile with the x-axis has some value α_0. At the three-quarter point Q we attach a little wheel with its axis of rotation along the line r. This wheel rolls on the (x, y) plane and forces the profile to move in the momentary direction m of $(A\text{–}B)$. In other words, when the point R moves along $y = f(x)$, the profile behaves as a bicycle of which the front-wheel at R is guided along $y = f(x)$ and the back-wheel at Q follows and describes the line L of Figure 4.2.1. If a profile carries out this motion, its circulation will remain zero. The line L we obtain in this way depends on the initial position α_0 of the profile. If we can find an initial position such that L is periodic in the x-direction, we have a simple characterization of the base motion of a periodically behaving sculling propeller.

Because the periodic line L is essential for the theory, we will sketch a proof of its existence due to F. Takens (private communication). Consider a position of the line segment $(Q\text{–}R)$ which is vertical above or below any point $R = (x^*, y^* = f(x^*))$ of the line $y = f(x)$. Then it is not difficult to show that by increasing x^* a little, the point Q will lag behind the point R when R moves along $y = f(x)$. From this it follows that, because $y = f(x)$ is uniquely valued, the point Q, when it is behind the point R, can never come in front of R when R moves along $y = f(x)$ in the positive x-direction.

Now consider in Figure 4.7.2 the line segment $(Q\text{–}R)$ in an arbitrary position while $R = (0, f(0))$, however with Q to the left of the y-axis. The point R moves along $y = f(x)$, starting at $(0, f(0))$ and ending one period further at $(2h, f(2h)) = (2h, f(0))$. The point Q then follows the point R with a velocity directed along $(Q\text{–}R)$. After this $(Q\text{–}R)$ assumes a new position $(Q^{(1)}\text{–}R^{(1)})$. This can be done for all possible starting positions of Q at the semi-circle I, the end positions $Q^{(1)}$ are lying at the semi-circle II. From the first part of this paragraph it follows that the vertical

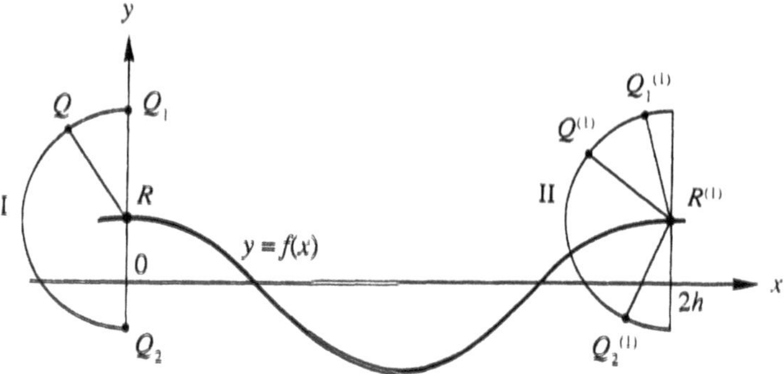

Fig. 4.7.2. Existence proof of periodic line L.

segments (Q_1–R) and (Q_2–R) obtain after one period new positions ($Q_1^{(1)}$–$R^{(1)}$) and ($Q_2^{(1)}$–$R^{(1)}$) which lie to the left of the vertical line $x = 2h$. Hence the semi-circle I is mapped on part of the semi-circle II. Besides we know that the points $Q^{(1)}$ at II have the same ordering as the corresponding points Q at I. Suppose this is not true, then, when we let all segments (Q–R) with Q at I move at the same time, two of them have to coincide at some moment and afterwards they would not carry out the same motion, which is impossible. This shows that the ordering of the points Q is preserved.

We now repeat the mentioned procedure any number of times by letting the point R move along $y = f(x)$ to points $(2nh, f(2nh)) = (2nh, f(0))$, $n = 2, 3, \ldots$ Then the original points Q_1 and Q_2 obtain positions $Q_1^{(n)}$ and $Q_2^{(n)}$ after each period. There are two possibilities, either the points $Q_1^{(n)}$ and $Q_2^{(n)}$ tend to each other or they remain apart and tend to two different positions in a monotonic way. In both cases the limit points yield starting positions of (Q–R) which are reproduced after one period.

Next we will derive the differential equations for the line L in terms of the prescribed path $y = f(x)$ of the pivotal point R of the profile and the differential equations for the wake H^* or H of the profile as defined for Regime (3).

We denote the coordinates of R by $(\xi, f(\xi))$, then ξ is a parameter for L and we denote the coordinates of Q by $x_Q = x_Q(\xi)$ and $y_Q = y_Q(\xi)$. The condition of Q moving along the unknown line L, while (A–B) is tangent to L, is

$$\frac{y'_Q(\xi)}{x'_Q(\xi)} = \frac{(f(\xi) - y_Q(\xi))}{(\xi - x_Q(\xi))} , \tag{4.7.6}$$

and the condition that the distance between Q and R is $(\frac{1}{2}\,l + b)$ is

$$\{\xi - x_Q(\xi)\}^2 + \{f(\xi) - y_Q(\xi)\}^2 = (\tfrac{1}{2}\,l + b)^2 . \tag{4.7.7}$$

Differentiating (4.7.7) with respect to ξ and eliminating $y_Q(\xi)$ and $y'_Q(\xi)$ from the resulting equation and from (4.7.6) and (4.7.7) yields for $x_Q(\xi)$ the non-linear differential equation

$$(\tfrac{1}{2}\,l+b)^2 x_Q' - (\xi - x_Q)^2 \\ - f'(\xi)(\xi - x_Q)\{(\tfrac{1}{2}\,l+b)^2 - (\xi - x_Q)^2\}^{1/2} = 0 \ . \tag{4.7.8}$$

Also we have the periodicity condition for L,

$$x_Q(2h) = 2h + x_Q(0) \ . \tag{4.7.9}$$

We know already by the foregoing that there exists at least one solution $x_Q = x_Q(\xi)$ of (4.7.8) under the condition (4.7.9). When we have calculated such a solution, we obtain $y_Q = y_Q(\xi)$ from (4.7.7). Then the line L is determined, by elimination of ξ, we write L as $y = g(x)$. Now we can characterize the base motion of the profile by its sliding tangently along L while touching L at its three-quarter chord point Q hence for $s = -\frac{1}{2}\,l$. Also we can calculate the angle $\alpha_0 = \alpha_0(x)$ which the profile carrying out the base motion, forms with the x-axis, when Q has the coordinates $(x, g(x))$. From now on we assume $L(y = g(x))$ and $\alpha_0(x)$ to be known.

The vorticity $\gamma_0(s,t)$ at the profile, belonging to the base motion, is by (4.7.3) and (4.7.5)

$$\gamma_0(s,t) = \omega(l-2s)\left(\frac{l+s}{l-s}\right)^{1/2} , \quad \omega = \frac{d}{dt}\,\alpha_0(x_Q) = \alpha_0'(x_Q)\,\frac{dx_Q}{dt} \ , \tag{4.7.10}$$

where dx_Q/dt follows from the prescribed velocity of the pivotal point R. By means of the law of Biot and Savart we can at each moment determine the velocity field induced by the base motion of the profile and we can calculate the trajectories passed through by the fluid particles. The trajectories $(x(t), y(t))$ of the fluid particles have to satisfy the two ordinary differential equations

$$\dot{x}(t) = v_x(x(t), y(t), t) \\ = -\frac{\omega}{2\pi}\int_{-l}^{l}(l-2s)\left(\frac{l+s}{l-s}\right)^{1/2}\frac{\{y - y_Q - (s+\frac{1}{2}\,l)\sin\alpha_0\}}{R^2}\,ds \ , \tag{4.7.11}$$

$$\dot{y}(t) = v_y(x(t), y(t), t) \\ = \frac{\omega}{2\pi}\int_{-l}^{l}(l-2s)\left(\frac{l+s}{l-s}\right)^{1/2}\frac{\{x - x_Q - (s+\frac{1}{2}\,l)\cos\alpha_0\}}{R^2}\,ds \ , \tag{4.7.12}$$

where

$$R^2 = \{x - x_Q - (s+\tfrac{1}{2}\,l)\cos\alpha_0\}^2 + \{y - y_Q - (s+\tfrac{1}{2}\,l)\sin\alpha_0\}^2 \ , \tag{4.7.13}$$

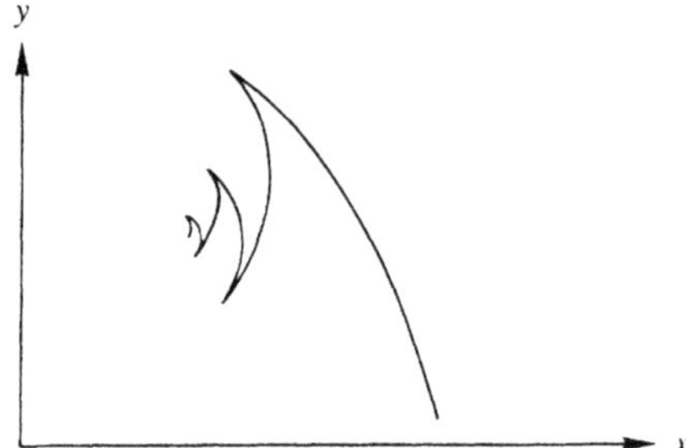

Fig. 4.7.3. Impression of the trajectory of one fluid particle, base motion.

and where we consider x_Q, y_Q and α_0 to be functions of t.

Now we reformulate the concept wake of the base motion. A point (x^*, y^*) belongs at time t^* to the wake, if there exists a solution $(\tilde{x}(t), \tilde{y}(t))$ of (4.7.12) and (4.7.13) with

$$x^* = \tilde{x}(t^*) \ , \qquad y^* = \tilde{y}(t^*) \ , \tag{4.7.14}$$

and if there exists $t' \leq t^*$ such that $(\tilde{x}(t'), \tilde{y}(t'))$ are the coordinates of the trailing edge at the moment t'. Hence, as has been said previously, the wake (H^* and H, Section 4.2) consists of those fluid particles which once have passed along the profile performing its base motion and left the profile at its trailing edge.

The numerical calculation of the wake is not difficult. We consider a fluid particle which at $t = t_j$ leaves the trailing edge and calculate its motion by (4.7.11) and (4.7.12). This has to be done for a large number of times $j = 1, 2, 3, \ldots$, the particles which left at these times the trailing edge then form the wake. In the neighbourhood of the trailing edge the wake is still in motion and was denoted by H^*, however when the profile moves on, the particles will come to rest and form far behind the profile the ultimate wake H which again is periodic.

It is interesting to look at the trajectory of an arbitrary fluid particle under influence of the base motion in case of the flat plate. An example is given in Figure 4.7.3. The sharp cusps occur at the moment that the three-quarter chord point Q of the flat profile of zero thickness passes a point of inflection of the line L, then the fluid comes to rest and starts to move in the inverse direction. Of course, the particles which have passed along the profile have the same type of motion. However, the wake, which is the line passing through these fluid particles, is smooth.

From the above it follows that besides the line $L(y = g(x))$ and the angle $\alpha_0(x)$, we also have to calculate the time-dependent part H^* of the wake and its ultimate periodic shape H far behind the profile.

Now, as we mentioned, we have to choose in the semi-linear theory an added motion of the profile in the neighbourhood of the base motion, which deviates only by a small amount of $O(\varepsilon)$ from it. For instance, we can define the added motion by a small periodic rotation $\alpha_1 = \alpha_1(x_Q)$ of the profile around the pivotal point R

(Figure 4.7.1). Of course, more general added motions are possible. The shed free vorticity which is of $O(\varepsilon)$ is then transported with the velocity $\dot{x}(t), \dot{y}(t))$ given by (4.7.11) and (4.7.12).

Suppose we have made a choice for $\alpha_1(x_Q)$, then we will need at the profile besides $\gamma_0(s,t)$ (4.7.10) which is of $O(\varepsilon^0)$, a small extra amount of vorticity $\gamma_1(s,t)$ of $O(\varepsilon)$, in order to let the fluid pass along it. Using the law of Biot and Savart we have the following singular integral equation for $\gamma_1(s,t)$

$$\frac{1}{2\pi} \oint_{-l}^{l} \frac{\gamma_1(\sigma,t)}{(s-\sigma)} \, d\sigma = -\alpha_1 V_Q + (s-c)\frac{d\alpha_1}{dt} - \tilde{v}_n \ , \tag{4.7.15}$$

where V_Q is the velocity of Q along L and $\tilde{v}_n$ is the still unknown normal component of the velocity at the profile induced by the shed free vorticity situated at the wake H^*. The shed free vorticity follows from the change per unit of time of the circulation around the profile and the velocity V_Q. Because our theory is linearized with respect to the small added motion, it is allowed within the accuracy of the theory to satisfy (4.7.15) at the profile (A–B) carrying out the base motion. By this the shed vorticity layer at H^* is connected to the trailing edge of the profile. In Section 6.10 we will continue this discussion and consider added motions which yield an optimum efficiency for the propulsion. This is carried out, however, for the more complicated case of two rigid flat profiles acting in each others neighbourhood and hence influencing each other.

4.8. Large-Amplitude Unsteady Propulsion, Rigid Wing of Finite Span

In this section we consider more closely the possibility of a base motion of a rigid flat wing of zero thickness and of finite span. The essence of this is discussed in Section 4.2 under Regime (4). We will describe how to find the contour of such wings for which a base motion in the sense of Figure 4.2.2 exists, and give some examples of these contours.

As we discussed in Section 4.2, in order to solve the above-mentioned non-linear problem, we have to consider the following linear problem. We start with a flat wing $\tilde{W}$ (Figure 4.8.1) which is symmetric with respect to the x-axis and which is also symmetric with respect to the y-axis. The wing has a given chord length distribution $c(y)$ and its span is denoted by s_0. The leading edge $l(y)$ of the wing and its trailing edge $t(y)$ are given by

$$x = l(y) \ , \qquad x = t(y) = l(y) + c(y) \ , \tag{4.8.1}$$

respectively. The normal velocity v_n at the wing is prescribed by

$$v_n = \omega x \ . \tag{4.8.2}$$

The wing is placed in a parallel flow with velocity U in the positive x-direction.

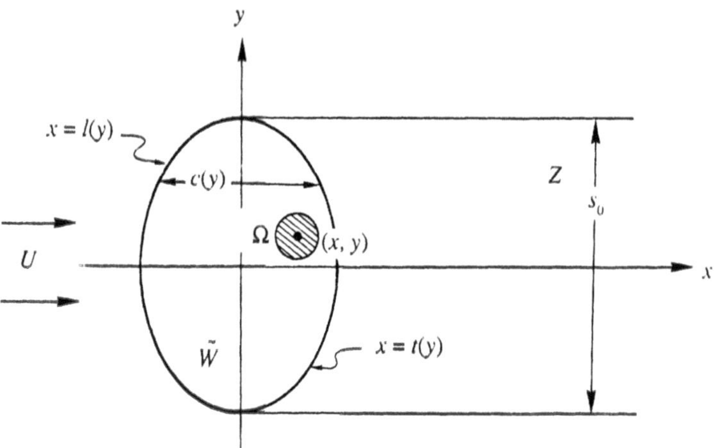

Fig. 4.8.1. Illustration of some quantities.

Now we have to change the shape of the leading edge to obtain a wing $\tilde{W}^*$ which has no trailing vorticity, hence of which the induced residence J becomes zero. The deformation of the leading edge will be defined by a deformation function $h(y)$, which is the local shift of the leading edge in the positive x-direction. The trailing edge then follows by keeping the chord length $c(y)$ the same. Hence the deformed wing has the same span s_0 and its leading edge and its trailing edge become

$$x = l(y) + h(y) \ , \qquad x = t(y) + h(y) = l(y) + c(y) + h(y) \ , \tag{4.8.3}$$

respectively. This means that its induced resistance J is a functional of $h(y)$

$$J = J(h) \ . \tag{4.8.4}$$

The question is now how to compute the induced resistance of $\tilde{W}$. Therefore, we introduce the region D as that part of the (x, y) plane which consists of the wing $\tilde{W}$ and its downstream region or wake Z of width s_0. Outside D the fluid velocity field is the gradient of a potential Φ which will be considered to be caused by a layer of velocity dipoles at D. The local strength of these dipoles is denoted by $\psi(x, y)$, which is the jump of the potential Φ across D and which can be considered as the stream function of the vorticity field $\vec{\gamma}(x, y) = (\gamma_x(x, y), \gamma_y(x, y))$ on D

$$\gamma_x = -\frac{\partial \psi}{\partial y} \ , \qquad \gamma_y = \frac{\partial \psi}{\partial x} \ . \tag{4.8.5}$$

In the wake Z, the dipole strength does not change in the downstream direction, which means when we temporarily use the notation for the undeformed wing

$$\psi(x, y) = \psi(t(y), y) \ , \qquad x > t(y) \ , \qquad |y| \leq \tfrac{1}{2}\, s_0 \ . \tag{4.8.6}$$

We remark that we can use here, just as in the lifting surface theory of Section 1.10 (see below (1.10.9)), the classical velocity dipole, because we are looking outside the dipole layer.

The Kutta condition yields that, on the planform the bound vorticity γ_y tends to zero at the trailing edge. Hence by (4.8.5)

$$\lim_{\varepsilon \to 0} \left\{ \frac{\partial \psi}{\partial x} (t(y) - \varepsilon, y) \right\} = 0 \ , \quad \varepsilon > 0 \ , \quad |y| < s_0 \ . \tag{4.8.7}$$

On the leading edge $x = l(y)$ and on the lines $y = \pm\frac{1}{2}s_0$ behind the tips of $\tilde{W}$, $\psi(x, y)$ has to be zero. Otherwise it follows from (4.8.5) that at these lines γ_x and (or) γ_y would behave like δ-functions of Dirac, in other words there would be a concentrated vortex at these lines which cannot be accepted for the following reasons. First, at the leading edge this is not in agreement with (4.8.2). Second, at the boundaries $y = \pm\frac{1}{2}s_0$ of the wake the concentrated vortex would have an infinite amount of kinetic energy around it per unit of length in the x-direction. Hence J would not be zero but infinite.

We will now discuss a method, as given by Sijtsma in [61], to find an integral equation for $\psi(x, y)$ which is interesting for its own sake because it avoids the Hadamard principal value. The potential $\Phi(x, y, z)$ of the fluid velocity $\vec{v}$ outside D, can be expressed in the strength $\psi(x, y)$ of the dipole layer by

$$\Phi(x, y, z) = \frac{1}{4\pi} \iint_D \psi(\xi, \eta) \frac{\partial}{\partial z} \left(\frac{1}{R} \right) d\xi \, d\eta \ , \tag{4.8.8}$$

where $R = \{(x - \xi)^2 + (y - \eta)^2 + z^2\}^{1/2}$. Then outside D we have for the z-component v_z of $\vec{v}$

$$\begin{aligned} v_z(x, y, z) &= \frac{\partial \Phi}{\partial z}(x, y, z) = \frac{1}{4\pi} \iint_D \psi(\xi, \eta) \frac{\partial^2}{\partial z^2} \left(\frac{1}{R} \right) d\xi \, d\eta \\ &= -\frac{1}{4\pi} \iint_D \psi(\xi, \eta) \left(\frac{\partial^2}{\partial x^2} + \frac{\partial^2}{\partial y^2} \right) \frac{1}{R} \, d\xi \, d\eta \\ &= -\frac{1}{4\pi} \iint_D \psi(\xi, \eta) \left(\frac{\partial^2}{\partial \xi^2} + \frac{\partial^2}{\partial \eta^2} \right) \frac{1}{R} \, d\xi \, d\eta \ . \end{aligned} \tag{4.8.9}$$

We exclude from the region of integration a region Ω (Figure 4.8.1) around the point (x, y) of D where we want to calculate the normal velocity component $v_n(x, y)$. The integration over Ω will be rewritten by means of partial integrations as follows

$$-\frac{1}{4\pi} \iint_\Omega \psi(\xi, \eta) \left(\frac{\partial^2}{\partial \xi^2} + \frac{\partial^2}{\partial \eta^2} \right) \frac{1}{R} \, d\xi \, d\eta$$

$$= -\frac{1}{4\pi} \iint\limits_{\Omega} \left[\frac{\partial}{\partial \xi} (\psi(\xi, \eta)) \frac{\partial}{\partial \xi} \frac{1}{R} + \frac{\partial}{\partial \eta} (\psi(\xi, \eta)) \frac{\partial}{\partial \eta} \frac{1}{R} \right] d\xi \, d\eta$$

$$- \frac{1}{4\pi} \int\limits_{\partial \Omega} \psi(\xi, \eta) \frac{\partial}{\partial n} \frac{1}{R} \, ds$$

$$= -\frac{1}{4\pi} \iint\limits_{\Omega} \frac{1}{R} \left(\frac{\partial^2}{\partial \xi^2} + \frac{\partial^2}{\partial \eta^2} \right) \psi(\xi, \eta) \, d\xi \, d\eta$$

$$+ \frac{1}{4\pi} \int\limits_{\partial \Omega} \frac{1}{R} \frac{\partial}{\partial n} \psi(\xi, \eta) \, ds - \frac{1}{4\pi} \int\limits_{\partial \Omega} \psi(\xi, \eta) \frac{\partial}{\partial n} \frac{1}{R} \, ds \, , \tag{4.8.10}$$

where $\vec{n}$ is the normal at $\partial\Omega$ lying in the (x, y) plane. In the right-hand side of (4.8.10) the limit $z \to 0$ can be taken, because the integration over Ω has only a singularity of order R^{-1} and the points of the line integrals along $\partial\Omega$ have a non-zero distance to the point (x, y) under consideration.

So by (4.8.2), (4.8.9) and (4.8.10) we find for $\psi(x, y)$ the equation

$$4\pi\omega x = - \iint\limits_{D \backslash \Omega} \frac{\psi(\xi, \eta) \, d\xi \, d\eta}{\{(x - \xi)^2 + (y - \eta)^2\}^{3/2}}$$

$$- \iint\limits_{\Omega} \frac{\left(\frac{\partial^2}{\partial \xi^2} + \frac{\partial^2}{\partial \eta^2} \right) \psi(\xi, \eta) \, d\xi \, d\eta}{\{(x - \xi)^2 + (y - \eta)^2\}^{1/2}}$$

$$- \int\limits_{\partial \Omega} \psi(\xi, \eta) \frac{\partial}{\partial n} \left\{ \frac{1}{\{(x - \xi)^2 + (y - \eta)^2\}^{1/2}} \right\} ds$$

$$+ \int\limits_{\partial \Omega} \frac{1}{\{(x - \xi)^2 + (y - \eta)^2\}^{1/2}} \frac{\partial \psi}{\partial n} (\xi, \eta) \, ds \, , \tag{4.8.11}$$

where Ω is any neighbourhood of (x, y), such that $\bar{\Omega} \subset D$.

The work applied to the fluid equals the increase of the kinetic energy of the fluid (Section 1.17). Hence we have that the induced resistance J equals the kinetic energy far behind the wing over a unit of length in the x-direction, when the wing moves through the fluid at rest. Far behind the wing we denote the disturbance potential by $\varphi(y, z)$. Then

$$J = \tfrac{1}{2} \rho \iint\limits_{-\infty}^{\infty} |\text{grad}\, \varphi|^2 \, dy \, dz = \tfrac{1}{2} \rho \iint\limits_{-\infty}^{\infty} \text{div}\, (\varphi \, \text{grad}\, \varphi) \, dy \, dz$$

$$= -\tfrac{1}{2}\rho \int_{-\frac{1}{2}s_0}^{\frac{1}{2}s_0} [\varphi]_-^+(y,0)\, \frac{\partial \varphi}{\partial z}(y,0)\, dy$$

$$= -\frac{\rho}{4\pi} \int_{-\frac{1}{2}s_0}^{\frac{1}{2}s_0} \psi(x_0,y) \left\{ \oint_{-\frac{1}{2}s_0}^{\frac{1}{2}s_0} \frac{\partial \psi}{\partial y}(x_0,\eta)\, \frac{d\eta}{(y-\eta)} \right\} dy \ , \qquad (4.8.12)$$

where x_0 is any value of x far behind the wing. The Cauchy principal value integral in the last expression of (4.8.12) follows from the fact that $\partial\psi/\partial y = -\gamma_x$ represents two-sided infinitely long vortex lines which induce the normal velocity $\partial\varphi/\partial z$ far behind the wing.

For each choice of the deformation function $h(y)$ (4.8.3) we have to solve (4.8.11) and to calculate (4.8.12). The problem is to find a $h(y)$ for which $J = 0$.

The numerical procedure used in [61], introduces a coordinate transformation in the (x, y) plane, such that the deformed wing becomes a square on which by a bi-quadratic spline the function ψ is approximated. The coefficients of the spline are calculated by a collocation method. The deformation function h is approximated by a finite series, by which J becomes a function of a finite number of variables, which is minimized by a quasi-Newton method.

We now give some results obtained by this method. In Figures 4.8.2 (a)–4.8.5 (a) we have drawn the starting wing shapes which are, besides symmetric with respect to the x-axis, also symmetric with respect to the y-axis. In Figures 4.8.2 (b)–4.8.5 (b) we have drawn the deformed shapes of which the induced resistance J is at least a factor 10^{-3} smaller than the value of J for the undeformed wings.

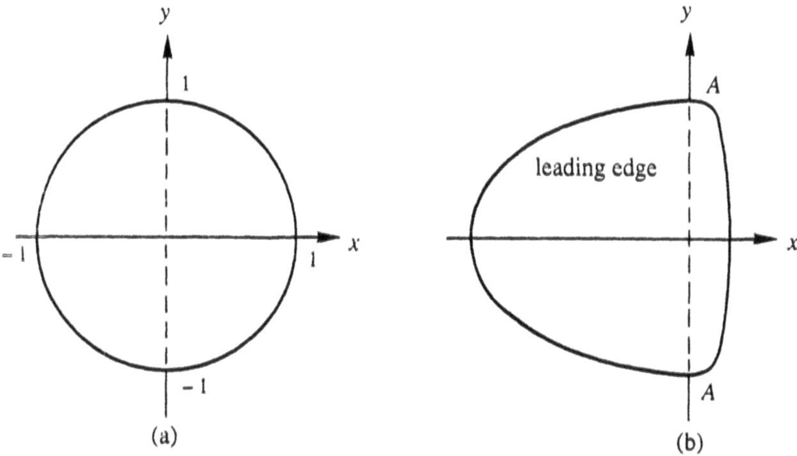

Fig. 4.8.2. (a) Circular wing, (b) deformed wing.

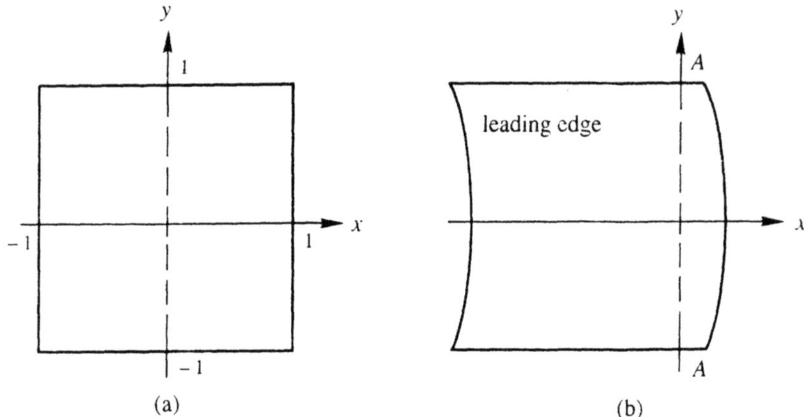

Fig. 4.8.3. (a) Square wing, (b) deformed wing.

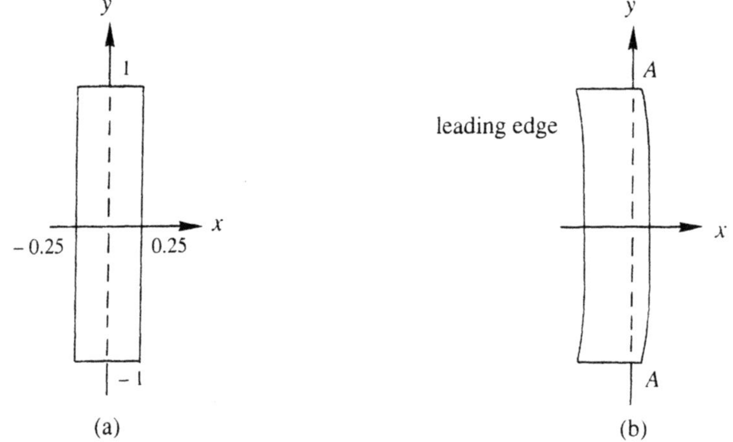

Fig. 4.8.4. (a) Rectangular wing, (b) deformed wing.

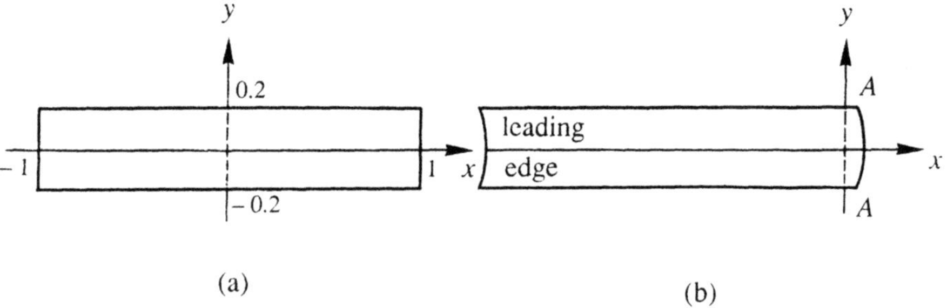

Fig. 4.8.5. (a) Rectangular wing, (b) deformed wing.

It follows from Figure 4.8.4 (b) that the line of tangency $(A\text{–}A)$ (Figure 4.2.2) lies in the neighbourhood of the three-quarter chord point of Section 4.7, where the 2-dimensional case of a wing with infinite span was considered. From Figure 4.8.5 (b) we find that the line $(A\text{–}A)$ is rather close to the trailing edge. This is in agreement with slender body theory, which predicts no vorticity shedding or $J = 0$ for $(A\text{–}A)$ at the trailing edge [61].

So we have found by numerical means that for the deformed wings base motions seem to exist. In order that a mean thrust will be delivered an added motion of $O(\varepsilon)$ has to be superimposed on the base motion. We remark that many of the considerations of the previous section with respect to the wing of infinite span can be applied "mutatis mutandis" to this case of a wing with a finite span. This will be left to the reader.

4.9. The Voith-Schneider Propeller

In technics there is up to now one important realization of an unsteady large-amplitude propulsion device namely the Voith-Schneider or vertical axis propeller, of which a scheme is drawn in Figure 4.9.1. Underneath a ship we imagine a horizontal circular disk which rotates about a vertical axis l through its centre C. The rotational velocity of the disk is ω. On the disk are mounted several vertical wing-like blades. These blades can perform oscillatory motions about vertical pivotal axes, denoted in Figure 4.9.1 by $l_1, \ldots, l_4$ which cut the circular disk in the pivotal points $Q_1, \ldots, Q_4$, respectively.

We now discuss the cylindrical surfaces described by the pivotal axes. We use a Cartesian coordinate system (x, y, z) in rest with respect to the fluid. The circular

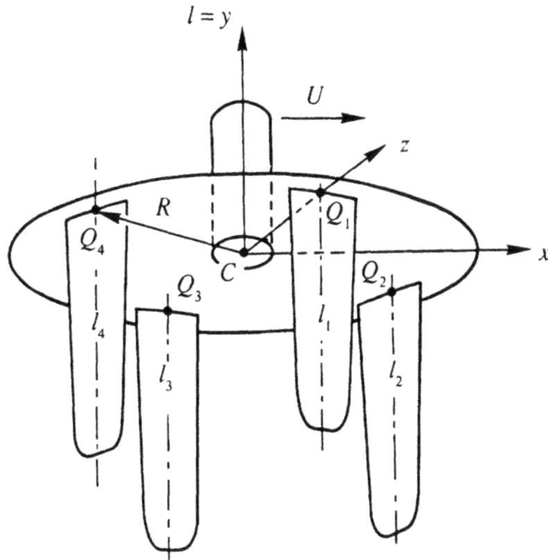

Fig. 4.9.1. Scheme of a four-bladed Voith-Schneider propeller, $t = 0$.

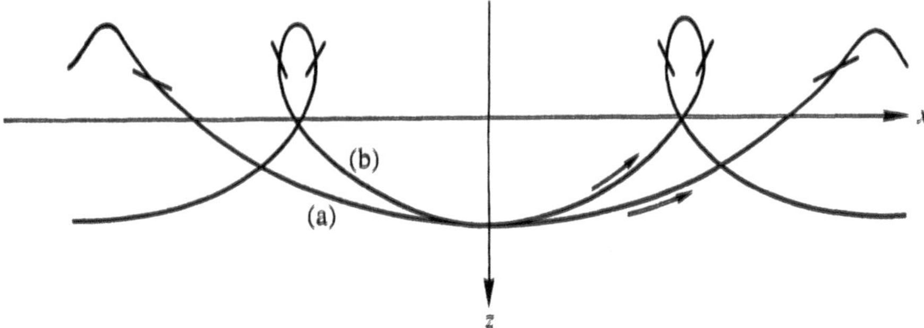

Fig. 4.9.2. Two types of cycloids, (a) $1 < \nu$, (b) $0.212 < \nu < 1$.

disk lies in the (x, z) plane and at time $t = 0$ the axis of rotation l of the propeller coincides with the y-axis. The propeller has a velocity U in the positive x-direction. We consider the path in the (x, z) plane of the pivotal point Q_1 which is assumed at $t = 0$ to be on the positive z-axis. We find

$$x = R(\nu\omega t + \sin\omega t) \ , \qquad z = R\cos\omega t \ , \qquad \nu = \frac{U}{\omega R} \ , \tag{4.9.1}$$

where R is the distance from the axes $l_1, \ldots, l_4$ to l. This path is a cycloid of which the character is drawn in Figure 4.9.2 for two values of ν. When $1 < \nu$, the blades carry out an asymmetrical swimming motion of a fish tail; when $0 < \nu < 1$, the cycloid intersects itself. When $\nu < 0.212$, more than one intersection per period occurs. In practice we have $\nu < 1$.

In order that the propeller provides a mean force in the x-direction, it is necessary that the blades have appropriate angles of incidence during their motion. This is accomplished by making them execute a suitable periodic oscillatory motion around their pivotal axes $l_1, \ldots l_4$, such as indicated in Figure 4.9.2. The way in which this motion of the blades about the axes is controlled mechanically, will not be discussed here, we refer to Mueller [50]. We only mention that, by turning the whole machinery, we can also turn the direction of the thrust. By this it is possible to steer a ship provided with this type of propeller, hence a rudder becomes superfluous.

When ν is sufficiently small, hence when the rotational velocity ω is high or the speed U of the ship is small, the cycloids passed through by the pivotal axes resemble more or less circles. Then when the chord lengths of the profiles are not too large with respect to the radii of curvature of the cycloids, it is possible to describe the working of the Voith-Schneider propeller by means of Regime (2) of Section 4.1. In this case, the shed free vorticity is situated on the cycloids. The boundary conditions on the wings then have to be satisfied on the projection of the wings on the cycloidal cylinders. In order to allow for the bottom of the ship we can take the span of the wings to which we apply our theory, equal to two times the span of the real wings of the propeller. Then the plane through the midspan of our mathematical model is a plane which by symmetry is not passed through by fluid particles, hence it can represent the bottom.

When the chord lengths of the profiles are not small with respect to the curvature of the cycloidal paths, it is possibly better to use Regime (3) in case the wings are slender and can be considered as parts of wings with infinite span. Then an exact base motion can be constructed for more wings as is mentioned at the end of the discussion of Regime (3). Also it is possible to give each wing of finite length a shape as is discussed in relation to Regime (4), although then an exact base motion does not seem to exist.

It is clear that the Voith-Schneider propeller gives room for many different approximate theoretical approaches. Of course, in the above-mentioned possibilities the propeller has to be lightly loaded.

In Section 6.12 we discuss the optimization of a Voith-Schneider propeller with many blades of finite span.

4.10. Some Remarks and Conclusions

We start with a short discussion of a possible realization of Regime (1.b) for the 2-dimensional case, hence of the small-amplitude propulsion by means of two profiles, one in front of the other. The theory is linear, the boundary conditions on the profiles have to be satisfied at the x-axis and the vorticities are assumed to be situated at the x-axis (Figure 4.10.1). Both profiles are rigid, flat and of zero thickness. The profiles are coupled by means of two rigid bars (b_1, s_1) and (s_1, a_2) which are so slender that they do not disturb the fluid and which are connected to each other at $x = s_1$ by means of a linear elastic hinge. The whole system is placed in a parallel flow with velocity U. The motion of the first profile (a_1, b_1) is prescribed in a suitable way while the profile (a_2, b_2) follows passively, hence its motion has to be calculated by means of the occurring pressures in the fluid. This motion is determined by the stiffness of the hinge, the length of (a_1, b_1), the values of s_1, a_2 and b_2, and the values of the mass of the second profile, the position of its centre of gravity and its moment of inertia around its centre of gravity. Then we can look for values of these quantities

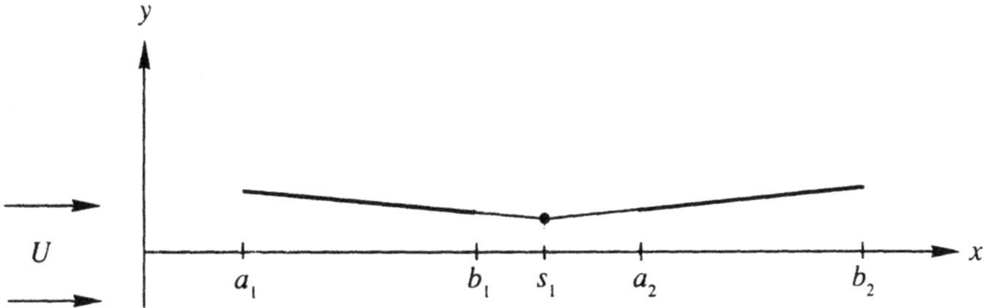

Fig. 4.10.1. Two flat profiles of zero thickness, coupled by an elastic hinge.

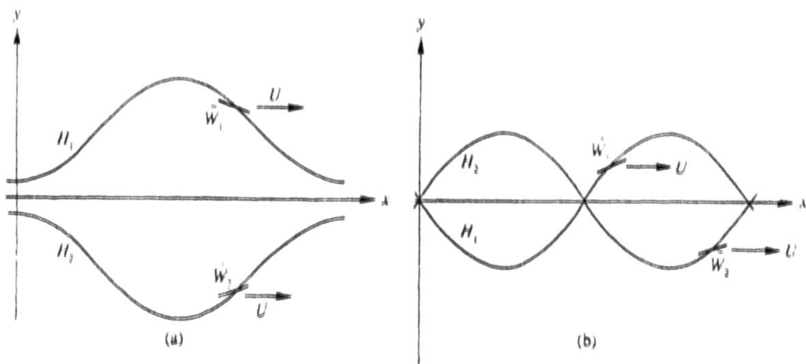

Fig. 4.10.2. Two profiles large-amplitude, (a) symmetric motion, (b) nearly symmetric motion.

such that the free vorticity of the first profile is cancelled by the shed free vorticity of the second one. This problem is in essence not more difficult than the one of the single profile as discussed in Sections 4.3–4.5, although it is much more complicated. For the theory and numerical results we refer to Sparenberg and de Vries [67].

In Section 4.1 we had to assume for a rational description of linearized large-amplitude propulsion (Regime (2)) that the wing $\tilde{W}$ is flexible. However, as we mentioned already in the previous section, when the chord length of $\tilde{W}$ is not too large, the wing can be taken rigid, without violating the assumptions of the linearization procedure too much. This has been done, for instance, by Wu [79], by de Graaf [20] and by Chopra [10]. In the work of Katz and Weihs [35] the profile is allowed to be elastic, it has a certain distributed bending stiffness by which it gives passively way to the hydrodynamic pressures, when its front part is forced to move in one way or another. Also in this paper the influence of the transport of the shed free vorticity by the profile motion and by the fluid velocities induced by the shed free vorticity itself is considered.

In principle, there are two different ways of approach to propulsive systems. The first one is to prescribe the motion of the propulsive system, or in case of the flexible profile the way in which it is forced. Then the mean value of the thrust and the efficiency can be calculated. This has been done in most research pertaining to biology such as [79], [10] and [35].

The second approach is, when we confine ourselves for instance to Regime (2), to prescribe the mean value of the thrust and the reference surface H, while we still do not specify the profile motions. Here we have the possibility to look for propulsion with optimum efficiency, because this optimum efficiency depends only on H as will be shown in the next chapter. Then the optimum motion of the wing $\tilde{W}$ is not uniquely determined and we can choose motions which meet requirements for a technical realization. The only constraint on the wing motions is that $\tilde{W}$ leaves behind at H the a priori calculated optimum free vorticity. This will be discussed in Section 5.5. An analogous treatment holds for Regime (3), however then we have to replace the reference surface H by the calculated wake of the base motion far behind the propeller.

In realizations of unsteady propulsion it seems recommendable to use more than one wing. For instance, $\tilde{W}_1$ and $\tilde{W}_2$, as is drawn in Figure 4.10.2 for the 2-dimensional case, which carry out motions which are reflections of each other with respect to the x-axis. This is favourable for two reasons. First, the avoidance of large time-dependent lateral forces, which cause unsteady moments at the propelled body. Second, the hydrodynamical interaction of the two wings can be important with respect to increasing the efficiency of the propeller as we will discuss in the next chapter.

Finally, we remark that in fact the screw propeller with blades of zero thickness (Section 3.3) can be considered to belong to Regime (2). The exact gliding of the blades along the helicoidal reference surfaces is the base motion and the function $h_Q^*(x, z)$ in (3.3.1) represents the added motion.

Chapter 5

Optimization Theory

The theory discussed in this chapter is intended to give insight in the best way of working of a device which produces lift, thrust or any other prescribed force action. We will develop a linear theory and assume the fluid to be inviscid and incompressible. In the last section we mention influences of the viscosity of water and of the non-linearity of the flow, which will be dealt with in more detail in Chapter 6.

First we have to define what will be called the best way of working of a device. We restrict ourselves to the minimization of a simple cost function, namely the kinetic energy losses per unit of time. For instance, we consider a screw propeller with a given diameter, number of blades, velocity of advance and rotational velocity, which has to deliver a prescribed thrust. The question is: what has to be the circulation distribution along the blades in order that the kinetic energy left behind is as small as possible. This special problem has been solved by Betz in his classical paper [5].

In this chapter we will derive necessary conditions for the optimum working of more general devices, including unsteady ones. The motion of the devices is assumed to be periodic. We demand a non-zero mean value with respect to time of the force action, otherwise the kinetic energy left behind can be made zero and we have a trivial optimum.

The constraints on the force actions can be rather general. For instance, it can be demanded, that a wing carrying out a flapping motion delivers a prescribed mean value of lift as well as a prescribed mean value of thrust. Then we can ask for its time-dependent circulation distribution which optimizes both force actions together. It will turn out that the free vorticity left behind by the optimum system can be characterized in a simple way.

It is allowed that the fluid through which the device moves is disturbed by a periodic velocity field of $O(\varepsilon)$. When the situation is favourable, energy can be extracted from this field in a useful way, by which the device needs less energy to perform its task. A disturbance velocity field, however, can also be unfavourable as we will discuss. Then, in spite of its energy extraction, the device needs more energy to generate the prescribed force action than in an undisturbed fluid.

When the viscosity of the water is neglected it will be seen that by increasing the size of a propeller its efficiency can be raised. This is the reason that in the theory of this chapter we have to make a choice of, for instance, the diameter of a screw propeller. In realistic optimization problems when the viscosity of the water is taken into account, it can happen that such a choice need not to be made. Then the diameter can be chosen in an optimum way so that the inviscid theoretical increase of the efficiency caused by an increase of the diameter will be annihilated by a decrease of the efficiency caused by the friction losses of the fast moving tips.

An important quantity for the judgement of a propeller is its quality number. This number gives information about a propeller which differs from the information given by its efficiency. It shows how much the working of the propeller resembles the working of the ideal propeller, hence if its geometry can be improved. We will discuss in Section 5.8 when such an improvement makes sense.

In our linearized theory we will neglect forces of $O(\varepsilon^2)$ with respect to the desired force action of $O(\varepsilon)$. The forces of $O(\varepsilon^2)$ are due to leading edge suction and to second order errors of first order forces caused by the assumption that the blade vorticity as well as the trailing vorticity are assumed to be situated at the reference surfaces. For a screw propeller, the leading edge suction forces are not important because they have only a small component in the direction of the thrust. When, however, a ducted propeller is considered, the suction forces acting at the leading edge of the shroud point in the direction of the thrust and can, in practice, be a non-negligible part of it. For the question of the possibility of calculating the efficiency accurate upto and including $O(\varepsilon)$ in spite of errors of the thrust of $O(\varepsilon^2)$ we refer to the text around (4.1.5).

The concept efficiency applies only to lifting surfaces which perform useful work. For instance, it applies to a screw propeller but not to a wing which has to generate a lift force which is nearly perpendicular to its translational velocity. Of course, we can say that such a wing is efficient when the amount of kinetic energy it leaves behind per unit of time is small, in other words, when its induced resistance is small.

5.1. Lifting Surface System

We have a Cartesian coordinate system (x, y, z) embedded in an inviscid and incompressible fluid. The fluid is at rest at infinity with respect to the coordinate system. Consider m immaterial geometric reference surfaces H_l

$$H_l(x, y, z) = 0 \ , \qquad l = 1, \dots, m \ , \tag{5.1.1}$$

with

$$H_l(x + b, y, z) = H_l(x, y, z) \ , \tag{5.1.2}$$

hence these surfaces are periodic with period b in the x-direction and are assumed to be sufficiently smooth. On each surface we have a locally orthogonal coordinate

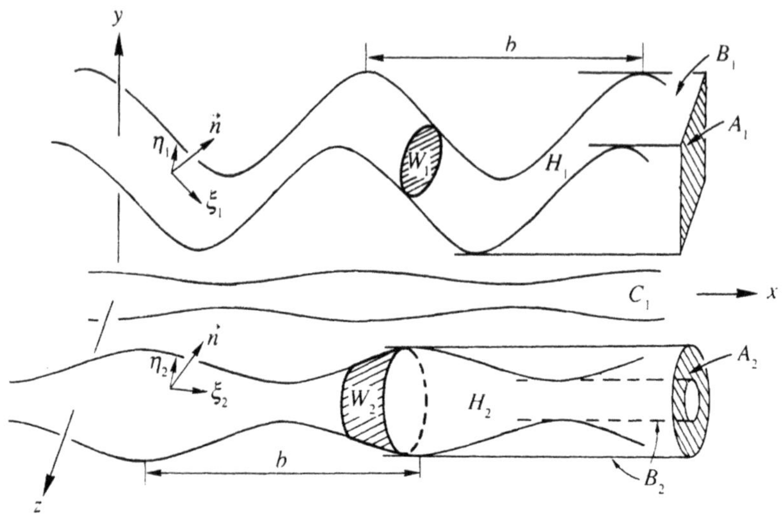

Fig. 5.1.1. A lifting surface system $m = 2, j = 1$.

system (ξ_l, η_l) (Figure 5.1.1) in such a way that an increase in ξ_l by a number b_l while η_l remains constant, makes that we obtain an equivalent point of H_l with respect to its periodicity. The regions of ξ_l and η_l are given by

$$-\infty < \xi_l < \infty \ , \qquad \eta_{0,l} \le \eta_l \le \eta_{1,l} \ , \qquad l = 1, \dots, j \ , \tag{5.1.3}$$

$$-\infty < \xi_l < \infty \ , \qquad \eta_{0,l} \le \eta_l < \eta_{1,l} \ , \qquad l = j+1, \dots, m \ , \tag{5.1.4}$$

where the half open intervals in (5.1.4) belong to closed surfaces as for instance H_2 in Figure 5.1.1. The lines

$$\eta_l = \text{const.} \ , \tag{5.1.5}$$

will form a one parameter family of curves on H_l, such that through each point of H_l passes one and only one such a line. Such locally orthogonal coordinate systems do exist as is proved by M. van der Put and F. Takens (private communication).

In order to introduce a + and a − side on H_l we consider the unit vectors $\vec{e}_\xi$ and $\vec{e}_\eta$ tangent to H_l and in the positive directions of ξ_l and η_l respectively. Then we agree that the unit normal $\vec{n} = \vec{e}_\xi \times \vec{e}_\eta$ points from the negative side H_l^- of H_l to its positive side H_l^+.

Next we have possibly flexible surfaces $\tilde{W}_l$ of finite extent. When these surfaces move exactly along the H_l, the fluid will not be disturbed when it was undisturbed before. Then we have the base motion as discussed in Sections 4.1 and 4.2. When the surfaces have to perform force actions, each of them will carry out an added motion in an ε neighbourhood of its reference surface H_l. Because our theory is linearized we identify the lifting surfaces as locations of vorticity with their projections on the H_l, hence with their planforms W_l.

The force actions of the W_l are represented by external force fields $\vec{F}_l$ of $O(\varepsilon)$, which are in the direction of the normal $\vec{n}$

$$\vec{F}_l = f_l(\xi_l, \eta_l, t)\, \vec{n}_l(\xi_l, \eta_l) \ , \qquad l = 1, \ldots, m \ , \tag{5.1.6}$$

where f_l are periodic scalar strenghts (per unit of area)

$$f_l(\xi_l + b_l, \eta_l, t + \tau) = f_l(\xi_l, \eta_l, t) \tag{5.1.7}$$

and τ is the time period of the motion of the profiles. These force fields give rise to the pressure jump

$$(p^+ - p^-) = [p]_-^+ = f_l(\xi_l, \eta_l, t) \ , \tag{5.1.8}$$

over the planforms W_l. Because our theory will be linear, the free vorticity $\vec{\gamma}_l$ which is shed by the force fields $\vec{F}_l$ remains where it is formed, hence at the surfaces H_l.

In the fluid we admit impermeable rigid bodies C_j which are also periodic with period b in the x-direction. These bodies can be simply connected, hence stretch from $x = -\infty$ towards $x = \infty$ (C_1, Figure 5.1.1) or they consist of periodically disconnected parts. When the lifting surfaces move through the fluid, these bodies are boundaries on which the normal component of the velocity has to vanish.

We now define a lifting surface system, as the periodic reference surfaces H_l, with the force fields $\vec{F}_l$, together with the impermeable bodies C_j. Sometimes we denote such a system simply by W_l.

Finally, we introduce the working region and the working area of a lifting surface system. The working region is the 3-dimensional region of space enclosed by the most narrow geometric cylindrical surfaces with generators parallel to the x-axis which enclose the H_l. The cross sections of these cylinders will be called the working area of the system. In Figure 5.1.1 the cylindrical surfaces are denoted by B_1 and B_2, where B_2 consists of two parts, one part outside H_2 and one part inside H_2. The working area consists of the regions denoted by A_1 and A_2, where A_2 is multiply connected.

5.2. Energy Extraction out of a Disturbed Fluid, One Wing

Consider an inviscid and incompressible fluid disturbed by a time-independent velocity field v_0^* of $O(\varepsilon)$

$$\vec{v}_0{}^*(x, y, z) = (v_{0x}^*(x, y, z) \ , v_{0y}^*(x, y, z) \ , v_{0z}^*(x, y, z)) \ , \tag{5.2.1}$$

which is supposed to be periodic in the x-direction

$$\vec{v}_0{}^*(x + b, y, z) = \vec{v}_0{}^*(x, y, z) \ . \tag{5.2.2}$$

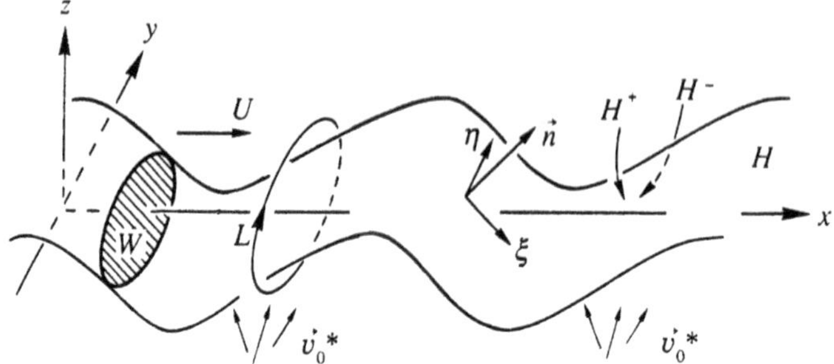

Fig. 5.2.1. Wing moving through disturbed fluid.

We want to discuss the optimum energy extraction out of this fluid by means of a lifting surface system.

In order to keep the discussion as simple as possible we start with one flexible lifting surface moving through the fluid in an ε neighbourhood of a reference surface H, with a constant velocity U in the positive x-direction (Figure 5.2.1). The discussion we give holds analogously for more lifting surfaces, while also one or more bodies of the type mentioned before are allowed to be present.

The wing has to extract as much as possible energy out of the velocity field by a suitable added motion. The questions we will consider are: how much energy can be extracted by W out of the fluid when a linearized theory is valid and how can this energy be characterized? First we give a mechanical reasoning and then show by an analytical verification that the results are correct.

When the wing (planform) moves along the reference strip H, it can extract energy only by forces of $O(\varepsilon^2)$ in the direction of its motion. These forces originate from the interaction of the bound vorticity of $O(\varepsilon)$ of the wing and the component of the velocity field $\vec{v}_0^*$ of $O(\varepsilon)$ normal to H. Hence in the linearized theory this normal component of $\vec{v}_0^*$ is the only thing that matters to the wing. Therefore, we replace the velocity field $\vec{v}_0^*$ by another one $\vec{v}_0$ which has the same normal component at H and of which the vorticity $\vec{\gamma}_0$ is entirely confined to H. In this case $\vec{\gamma}_0$ has two components

$$\vec{\gamma}_0 = \left(\gamma_{0\xi}(\xi,\eta), \gamma_{0\eta}(\xi,\eta)\right) \quad . \tag{5.2.3}$$

This can be achieved by solving the following Neumann problem for the velocity potential Φ_0

$$\left(\frac{\partial^2}{\partial x^2} + \frac{\partial^2}{\partial y^2} + \frac{\partial^2}{\partial z^2}\right)\Phi_0(x,y,z) = 0 \ , \qquad (x,y,z) \notin H \ , \tag{5.2.4}$$

$$\frac{\partial \Phi_0}{\partial n}(x,y,z) = \vec{v}_0^* \cdot \vec{n} \ , \qquad (x,y,z) \in H \ , \tag{5.2.5}$$

where $\vec{n}$ is the unit normal at H. This potential can be made "uniquely valued" by demanding that the circulation along any contour L (Figure 5.2.1) around H is zero.

Then the velocity field $\vec{v}_0$ becomes

$$\vec{v}_0 = \text{grad}\ \Phi_0 \ . \tag{5.2.6}$$

The kinetic energy of the fluid per period in the x-direction has changed, because in general the new field $\vec{v}_0$ is different from the old one $\vec{v}_0^{\,*}$. It is the kinetic energy of the field $\vec{v}_0$ which can be extracted entirely by a flexible wing. The reason is that such a wing, when it deforms suitably, can shed at H any vorticity for which the circulation around H is zero. Hence it can annihilate $\vec{\gamma}_0$ (5.2.3) and bring to rest the velocity field $\vec{v}_0$. We remark that when we had not taken the circulation around H equal to zero, the kinetic energy of $\vec{v}_0$ would have been infinite.

The kinetic energy belonging to $\vec{v}_0$ is smaller than or at most equal to the kinetic energy of the original field $\vec{v}_0^{\,*}$. This is clear from the mechanical point of view, the wing W cannot extract more energy out of the fluid than is present in it.

We still give two somewhat different formulations of the principal result. First, *the optimum wing W has to leave behind free vorticity which far behind W induces a velocity field of which the normal component at H is opposite to the normal component of $\vec{v}_0^{\,*}$*. Second, *the optimum wing W has to leave behind that vorticity on H which is needed when H is, as a rigid and impervious surface of zero thickness, embedded in the velocity field $\vec{v}_0^{\,*}$*.

Next we give an analytic discussion of the subject. We suppose that the kinetic energy E_0^* of the original field $\vec{v}_0^{\,*}$ per period b in the x-direction, is finite

$$E_0^* = \tfrac{1}{2}\,\rho \int\int_{-\infty}^{\infty}\int_0^b (\vec{v}_0^{\,*})^2\, dx\, dy\, dz < \infty \ . \tag{5.2.7}$$

We want to utilize this kinetic energy as much as possible, by a suitable action of the wing, hence by adding a periodic velocity field grad Φ_1 of which the vorticity is confined to H. Note that we are looking again far behind the wing. Then we have to choose Φ_1, so that we minimize

$$\tilde{E}(\Phi_1) = \tfrac{1}{2}\,\rho \int\int_{-\infty}^{\infty}\int_0^b \left(\vec{v}_0^{\,*} + \text{grad}\ \Phi_1\right)^2\, dx\, dy\, dz \ . \tag{5.2.8}$$

Assuming that Φ_1 is the optimum potential, we change Φ_1 into $\Phi_1+\delta\Phi_1$, where $\delta\Phi_1$ is a periodic disturbance potential of which the vorticity is also confined to H. Then we consider the first variation $\delta\tilde{E}(\delta\Phi_1)$, which is that part of $\tilde{E}(\Phi_1+\delta\Phi_1) - \tilde{E}(\Phi_1)$ which is linear in $\delta\Phi_1$

$$\delta\tilde{E}(\delta\Phi_1) = \rho \int\int_{-\infty}^{\infty}\int_0^b \left(\vec{v}_0^{\,*} + \text{grad}\ \Phi_1\right)\cdot \text{grad}\ \delta\Phi_1\, dx\, dy\, dz \ . \tag{5.2.9}$$

A necessary condition for Φ_1 to be optimum is that $\delta\tilde{E}(\delta\Phi_1)$ is equal to zero for all admissible $\delta\Phi_1$.

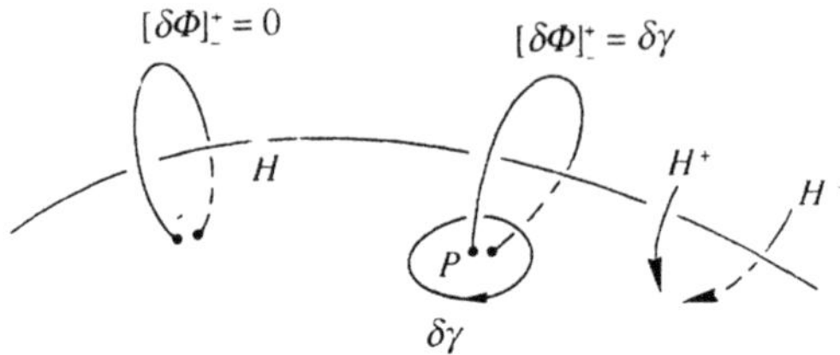

Fig. 5.2.2. The disturbance potential $\delta\Phi_1$.

Because the velocity field is without divergence we can write

$$\delta\tilde{E}(\delta\Phi_1) = \rho \int_{-\infty}^{\infty}\int \int_0^b \operatorname{div}\ \{(\vec{v}_0^{\,*} + \operatorname{grad}\Phi_1)\,\delta\Phi_1\}\ dx\,dy\,dz\ . \tag{5.2.10}$$

The boundary of the region of integration consists of the planes $x = 0$ and $x = b$ and the intermediate part of H. When we convert the volume integral into an integral over this boundary, it is seen by the periodicity of $(\vec{v}_0^{\,*} + \operatorname{grad}\Phi_1)\cdot\delta\Phi_1$ that the planes $x = 0$ and $x = b$ do not contribute. Hence there remains an integration over $H(b)$, which denotes one period of H with $0 \le x \le b$. We find

$$\delta\tilde{E}(\delta\Phi_1) = -\rho \iint_{H(b)} \{(\vec{v}_0^{\,*} + \operatorname{grad}\Phi_1)\cdot\vec{n}\}\,[\delta\Phi_1]_-^+\ dS\ , \tag{5.2.11}$$

where $[\delta\Phi_1]_-^+$ is the jump of $\delta\Phi_1$ over H and dS is an element of area. We used the facts that, because of the preservation of fluid, $(\vec{v}_0^{\,*} + \operatorname{grad}\Phi_1)\cdot\vec{n}$ is continuous over H and that $\vec{n}$ points from H^- towards H^+.

The quantity $[\delta\Phi_1]_-^+$ is still arbitrary, we choose it such that it is zero at $H(b)$ with the exception of a small neighbourhood of area ΔS of an arbitrary point P of $H(b)$. This is most easily effectuated by taking for $\delta\Phi_1$ the potential of a small vortex ring of strength $\delta\gamma$ at $H(b)$ around P (Figure 5.2.2) and around points which are equivalent to P with respect to the periodicity of H. Then $[\delta\Phi_1]_-^+ = \delta\gamma$ on ΔS. Hence because in the optimum case $\delta\tilde{E}(\delta\Phi_1) = 0$, it follows from (5.2.11)

$$\frac{\partial\Phi_1}{\partial n}(P) = \operatorname{grad}\Phi_1(P)\cdot\vec{n}(P) = -\vec{v}_0^{\,*}(P)\cdot\vec{n}(P)\ , \tag{5.2.12}$$

for any point P of H. Comparing (5.2.12) with (5.2.5), we find

$$\Phi_1 = -\Phi_0\ . \tag{5.2.13}$$

In the previous mechanical reasoning it was argued that the maximum amount of energy which could be extracted out of the fluid was the kinetic energy $E(\Phi_0)$

belonging to Φ_0. Hence by (5.2.13) we have to prove

$$E_0^* - \tilde{E}(\Phi_1) = E(\Phi_0) = \tfrac{1}{2}\,\rho \int\!\!\int_{-\infty}^{\infty}\!\int_0^b (\text{grad }\Phi_1)^2\, dx\, dy\, dz \ . \tag{5.2.14}$$

Subtracting (5.2.8) from (5.2.7) we find that in order to prove (5.2.14), we have to show that

$$-\int\!\!\int_{-\infty}^{\infty}\!\int_0^b \vec{v}_0^{\,*} \cdot \text{grad }\Phi_1\, dx\, dy\, dz = \int\!\!\int_{-\infty}^{\infty}\!\int_0^b (\text{grad }\Phi_1)^2\, dx\, dy\, dz \ . \tag{5.2.15}$$

The left-hand side of (5.2.15) can be written as

$$-\int\!\!\int_{-\infty}^{\infty}\!\int_0^b \text{div }(\Phi_1 \vec{v}_0^{\,*})\, dx\, dy\, dz = \int_{H(b)} \vec{v}_0^{\,*} \cdot \vec{n}\, [\Phi_1]_-^+\, dS \tag{5.2.16}$$

and the right-hand side as

$$\int\!\!\int_{-\infty}^{\infty}\!\int_0^b \text{div }(\Phi_1 \text{grad }\Phi_1)\, dx\, dy\, dz = -\int_{H(b)} \frac{\partial \Phi_1}{\partial n}\, [\Phi_1]_-^+\, dS \ . \tag{5.2.17}$$

Equality of (5.2.16) and (5.2.17) then follows from (5.2.12), hence (5.2.14) is shown to be correct. It is seen by (5.2.14), (5.2.15) and (5.2.16) that the extractable kinetic energy per period of length b can be written as

$$E(\Phi_0) = \tfrac{1}{2}\,\rho \int_{H(b)} \vec{v}_0^{\,*} \cdot \vec{n}\, [\Phi_1]_-^+\, dS \ . \tag{5.2.18}$$

The jump of the velocity potential $[\Phi_1]_-^+$ is directly coupled to the bound vorticity of the wing or to the external force field, perpendicular to H, which can represent the wing. By this it is possible to interpret (5.2.18) in two ways. First, as the work delivered by the forces which are tangential to H. Second, as the work carried out by the perpendicular forces. This is left to the reader. In fact (5.2.18) could have been written down already in the mechanical considerations of the first part of this section.

From (5.2.14) it follows, because $\tilde{E}(\Phi_1) > 0$, that

$$E_0^* > E(\Phi_0) \ , \tag{5.2.19}$$

as we expected already. This inequality is closely related to Kelvin's minimum energy principle ([4], p. 384).

5.3. Energy Extraction out of a Disturbed Fluid, Many Wings

In this section we consider the question, what is the amount of kinetic energy per period b which can be extracted out of the periodic disturbance field $\vec{v}_0^*$ by means of an *unlimited* number of flexible lifting surface systems. The only condition on these lifting surfaces is that their reference surfaces H_l have to be situated within a working region defined by the geometrical cylinder B

$$B: \quad -\infty < x < \infty \ , \qquad (y, z) \in A \ , \tag{5.3.1}$$

where the working area A is some closed region in the (y, z) plane.

In the optimum case, as we have seen in the previous section, the vorticity shed by the flexible wings moving in the neighbourhood of the H_l is such that the H_l can be considered as rigid impervious surfaces. Hence by crowding the working region B more and more with reference surfaces, we can cut B into smaller and smaller disconnected regions, say into "cubes" of volumes of $O(\tilde{\varepsilon}^3)$ where $\tilde{\varepsilon}$ is the small length of an edge of a "cube". We remark that this geometrical $\tilde{\varepsilon}$ has no connection with the ε of the linearization procedure of the hydrodynamical theory.

Consider such a small cube in which we possibly have continuous free vorticity $\vec{\gamma}_0^*$ which belongs to the disturbance velocity field $\vec{v}_0^*$. Then we will determine by dimension analysis the amount of kinetic energy E^* in this cube. The kinetic energy depends on ρ, $\tilde{\varepsilon}$ and $\vec{\gamma}_0^*$. From Appendix E it follows that we have to consider the following relation between the dimensions of these quantities

$$[E^*] = [\rho]^{\alpha_1} \cdot [\tilde{\varepsilon}]^{\alpha_2} \cdot [\vec{\gamma}_0^*]^{\alpha_3} \ , \tag{5.3.2}$$

where

$$[E^*] = [m]\,[l]^2\,[t]^{-2} \ , \qquad [\rho] = [m]\,[l]^{-3} \ ,$$

$$[\tilde{\varepsilon}] = [l] \ , \qquad [\vec{\gamma}_0^*] = [t]^{-1} \ . \tag{5.3.3}$$

Substituting (5.3.3) into (5.3.2) and comparing the exponents of $[l]$, $[m]$ and $[t]$ respectively, we find $\alpha_1 = 1$, $\alpha_2 = 5$ and $\alpha_3 = 2$. Hence we obtain the result

$$E(\tilde{\varepsilon}) = O(\tilde{\varepsilon}^5) \ , \qquad \tilde{\varepsilon} \to 0 \ . \tag{5.3.4}$$

Next we consider the case that inside a small cube is a sheet of vorticity, hence vorticity concentrated on a surface. Then the dimension of $\vec{\gamma}_0^*$ differs from the one given in (5.3.3); it becomes

$$[\vec{\gamma}_0^*] = [l]\,[t]^{-1} \ . \tag{5.3.5}$$

Repeating the reasoning given above, we find in this case $\alpha_1 = 1$, $\alpha_2 = 3$ and $\alpha_3 = 2$. Hence now we have

$$E^* = O(\tilde{\varepsilon}^3) \ , \qquad \tilde{\varepsilon} \to 0 \ . \tag{5.3.6}$$

The case that the vorticity is concentrated on a line has no meaning with respect to kinetic energy considerations. The reason is that such a vortex line has per unit of length in each tube around it, however narrow it may be, a non-physical infinite amount of kinetic energy.

So we have to discuss the two cases: (1) continuous vorticity and (2) vortex sheets.

First, case (1) with continuous vorticity. When we consider a part $B(b)$ of length b of the cylinder B, we can, by means of the unlimited number of impervious and rigid reference surfaces, divide this space into more and more disconnected cubes, the number of which becomes $O(\tilde{\varepsilon}^{-3})$. Because each cube has by (5.3.4) a kinetic energy content of $O(\tilde{\varepsilon}^{5})$, the kinetic energy content of $B(b)$ becomes

$$O(\tilde{\varepsilon}^{-3}) \cdot O(\tilde{\varepsilon}^{5}) = O(\tilde{\varepsilon}^{2}) \to 0 \ , \qquad \tilde{\varepsilon} \to 0 \ . \tag{5.3.7}$$

In other words, we find for case (1) that by the wings W_l in the optimum case, the fluid can be set at rest in the region $B(b)$ when $\tilde{\varepsilon} \to 0$.

Second, case (2) when we have for instance one vortex sheet G in the cylinder B. Then we have to divide only a small part of B of width $\tilde{\varepsilon}$ around G into small cubes with edges of length $\tilde{\varepsilon}$. Here we need a number of cubes of $O(\tilde{\varepsilon}^{-2})$ within $B(b)$. Because each sequestered cube now has by (5.3.6) a kinetic energy content of $O(\tilde{\varepsilon}^{3})$, the kinetic energy content of $B(b)$ becomes

$$O(\tilde{\varepsilon}^{-2})O(\tilde{\varepsilon}^{3}) = O(\tilde{\varepsilon}) \to 0 \ , \qquad \tilde{\varepsilon} \to 0 \ . \tag{5.3.8}$$

Hence also here we can in the optimum case, set the fluid at rest in $B(b)$ by means of optimum wing systems W_l, when also the boundary of B is taken as a reference surface. This holds also for any finite number of vortex sheets within B.

Note that in general it is not possible to let a reference surface H of a flexible wing W coincide with the vortex sheet G and then choose the bound vorticity of W such that it annihilates the vorticity on G. The reason is that the circulation around G inside the working region B need not be zero. So a wing W which always leaves behind free vorticity with total circulation equal to zero, cannot annihilate the vorticity on G in this case. When, however, the circulation of the sheet G within B is equal to zero, then we can extract its energy by means of a single wing of which the reference surface coincides with G.

We remark that the above-mentioned kinetic energy extraction can also be carried out by only one flexible wing. However, then its reference surface has to become extremely complicated because it has to divide the space into smaller and smaller sequestered cubes or other small disconnected regions.

Now we go on with the cylinder B, placed in the disturbed fluid with periodic velocity field $\vec{v}_0^{\,*}$. Inside B we assume continuous vorticity and a finite number of vortex sheets. Then by the shed free vorticity of the unlimited number of optimum W_l or, what is the same, by the presence of the impervious reference surfaces H_l, the component of the resulting velocity field, normal to the boundary ∂B of B, has to vanish. The field outside the cylinder induced by the shed vorticity at the H_l has a uniquely valued potential Φ_1 which has to satisfy

$$\frac{\partial \Phi_1}{\partial n} = -\vec{v}_0^{\,*} \cdot \vec{n} \ , \qquad (x,y,z) \in \partial B \ , \tag{5.3.9}$$

where $\vec{n}$ is the unit normal at ∂B pointing out of B. This potential follows again from the solution of a Neumann problem when we demand that the circulation around B is zero.

From the foregoing it follows that the extractable kinetic energy E has the value

$$E = \tfrac{1}{2}\rho \Big\{ \int_{V_0(b)} (\vec{v}_0^{\,*})^2 \, d\text{Vol} - \int_{V_0(b)} (\vec{v}_0^{\,*} + \text{grad}\ \Phi_1)^2 \, d\text{Vol} + \int_{V_i(b)} (\vec{v}_0^{\,*})^2 d\text{Vol} \Big\} \ , \tag{5.3.10}$$

where

$$V_0(b): \quad 0 \le x \le b \ , \qquad (x,y,z) \notin B \ ,$$
$$V_i(b): \quad 0 \le x \le b \ , \qquad (x,y,z) \in B \ . \tag{5.3.11}$$

Following an analogous reasoning as we used below (5.2.15) and by (5.3.9) we can write the extractable energy (5.3.10) as

$$E = \tfrac{1}{2}\rho \Big\{ -\int_{\partial B(b)} \vec{v}_0^{\,*} \cdot \vec{n}\, \Phi_1 \, dS + \int_{V_i(b)} (\vec{v}_0^{\,*})^2 \, d\text{Vol} \Big\} \ , \tag{5.3.12}$$

where

$$\partial B(b): \quad 0 \le x \le b \ , \qquad (x,y,z) \in \partial B \ . \tag{5.3.13}$$

It can be proved analogously as in the previous section that $\Phi_0 = -\Phi_1$ is the potential of the velocity field $\vec{v}_0$ outside B, by which we can replace $\vec{v}_0^{\,*}$ outside B without changing the maximum amount of extractable kinetic energy.

5.4. The Variational Problem for Lifting Surface Systems

Our object will be the optimization of lifting surface systems as defined in Section 5.1. The reference surfaces H_l and the impermeable rigid periodical bodies C_j are prescribed, while the intensities $f_l(\xi_l, \eta_l, t)$ of the force fields (5.1.6) which represent the lifting surfaces have to be optimized. For simplicity we start with the case that only one reference surface H with coordinates ξ, η is present (Figure 5.2.1). The fluid is assumed to be disturbed by a time-independent velocity field $\vec{v}_0^{\,*}$ of $O(\varepsilon)$. We assume again that H and $\vec{v}_0^{\,*}$ are periodic with period b in the x-direction and that the force field repeats its action after each period of length.

When the wing W of finite extent has passed along, it has left behind free vorticity $\vec{\gamma}$ at H by which it alters the kinetic energy, which far before the arrival of W belonged to $\vec{v}_0{}^*$ alone. The resulting kinetic energy cannot be used any more and should be made as small as possible. We introduce the velocity potential $\Phi(x, y, z)$ which belongs to the velocity field induced by $\vec{\gamma}$. This potential is independent of time because the wing is supposed to be already at a large distance. The kinetic energy E left behind per period of length has the value

$$E(\Phi) = \tfrac{1}{2}\,\rho \int\int_{-\infty}^{\infty}\int_{0}^{b} \{\vec{v}_0{}^* + \text{grad}\,\Phi\}^2 \; dx\, dy\, dz \ . \tag{5.4.1}$$

We remark that in expression (5.4.1) the behaviour of the wing is represented only by the potential Φ of the disturbance velocities caused by the shed vorticity $\vec{\gamma}$. From this it follows that any wing which sheds the same free vorticity, has the same losses of kinetic energy. Hence we can, for instance, replace the lifting surface, with respect to the linearized optimization theory, also by a lifting line with a suitable circulation distribution. This does not mean that the lifting surface has to be slender, but only that the optimization theory is linear. Using lifting lines sometimes facilitates thinking about a problem, see for instance Sections 6.3 and 6.13.

The kinetic energy E (5.4.1) has to be minimized under some constraints. For instance, we can demand that the mean value of the thrust with respect to time, reckoned positive in the positive x-direction, must have a prescribed value $\bar{T}$. This condition can by (5.1.6) be written as

$$\frac{1}{\tau}\int_0^{\tau}\int_H f(\xi,\eta,t)\cos_{(n,x)}(\xi,\eta)\; dS\; dt = -\bar{T} \ , \tag{5.4.2}$$

where $\cos_{(n,x)}$ is the cosine of the angle between the normal $\vec{n}$ at the reference surface H (Figure 5.1.1) and the positive x-direction, τ is the time in which the wing passes along a period b and dS is an element of area of H. The minus sign at the right-hand side of (5.4.2) occurs because the external force field $\vec{F} = f(\xi,\eta,t)\vec{n}$ acts on the fluid, hence by action equals reaction, minus this force field acts on the wing. In (5.4.2) only the area of the planform of the wing, which is a finite part of H, contributes to the integral.

Using the periodicity of the external force field and of H, we can write (5.4.2) as

$$\frac{1}{\tau}\int_0^{\tau}\sum_{m=-\infty}^{\infty}\int_{H(b)} f(\xi,\eta,t+m\tau)\cos_{(n,x)}(\xi,\eta)\; dS\; dt = -\bar{T} \ , \tag{5.4.3}$$

where $H(b)$ is one period of H, hence

$$\frac{1}{\tau}\int_{-\infty}^{\infty}\int_{H(b)} f(\xi,\eta,t)\cos_{(n,x)}(\xi,\eta)\; dS\; dt = -\bar{T} \ . \tag{5.4.4}$$

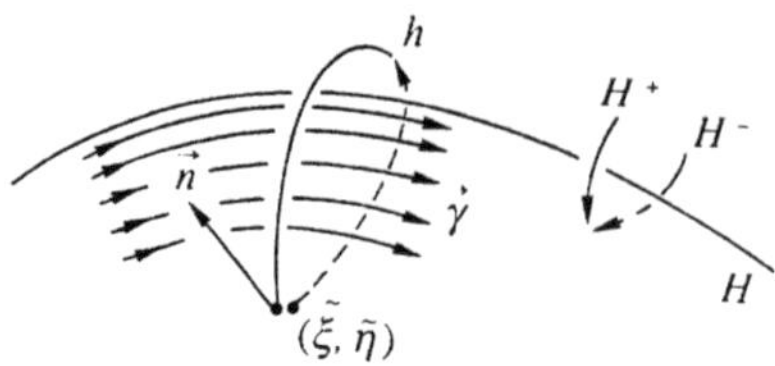

Fig. 5.4.1. The potential jump $[\Phi]_-^+$ across H.

Only a finite interval of time contributes to this integral, namely the interval during which W passes $H(b)$. First we carry out the integration with respect to t for a fixed point $(\tilde{\xi}, \tilde{\eta})$ at H. We consider a closed contour h (Figure 5.4.1) which pierces H perpendicularly (not necessary) at the point $(\tilde{\xi}, \tilde{\eta})$ and further has no contact with H. Then when the wing or, what is the same, the external force field has passed by, it follows from (1.3.25) that the circulation along h has been increased by the amount

$$-\frac{1}{\rho}\int_{-\infty}^{\infty} f(\tilde{\xi}, \tilde{\eta}, t)\, dt \ . \tag{5.4.5}$$

However, this increase is nothing but the jump in potential $[\Phi]_-^+$ at the point $(\tilde{\xi}, \tilde{\eta})$ of H. Hence we can write condition (5.4.4) as

$$\frac{\rho}{\tau}\int_{H(b)} [\Phi]_-^+(\xi, \eta)\cos_{(n,x)}(\xi, \eta)\, dS = \bar{T} \ . \tag{5.4.6}$$

The jump $[\Phi]_-^+$ at the point $(\tilde{\xi}, \tilde{\eta})$ on H is also equal to the amount of shed vorticity $\vec{\gamma}$ which lies "between" $(\tilde{\xi}, \tilde{\eta})$ and that boundary of H which is enclosed by h. The direction of $\vec{\gamma}$ is coupled with a right-hand screw to the orientation of the contour h which is from the H^- side towards the H^+ side. Here we have for $f(\xi, \eta, t) > 0$, $[\Phi]_-^+ < 0$, then $\vec{\gamma}$ is opposite to its indicated direction in Figure 5.4.1.

Formula (5.4.6) represents a constraint on the periodic potential functions Φ (x, y, z) which are admitted in the variational problem to minimize the lost kinetic energy. By replacing for instance the function $\cos_{(n,x)}(\xi, \eta)$ by $\cos_{(n,y)}(\xi, \eta)$, which is the cosine between the normal $\vec{n}$ at H and the positive y-direction (Figure 5.2.1), we put a demand on $\Phi(x, y, z)$ so that a mean force in the y-direction or in other words a mean lift $\bar{L}$ is delivered

$$\frac{\rho}{\tau}\int_{H(b)} [\Phi]_-^+(\xi, \eta)\cos_{(n,y)}(\xi, \eta)\, dS = \bar{L} \ . \tag{5.4.7}$$

Demanding (5.4.6) and (5.4.7) simultaneously, we ask for a potential which delivers a mean thrust $\bar{T}$ as well as a mean lift $\bar{L}$ in an optimum way. By this it is possible

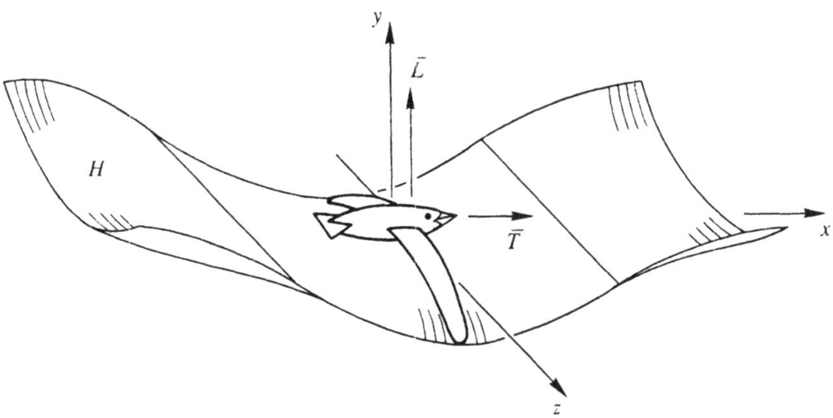

Fig. 5.4.2. Impression of the reference surface of the wings of a bird.

to calculate the least power needed by a bird when the reference surface of its wings is prescribed, for instance by taking a moving picture of its flight (Figure 5.4.2).

More generally, we can put M constraints on the potential Φ and choose instead of the cos functions other periodic functions $g_\nu(\xi,\eta)$, $\nu = 1,\ldots,M$,

$$\int_{H(b)} [\Phi]_-^+(\xi,\eta)\, g_\nu(\xi,\eta)\, dS - \frac{T}{\rho} F_\nu = 0 \ , \qquad \nu = 1,\ldots,M \ , \tag{5.4.8}$$

where the F_ν are the demanded force actions.

Also we can consider the case that we have m lifting surfaces, hence m reference surfaces H_l, $l = 1,\ldots,m$. Then we have to replace (5.4.8) by

$$\sum_{l=1}^{m} \int_{H_l(b)} [\Phi]_{-,l}^+(\xi_l,\eta_l)\, g_{l\nu}(\xi_l,\eta_l)\, dS - \frac{T}{\rho} F_\nu = 0 \ , \quad \nu = 1,\ldots,M \ , \tag{5.4.9}$$

where $[\Phi]_{-,l}^+$ denotes the values of $(\Phi^+ - \Phi^-)$ at H_l. By choosing the functions $g_{l\nu}(\xi_l,\eta_l) = 0$ for a number of values of l, the corresponding wings are not subject to constraint ν. Hence we can also put constraints on the action of selected groups of wings.

We assume that the constraints are not contradictory. Consider for instance a screw propeller with given angular velocity ω and velocity of advance U. Then in the linearized theory the thrust T and the moment M around the axis of rotation satisfy

$$TU = M\omega \ . \tag{5.4.10}$$

Hence T and M cannot be prescribed independently of each other. Equation (5.4.10) follows from an energy consideration because both T and M are $O(\varepsilon)$ while the lost kinetic energy is $O(\varepsilon^2)$. Another way of deriving (5.4.10) is by means of a consideration of the geometry of the reference surfaces of the blades. The force on a blade element is in the linearized theory perpendicular to the reference surface. Its components in the x-direction and in the direction of the rotational velocity are geometrically coupled by the shape of the helicoidal reference surface in such a way that (5.4.10) arises.

5.5. Necessary Condition for an Optimum

In this section we will give necessary conditions for the minimization of the kinetic energy $E(\Phi)$ (5.4.1) with respect to the velocity potential $\Phi(x, y, z)$. We consider, as in Section 5.1, m reference surfaces H_l ($l = 1, \ldots, m$) and n periodic bodies C_j ($j = 1, \ldots, n$). On Φ we put M constraints of the type (5.4.9). Then instead of E we can consider the functional

$$J(\Phi) = E(\Phi) + \sum_{\nu=1}^{M} \lambda_\nu \left\{ \sum_{l=1}^{m} \int_{H_l(b)} [\Phi]^{+}_{-,l}(\xi_l, \eta_l)\, g_{l\nu}(\xi_l, \eta_l)\, dS - F_\nu \right\}, \tag{5.5.1}$$

where the λ_ν are still unknown numbers, the Lagrange multipliers. We can now treat the functional $J(\Phi)$ just as if there are no constraints on Φ, see for instance [16].

Suppose $\Phi(x, y, z)$ is the optimum potential we are looking for. Then we replace Φ by $\Phi + \delta\Phi$, where the perturbation potential $\delta\Phi(x, y, z)$ has to be periodic with period b in the x-direction and has to satisfy the condition of tangential flow to the periodic bodies C_j. We now have to determine the first variation δJ of J. Hereto we start with the first variation δE of E which can be written as (see (5.2.10))

$$\delta E(\delta\Phi) = \rho \iint_{-\infty}^{\infty} \int_{0}^{b} \text{div}\ \{(\vec{v}_0^{\,*} + \text{grad}\ \Phi)\delta\Phi\}\ dx\, dy\, dz\ . \tag{5.5.2}$$

In (5.5.2) the integration has to be carried out over the slab of space $0 \leq x \leq b$, $-\infty < y, z < \infty$, minus the region cut out by the bodies C_j.

The boundary of this region of integration consists of the $H_l(b)$ ($l = 1, \ldots, m$), the planes $x = 0$ and $x = b$ minus the parts where the C_j cut these planes and the relevant parts of the boundaries ∂C_j of the C_j. When we convert the volume integral in (5.5.2) into an integral over this boundary, it is seen that besides the planes $x = 0$ and $x = b$ also the boundaries ∂C_j yield no contribution, because the flow is tangent to the C_j. Hence there remain the integrations over both sides of the $H_l(b)$, we find analogously to (5.2.11)

$$\delta E(\delta\Phi) = -\rho \sum_{l=1}^{m} \int_{H_l(b)} \{(\vec{v}_0^{\,*} + \text{grad}\ \Phi) \cdot \vec{n}\}\ [\delta\Phi]^{+}_{-,l}\ dS\ , \tag{5.5.3}$$

because $(\vec{v}_0^* + \text{grad}\,\Phi) \cdot \vec{n}$ is continuous across the H_l and $\vec{n}$ points from H_l^- towards H_l^+.

In the case of an optimum Φ the first variation $\delta J(\delta\Phi)$ of $J(\Phi)$ (5.5.1) has to vanish hence with (5.5.3)

$$\delta J(\delta\Phi) = -\rho \sum_{l=1}^{m} \int_{H_l(b)} \{(\vec{v}_0^* + \text{grad}\,\Phi) \cdot \vec{n}\}\, [\delta\Phi]_{-,l}^{+}\, dS$$

$$+ \sum_{\nu=1}^{M} \lambda_\nu \sum_{l=1}^{m} \int_{H_l(b)} g_{l\nu}(\xi_l, \eta_l)\, [\delta\Phi]_{-,l}^{+}\, dS = 0 \ . \qquad (5.5.4)$$

As in Section 5.2 the jump $[\delta\Phi]_-^+$ can be chosen to be zero everywhere, with the exception of the neighbourhood of some arbitrary point P at $H_l(b)$ (Figure 5.2.2), then we find from (5.5.4)

$$\frac{\partial\Phi}{\partial n}(\xi_l, \eta_l) = -\vec{v}_0^* \cdot \vec{n}(\xi_l, \eta_l) + \frac{1}{\rho} \sum_{\nu=1}^{M} \lambda_\nu g_{l\nu}(\xi_l, \eta_l) \ , \quad l = 1, \ldots, m \ . \qquad (5.5.5)$$

This relation is a necessary condition for the normal component $\partial\Phi/\partial n$ of the velocity induced at each point of the reference surfaces H_l by the free vorticity shed by the optimum wings. Hence we have to solve for the calculation of Φ a Neumann problem (also $\partial\Phi/\partial n = 0$ at ∂C_j), while the Lagrange multipliers λ_ν are still unknown. Having solved this problem, we can determine the λ_ν by the conditions (5.4.9).

From the necessary conditions (5.5.5) for an optimum it follows again that we could have replaced the original disturbance field $\vec{v}_0^*$ by the field $\vec{v}_0$ as defined in Section 5.2, hence with its vorticity $\vec{\gamma}_0$ confined to the H_l and with a potential Φ_0 (5.2.6) defined outside the H_l. Also it follows from (5.5.5) that we can divide the action of the lifting surface system into two parts, in other words the potential Φ can be split into two parts

$$\Phi = \Phi_1 + \Phi_2 \ , \qquad (5.5.6)$$

which we will discuss separately.

In the first place the lifting surfaces W_l have to "clean" their reference surfaces H_l from the vorticity $\vec{\gamma}_0$ belonging to the modified disturbance velocity field $\vec{v}_0$ of which the kinetic energy can be extracted. The corresponding part Φ_1 of Φ follows from

$$\frac{\partial\Phi_1}{\partial n}(\xi_l, \eta_l) = -\vec{v}_0^* \cdot \vec{n}(\xi_l, \eta_l)$$

$$= -\frac{\partial\Phi_0}{\partial n}(\xi_l, \eta_l) \ , \qquad l = 1, \ldots, m \ , \qquad (5.5.7)$$

where Φ_0 and Φ_1 have the same meaning as in Section 5.2, where just like here $\Phi_0 = -\Phi_1$.

Also here we have to demand that the circulation vanishes around each H_l ($l = 1, \ldots, m$) separately, otherwise it is not possible for the wings to extract the kinetic energy from $\vec{v}_0$. In performing a motion suitable for this task, force actions have to be exerted on the fluid and their reactions act at the W_l. These reactions of the fluid will in general contribute already to the constraints (5.4.9) put on the lifting surface system. Whether this contribution will be favourable or not, will be discussed in the last but one paragraph of this section.

The second part Φ_2 of Φ belongs to that part of the motion of the W_l, which is carried out along their "cleaned" reference surfaces H_l. These motions have to supply the deficiency in the demanded force actions (5.4.9).

$$\frac{\partial \Phi_2}{\partial n}(\xi_l, \eta_l) = \frac{1}{\rho} \sum_{\nu=1}^{M} \lambda_\nu \, g_{l\nu}(\xi_l, \eta_l) \ , \qquad l = 1, \ldots, m \ , \tag{5.5.8}$$

where in this Neumann problem again the circulations around the H_l have to be zero. By the linearity of the problem we can express Φ_2 as a linear combination of potentials each multiplied by an unknown λ_ν ($\nu = 1, \ldots, M$), then the λ_ν have to be chosen so that $\Phi = \Phi_1 + \Phi_2$ satisfies the M conditions (5.4.9).

Now we consider the kinetic energy E_p put into the fluid per period of length b of the motion by the optimum lifting surface system W_l. This energy is the difference between the kinetic energy in the fluid over one period far downstream of the W_l and the kinetic energy which was present in the fluid at that period long before the W_l arrived, hence

$$E_p = \tfrac{1}{2}\rho \int\!\!\int_{-\infty}^{\infty} \int_0^b \left(\vec{v}_0^{\,*} + \text{grad}\, \Phi_1 + \text{grad}\, \Phi_2\right)^2 \, dx\, dy\, dz$$
$$- \tfrac{1}{2}\rho \int\!\!\int_{-\infty}^{\infty} \int_0^b (\vec{v}_0^{\,*})^2 \, dx\, dy\, dz \ . \tag{5.5.9}$$

It is not difficult to show by using formulas analogous to those below (5.2.14) that we can write E_p also in the form

$$E_p = \tfrac{1}{2}\rho \int\!\!\int_{-\infty}^{\infty} \int_0^b \{(\text{grad}\, \Phi_2)^2 - (\vec{v}_0)^2\} \, dx\, dy\, dz \ . \tag{5.5.10}$$

This expression can be reduced as before to an integral over the reference surfaces. Hence all disturbance fields $\vec{v}_0^{\,*}$ with the same normal component at the H_l are equivalent with respect to our optimization problem in the linearized theory.

It is elucidating to apply the previous results to the optimization of the flight of the bird of Figure 5.4.2. For the sake of simplicity, we take $\vec{v}_0^{\,*} = 0$ and hence $\Phi_1 = 0$. Condition (5.5.8) yields

$$\frac{\partial \Phi_2}{\partial n}(\xi,\eta) = \frac{1}{\rho}\left\{\lambda_1 \cos_{(n,x)}(\xi,\eta) + \lambda_2 \cos_{(n,y)}(\xi,\eta)\right\} . \tag{5.5.11}$$

This means that the optimum vorticity left behind by the bird consists of two types. First, vorticity which is induced on H when H as a rigid and impervious surface is translated in the x-direction, say with velocity one. This type "takes care" of the mean thrust. Second, vorticity which is induced when H is translated in the y-direction, say also with velocity one. This type "takes care" of the mean lift. The combination of these two vorticity types, multiplied by λ_1 and λ_2 respectively which follow from (5.4.6) and (5.4.7), has to be left behind by the bird in its optimum flight.

Finally, we discuss the usefulness of the kinetic energy of the equivalent disturbance velocity field $\vec{v}_0$. For this purpose we have to consider the force actions needed to induce Φ_1 or to "clean" the reference surfaces H_l from their vorticity $\vec{\gamma}_0$. Suppose the reactions of these forces are in the desired direction of the forces to be delivered by the wings. Then the values of $|\lambda_\nu|$ will become smaller than they have to be without the disturbance field $\vec{v}_0$. This is favourable for two reasons. First, the kinetic energy is extracted out of the field $\vec{v}_0$ and becomes available. Second, the energy needed for the smaller values of $|\lambda_\nu|$ is less than the energy needed to satisfy the force constraints in an undisturbed fluid. In the other case, when the reactions of the force actions needed to create Φ_1 are or are for an important part opposite to the demanded force actions of the wings, the $|\lambda_\nu|$ or some of them have to be larger than in the undisturbed case. Then, although energy is extracted from the velocity field $\vec{v}_0$, the extra energy needed for these larger values of the $|\lambda_\nu|$ can become so large that it nullifies or makes negative the profit of the energy extraction. In the latter case the disturbance field $\vec{v}_0$ is unfavourable.

In the next section we will show by a simple example the possibility of favourable or unfavourable disturbance velocity fields.

5.6. Influence of a Disturbance Velocity Field

A screw propeller is in general placed behind the stern of a ship. The water, flowing along the hull, is dragged by its viscosity in the direction of motion of the ship. Hence behind the ship it will have a mean disturbance velocity in the direction of the thrust of the propeller. We will make it plausible in this section that this is favourable for the efficiency of the propeller. More specifically we can say that, with respect to the blades of the ship screw, the mean disturbance velocity of the water has a component in the direction of their lift. We can simplify the problem, without loosing its essence, by considering instead of the blades which move along their helicoidal reference surfaces, a flat lift-producing wing through a fluid with an upward disturbance velocity.

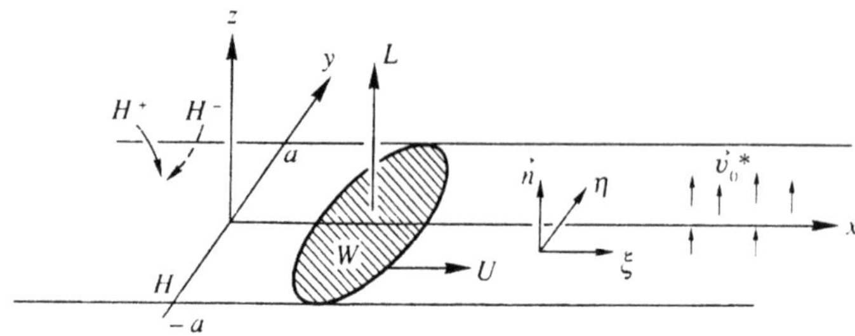

Fig. 5.6.1. Wing moving along flat strip.

We consider an optimum wing moving with constant velocity U in the positive x-direction along a flat strip H; $-\infty < x < \infty$, $|y| < a$, $z = 0$ (Figure 5.6.1). The coordinate system (ξ, η) of H can be identified in this case with (x, y). The disturbance velocity field $\vec{v}_0{}^*$ is assumed to be constant and has the form

$$\vec{v}_0{}^* = (0, 0, v_{0z}^*) \ , \qquad v_{0z}^* = O(\varepsilon) \ . \tag{5.6.1}$$

When the wing has passed and is far away in the positive x-direction, all phenomena are independent of x. Hence for the period b in the x-direction, we can take any number, for instance $b = 1$. We demand that the wing delivers a lift force L in the positive z-direction.

The potential Φ_0 of the modified disturbance velocity field $\vec{v}_0$ satisfies by (5.2.5)

$$\frac{\partial \Phi_0}{\partial z} = v_{0z}^* \ , \quad -\infty < x < \infty \ , \quad |y| < a \ , \quad z = 0 \ . \tag{5.6.2}$$

Because, as we mentioned, the problem is independent of the x-coordinate, the vorticity $\vec{\gamma}_0$ at H is parallel to the x-axis and will be denoted by the scalar function $\gamma_0(y)$. Using a formula analogous to (1.1.16) we form for $\gamma_0(y)$ the singular integral equation

$$-\frac{1}{2\pi} \oint_{-a}^{a} \frac{\gamma_0(\eta)}{(\eta - y)} \, d\eta = v_{0z}^* \ , \qquad |y| < a \ , \tag{5.6.3}$$

where $\gamma_0(y)$ is positive with a right-hand screw in the positive x-direction. Further, because the total circulation around H has to be zero, we demand

$$\int_{-a}^{a} \gamma_0(y) \, dy = 0 \tag{5.6.4}$$

and we tolerate for $\gamma_0(y)$ in connection with the integrability of the kinetic energy, singularities at $y = \pm a$ of the square root type. The solution of (5.6.3) is discussed in Appendix A, (A.3.14) with $c_2 = 0$, we find

$$\gamma_0(y) = -2 v_{0z}^* y \, (a^2 - y^2)^{-1/2} \ , \qquad |y| < a \ . \tag{5.6.5}$$

We have one wing and one constraint on the lift, hence in (5.4.9) $m = M = 1$ and $g_{11}(\xi, \eta) = \cos_{(n,y)} = 1$. By this, the necessary condition (5.5.5) for the potential Φ, "added" by the wing to the disturbance flow and far behind the wing, becomes

$$\frac{\partial \Phi}{\partial z} = -v_{0z}^* + \lambda \ , \quad |y| < a \ , \quad z = 0 \ , \tag{5.6.6}$$

where we have absorbed the constant ρ^{-1} in the Lagrange multiplier λ. The vorticity γ which is shed by the wing and which belongs to Φ (5.6.6) follows directly from (5.6.5), because both (5.6.2) and (5.6.6) have constants at their right-hand side. So replacing in (5.6.5) v_{0z}^* by $(-v_{0z}^* + \lambda)$ we obtain

$$\gamma(y) = -2\,(-v_{0z}^* + \lambda)\, y\, (a^2 - y^2)^{-1/2} \ , \qquad |y| < a \ . \tag{5.6.7}$$

From (5.6.7) we find by $[\Phi]_-^+(y) = \int_{-a}^{y} \gamma(\eta)\, d\eta$,

$$[\Phi]_-^+(y) = 2\,(-v_{0z}^* + \lambda) \int_{-a}^{y} \frac{\eta\, d\eta}{(a^2 - \eta^2)^{1/2}} = -2\,(-v_{0z}^* + \lambda)(a^2 - y^2)^{1/2} \ . \tag{5.6.8}$$

Because $[\Phi]_-^+(y)$ equals the circulation of the wing W at the place y, we have recovered the well-known elliptic circulation distribution of a wing with the lowest induced resistance.

In order to calculate the still unknown Lagrange multiplier λ we use (5.4.9) which reduces in this case to

$$\int_{-a}^{a} [\Phi]_-^+(y)\, dy = \frac{L}{\rho\, U} \ , \tag{5.6.9}$$

because here we have taken $b = 1$ and hence the time period τ becomes U^{-1}. Substitution of (5.6.8) into (5.6.9) yields for λ the value

$$\lambda = \frac{-L}{\rho\, \pi U a^2} + v_{0z}^* \ . \tag{5.6.10}$$

We now calculate the kinetic energy per unit of length in the x-direction which remains in the fluid after the wing has passed by and we have again changed from $\vec{v}_0^{\,*}$ to $\vec{v}_0$. Using (5.6.2), (5.6.6), (5.6.8) and the analogous formula of (5.6.8) for $[\Phi_0]_-^+$ we obtain for this energy

$$-\tfrac{1}{2}\,\rho \int_{-a}^{a} [\Phi + \Phi_0]_-^+ \frac{\partial}{\partial n}(\Phi + \Phi_0)\, dy = \tfrac{1}{2}\,\pi \rho\, a^2 \left(\frac{-L}{\rho\, \pi U a^2} + v_{0z}^* \right)^2 \ . \tag{5.6.11}$$

The energy needed to move the wing over one unit of length in the x-direction is this amount minus the energy belonging to the modified velocity field $\vec{v}_0$ which was

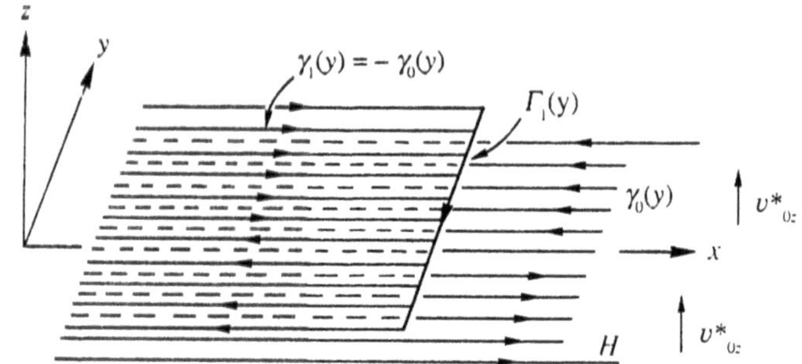

Fig. 5.6.2. Extraction of kinetic energy out of $\vec{v}_0$; $v^*_{0z} > 0$.

already present in the fluid. This amount follows from (5.6.11) by replacing at the right-hand side the expression between brackets by v^*_{0z}.

We find for the needed energy

$$\frac{L^2}{2\rho\,\pi U^2 a^2} - \frac{L v^*_{0z}}{U} \; . \tag{5.6.12}$$

We see that the first term is the energy needed when the fluid is undisturbed. The second term reduces this energy when $v^*_{0z} > 0$, hence when v^*_{0z} is in the direction of the desired lift and augments this energy when $v^*_{0z} < 0$ hence in the direction opposite to the desired lift.

We now compare these results with the remarks of the last but one paragraph of the previous section, especially the influence of the direction of the force actions connected to the part $\Phi_1 = -\Phi_0$ (5.5.7) of the velocity potential Φ, which is needed to extract the kinetic energy of $\vec{v}_0$. In order to calculate these force actions we replace the wing by a straight lifting line parallel with the y-axis, with bound vorticity $\Gamma_1(y)$ (Figure 5.6.2). As has been argued in Section 5.4, this does not mean that the wing has to be slender but only that we consider a linearized theory. The bound vorticity $\Gamma_1(y)$ has in order to extract the kinetic energy out of the velocity field $\vec{v}_0$, to shed free vorticity $\gamma_1(y)$ opposite to the vorticity $\gamma_0(y)$ of $\vec{v}_0$, clearly

$$\Gamma_1(y) = -2\, v^*_{0z}(a^2 - y^2)^{1/2} \; ,$$

$$\gamma_1(y) = \frac{d\Gamma_1}{dy} = 2\, v^*_{0z}\; fracy(a^2 - y^2)^{1/2} = -\gamma_0(y) \; , \tag{5.6.13}$$

where $\Gamma_1(y)$ is reckoned positive with a right-hand screw in the positive y-direction. In Figure 5.6.2 we have given, in spite of this agreement, the occurring rotation of the vorticity with a right-hand screw. Far behind the lifting line the free vorticity γ_1 creates the potential $\Phi_1 = -\Phi_0$.

When the bound concentrated vortex Γ_1 moves with the velocity U in the positive x-direction, it delivers for the case $v^*_{0z} > 0$ of Figure 5.6.2, two force components.

First, a force component $\tilde{T}$ of $O(\varepsilon^2)$ in the positive x-direction of magnitude

$$\tilde{T} = -\rho(\tfrac{1}{2}\, v_{0z}^*) \int_{-a}^{a} \Gamma_1(y)\, dy = \tfrac{1}{2}\, \pi\rho\, a^2 (v_{0z}^*)^2 = O(\varepsilon^2) \ , \tag{5.6.14}$$

which assists in moving the wing and by which energy is extracted out of the fluid. Here we have to use $\frac{1}{2}\, v_{0z}^*$ in the calculation of $\tilde{T}$ because behind $\Gamma_1(y)$ the vorticity has disappeared by the addition of $\gamma_1(y)$ so that only a half infinite strip with $\gamma_0(y)$ remains in front of the lifting line. Second, a lift force $\tilde{L}$ in the positive z-direction of magnitude

$$\tilde{L} = -\rho\, U \int_{-a}^{a} \Gamma_1(y)\, dy = \rho\, \pi U a^2 v_{0z}^* = O(\varepsilon) \ , \tag{5.6.15}$$

which contributes to the prescribed lift force L. When v_{0z}^* is sufficiently large it is even possible that $\tilde{L}$ is larger than L, then we do not need to extract all the kinetic energy out of the modified disturbance velocity field $\vec{v}_0$.

The above discussion shows why a sea gull can fly along a range of dunes without moving its wings, when a transverse wind causes an upward flow. It uses the thrust $\tilde{T}$ (5.6.14) to overcome its resistance.

When $v_{0z}^* < 0$ also $\Gamma_1(y)$ changes sign, then although $\tilde{T}$ (5.6.14) remains positive, $\tilde{L}$ (5.6.15) becomes negative, this is the unfavourable situation. The energy needed to deliver the lift force $L - \tilde{L} > L$ in the now undisturbed fluid, minus $\tilde{T}U$, is larger than the energy needed to deliver L in the undisturbed fluid, see (5.6.12).

It is clear that, as we mentioned in the beginning of this section, the same happens to the blades of a screw propeller. Then the flat strip H we used here, has to be replaced by helicoidal strips, however this does not change in essence the phenomena. So from the point of view of efficiency it is recommendable to place the screw behind the stern of a ship. This holds also for an unsteady propeller.

5.7. Classes of Lifting Surface Systems

Up to now we have introduced lifting surface systems and have discussed their optimization in an inviscid and incompressible fluid. The reference surfaces H_l were given and we determined the circulation distributions or the force fields $\vec{F}_l$ (5.1.6) of the wings W_l moving along them, in such a way that certain constraints (5.4.9) are satisfied and that the kinetic energy left behind per period is minimum. Now we will give more freedom to the admitted lifting surface systems; to this end we introduce classes of these systems. It is again allowed that the fluid is disturbed by a velocity field $\vec{v}_0{}^*(x, y, z)$ of $O(\varepsilon)$, periodic in the x-direction.

A class of periodically moving lifting surface systems consists of all those systems which meet a certain number of conditions. There are four principal properties which systems belonging to one and the same class will have in common. They will have:

1. the same mean velocity of advance,
2. the same shortest period,
3. reference surfaces which can be embedded in the same cylindrical region B, called working region of the class, outside of which there is allowed a fixed set of rigid periodic bodies,
4. the same prescribed mean force actions.

Additional conditions with respect to the geometry of the reference surfaces can be imposed. For instance we can demand that these surfaces form local angles with the x-axis which are smaller than or equal to a prescribed value. This means that the reference surfaces are not allowed to be too steep.

We will consider a "best" lifting surface of a class. This is a system which leaves behind the "greatest lower bound" of the kinetic energy of all systems of that class.

We explain why we use the concept "greatest lower bound" and not the concept "minimum". Not each class has one or more best lifting surface systems. Assume for simplicity that the fluid is undisturbed, hence $\vec{v}_0^* = 0$ and consider the class of screw propellers with (1) the same velocity of advance, (2) the same shortest period, (3) the same diameter, (4) the same thrust and no other conditions. The rotational velocity is such that first it accelerates slightly and then decelerates slightly, within one period. Then we can increase the mean rotational velocity without changing its shortest period. By this the reference surfaces will crowd more and more the working region and as we will see later on, the efficiency of the propeller will increase. Hence no finite mean rotational velocity can appear as the result of an optimization process and hence there is no best propeller in this class. This is the reason that in the definition of best lifting surface system in a class we used the expression "greatest lower bound" of the kinetic energy, because that always exists. Of course, this example only makes sense in our linearized theory and with the neglect of viscosity and cavitation.

We next introduce the concept "ideal lifting surface system" connected to a class C. To this end we consider another class C_0 which has only the four principal properties of C, as mentioned in the second paragraph of this section. Hence C_0 can be more comprehensive than C, because C can have additional constraints on the geometry of its reference surfaces. We now will call a best lifting surface system of C_0, if it exists, an ideal lifting surface system connected to C.

Suppose we take some set of reference surfaces H_l which are compatible with C_0, hence they are within the working region and have the correct period. Then we can find an optimum wing system W_l for these H_l by the method described before. This system belongs to the class C_0. We cannot expect that W_l is the best system of C_0 or in other words is an ideal system connected to C, because in general we will need more reference surfaces to achieve this. Choose another set $\tilde{H}_l$ which is compatible with C_0 and look at the combined set $H_l \cup \tilde{H}_l$. Again we can find an optimum wing system by our previous method, which now belongs to $H_l \cup \tilde{H}_l$. This latter system, being possibly not ideal, is more efficient than the first one because it has a greater freedom for the distribution of its optimum vorticity. Now we can formulate the following criterion:

When for each combination of a fixed set H_l with any other set $\tilde{H}_l$ no vorticity is needed in the optimum case on the set $\tilde{H}_l$, it follows that the set H_l itself is able to yield an ideal system.

We will discuss this subject in some more detail. In order to find an ideal lifting surface system, we follow the same method as in Section 5.3, where we considered the energy extraction by means of lifting surfaces W_l moving along reference surfaces H_l which have to remain within the working region B. Also now we have a periodic initial disturbance field $\vec{v}_0^*$ in space. Then we crowd the cylinder B more and more with periodic reference surfaces H_l and optimize each time the lifting surface systems under the constraint of some prescribed force action. In imitation of (5.5.6) we split the originating potentials Φ into Φ_1 and Φ_2, where Φ_1 belongs to vorticity which annihilates the normal component of $\vec{v}_0^*$ at the H_l. Hence, as before, the effect of Φ_1 outside B is that in the limit of very many H_l it compensates also the normal component of $\vec{v}_0^*$ at the boundary ∂B of B,

It is clear that an ideal system does not exist when we have in the working region B continuously distributed vorticity $\vec{\gamma}_0^*$ or vortex sheets of which the total circulation of each one apart, is not zero. Then we have, as we discussed in Section 5.3, to divide B into very small sequestered cubes by means of an indefinitely increasing number of reference surfaces. However, we can still use the expression ideal system but then in a symbolic way.

The potential Φ_2 takes care of the prescribed force actions and can be considered to be created by lifting surfaces moving in an undisturbed fluid, because by the potential Φ_1 the fluid inside B has been brought to a standstill in the limit case of an "infinite" number of reference surfaces. The force action to be delivered by Φ_2 depends on the force action already delivered by Φ_1 as we discussed in Sections 5.5 and 5.6. It turns out, as we will show in Section 5.9 by means of an example, that for the creation of an optimum Φ_2 we only need a finite number of reference surfaces.

5.8. Quality Number

We now introduce an important quantity belonging to a lifting surface system W_l which is a member of a certain class C, namely its quality number q. Because we are interested here in optimization, we will consider especially this number with respect to optimum systems although it can be defined easily for any system. This number indicates the possibility of improving a lifting surface system by changing the geometry of its reference surfaces. For instance, the influence of the application of end plates to the blade tips of a screw propeller. The practical effectiveness of such a change, however, has to be judged by taking into account also the efficiency η of the propeller as we will discuss below.

The kinetic energy E_i denotes the kinetic energy which is added to the fluid per period of length b by an ideal lifting surface system connected to the class C, in the sense as defined in the previous section. This amount is by (5.5.10) equal to

$$E_i = \frac{1}{2}\rho \int\int_{-\infty}^{\infty} \int_0^b \{(\text{grad}\,\tilde{\Phi}_2)^2 - (\vec{v}_0)^2\}\,dx\,dy\,dz \ , \tag{5.8.1}$$

where $\tilde{\Phi}_2$ is the potential which takes care of the demanded force action of the ideal system.

The kinetic energy E_p put into the fluid per period b by the optimum lifting surface system W_l is given by (5.5.9) or (5.5.10), where Φ_1 and Φ_2 there belong to W_l.

It can happen theoretically that E_p or E_i or both are negative which means that more energy is extracted out of the fluid than is needed for the force action. In practical situations of, for instance, ship propulsion this does not occur and we assume $E_p > 0$ and $E_i > 0$. Note that in the case of the already mentioned sea gull flying along a range of dunes or of a sail plane flying in thermals, this need not be true.

We define the quality number by

$$q = \frac{E_i}{E_p} \leq 1 \ , \qquad E_p > 0 \ , \qquad E_i > 0 \ , \tag{5.8.2}$$

where the first inequality follows directly from the meaning of E_i.

In the above general case the quality number q will depend in general on the magnitude of the prescribed force actions. This is caused by the presence of the disturbance velocity field $\vec{v}_0^{\,*}$. When we consider however the case that the fluid is initially undisturbed, we have

$$\vec{v}_0^{\,*} = \vec{v}_0 = 0 \ , \qquad \Phi_1 = \text{constant } (= 0) \ , \tag{5.8.3}$$

where the latter equality follows from (5.5.7) or from the fact that no energy can be extracted out of the undisturbed fluid. Then we find from (5.8.1) and (5.5.9) or (5.5.10)

$$q = \frac{E_i}{E_p} = \frac{\int\int_{-\infty}^{\infty} \int_0^b (\text{grad}\,\tilde{\Phi}_2)^2\,dx\,dy\,dz}{\int\int_{-\infty}^{\infty} \int_0^b (\text{grad}\,\Phi_2)^2\,dx\,dy\,dz} \ . \tag{5.8.4}$$

When now we multiply the prescribed force actions by the same factor, $\tilde{\Phi}_2$ and Φ_2 will be multiplied by this factor and hence q depends only on the ratio of these force actions.

Making a further simplification, namely that there is only one constraint on the force action, for instance a prescribed thrust of a propeller or a prescribed lift of a wing, we have that both $\tilde{\Phi}_2$ and Φ_2 are proportional to the magnitude of this thrust or lift. Then q depends by means of E_i on the admitted class C and by means of E_p on the geometry of the prescribed reference surfaces of the optimum system, *it does not depend on the prescribed thrust or lift anymore*.

Finally, we consider only propellers. Then we can speak of the efficiency η of the propeller, which can be a ship screw or an unsteady propeller. We now discuss

Table 5.8.1. Four main possibilities of q and η.

	$q \approx 1$	$q < 1$
$\eta \approx 1$	I	II
$\eta < 1$	III	IV

Table 5.8.1. In our linearized theory, for the type of propulsion we consider in this chapter, we have $\eta = 1 - O(\varepsilon)$. Hence it seems that always $\eta \approx 1$. However, the linearized theory is most probably not devoid of sense for, let us say, η is about 0.7. For this kind of values we write $\eta < 1$.

The efficiency is about one in the cases I and II of Table 5.8.1. This means that the propeller cannot be improved substantially. In case III, q is nearly one. Hence although η is low, the propeller cannot be improved significantly because its energy loss per period is nearly equal to the energy loss of an ideal propeller connected to class C ($E_p \approx E_i$).

In case IV we have $q < 1$ and $\eta < 1$. Then we can try to improve, from the potential point of view, the efficiency of the propeller by changing its reference surfaces while it still remains within the class C or by enlarging this class by for instance demanding only the four properties $1, \ldots, 4$ given in the second paragraph of this section. In Section 6.4 we give an application of the quality number in relation with the performance of screw propellers with or without end plates.

Using the quality number, we can write the efficiency of a propeller as

$$\eta = \frac{b\,\bar{T}}{b\,\bar{T} + E_p} = \left(1 + \frac{E_i}{q\,b\,\bar{T}}\right)^{-1} , \tag{5.8.5}$$

where $\bar{T}$ is the mean value of the thrust. Note that for a given velocity of advance, a given period, a prescribed working region and no disturbance velocities, E_i = const. $\bar{T}^2$.

We remark that the quality number q of a propeller in a semi-linear theory depends on the definition of its working region. A realistic way seems at first sight to be: the working region is the most narrow cylinder which can contain the moving blades of a propeller. This is literally the space which the propeller needs for its functioning. However, then it is possible that by the non-linear base motion the shed vorticity of the added motion comes outside this cylinder and then it can happen that $q > 1$. This we will show in Section 7.5.

However, when we take as the working region the most narrow cylinder (if this one exists), which can contain the shed free vorticity, then $q \leq 1$.

In case of a linear theory, the two mentioned possibilities are the same because the vorticity remains at the place where it is shed from the blades.

5.9. An Ideal Propeller

In this section we consider in detail an example of an ideal propeller. Consider the circular cylindrical region B (Figure 5.9.1)

$$B: \quad -\infty < x < \infty \ , \qquad y^2 + z^2 \leq R^2 \ , \tag{5.9.1}$$

which is assumed to be the working region of a class C of lifting surfaces as discussed in Section 5.7. The reference surfaces H_l will have a period b in the x-direction, the lifting surfaces have a mean velocity of advance U and have to deliver together a mean thrust $\bar{T}$. We impose no other conditions on C, so the class C_0 is equal to C. For simplicity we assume the fluid to be undisturbed, hence $\vec{v}_0{}^* = 0$. We will construct an ideal propeller for this class.

A set H_l (l = 1, 2) of two reference surfaces H_1 and H_2 is sufficient for our purpose. The surface H_1 is the cylinder which is the boundary of the working region B

$$H_1 = \partial B : \quad y^2 + z^2 - R^2 = 0 \ . \tag{5.9.2}$$

The surface H_2 is more complicated. First, it is periodic in the x-direction with period b and it is axi-symmetric with respect to the x-axis. Second, it contracts up to the x-axis, then expands up to H_1, then contracts again up to the x-axis and so on.

The condition for the potential $\Phi(x, y, z)$ belonging to the shed vorticity of the propeller, follows from the general formula (5.5.5). We have two reference surfaces H_1 and H_2 and only one condition for the thrust delivered by both together. Hence $m = 2$, $M = 1$ and $g_{l1} = \cos_{(n,x)}(\xi_1, \eta_1)$, so we find

$$\frac{\partial \Phi}{\partial n}(\xi_l, \eta_l) = \lambda \cos_{(n,x)}(\xi_l, \eta_l) \ , \qquad l = 1, 2 \ , \tag{5.9.3}$$

where $\cos_{(n,x)}(\xi_l, \eta_l) = 0$. Condition (5.9.3) can be interpreted as follows. The potential Φ can be created by translating the two reference surfaces as rigid and

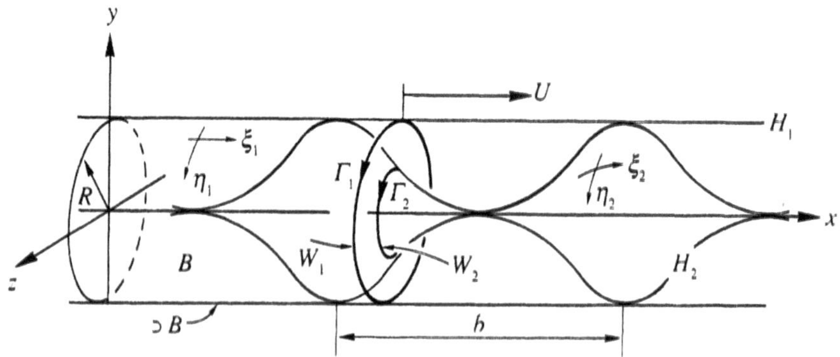

Fig. 5.9.1. A possible "realization" of an ideal thrust producing system.

impermeable surfaces with the velocity λ in the positive x-direction. The vorticity needed on the impermeable H_1 and H_2 for this translation is the optimum free vorticity left behind by the propeller. The constant λ follows from the condition that the mean value of the thrust has to be $\bar{T}$ in the positive x-direction. It will turn out that λ has to be negative, so in fact we have to translate H_1 and H_2 in the negative x-direction, which is in agreement with the direction of the slip stream caused by the propeller.

It is clear that by translating H_1 and H_2 as impermeable rigid surfaces, the fluid in the cut off compartments inside H_1 will translate also with the same velocity λ. Hence H_2 will not carry any vorticity within H_1. Outside H_1 we have the fluid in rest, hence the vorticity on H_1 is constant, circular and perpendicular to the x-axis.

We now consider two flexible ring wings W_1 and W_2, moving along H_1 and H_2, respectively. Without restricting generality we can in our linearized theory represent W_1 and W_2 by two circular concentrated bound vortices of strength Γ_1 and Γ_2 (see below (5.4.1)). The positive directions of rotation of Γ_1 and Γ_2 are with a right-hand screw denoted in Figure 5.9.1. We assume that the velocity of Γ_1 is U and that Γ_2 remains in the plane of Γ_1.

The thrust is entirely delivered by Γ_2, because Γ_1 cannot produce a force of $O(\varepsilon)$ in the x-direction. The process is as follows. The strength of Γ_1 increases linearly with x or with time, hence it sheds the mentioned circular vorticity at H_1. This, however, cannot go on indefinitely because then the strength of Γ_1 would increase beyond all bounds and the propulsion system would not be periodic. It is here that Γ_2 appears on stage. First, as we saw before, it is not allowed that Γ_2 sheds any vorticity on H_2, inside H_1. Hence its strength has to be constant at parts of H_2 which are not in contact with H_1. Second, when Γ_2 has to deliver a thrust, it must have different strengths at the positive and negative slopes of H_2. These two properties can be reconciled with each other. When Γ_2 arrives at the contact circle of H_2 with H_1, it changes sign instantaneously and leaves behind a concentrated vortex at H_1. However, when Γ_1 changes its magnitude at exactly the same place, it cancels this concentrated vortex. By this change, Γ_1 has not to grow beyond all bounds and can be made periodic. The bound vorticity Γ_2 can however also change its sign inside H_1 at the points where H_2 touches the x-axis, there its radius and hence its length is zero and no vorticity is left behind. In Figure 5.9.2 we have drawn possible strengths of Γ_1 and Γ_2 as a function of x. We note that it is not allowed to leave behind concentrated free vorticity because the kinetic energy around a concentrated vortex is infinite. Hence the efficiency of the propeller would become zero, anyhow theoretically.

We now determine the Lagrange multiplier λ which, as we mentioned before, can be interpreted as the homogeneous velocity of the fluid inside H_1. Hence by the theorem of momentum (see fourth paragraph of Section 1.15) we find

$$\rho\,(U\pi R^2)\lambda = -\bar{T}\ , \qquad \lambda = \frac{1}{\rho\, U}\cdot\frac{-\bar{T}}{\pi R^2} < 0\ . \tag{5.9.4}$$

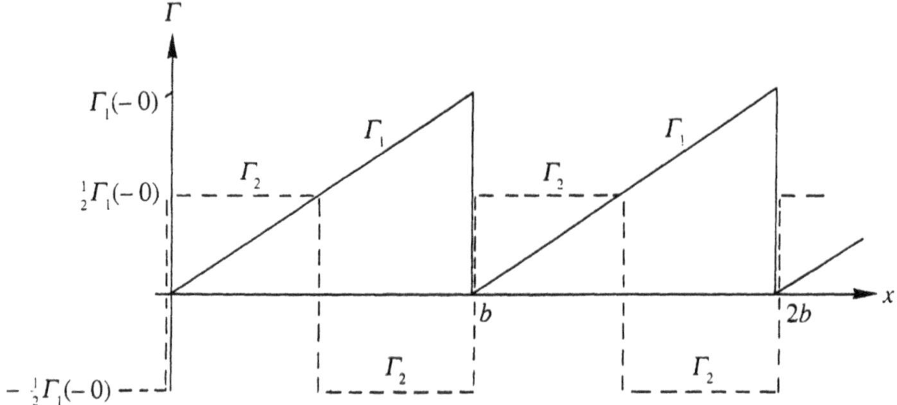

Fig. 5.9.2. Bound vorticity Γ_1 (drawn line) and Γ_2 (dashed line).

The left-hand side of the first equation is the momentum added to the fluid per unit of time and its right-hand side is the force exerted on the fluid, which is the reaction of the thrust.

From its construction it follows that this propulsion device is ideal in the sense of Section 5.7. Namely, when any other reference surface is added, which is inside H_1 and translates with H_1 and H_2, it will not change the flow. Hence the free vorticity shed by the supplemented optimum propeller remains the same as before. This means that the efficiency of the original one cannot be raised. We remark that the surface H_2 is not uniquely determined, each reference surface which divides the interior of H_1 into disconnected parts, will do.

When we look at the velocity of the homogeneous wake behind our ideal propeller, we see that it is the same as the velocity field (2.1.13) behind an actuator disk with a constant load $f(y,z) = \bar{T}/\pi R^2 = \text{const.}$, when it has the same velocity of advance U and is described by a linear theory. However, the system discussed here has the highest efficiency in comparison with each system of lifting surfaces of the considered class. Hence also the efficiency of an actuator disk with a constant load as given in (2.2.7) yields an upperbound for any conceivable propeller of the class under consideration. This is the reason that the actuator disk with a constant load acting in an undisturbed fluid often is called an ideal propeller.

We remark that the choice of a circular working area A, which is the cross section of the working region B, is of no importance in the above. All considerations remain valid, mutatis mutandis, for any other shape of A. For instance, we have to replace in (5.9.4) πR^2 by the area of A.

When there is a disturbance velocity field $\vec{v}_0^{\,*}$, we need in general infinitely many reference surfaces to extract the kinetic out of the working region B, as we discussed in Section 5.7. After this, the demand on the force action was satisfied by translating the reference surfaces H_1 and H_2, considered to be rigid and impervious, in the

x-direction by which a homogeneous wake with a constant velocity occurs. We have demonstrated in this section that the potential Φ_2 needed for this can be obtained by a finite number of reference surfaces. This Φ_2 has to develop a thrust of such a magnitude that, together with the force in the x-direction which originates from bringing the field $\vec{v}_0{}^*$ to a standstill inside B, the total demanded thrust $\bar{T}$ occurs.

We now calculate the mean value of the force acting at the fluid in the x-direction needed to annihilate $\vec{v}_0{}^*$ inside B. In case of the ideal propeller, it follows from our previous considerations that the resulting velocity field is such that its component normal to the boundary ∂B of the working region B, is zero. This can be achieved in the following way, where we again use the configuration of Figure 5.9.1.

Consider a half infinite cylinder which is rigid and impervious, has radius R and stretches along the x-axis, $-\infty < x \leq Vt$, hence it moves with some velocity V in the positive x-direction along ∂B. Then the only force in the x-direction is the suction force at the leading edge of the cylinder. This force is of $O(\varepsilon^2)$. The momentum I_x in the x-direction of the fluid per period b of the velocity field $\vec{v}_0{}^*$ is, however, of $O(\varepsilon)$. Hence the leading edge suction cannot alter this momentum. So when the leading edge of the cylinder is far away in the positive x-direction, we still have per period b the same amount of momentum I_x.

Next we have to calculate in this situation the velocity field outside and inside the cylinder. Outside the cylinder we have the total velocity field

$$\vec{v}_0{}^* - \vec{v}_0 \ , \qquad r > R \ , \tag{5.9.5}$$

where $\vec{v}_0$ is the velocity field which occurs in the last paragraph of Section 5.3. It is calculated by the Neumann problem (5.3.9) for its potential $\Phi_0 = -\Phi_1$. Inside the cylinder we have to solve also a Neumann problem with the same boundary values at $r = R$. We denote the resulting velocity field by $\vec{w}_0$. Then inside the cylinder we have the velocity field

$$\vec{v}_0{}^* - \vec{w}_0 \ , \qquad r < R \ . \tag{5.9.6}$$

Hence the total amount of momentum I_x can be written as

$$I_x = \iint\limits_{-\infty}^{\infty} \int\limits_0^b v^*_{0,x} \, dx \, dy \, dz = \iint\limits_{r<R} \int\limits_0^b (v^*_{0,x} - w_{0,x}) \, dx \, dy \, dz$$
$$+ \iint\limits_{r>R} \int\limits_0^b (v^*_{0,x} - v_{0,x}) \, dx \, dy \, dz \ , \tag{5.9.7}$$

where the second equality originates from the considerations given by means of the half infinite impervious cylinder moving along ∂B. Now the first amount of momentum at the right-hand side is annihilated. This means that by the momentum theorem there is needed a mean resultant force K in the positive x-direction, of which the magnitude follows from

$$\frac{b}{U}K = -\rho \iint\limits_{r<R} \int\limits_{0}^{b} (v_{0,x}^{*} - w_{0,x})\, dx\, dy\, dz \ , \tag{5.9.8}$$

acting at the fluid to bring the fluid at rest for $r < R$.

By this the potential Φ_2 has to deliver a thrust

$$\bar{T} + K \ . \tag{5.9.9}$$

A very simple example occurs for $\vec{v}_0^{\,*} = (c, 0, 0)$, $r < R$, $\vec{v}_0^{\,*} = (0, 0, 0)$, $r > R$, for some constant c. Then

$$K = -\pi \rho\, c U R^2 \ . \tag{5.9.10}$$

This is favourable for $c > 0$ and unfavourable for $c < 0$. Hence we arrive again at the result that it is favourable when the disturbance velocity is in the direction of the demanded thrust.

5.10. Comparison of the Efficiency of Optimum Propellers by Inspection

In this section we compare optimum propulsion systems which have prescribed reference surfaces. We will show that sometimes it is possible, by a simple inspection of the shape of these surfaces, to judge which one can be expected to yield a propeller with the highest efficiency, from the potential point of view.

When we compare two propellers, say I and II, of which II has besides the reference surfaces of I still other reference surfaces, then the quality number of II will be larger than the quality number of I. This follows directly, as we mentioned before, from the optimization theory because in case of propeller II there is more freedom to arrange the shed free vorticity favourably. Then by equal thrust, propeller II will have a higher efficiency than propeller I. An example of this is the application of end plates to the tips of the blades of a screw propeller. Then we have to compare simple helicoidal reference surfaces and the same surfaces but now with a strip attached to their outer rim (Section 6.4).

The comparison becomes more uncertain when the reference surfaces have not such a simple relation. Then, however, we can formulate a rather vague but useful statement, which says:

Given two propulsion systems which have the same mean velocity of advance U, the same working region, the same value of thrust $\bar{T}$ and length period b. Then the propeller of which the reference surfaces, considered to be impervious and rigid, hamper most in the working region a flow parallel to the mean velocity of advance, will have the highest efficiency.

The mentioned parallel flow is the flow of velocity λ, discussed for instance below (5.9.3), but is now taken relatively to the reference surfaces, which are here considered to be fixed in space.

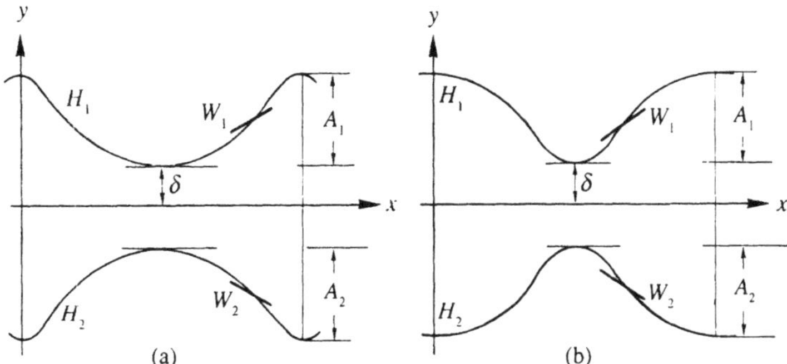

Fig. 5.10.1. Two different arrangements of geometrically congruent shapes of reference surfaces, (a) $h = 0.0375$, (b) $h = -0.0375$.

The examples we discuss are 2-dimensional. Each propulsion device consists of two flexible profiles W_1 and W_2, moving in an ε neighbourhood of the two reference surfaces H_1 and H_2, respectively, which are given by

$$H_l : \quad y = \pm 0.15 \mp h \pm \delta \pm 0.15 \cos 2\pi x \pm h \cos 4\pi x \ , \quad l = 1,2 \ , \quad (5.10.1)$$

where the upper signs belong to $l = 1$ and the lower signs to $l = 2$. The cross sections of the H_l are drawn in Figure 5.10.1 for two specific choices of h, namely $h = 0.0375$ and $h = -0.0375$. The parameter δ is still variable. The two figures are clearly related. The surface H_1 of Figure 5.10.1 (b) originates from H_1 of Figure 5.10.1 (a) by a reflection and a translation.

We now discuss the quality number q (5.8.2) of these two types of propellers. As their working area per unit of span we take the two parts A_1 and A_2 covered by the amplitude of H_1 and H_2, separated by a gap of width 2δ. Hence this area equals 0.6. The calculated values of q are given in Figure 5.10.2, as a function of the minimum distance 2δ between the two reference surfaces for the two cases (a) and (b) of Figure 5.10.1 and also for a case (c) where H_1 and H_2 have a simple cosine shape with $h = 0$ in (5.10.1). Note that in all three cases the working area per unit of span is the same.

First we discuss the cases (a) and (b) with $h = 0.0375$ and $h = -0.0375$, respectively, and try to predict which one has the highest quality number. It is clear that for large values of δ there is no interaction between the wings W_1 and W_2 and between the vortex sheets H_1 and H_2 in each case (a) and (b). Furthermore, the geometry of H_1 and H_2 is the same in both cases, hence there will be no difference in their quality number q for $\delta \to \infty$. When δ becomes smaller, interaction will occur between H_1 and H_2. Then the parallel flow of velocity λ is hampered in the working region B_1 besides by H_1 also by H_2 and analogously for B_2. Hence it is clear that for smaller values of δ, in the case of optimum propellers, q increases for configuration (a) as well as for (b). For $\delta = 0$, the working regions B_1 and B_2 have periodically

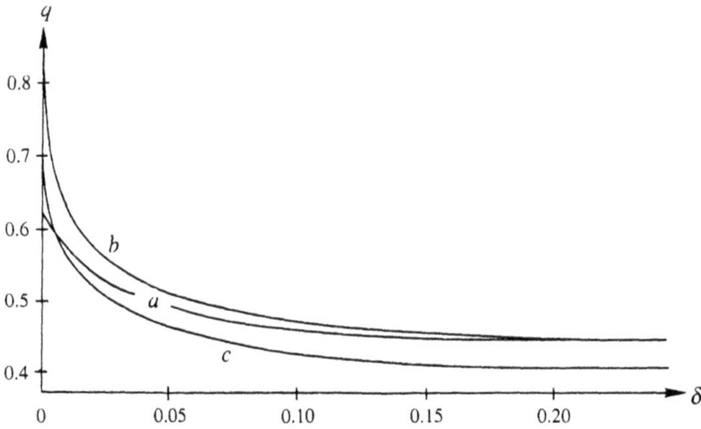

Fig. 5.10.2. The quality number q as a function of the minimum distance 2δ of the two wings, for three different sets of H_1 and H_2.

situated boundary points in common and hence cut off finite regions of the (x, y) plane. In case (b) more "fluid is enclosed" in between H_1 and H_2 than in case (a). Hence in case (b) the parallel flow seems to be hampered more than in case (a) and it can be expected that q for $\delta = 0$ is larger in case (b) than in case (a); moreover, it is likely that this holds for all values of δ. These conclusions are confirmed by the calculated results given in Figure 5.10.2.

Next we consider case (c) where $h = 0$, in comparison with the cases (a) and (b). We repeat that the working area of case (c) is equal to the working area of (a) and (b). For large values of δ it is conceivable that the more sharply indented H_1 and H_2 of cases (a) and (b) hamper the parallel flow inside B_1 and B_2 more than the more gently waving pure cosine of case (c) will do, although they have the same amplitude. This is in agreement with Figure 5.10.2, where for large values of δ the quality number q of (c) is less than the value of q for the cases (a) and (b). When $\delta = 0$, however, the two pure cosines of case (c) "enclose more fluid" than the surfaces H_1 and H_2 in case (a) and less than the surfaces H_1 and H_2 belonging to case (b). So it seems predictable that for sufficiently small values of δ the quality number of case (c) is in between those of case (a) and case (b). This is also confirmed by the results given in Figure 5.10.2 [20].

The optimum efficiency of these propulsion systems follows from their quality number q as given in Figure 5.10.2. Using (5.8.5) we can write the efficiency of the three types of propellers under consideration as

$$\eta = \frac{\bar{T}}{\bar{T} + E_p} = \left(1 + \frac{E_i}{q\bar{T}}\right)^{-1} , \tag{5.10.2}$$

where $\bar{T}$ is the thrust per unit of span and in this case the period of length of the propeller motion is 1 and E_p and E_i are the energy losses per unit of length of a propeller and of the ideal propeller respectively, both in an undisturbed fluid. We have shown that in this case an ideal propeller can be represented by an actuator disk with a constant load hence we can use (2.1.15). However, this formula gives the kinetic energy per unit of time. Because, as we mentioned before, the working area per unit of span is here 0.6, we have to take in (2.1.15) $f(y, z) = \bar{T}/0.6$ = const. After a simple reduction we find $E_i = \bar{T}^2/1.2\rho U^2$ and hence

$$\eta = \left(1 + \frac{\bar{T}}{1.2\,\rho\,q\,U^2}\right)^{-1} , \tag{5.10.3}$$

where $\bar{T}$ is the thrust per unit of span.

Equation (5.10.3) gives an addition to the discussion of Table 5.8.1 about the relation between η and q. When $\bar{T}$ is sufficiently small, the efficiency will be high in general and q has not so much influence on the efficiency. However, when $\bar{T}$ becomes larger, then it is important to have q close to one. In other words, then we need a propeller of which the reference surfaces have a good shape in order to obtain a relatively better efficiency.

In the same way as we did in this section for the unsteady propulsion, we can compare other propulsion systems. For instance, consider two optimum ship screws, with the same values of the velocity of advance U, diameter, hub radius, thrust $\bar{T}$ and rotational velocity, but differing in their number of blades. Then the one with the highest number of blades will have the highest value of q and hence the highest efficiency. Analogously, from two optimum screw propellers differing only in their rotational velocity, the one with the highest rotational velocity will have the highest value of q and hence the highest efficiency.

Again we have to realize that viscosity effects of the fluid have been left out of consideration.

5.11. On the "Optimization" of a Rigid Lifting Surface in a Disturbed Fluid

Since a screw propeller generally works in the disturbed flow behind a ship, it seems of interest to investigate the possibility of energy extraction out of the inhomogeneous inflow at the propeller disk. Up to now we have assumed that the lifting surfaces which extracted energy out of a disturbance field were flexible and moving along arbitrarily curved reference surfaces. A screw propeller, however, has undeformable blades which are mostly connected rigidly to the hub. In order to keep the considerations as simple as possible we discuss in this section a rigid planar wing translating through a disturbed fluid. The same reasoning can be used for the non-planar blades of the ship screw moving along their helicoidal reference surfaces.

The reference strip H along which the wing W moves with velocity U is defined by (Figure 5.11.1)

$$H: \quad -\infty < x < \infty \ , \quad y = 0 \ , \quad |z| < a \ . \tag{5.11.1}$$

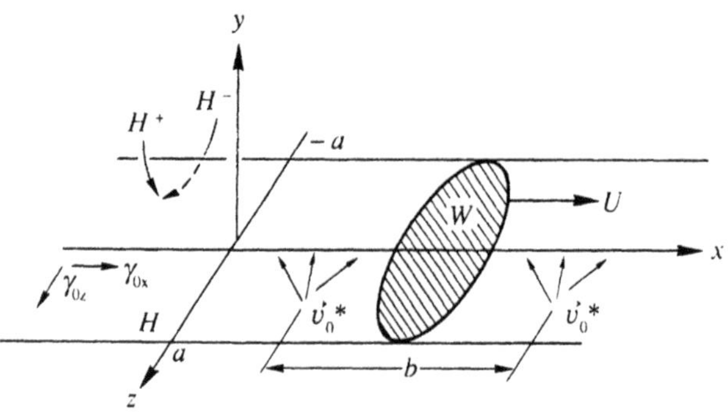

Fig. 5.11.1. Energy extraction by means of rigid wing W.

As before we have a periodic disturbance velocity field $\vec{v}_0{}^*$, which we replace by $\vec{v}_0$ having the same normal component at H and of which the vorticity $\vec{\gamma}_0(x, z) = (\gamma_{0x}(x, z), \gamma_{0z}(x, z))$ is confined to H while the total circulation around H is zero. This vorticity field can be expanded into a Fourier series

$$\gamma_{0x}(x, z) = \alpha_{x0}(z) + \sum_{n=1}^{\infty} \left\{ \alpha_{xn}(z) \cos \frac{2\pi n x}{b} + \beta_{xn}(z) \sin \frac{2\pi n x}{b} \right\} , \quad (5.11.2)$$

$$\gamma_{0z}(x, z) = \alpha_{z0}(z) + \sum_{n=1}^{\infty} \left\{ \alpha_{zn}(z) \cos \frac{2\pi n x}{b} + \beta_{zn}(z) \sin \frac{2\pi n x}{b} \right\} . \quad (5.11.3)$$

A vorticity field is free of divergence, div $\vec{\gamma}_0(x, z) = 0$, hence

$$\frac{\partial}{\partial z}\alpha_{z0}(z) = 0 , \quad \frac{\partial}{\partial z}\alpha_{zn}(z) = -\frac{2\pi n}{b}\beta_{xn}(z) ,$$
$$\frac{\partial}{\partial z}\beta_{zn}(z) = \frac{2\pi n}{b}\alpha_{xn}(z) . \quad (5.11.4)$$

Because outside H there is no vorticity, we have to take

$$\alpha_{z0} = 0 , \quad \alpha_{zn}(\pm a) = 0 , \quad \beta_{zn}(\pm a) = 0 , \quad (5.11.5)$$

otherwise $\vec{\gamma}_0(x, z)$ would have concentrated vortices in the x-direction along the edges of the strip and the kinetic energy of $\vec{v}_0$ would be infinite per unit of length.

The velocity potential Φ_0 of the modified velocity field $\vec{v}_0$ is likewise expanded in a Fourier series

$$\Phi_0(x, y, z) = \Phi_{00c}(y, z) + \sum_{n=1}^{\infty} \left\{ \Phi_{0nc}(y, z) \cos \frac{2\pi n x}{b} + \Phi_{0ns}(y, z) \sin \frac{2\pi n x}{b} \right\} . \quad (5.11.6)$$

The jump $[\Phi_0]^+_-$ of this potential over H can be written as

$$[\Phi_0]^+_-(x,z) = -\int_z^a \alpha_{x0}(s)\,ds$$

$$-\sum_{n=1}^{\infty}\int_z^a \left\{\alpha_{xn}(s)\cos\frac{2\pi nx}{b} + \beta_{xn}(s)\sin\frac{2\pi nx}{b}\right\}\,ds$$

$$= -\int_z^a \alpha_{x0}(s)\,ds$$

$$+\frac{b}{2\pi}\sum_{n=1}^{\infty}\frac{1}{n}\left\{\beta_{zn}(z)\cos\frac{2\pi nx}{b} - \alpha_{zn}(z)\sin\frac{2\pi nx}{b}\right\}$$

$$\stackrel{\text{def}}{=} [\Phi_0]^+_{-0c}(z) + \sum_{n=1}^{\infty}\left\{[\Phi_0]^+_{-nc}(z)\cos\frac{2\pi nx}{b} + [\Phi_0]^+_{-ns}\sin\frac{2\pi nx}{b}\right\}, \quad (5.11.7)$$

where we used (5.11.4) and (5.11.5). Then we can write the kinetic energy E, belonging to Φ_0 over one period b in the x-direction, as

$$E = -\tfrac{1}{2}\rho\int_{-a}^{a}\int_0^b \left\{[\Phi_0]^+_-(x,z)\frac{\partial}{\partial y}\,\Phi_0(x,0,z)\right\}\,dx\,dz$$

$$= -\tfrac{1}{4}\,b\rho\int_{-a}^{a}\Bigg[2\left\{[\Phi_0]^+_{-0c}(z)\,\frac{\partial}{\partial y}\,\Phi_{00c}(0,z)\right\}$$

$$+\sum_{n=1}^{\infty}\left\{[\Phi_0]^+_{-nc}(z)\,\frac{\partial}{\partial y}\,\Phi_{0nc}(0,z) + [\Phi_0]^+_{-ns}(z)\frac{\partial}{\partial y}\,\Phi_{0ns}(0,z)\right\}\Bigg]\,dz$$

$$\stackrel{\text{def}}{=} E_0 + \sum_{n=1}^{\infty}(E_{cn} + E_{sn}) \quad . \quad (5.11.8)$$

By this we have split the kinetic energy E, which can be extracted entirely by a flexible wing because it belongs to the velocity field $\vec{v}_0$ "into" parcels E_n which belong to the different terms between braces in the Fourier expansion (5.11.6).

Next we consider the rigid wing W moving with velocity U along H. We introduce the coordinates ξ, η and ζ which are in the direction of x, y and z, which form a reference system (ξ,η,ζ) which translates with W. Because $\vec{\gamma}_0(x,z)$ has a period b in the x-direction we write for the vorticity $\vec{\gamma}_W = (\gamma_{W\xi},\gamma_{W\zeta})$ on W and shed by W

$$\gamma_{W\xi}(\xi,\zeta,t) = \tilde{\alpha}_{\xi 0}(\xi,\zeta)$$

$$+\sum_{n=1}^{\infty}\left\{\tilde{\alpha}_{\xi n}(\xi,\zeta)\cos\frac{2\pi nUt}{b}+\tilde{\beta}_{\xi n}(\xi,\zeta)\sin\frac{2\pi nUt}{b}\right\}\ , \tag{5.11.9}$$

$$\gamma_{W\zeta}(\xi,\zeta,t)=\tilde{\alpha}_{\zeta 0}(\xi,\zeta)$$
$$+\sum_{n=1}^{\infty}\left\{\tilde{\alpha}_{\zeta n}(\xi,\zeta)\cos\frac{2\pi nUt}{b}+\tilde{\beta}_{\zeta n}(\xi,\zeta)\sin\frac{2\pi nUt}{b}\right\}\ . \tag{5.11.10}$$

We suppose that W moves just above H, hence just above the vorticity $\vec{\gamma}_0$ and keep in mind that we have to take the limit of vanishing distance between W and $\vec{\gamma}_0$. The vorticity $\vec{\gamma}_W$ caused by W induces a velocity potential $\tilde{\Phi}(\xi,\eta,\zeta,t)$. The pressure jump across W follows from the linearized unsteady Bernoulli equation (1.2.12)

$$[p]_-^+(\xi,\zeta,t)=-\rho U\left[\frac{\partial\tilde{\Phi}}{\partial\xi}\right]_-^+(\xi,0,\zeta,t)-\rho\left[\frac{\partial\tilde{\Phi}}{\partial t}\right]_-^+(\xi,0,\zeta,t)\ . \tag{5.11.11}$$

The potential Φ_0 induced by the already existing vorticity $\vec{\gamma}_0$ at H, being just below the wing does not contribute to the jump of pressure across W.

Extraction of energy out of the fluid is effectuated by the component in the x-direction of the force exerted by the fluid at W. Suppose the shape of W is given by

$$\eta=y=f(\xi,\zeta)=O(\varepsilon)\ ,\qquad(\xi,\zeta)\in W\ , \tag{5.11.12}$$

where, because the theory is linearized, the planform of the wing is also denoted by W. Then the contribution of an elementary area $d\xi\ d\eta$ of W to the mean value with respect to time, of the force of $O(\varepsilon^2)$, in the x-direction becomes

$$\frac{U}{b}\int_0^{b/U}\left\{[p]_-^+(\xi,\zeta,t)\,\frac{\partial f}{\partial\xi}(\xi,\zeta)\,d\xi\,d\zeta\right\}dt\ . \tag{5.11.13}$$

It is clear from (5.11.9) and (5.11.11) that we can develop $[p]_-^+$ in a Fourier series with respect to t. It follows from (5.11.13) that the elementary area $d\xi\ d\eta$ can contribute only a non-zero value to the mean force in the x-direction by means of the time-independent part of $[p]_-^+$. Because $[\partial\Phi/\partial t]_-^+$ does not have such a part, we have to use the time-independent part of $[\partial\Phi/\partial\xi]_-^+$ which reads by (5.11.9)

$$-\int_\zeta^a\frac{\partial}{\partial\xi}\tilde{\alpha}_{\xi 0}(\xi,s)\,ds=\int_\zeta^a\frac{\partial}{\partial\zeta}\tilde{\alpha}_{\zeta 0}(\xi,s)\,ds=-\tilde{\alpha}_{\zeta 0}(\xi,\zeta)\ , \tag{5.11.14}$$

where we used div $\vec{\gamma}_W = 0$. Hence we can write (5.11.13) as

$$\frac{\rho U^2}{b} \int_0^{b/U} \left\{ \tilde{\alpha}_{\zeta 0}(\xi,\zeta) \frac{\partial f}{\partial \xi}(\xi,\zeta)\, d\xi\, d\zeta \right\} dt$$

$$= \rho U \tilde{\alpha}_{\zeta 0}(\xi,\zeta) \frac{\partial f}{\partial \xi}(\xi,\zeta)\, d\xi\, d\zeta \ . \tag{5.11.15}$$

When in one way or another the wing leaves behind vorticity of which the x-component has one of the strengths

$$-\alpha_{x0}(z) \ , \quad -\alpha_{xn}(z) \cos \frac{2\pi n x}{b} \ , \quad -\beta_{xn}(z) \sin \frac{2\pi n x}{b} \ , \tag{5.11.16}$$

then the corresponding parcel of kinetic energy E_0, E_{cn} or E_{sn} $(n \neq 0)$ would disappear from E (5.11.8).

First we consider separately the case that only the kinetic energy E_0 (5.11.8) is present. This energy can be extracted entirely by a rigid wing when we demand that its circulation distribution $\Gamma_0(z)$ has the form

$$\Gamma_0(z) = -\int_{-a}^{z} \alpha_{x0}(s)\, ds \ , \tag{5.11.17}$$

which leaves behind free vorticity of the strength $d\Gamma_0/dz = -\alpha_{x0}(z)$, which annihilates $\alpha_{x0}(z)$ (5.11.2). Now we can absorb this energy entirely by the inner parts of the wing by choosing the distribution of the bound vorticity in chordwise direction so that we have a smooth flow at the leading edge, for instance by taking the bound vorticity equal to zero at this edge. Then we have no suction force and we calculate the camber of the wing by (3.3.25) or (3.5.24). It is of course also possible by admitting a square root singularity at the leading edge to extract part of E_0 by suction forces.

The parcels E_{cn} and E_{sn} $(n \neq 0)$ of the kinetic energy can only be influenced by shed vorticity which is periodic in the x-direction, hence which in the (ξ, η, ζ) coordinate system is created by vorticity at W which is periodic in time. The corresponding periodic pressure jumps, however, yield by (5.11.13) no mean value of a force in the x-direction at inner parts of W, which means that the inner parts of W cannot contribute to the energy extraction of E_{cn} and E_{sn}. In other words this has to be done by the suction forces at the leading edge.

We will discuss this somewhat further. The action of the wing W is split into separate parts by considering separate wings with the same planform as W however working in differently disturbed fluids.

(i) Wing W_0 which is the same as the original wing W, moves through a fluid which has at H the vortex field $\alpha_{x0}(z)$ (5.11.2) in the x-direction. W_0 has a camber such that its circulation distribution equals $\Gamma(z)$ (5.11.17) and hence it sheds free vorticity which annihilates $\alpha_{x0}(z)$.

(ii) Wings W_n ($n = 1, 2, \ldots$), these are the flat impermeable planforms of W at H, each moves through a fluid which has at H the vortex field ((5.11.2), (5.11.3))

$$\left\{\alpha_{xn}(z)\cos\frac{2\pi nx}{b} + \beta_{xn}\sin\frac{2\pi nx}{b},\right.$$

$$\left.\alpha_{zn}(z)\cos\frac{2\pi nx}{b} + \beta_{zn}\sin\frac{2\pi nx}{b}\right\} . \tag{5.11.18}$$

They shed free vorticity which we denote by

$$\left\{\alpha^*_{xn}(z)\cos\frac{2\pi nx}{b} + \beta^*_{xn}\sin\frac{2\pi nx}{b},\right.$$

$$\left.\alpha^*_{zn}(z)\cos\frac{2\pi nx}{b} + \beta^*_{zn}\sin\frac{2\pi nx}{b}\right\} . \tag{5.11.19}$$

Wing W_0 (= W) extracts the kinetic energy E_0 (5.11.8) entirely, while each W_n extracts a certain unknown, however calculable, amount of energy E^*_n out of $(E_{cn} + E_{sn})$. The wings ($n = 1, 2, \ldots$) have to extract, because they are flat, energy by means of their suction forces at the leading edge, however, this could not be done otherwise as we discussed in relation with (5.11.13). When we now let move the original wing W (= W_0) along the total vortex field (5.11.2), (5.11.3) it will, by the linearity of the problem, subtract the same amounts E_0 and E^*_n from each of the energy parcels as the separate wings $W_0, W_1, W_2, \ldots$, would do.

We now consider a rigid wing which has to deliver a certain mean value $\bar{L}$ of lift. Then we can partially follow the directives, given in the optimization theory below (5.5.6). First we clean as much as possible the reference surface H from the vorticity $\vec{\gamma}_0(x, z)$ of the modified disturbance field $\vec{v}_0$. The vorticity $\alpha_{x0}(z)$ can be annihilated entirely, hence the kinetic energy parcel E_0 can be absorbed. The other periodic vorticity components (5.11.18) will be diminished to a certain extent, because the suction forces of the flat wings W_n extract some energy out of the parcels $(E_{cn} + E_{sn})$.

The mean value of the lift which originates from this energy extraction is caused only by the extraction of E_0, because the other lift forces of the W_n ($n = 1, 2, \ldots$) are periodic with mean value equal to zero. This lift L_0 follows directly from (5.11.17) and has the value

$$L_0 = \rho U \int_{-a}^{a} \Gamma_0(z)\, dz . \tag{5.11.20}$$

It is not difficult to show that L_0 is positive when the y-component of $\vec{v}_0$ is predominantly in the positive y-direction. This means that the wing has to deliver an additional lift L of magnitude

$$L = \bar{L} - L_0 . \tag{5.11.21}$$

Because now the "optimization" procedure is no longer exact, we can take for instance a wing which delivers the lift L (of (5.11.21) in an undisturbed fluid. When we add the camber of this wing to the camber of W_0, the new wing created in this manner will deliver the prescribed mean lift $\bar{L}$.

We do not know if all the energy can be extracted out of the energy parcels $(E_{cn} + E_{sn})$ by wings of finite chord length. In the next section we make it plausible that it can be done by wings of "infinite chord length". It is also proved there that in the 2-dimensional case it cannot be done by wings of finite chord length. So it is at least plausible that in order to extract all the energy E_n $(n = 1, 2, \dots)$ by means of a rigid wing, this wing has to have an infinite chord length.

The above considerations hold nutatis mutandis, for the rigid blades of a screw propeller, moving along helicoidal reference surfaces in the disturbance velocity field of the wake behind a ship. Then the desired mean value of the thrust is coupled geometrically to the lift forces of the blades. We remark that the friction caused by the viscosity of water will be detrimental for the gain predicted by potential theory for blades with very large chord lengths. Furthermore, the extraction of energy by means of suction forces at the sharp leading edges of the blades does not seem to be very reliable, small disturbances can cause separation of the flow by which the magnitude of the suction forces will fall off. *Hence the most important energy to recover seems to be E_0, which can be done by blades with finite chord lengths while no suction forces are needed for that purpose.*

5.12. Optimum Energy Extraction by a Rigid Wing

Consider a flat wing W moving along a reference strip H (Figure 5.11.1) of width $2a$. On H we have vorticity $\vec{\gamma}_0(x,z)$ of which the total circulation is zero and which is periodic with period b in the x-direction. For the sake of simplicity we assume that the wing is a rectangle of width $2a$ and chord length l with $l \gg 2a$ and $l \gg b$, which moves with a velocity U in the positive x-direction.

The component of the velocity induced by $\vec{\gamma}_0(x,z)$, normal to W has to be cancelled by vorticity $\vec{\gamma}_W$ at W. This vorticity will be split into three parts

$$\vec{\gamma}_W(x,z,t) = \vec{\gamma}_{W1}(x,z,t) + \vec{\gamma}_{W2}(x,z,t) + \vec{\gamma}_{W3}(x,z,t) \ , \tag{5.12.1}$$

where for the description of the vorticity at W we now use the coordinates (x,z) instead of those of a coordinate system which moves with W.

The first part we choose as

$$\vec{\gamma}_{W1}(x,z,t) = -\vec{\gamma}_0(x,z) \ , \qquad (x,z) \in W \ , \tag{5.12.2}$$

and state that this vorticity is shed at the trailing edge of W when the chord length l is sufficiently large. The correctness of this will be checked. Then this free vorticity annihilates the vorticity $\vec{\gamma}_0(x,z)$ at H behind the wing.

This vorticity $\vec{\gamma}_{W1} = -\vec{\gamma}_0$ needs a correction for two reasons. First, it is not free of divergence at the leading edge of W. Second, it cannot cancel all the normal velocities induced at W by $\vec{\gamma}_0$, because in front of the leading edge we have still vorticity at H of which the induced normal velocities at W are not compensated by those of $\vec{\gamma}_{W1}$. The first shortcoming will be repaired by the vortex system $\vec{\gamma}_{W2}$, which can be defined in a fixed region of W behind the trailing edge and of which the leading edge itself is part of its boundary. This vortex system can be chosen in an infinite number of ways. The second shortcoming will be corrected by choosing $\vec{\gamma}_{W3}$ such that the normal velocities at W induced by the vorticity $\vec{\gamma}_0$ in front of W and by the vorticity $\vec{\gamma}_{W2}$ are compensated. These normal velocities, however, tend to zero as

$$O(|x - x_l|^{-2}) \ , \qquad (x - x_l) \to -\infty \ , \tag{5.12.3}$$

where x_l is the x-coordinate of the leading edge. This relation can be shown to be true by an asymptotic expansion of the velocities induced by some representative vortex system.

We now show that also $\vec{\gamma}_{W3}$ tends to zero for $(x - x_l) \to -\infty$. Consider a region $\tilde{W}$ of W of which the length is large with respect to the span $2a$ of W. This region is situated at a large distance $\tilde{l} \gg 2a$ from the leading edge as well as from the trailing edge. Suppose that $|\vec{\gamma}_{W3}|$ assumes finite values larger than c with $c > 0$, on $\tilde{W}$ no matter how long $\tilde{W}$ is and how far $\tilde{W}$ is situated behind the leading edge. On $\tilde{W}$ the velocities induced by the vorticity in front of the leading edge and by $\vec{\gamma}_{W2}$ have disappeared when $\tilde{l}$ is sufficiently large. This means that the time periodic $\vec{\gamma}_{W3}$ on the region $\tilde{W}$ is not allowed to induce a non-zero velocity component normal to W, because there $\vec{\gamma}_{W1} = -\vec{\gamma}_0$ already compensates $\vec{\gamma}_0$ situated at H. This can happen only when $\vec{\gamma}_{W3} = 0$ at $\tilde{W}$. Namely its circulation around $\tilde{W}$ is zero because a wing can create only such vorticity. Hence the potential Φ_{W3} of its induced velocities outside $\tilde{W}$ is a uniquely valued function of which the derivative normal to $\tilde{W}$ is zero and which can be taken equal to zero for $(y^2 + z^2) \to \infty$. This means that $\Phi_{W3} = 0$ and then the same holds for $\vec{\gamma}_{W3}$. Hence when the chord length l of the wing is large enough, it is clear that in the neighbourhood of the trailing edge $\vec{\gamma}_{W1} = -\vec{\gamma}_0$ dominates by far and this confirms our statement directly below (5.12.2).

The result is that when the wing is flat and has a very large, in fact an infinite, chord length it annihilates the vorticity $\vec{\gamma}_0$ at H and extracts by its suction forces at the leading edge all the kinetic energy belonging to the modified disturbance field $\vec{v}_0$ out of the fluid.

The question remains if a flat wing with finite chord length can extract all kinetic energy belonging to vorticity with Fourier coefficients $\alpha_{xn}(z)$, $\beta_{xn}(z)$, $\alpha_{zn}(z)$ and $\beta_{zn}(z)$, $n \neq 0$ (5.11.2), (5.11.3) out of the fluid. That this is not plausible can be seen by the following calculation.

We consider a 2-dimensional problem. The Cartesian coordinate system (x, y) is embedded in an incompressible and inviscid fluid. The fluid has an incoming velocity U in the positive x-direction (Figure 5.12.1). Transported with the fluid is a vortex layer $\gamma_0(x, t)$ of strength

$$\gamma_0(x, t) = Ae^{i\nu(x - Ut)} \ . \tag{5.12.4}$$

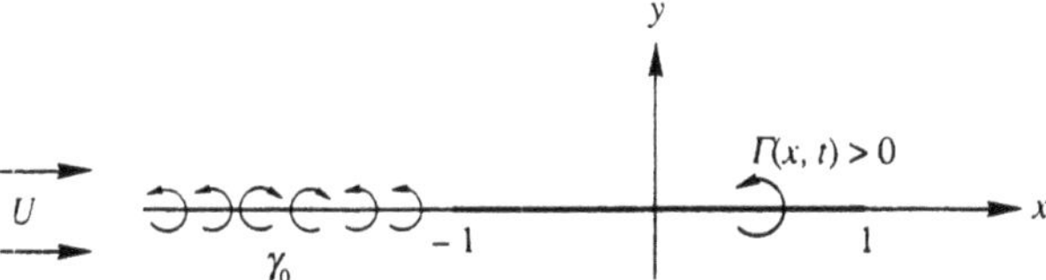

Fig. 5.12.1. Two-dimensional profile with incoming vortex layer γ_0.

This problem can be considered as a special case of Figure 5.9.1 for a very wide strip H hence for a large value of a. Then we neglect in a finite neighbourhood of the edges H at $z = a$ and $z = -a$, vorticity with a non-zero x-component needed to satisfy the condition div $\vec{\gamma}_0 = 0$. However, the amount of kinetic energy per unit of length in the x-direction added by this vorticity component can be made arbitrary small with respect to the total amount of energy connected to γ_0 (5.12.4), when the width $2a$ of the strip H is sufficiently large. So the results of this 2-dimensional calculation seem to have a bearing on the 3-dimensional case.

That our profile has the chord length 2 is not a restriction of generality, it is only a choice of the unit of length. Also that the profile is chosen to be flat is not a restriction, because we already discussed (see (5.11.13)) that a profile with camber cannot extract more energy out of a periodic vorticity field than a flat profile.

We assume that for some value ν it is possible to annihilate the free vorticity (5.12.4) behind the profile and we will arrive at a contradiction. If there is no vorticity behind the profile, the vorticity $\Gamma(x,t)$ at the profile has, in order that the flow is tangent to the profile, to satisfy

$$\oint_{-1}^{1} \frac{\Gamma(\xi,t)}{(\xi - x)}\, d\xi = -Ae^{-i\nu Ut} \int_{-\infty}^{-1} \frac{e^{i\nu\xi}}{(\xi - x)}\, d\xi \ , \quad -1 < x < 1 \ . \tag{5.12.5}$$

Writing $\Gamma(x,t) = \Gamma(x)e^{-i\nu Ut}$ we obtain

$$\oint_{-1}^{1} \frac{\Gamma(\xi)}{(\xi - x)}\, d\xi = -A \int_{-\infty}^{-1} \frac{e^{i\nu\xi}}{(\xi - x)}\, d\xi \ , \quad -1 < x < 1 \ . \tag{5.12.6}$$

The solution of this equation which satisfies the Kutta condition at the trailing edge, follows from Appendix A ((A.3.1) and (A.3.9))

$$\Gamma(x) = \frac{A}{\pi^2}\left(\frac{1-x}{1+x}\right)^{1/2} \oint_{-1}^{1} \left(\frac{1+\xi}{1-\xi}\right)^{1/2} \left\{ \int_{-\infty}^{-1} \frac{e^{i\nu\sigma}}{(\sigma - \xi)}\, d\sigma \right\} \frac{d\xi}{(\xi - x)} \ . \tag{5.12.7}$$

The condition that no free vorticity is present behind the profile reads

$$\gamma_0(-1,t) = Ae^{i\nu(-1-Ut)} = \frac{1}{U}\, \frac{d}{dt} \int_{-1}^{1} \Gamma(x)e^{-i\nu t}\, dx \ , \tag{5.12.8}$$

hence

$$\int_{-1}^{1} \Gamma(x)\, dx = \frac{iA}{\nu} e^{-i\nu} \ . \tag{5.12.9}$$

Substituting (5.12.7) into (5.12.9) we find, after a rearrangement of the integrations that the following condition must hold

$$\frac{1}{\pi^2} \int_{-\infty}^{-1} d\sigma\, e^{i\nu\sigma} \int_{-1}^{1} d\xi \left(\frac{1+\xi}{1-\xi}\right)^{1/2} \frac{1}{(\sigma-\xi)} \cdot \int_{-1}^{1} \left(\frac{1-x}{1+x}\right)^{1/2} \frac{dx}{(\xi - x)}$$

$$= \frac{i}{\nu}\, e^{-i\nu} \ . \tag{5.12.10}$$

Using the integrals ([23] II),

$$\oint_{0}^{\pi} \frac{\cos nx}{\cos x - \cos\alpha}\, dx = \pi \frac{\sin n\alpha}{\sin\alpha} \ , \quad n = 0, 1, 2, \dots \ , \tag{5.12.11}$$

and

$$\int_{a}^{b} \frac{dx}{(cx+d)\{(x-a)(b-x)\}^{1/2}} = \frac{\pi}{\{(ac+d)(bc+d)\}^{1/2}} \ , \tag{5.12.12}$$

$(ac+d) > 0$, $(bc+d) > 0$, we write condition (5.12.10) as

$$-\int_{-\infty}^{-1} e^{i\nu\sigma} \left\{1 + \frac{(1+\sigma)}{(\sigma^2-1)^{1/2}}\right\} d\sigma = \frac{i}{\nu}\, e^{-i\nu} \ . \tag{5.12.13}$$

Multiplication of the real parts of (5.12.13) by $\sin\nu$ and of the imaginary parts by $\cos\nu$ and addition of the two results, yields the condition

$$-\int_{-\infty}^{-1} \left\{1 + \frac{(1+\sigma)}{(\sigma^2-1)^{1/2}}\right\} \sin\nu(1+\sigma)\, d\sigma$$

$$= \int_{0}^{\infty} \left\{ 1 - \left(\frac{\xi}{\xi+2}\right)^{1/2} \right\} \sin\nu\xi\, d\xi = \frac{1}{\nu} \ , \tag{5.12.14}$$

where the first equality follows from the substitution $\sigma = -\xi - 1$. By two times partial integration we rewrite this condition as

$$\int_{0}^{\infty} \sin\nu\xi \frac{(2\xi+1)}{\xi^{3/2}(\xi+2)^{5/2}}\, d\xi = 0 \ . \tag{5.12.15}$$

However, because the factor of $\sin \nu\xi$ in the integrand is a monotonic decreasing function, it follows that (5.12.15) cannot be satisfied for any value of $\nu > 0$. This means that the profile cannot annihilate the incoming periodic vorticity. Hence our assumption stated above (5.12.5) is not correct.

We can state our result otherwise. When the fluid with the vorticity is in "rest" and the profile of which the chord length is finite, moves with the velocity U in the negative x-direction, it cannot extract by means of its suction forces at the leading edge, all the kinetic energy out of the fluid.

5.13. Some Additional Remarks

As we mentioned already in this chapter, there are two major shortcomings in our optimization theory, namely we neglected non-linear effects such as the deformation of the shed vorticity sheets and we neglected the influence of the viscosity of water. We will discuss some possibilities to correct partially these imperfections.

First we mention a non-linear correction with respect to the shed vorticity sheets. Consider a screw propeller with N blades working in an inviscid and undisturbed fluid. Hence in notation of this chapter $\Phi_0 = -\Phi_1 = 0$. Because we consider a propeller, only the thrust is prescribed and the optimization condition (5.5.8) for the potential $\Phi = \Phi_2$ (5.5.6) of the shed vorticity far behind the propeller becomes

$$\frac{\partial \Phi}{\partial n}(\xi_l, \eta_l) = \frac{\lambda}{\rho} \cos_{(n,x)}(\xi_l, \eta_l) \ , \quad l = 1, \ldots, N \ . \tag{5.13.1}$$

This condition has to be satisfied in the linearized theory at the reference surfaces H_l of the blades (3.1.1) (with $t = 0$)

$$H_l : \quad \varphi + ax = \frac{2\pi l}{N} \ , \quad a = \frac{\omega}{U} \ , \quad \omega > 0 \ , \quad l = 1, \ldots, N \ . \tag{5.13.2}$$

The Lagrange multiplier λ in (5.13.1) follows, as we discussed in Section 5.5, from the prescribed thrust of the propeller. We can interpret (5.13.1) by stating that far behind the propeller the reference surfaces H_l, and hence the shed free vorticity layers, translate with a velocity $|\lambda|/\rho$ backwards with respect to the ship. Just behind the propeller the free vorticity is situated only on half infinite reference surfaces. This means that just behind the propeller the magnitude of the induced velocity in axial-direction, which we denote by w_x, will be about

$$w_x = \tfrac{1}{2} \frac{|\lambda|}{\rho} \ . \tag{5.13.3}$$

Hence instead of the reference surfaces H_l given in (5.13.2), we can better use the reference surfaces $\tilde{H}_l$ given by

$$\tilde{H}_l : \quad \varphi + \tilde{a}x = \frac{2\pi l}{N} \ , \quad \tilde{a} = \frac{\omega}{(U + w_x)} \ , \quad \omega > 0 \ , \quad l = 1, \ldots, N \ , \tag{5.13.4}$$

as the surfaces on which the shed free vorticity is situated. Also on these surfaces we can place the planforms of the blades, where their bound vorticity is present.

For this new geometrical configuration we again carry out the optimization procedure and so on, until the results do not change anymore. In this way we come more close to the correct value of the lost kinetic energy E_{kin}.

It should be noted that the contraction of the flow in the neighbourhood of the propeller is not taken into account in this way, for this we refer to [17]. Also we have to be aware of the fact that the conservation of energy, which holds in the exact non-linear theory and in the linearized theory of an inviscid and incompressible fluid (Section 1.17) is no longer valid, because here we are "in between" these two theories.

We now discuss a simple method to determine approximately the influence of the viscosity of water on the efficiency of an "optimum" simple screw propeller. First, we optimize the propeller in the inviscid fluid under the condition that it delivers a demanded thrust, in the way as we described in this chapter. Then we know the lost kinetic E_{kin1} per unit of length and the circulation distribution $\Gamma_1(r)$ around the blades as a function of the radius r. Now we can distribute this $\Gamma_1(r)$ as bound vorticity along the chords and also choose chord lengths $c(r)$, both in relation to the avoidance of cavitation. Next we use a more or less experimental resistance law so that we can calculate the viscous resistance $D(r)$ per unit of span, of the chords $c(r)$ which are placed in a fluid velocity $V(r) = \{(U^2 + w_x)^2 + \omega^2 r^2\}^{1/2}$ (w_x from (5.13.3)). Then the viscous energy loss per unit of length in the direction of motion becomes for a screw with N blades

$$E_{visc1} = \frac{N}{U} \int_{r_1}^{r_2} V(r) D(r)\, dr \ , \tag{5.13.5}$$

where r_1 and r_2 are the inner and the outer radius of a blade. Also we can calculate the integral $Q_{visc1} > 0$ over the components of $D(r)$ in the direction of the screw axis, of the resistance of the chords or profiles.

Now we have to repeat these calculations, however, we take for the demanded thrust the value

$$T_2 = T + Q_{visc1} \ . \tag{5.13.6}$$

Then we arrive at new values E_{kin2}, E_{visc2} and Q_{visc2}. This can be continued until we arrive at values E_{kin} and E_{visc} which do not change anymore. The corrected efficiency for the "optimum" propeller in a viscous fluid then becomes

$$\eta = \frac{TU}{TU + E_{kin} + E_{visc}} \ . \tag{5.13.7}$$

This we call "classical optimization in relation to viscosity". As follows from the above, we use here circulation distributions $\Gamma_1(r), \Gamma_2(r), \ldots$, which follow from the inviscid optimization.

Another procedure called "optimization including viscosity" is much more complicated. There a formula for E_{visc} is derived which depends on the still unknown optimum circulation distribution $\Gamma(r)$ of the blades. This can be done because the value of $\Gamma(r)$ yields a chord length at r in order to avoid cavitation. By this the constraint on the demanded thrust becomes dependent on $\Gamma(r)$. Then we can minimize, with respect to $\Gamma(r)$, the *sum* $(E_{\text{kin}} + E_{\text{visc}})$, under the constraint of the demanded thrust. Then the efficiency follows from (5.13.7).

This latter method has, in comparison with the first method, the tendency to reduce slightly the circulation at the tips of the blades and by this the chord lengths in that region. The reason is that the tips of the blades have a high velocity with respect to the water hence their viscous resistance is relatively high. Then in order that the desired thrust is delivered, the circulation and hence the chord lengths have to be somewhat larger at the inner radii.

It turns out, however, that both methods yield nearly the same values for the influence of the viscosity on the efficiency. This is caused by the fact that for a realistic screw propeller the kinetic energy losses are much higher than the viscous losses, so the viscosity gives rise only to a relatively small decrease of the efficiency.

In Section 6.6 we will discuss the influence of viscosity on the span of end plates at the tips of the blades of a screw propeller.

We refer to de Jong [33] for an extensive treatment of the influence of the viscosity of water on optimum screw propellers. Potze discusses in [54] the influence of viscosity on the optimum unsteady propulsion by means of two wings.

Next we mention a difficulty which arises with respect to the optimization of a screw propeller surrounded by a duct. This configuration cannot be treated in a straightforward way by our optimization theory. The reason is that the profiles of the duct are prescribed and are anyhow fixed in space. Hence this yields a mixed problem of which the geometry is partly prescribed. We remark that a ring propeller of which the ring rotates with the blades is easily handled by our theory, analogously to the propeller with end plates. (Sections 6.1–6.6).

In this chapter it followed from the linear optimization theory that, for instance in an undisturbed fluid, the optimum free vorticity sheets far behind a propeller have to translate like rigid impermeable surfaces backwards. So at first sight there seems to be no reason for these sheets to change their shape and to roll up. However, such a "translation" would cause a square root singularity of the vorticity at the edges of the sheets, which would cause a suction force at these edges. Because there is no rigid material which can sustain these forces, in reality the sheets will roll up just as in Section 1.22 where at $t = 0$ the half infinite material plate was annihilated (below (1.22.6)). In fact, this roll up will start right behind the blades or wings of the propeller.

A remark has to be made with respect to the optimization theory of unsteady propellers of Regimes (3) and (4) of Section 4.2. There the shed free vorticity cannot be placed on reference surfaces of the wings as we did in the previous sections. In this case the shed vorticity has to be placed on the calculated wakes which belong to the base motion of the wings, as defined in Section 4.2. So in order to determine the optimum shed vorticity in this case, we have to translate the periodic wakes far behind the propeller as rigid and impermeable surfaces in the direction opposite to the one in which the ship moves. When in this way the optimum shed vorticity is found, we have to calculate via the non-periodic still moving part of the wake which stretches up to the wings, the time-dependent circulation around the wings which inversely yields this optimum shed vorticity. We return to this problem in Section 6.10.

Finally, we refer to the work of Coney [13], who discusses the optimization of contra-rotating propellers, vane wheel propulsors and propellers with a pre-swirl stator, of which the blades are represented by lifting lines. The optimization is carried out by means of a purely numerical iterative vortex lattice method, which takes into account the deformation of the free vortex sheets.

Chapter 6

Applications of Optimization Theory

We now discuss in more detail a number of problems concerned with the optimization of propulsion devices. These problems are such that they can be treated with the theory of the previous chapter. Our considerations are directed towards the determination of the shed free vorticity of the optimum propellers from which follows the span-wise distribution of the circulation around their blades or wings.

The first subject is the screw propeller with or without end plates at the tips of its blades. The influence of the shape of the generator lines of the propeller on the quality number q is considered. Also discussed is the combined optimization of the rotational velocity of the screw and the span of the end plates in relation to the viscosity of water.

The second problem is the optimization of a screw propeller with a hub of finite length of which the radius is not small with respect to the radius of the propeller. It is shown that as well in this case we can find the optimum free vorticity far behind the propeller.

Then we treat unsteady propulsion by means of one or two flexible wings of finite span which carry out a large-amplitude motion (Regime (2) of Section 4.1). We calculate the quality numbers of this type of propellers in the optimum case.

Also we consider the large-amplitude unsteady propulsion by means of two rigid wings. In the 2-dimensional case that these wings have infinite span, we discuss their base motion and their added motion. The theory is then applied to a more or less realistic situation and the results are compared with the results from the previous case (Regime (2) of Section 4.1). It follows that the tip vorticity of wings of finite span can have an important influence.

Next we discuss the optimization of a simple model with an infinite number of blades of finite span of the Voith-Schneider propeller. From this we obtain the information that the thrust has to be divided equally over the blades in the front position and the blades in the aft position, which is in agreement with experiments.

Finally, the optimization of the sails of a yacht sailing close to wind is considered. We calculate the optimum circulation distribution along the sails, under the constraint of a prescribed heeling moment.

6.1. Screw Propeller with or without End Plates, Basic Notations

We consider the optimization of a screw propeller of which the blades can be provided with end plates at their tips. We assume that these end plates are present, however, the theory holds equally well for propellers without end plates. In the first instance we neglect the viscosity of water and consider its influence at a later stage. The theory we will discuss is based, with the exception of the considerations about the disturbance velocity field, on de Jong and Sparenberg [33] and on de Jong [34].

For our geometrical considerations we use a Cartesian coordinate system (x, y, z) and a cylindrical one (x, r, θ) (Figure 6.1.1). We consider an N-bladed screw propeller which rotates with the angular velocity ω in the positive θ-direction and which moves with the velocity U in the positive x-direction. The hub of the propeller is the two-sided infinite cylinder $r = r_h$.

The reference surface of a blade with its end plate is described as follows. Consider (Figure 6.1.1), say at $t = 0$, in the (x, y) plane two generator lines g_1 and g_2, given by

$$g_1 : \quad x = f_1(y) \ , \quad 0 \le x \le x_1 \ , \quad r_h \le y \le r_1 \ , \quad f_1(r_h) = 0 \ , \tag{6.1.1}$$

$$g_2 : \quad y = f_2(x) \ , \quad x_2 \le x \le x_3 \ , \quad r_2 \le y \le r_3 \ , \tag{6.1.2}$$

where f_1 and f_2 are sufficiently smooth functions. We demand that g_2 passes through the point $(x, y) = (x_1, r_1)$ of g_1. The lines g_1 and g_2 are now translated in the positive x-direction with the velocity U and are at the same time rotated with the angular velocity ω in the positive θ-direction. By doing this, g_1 and g_2 will describe the reference surfaces H_1 and H_2 respectively, of the screw blade and of the end plate under consideration. If the translation and the rotation "started at $x = -\infty$", H_1 and H_2 stretch from $x = -\infty$ towards $x = \infty$. The surface of the hub is denoted by H_3.

When the propeller has N blades, we sometimes add to a symbol belonging to a blade or an end plate a second index $j = 1, \ldots, N$. The reference surfaces coincide with each other after a rotation over an angle $2\pi j/N$ $(j = 1, \ldots, N)$ around the x-axis. We find for these reference surfaces

$$H_{1,j} : \quad \theta + a(f_1(r) - x) = \frac{(j-1)2\pi}{N} \ ,$$

$$r_h \le r \le r_1 \ , \quad a = \frac{\omega}{U} \ , \quad j = 1, \ldots, N \ , \tag{6.1.3}$$

$$H_{2,j} : \quad \theta - ax = \frac{(j-1)2\pi}{N} - a\tilde{x} \ ,$$

$$r = f_2(\tilde{x}) \ , \quad x_2 \le \tilde{x} \le x_3 \ , \quad j = 1, \ldots, N \ , \tag{6.1.4}$$

$$H_3 : \quad r = r_h \ . \tag{6.1.5}$$

The radius r_p of the propeller is defined as the greatest value of r which is attained by points situated at the reference surfaces.

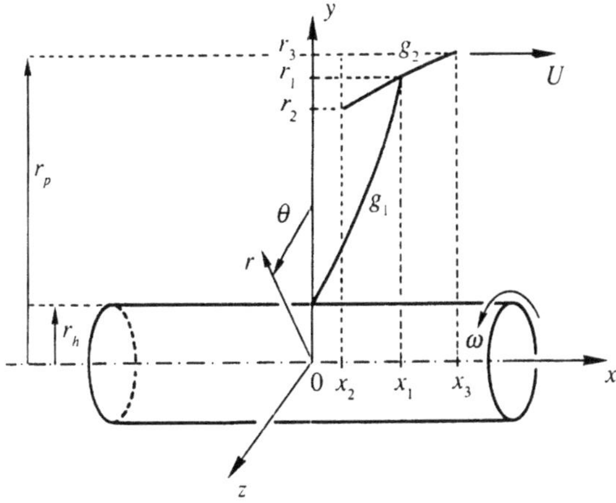

Fig. 6.1.1. Hub and generator lines, g_1 of blade and g_2 of end plate.

End plates can be “one-sided” or “two-sided”. A one-sided end plate has a generator line g_2 (6.1.2) with $x_2 = x_1$ or $x_3 = x_1$, while for a two-sided end plate $x_2 < x_1 < x_3$.

On the surfaces $H_{i,j}$ ($i = 1, 2; j = 1, \ldots, N$) and H_3 unit normals $\vec{n}_{i,j}$ and $\vec{n}_3$ are introduced. Their physical components in the local cylindrical coordinate directions $(\vec{c}_x, \vec{c}_r, \vec{c}_\theta)$ follow from (B.2.7)

$$\vec{n}_{1,j} = \frac{(-ar, arf_1'(r), 1)}{[1 + a^2r^2\{1 + (f_1'(r))^2\}]^{1/2}} , \tag{6.1.6}$$

$$\vec{n}_{2,j} = \frac{\left(-arf_2'\left(x - \frac{\theta}{a} + \frac{(j-1)2\pi}{aN}\right), ar, f_2'\left(x - \frac{\theta}{a} + \frac{(j-1)2\pi}{aN}\right)\right)}{\left[a^2r^2 + (1 + a^2r^2)\left\{f_2'\left(x - \frac{\theta}{a} + \frac{(j-1)2\pi}{aN}\right)\right\}^2\right]^{1/2}} , \tag{6.1.7}$$

$$\vec{n}_3 = (0, 1, 0) . \tag{6.1.8}$$

It follows from these formulas that, in case of “reasonably” chosen generator lines g_1 and g_2, $\vec{n}_{1,j}$ has a component in the direction of increasing values of θ and that $\vec{n}_{2,j}$ and $\vec{n}_3$ have a component in the direction of increasing values of r. We define the + side $H^+_{i,j}$ of $H_{i,j}$ to be oriented in the direction of $\vec{n}_{i,j}$, the – side $H^-_{i,j}$ is the opposite side.

Next we want to have at our disposal a measure which gives some information about the extent to which the fluid “outside” the end plate reference surfaces $H_{2,j}$ ($j = 1, \ldots, N$) is being separated from the fluid “inside” these surfaces. Hereto we introduce the ratio of covering k of the end plates. Suppose the helix through the point $(x, r, \theta) = (x_2, r_2, 0)$ (Figure 6.1.1) cuts the plane $x = 0$ in the point

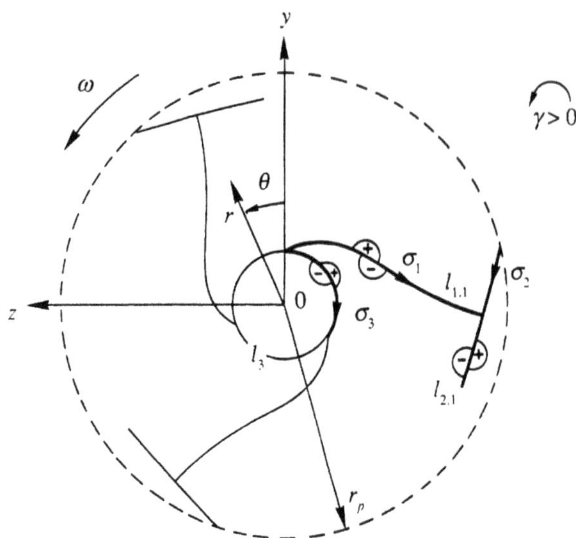

Fig. 6.1.2. Lifting lines $l_{1,1}$, $l_{2,1}$, l_3 and length parameters σ_i of a three-bladed screw propeller.

$(x, r, \theta) = (0, r_2, \theta_2)$ and the helix through $(x_3, r_3, 0)$ cuts that plane in $(0, r_3, \theta_3)$. For the N-bladed screw propeller with end plates the ratio of covering k is defined by

$$k = \frac{N(\theta_3 - \theta_2)}{2\pi} \ . \tag{6.1.9}$$

A screw propeller without end plates has $k = 0$. Here we will consider only screws with $0 \le k \le 1$.

Finally, we discuss the lifting line system which, in the linear theory, can represent the screw propeller. We take the lifting lines $l_{1,j}$, $l_{2,j}$ $(j = 1, \ldots, N)$ and l_3 as the intersection of the reference surfaces $H_{1,j}$, $H_{2,j}$ and H_3 with the (y, z) plane (Figure 6.1.2). In this way the hub is represented by the circular lifting line l_3 and can be treated in the same way as the screw blades and the end plates as will follow from the boundary conditions which we have to put in the optimum case on the velocity potential, see below (6.2.7). We could have taken the lifting lines in many other ways, for instance along the generator lines g_1 and g_2 and along $y = r_h$ in the (x, y) plane. However, the choice we made here, seems to be more palpable.

We make two remarks. First, the hub can also be considered as a periodic body as in Figure 5.1.1, where it is denoted by C_1. In fact, the admittance of these bodies in the optimization theory was suggested by the two-sided infinite cylinder as a model of a hub. Second, the curved lifting lines $l_{i,j}$ $(i = 1, 2;\ j = 1, \ldots, N)$ and l_3 have one drawback as we mentioned at the end of Section 1.1, namely they induce infinite velocities on themselves. Also the free vorticity shed by the lifting lines, induces by its component tangent to these lines infinite velocities on them. So we cannot, as we did in relation to the straight bound vortex of Figure 5.6.2, calculate the forces of $O(\varepsilon^2)$ on the curved lifting lines in a simple way.

For the N-bladed screw, along a relevant $1/N$-th part of the lifting line system, for which we take $j = 1$, length parameters are introduced. Along the lifting line l_3 representing the hub, the length parameter σ_3 is increasing from the point $(r, \theta) = (r_h, 0)$ to $(r, \theta) = (r_h, -2\pi/N)$. Along the lifting line $l_{1,1}$ representing the blade, we take the length parameter σ_1 increasing from the point $(r, \theta) = (r_h, 0)$ to $(r, \theta) = (r, -ax_1)$, $(a = \omega/U)$. If end plates are present, the length parameter σ_2 is used, increasing from the point $(r, \theta) = (r_2, -ax_2)$ to $(r, \theta) = (r_3, -ax_3)$. Start and end values of σ_i are denoted by $\sigma_{i,s}$ and $\sigma_{i,e}$, respectively. The value of σ_2 corresponding to the point which have $l_{1,1}$ and $l_{2,1}$ in common, is denoted by $\sigma_{2,bl}$.

The bound vorticity $\Gamma_i(\sigma_i)$ $(i = 1, 2, 3)$ which is the circulation around l_i at the place σ_i is taken with a right-hand screw in the positive σ_i-direction. The free vorticity $\gamma_i(\sigma_i)$ situated at H_i $(i = 1, 2, 3)$ along the helices at these surfaces, is reckoned positive when, with a right-hand screw, it has a positive component in the positive x-direction. We take the free vorticity $\gamma_i(\sigma_i)$ per unit of length in the σ_i-direction $(i = 1, 2, 3)$, which however is not perpendicular to the helicoidal vortex lines. Then between the bound vorticity $\Gamma_i(\sigma_i)$ and the free vorticity $\gamma_i(\sigma_i)$ we have the relations

$$\gamma_i(\sigma_i) = \frac{d\Gamma_i}{d\sigma_i}(\sigma_i) , \qquad (i = 1, 2, 3) . \tag{6.1.10}$$

A propeller cannot leave behind a concentrated free vortex because then its efficiency would be zero. Hence we find that the circulation at the tip of the lifting line representing the blade, has to be divided over the lifting lines representing the two end plate halves.

6.2. Optimization of the Screw Propeller

We consider the N-bladed propeller acting behind a ship in an inviscid and incompressible fluid. Then when its blades and end plates move along their reference surfaces $H_{i,j}$ $(i = 1, 2;\ j = 1, \ldots, N)$ they encounter a disturbance velocity field $\vec{v}_0^{\,*}$ for which as an approximation we can take the nominal wake of the hull. The component of $\vec{v}_0^{\,*}$ normal to the two-sided infinite hub $r = r_h$ can be taken equal to zero. This disturbance field is assumed to be independent of the x-coordinate.

Moving along a helicoidal reference surface a rigid screw blade with its end plates periodically encounters a disturbance field which has the same meaning for this blade as the periodicity of $\vec{v}_0^{\,*}$ in Section 5.11 for the flat rigid wing. Fourier expansions can be made here entirely analogous to those made in that section. It follows that the kinetic energy "parcels" belonging to the Fourier components $\cos nax$ and $\sin nax$ for $n \geq 1$ can be extracted only by means of suction forces at the leading edges of blades and end plates. This seems however not very reliable because these leading edges are rather sharp.

When we consider the kinetic energy which can be extracted by the blades, we first can replace (Section 5.2) $\vec{v}_0^*(x,r,\theta)$ by the equivalent velocity field $\vec{v}_0(x,r,\theta)$ of which the vorticity $\vec{\gamma}_0$ is confined to the reference surfaces and which has the same normal velocity components at these surfaces as $\vec{v}_0^*$. The total circulation around the reference surfaces has to be chosen equal to zero. Then we can consider from $\vec{\gamma}_0$ only that part which corresponds to $\alpha_{x0}(z)$ in (5.11.2), which we denote here by $\gamma_{00}(\sigma_i)$ and which consists of vorticity placed along the helicoidal lines at the reference surfaces.

From $\gamma_{00}(\sigma_i)$ we can calculate the velocity field $\vec{v}_{00}(x,r,\theta)$ and its velocity potential $\Phi_{00}(x,r,\varphi)$. The kinetic energy belonging to $\vec{v}_{00}(x,r,\theta)$ can be extracted entirely by the inner parts of the propeller blades and the end plates. In the following we neglect the higher order Fourier components of $\vec{\gamma}_0$ and consider only the vorticity field $\gamma_{00}(\sigma_i)$ at the reference surfaces and its velocity field $\vec{v}_{00}(x,r,\theta)$ with the potential $\Phi_{00}(x,r,\theta)$.

In order that the propeller is optimum in the disturbance field $\vec{v}_{00}(x,r,\theta)$, we have for the potential Φ of the velocity field induced by the shed vorticity far behind the propeller, the condition

$$\frac{\partial \Phi}{\partial n} = -\frac{\partial \Phi_{00}}{\partial n} + \frac{\lambda}{\rho} \cos_{(n,x)} , \tag{6.2.1}$$

at the reference surfaces $H_{i,j}$ ($i = 1,2; j = 1,\ldots,N$) and H_3. This equation follows from (5.5.5), with one Lagrange multiplier for the constraint on the thrust, while the function $\cos_{(n,x)}$ follows from a comparison of (5.4.9) with (5.4.4).

It follows again from (6.2.1) that we can split Φ into two parts

$$\Phi = \Phi_1 + \Phi_2 , \qquad \Phi_1 = -\Phi_{00} . \tag{6.2.2}$$

The first part $\Phi_1 = -\Phi_{00}$ "cleans" the reference surfaces by annihilating the free vorticity $\gamma_{00}(r)$ on these surfaces. This can be carried out by a circulation distribution $\Gamma_{i,00}$ of the lifting lines of the value (see (6.1.10))

$$\Gamma_{i,00}(\sigma_i) = -\int_{\sigma_{i,s}}^{\sigma_i} \gamma_{00}(\tilde{\sigma}_i)\, d\tilde{\sigma}_i , \qquad i = 1,2,3 . \tag{6.2.3}$$

The second part Φ_2 of the potential belongs to the velocity field caused by translating, through the undisturbed fluid, the reference surfaces as rigid impermeable surfaces, with the velocity λ/ρ in the x-direction. So when we solve the boundary value problem

$$\Delta \tilde{\Phi}_2 = 0 , \tag{6.2.4}$$

$$\frac{\partial \tilde{\Phi}_2}{\partial n_{i,j}} = (\vec{n}_{i,j} \cdot \vec{e}_x) , \quad \text{on} \quad H_{i,j} , \quad (i = 1,2 \;; j = 1,\ldots,N) , \tag{6.2.5}$$

$$\frac{\partial \tilde{\Phi}_2}{\partial n_3} = 0 \ , \qquad \text{on } H_3 \ , \tag{6.2.6}$$

where Δ is the Laplace operator and $\vec{e}_x$ is the unit vector in the x-direction, we find

$$\Phi_2 = \frac{\lambda}{\rho} \tilde{\Phi}_2 \ . \tag{6.2.7}$$

We remark that by (6.2.6) indeed the boundary condition is satisfied on the hub by the former introduction of the lifting line l_3.

It follows from Section 1.20 that the velocity fields caused by Φ_1 as well as by Φ_2, are perpendicular to the helicoidal lines.

The total circulation $\Gamma_i(\sigma_i)$ of the lifting line l_i $(i = 1, 2)$ can be split also into two parts

$$\Gamma_i(\sigma_i) = \Gamma_{i,00}(\sigma_i) + \Gamma_{i,2}(\sigma_i) \ , \tag{6.2.8}$$

where $\Gamma_{i,00}$ (6.2.3) takes care of the extraction of the kinetic energy of $\vec{v}_{00}(x, r, \theta)$ and

$$\Gamma_{i,2}(\sigma_i) = -[\Phi_2]^+_- = -\frac{\lambda}{\rho} [\tilde{\Phi}_2]^+_- \ . \tag{6.2.9}$$

By the law of Joukowski we find for the thrust T $(= O(\varepsilon))$

$$\begin{aligned} T &= N\rho\,\omega \sum_{i=1}^{2} \int_{l_i} r\Gamma_i(\sigma_i) \frac{(\vec{r} \cdot \vec{t}_i)}{r} \, d\sigma_i \\ &= N\rho\,\omega \left[\sum_{i=1}^{2} \int_{l_i} \Gamma_{i,00}(\sigma_i) \, (\vec{r} \cdot \vec{t}_i) \, d\sigma_i - \frac{\lambda}{\rho} \sum_{i=1}^{2} \int_{l_i} [\tilde{\Phi}_2]^+_- \cdot (\vec{r} \cdot \vec{t}_i) \, d\sigma_i \right], \end{aligned} \tag{6.2.10}$$

where $\vec{r}$ is the vector in the (y, z) plane pointing from the origin towards the point σ_i of l_i and $\vec{t}_i(\sigma_i)$ is the unit tangent vector to l_i at σ_i, pointing in the direction of increasing σ_i. From (6.2.10) follows the Lagrange multiplier λ, when T is prescribed.

In our linear theory the power needed to rotate the screw propeller when only the disturbance velocity field $\vec{v}_{00}(x, r, \theta)$ is present, is equal to $UT = O(\varepsilon)$, plus the kinetic energy $E_2 = O(\varepsilon^2)$ far behind the propeller over a length U in the x-direction which is caused by Φ_2, minus the absorbed kinetic energy $E_1 = O(\varepsilon^2)$ of $\vec{v}_{00}(x, r, \theta)$ also over this length.

First we give a formula for E_1

$$\begin{aligned} E_1 &= \tfrac{1}{2}\rho \int_{-\infty}^{\infty}\!\!\int \int_{0}^{U} (\vec{v}_{00})^2 \, dx \, dy \, dz \\ &= -\tfrac{1}{2}\rho N \sum_{i=1}^{2} \int_{H_i(0 \le x \le U)} \frac{\partial \Phi_{00}}{\partial n_i} \cdot [\Phi_{00}]^+_- \, dS_i \ , \end{aligned} \tag{6.2.11}$$

where dS_i are elements of area of H_i,

$$dS_1 = \frac{\{1+a^2r^2(1+f_1'(r)^2)\}^{1/2}}{\left(1+a^2r^2f_1'(r)^2\right)^{1/2}}\,dx\,d\sigma_1 = f_3(\sigma_1)\,dx\,d\sigma_1\ , \tag{6.2.12}$$

$$dS_2 = \frac{\{a^2r^2+(1+a^2r^2)f_2'(x-\theta/a)^2\}^{1/2}}{\left(a^2r^2+f_2'(x-\theta/a)^2\right)^{1/2}}\,dx\,d\sigma_2 = f_4(\sigma_2)\,dx\,d\sigma_2\ . \tag{6.2.13}$$

In (6.2.12) and (6.2.13) we introduced $f_3(\sigma_1)$ and $f_4(\sigma_2)$, which is possible because when σ_i = const. it follows that r = const. and hence $(x-\theta/a)$ = const. at some H_i. Also $\partial\Phi_{00}/\partial n_i$ and $[\Phi_{00}]_-^+$ depend on σ_i only, hence we can carry out the integration over x in the right-hand side of (6.2.11) simply by multiplying the integral over σ by U, we find

$$E_1 = -\tfrac{1}{2}\,\rho\,NU\left\{\int_{l_1}\frac{\partial\Phi_{00}}{\partial n_1}\cdot[\Phi_{00}]_-^+ f_3(\sigma_1)\,d\sigma_1 + \int_{l_2}\frac{\partial\Phi_{00}}{\partial n_2}\cdot[\Phi_{00}]_-^+\,f_4(\sigma_2)\,d\sigma_2\right\}\ . \tag{6.2.14}$$

In an analogous way we can treat E_2, and we obtain by (6.2.5)

$$\begin{aligned} E_2 &= \tfrac{1}{2}\,\frac{\lambda^2}{\rho}\int_{-\infty}^{\infty}\int\int_0^U(\text{grad}\,\tilde{\Phi}_2)^2\,dx\,dy\,dz \\ &= -\tfrac{1}{2}\,\frac{\lambda^2}{\rho}N\sum_{i=1}^{2}\int_{H_i(0\le x\le U)}(\vec{n}_i\cdot\vec{e}_x)\,[\tilde{\Phi}_2]_-^+\,dS_i \\ &= \tfrac{1}{2}\,\frac{\lambda^2}{\rho}\omega N\sum_{i=1}^{2}\int_{l_i}(\vec{r}\cdot\vec{t}_i)\,[\tilde{\Phi}_2]_-^+\,d\sigma_i\ , \end{aligned} \tag{6.2.15}$$

where we changed from $(\vec{n}_i\cdot\vec{e}_x)$ to $(\vec{r}\cdot\vec{t}_i)$, which also occurs in (6.2.10).

The efficiency of the propeller then becomes

$$\eta = \frac{UT}{UT+E_2-E_1}\ . \tag{6.2.16}$$

From now on we assume that the propeller works in an *undisturbed* fluid, then E_1 is equal to zero. E_2 can be rewritten by substituting λ from (6.2.10), where now $\Gamma_{i,00}=0$, into (6.2.12)

$$E_2 = T^2\left\{2\rho\,\omega N\sum_{i=1}^{2}\int_{l_i}[\tilde{\Phi}_2]_-^+\,(\vec{r}\cdot\vec{t}_i)\,d\sigma_i\right\}^{-1}\ . \tag{6.2.17}$$

The kinetic energy loss E_{ac} of an actuator disk with the same thrust T, the same working area $\pi(r_p^2 - r_h^2)$ and the same velocity U as our screw propeller, follows from (2.1.15)

$$E_{ac} = T^2\{2\rho\, U\pi(r_p^2 - r_h^2)\}^{-1} \ . \tag{6.2.18}$$

Then the quality number of the screw propeller becomes

$$q = \frac{E_{ac}}{E_2} = \frac{N\omega}{U\pi(r_p^2 - r_h^2)} \sum_{i=1}^{2} \int_{l_i} [\tilde{\Phi}_2]_-^+ \, (\vec{r} \cdot \vec{t}_i) \, d\sigma_i \leq 1 \ . \tag{6.2.19}$$

Using the quality number, the efficiency of the propeller in the undisturbed fluid can be written as

$$\eta = \frac{UT}{UT + E_2} = \left\{1 + \frac{T}{2\rho\, U^2\pi(r_p^2 - r_h^2)q}\right\}^{-1} \ . \tag{6.2.20}$$

One point of interest in the following will be, can we find shapes of the generator lines l_i $(i = 1, 2)$, which yield favourable values for the quality number q.

6.3. Some Aspects of Optimum Screw Propellers

First we discuss in this section a useful symmetry property of optimum screw propellers. Consider two propellers A and B (Figure 6.3.1) with the *same rotational velocity* ω around the x-axis, while each is moving in the positive x-direction with the velocity U. The propellers A and B are such that for a suitable position in the (y, z) plane the lifting lines *are each others image by a reflection in the y-axis.* Along the lifting lines l_1 and l_2 the length parameters σ_1 and σ_2 are used as denoted in Figure 6.3.1.

When the optimization method is applied to these two propellers we have, in order to find their optimum shed vorticity, to translate the reference surfaces of their blades and end plates as impermeable and rigid surfaces in the x-direction with some velocity λ. Now it is easily seen that in both cases the "same" free vorticities occur on the reference surfaces H_A of propeller A as on the reference surfaces H_B of propeller B. In fact rotating H_B around the y-axis over π radians it can be made to coincide with H_A. Then changing the sign of the translational velocity and of the vorticity, H_B becomes identical to H_A. Hence indeed the optimum vortices have to be the same in both cases

$${}_A\gamma_i(\sigma_i) = {}_B\gamma_i(\sigma_i) \ , \qquad i = 1, 2 \ . \tag{6.3.1}$$

Then also the optimum circulation distributions of the lifting lines have to be equal, as well as the quality numbers and the efficiencies when the thrust is prescribed to be the same.

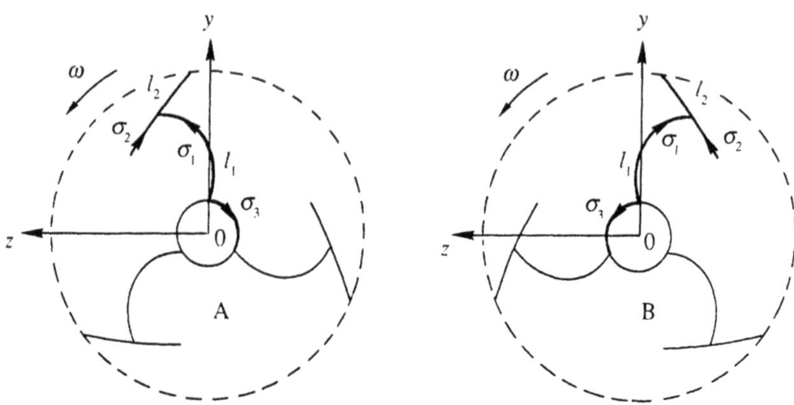

Fig. 6.3.1. Lifting lines of screw propellers A and B, three blades.

We now give three applications of this symmetry property. First: consider an optimum screw propeller with generator lines (Figure 6.1.1):

$$g_1 : \quad x = 0 \ , \quad r_h \le y \le r_1 \ , \tag{6.3.2}$$

$$g_2 : \quad y = f_2(x) \ , \quad x_2 \le x \le x_3 \ , \tag{6.3.3}$$

with $x_3 = -x_2$ and $f_2(-x) = f_2(x)$. It follows that the lifting line l_1 representing the blade is part of the y-axis and the lifting line l_2 representing an end plate is in the (y, z) plane symmetric with respect to the y-axis. Then (6.3.1) implies that the optimum free vorticity of the end plate is symmetrical in the sense

$$\gamma_2(\sigma_2) = \gamma_2(\sigma_{2,e} - \sigma_2) \ , \tag{6.3.4}$$

where, without loss of generality, we have taken the start value of σ_2 equal to zero.

Second: by our symmetry property we can obtain insight in the influence of a rake of the propeller blades on the optimum efficiency. Consider a screw with rake angle α, and end plates which are part of the cylinder $r = r_1 = r_p$

$$g_1 : \quad x = y \tan\alpha \ , \quad r_h \le y \le r_1 \ , \tag{6.3.5}$$

$$g_2 : \quad y = r_1 \ , \quad \xi = 1 \ , \tag{6.3.6}$$

where we introduced

$$\xi = (x_3 - x_1)/(x_1 - x_2) \ , \qquad 0 \le \xi \le \infty \ , \tag{6.3.7}$$

which will be called the end plate ratio. Then by our symmetry property such an optimum propeller has the same efficiency as one with rake angle $-\alpha$. This means that the quality number q or the efficiency η is a symmetric function of α with respect to $\alpha = 0$. Hence a small rake angle α will cause only very small changes of $O(\alpha^2)$ in the quality number and hence in the efficiency of a screw propeller.

Third: it follows almost directly from the symmetry property that the quality number is independent of the direction of the rotational velocity ω.

Next we treat some properties of the shed free vorticity of an optimum screw propeller. It follows from Section 1.20, as we mentioned before, that in the linearized theory, far behind a screw propeller the velocities induced by the shed free vorticity are perpendicular to the helices. Hence also in the case of the optimum free vorticity, these velocities are perpendicular to the helices at the reference surfaces and because in the optimum case these surfaces can be considered to be rigid and impervious, they are tangent to these surfaces. We denote the magnitude of these velocities at the + or − side of a reference surface H_i by

$$v_i^+(\sigma_i) \ , \qquad v_i^-(\sigma_i) \ , \tag{6.3.8}$$

respectively, they are reckoned positive when they have a positive component in the positive σ_i-direction. Then for the vorticity $\gamma_i(\sigma_i)$, which at the end of Section 6.1 is defined per unit of length in the σ_i-direction, we have

$$\gamma_i(\sigma_i) = (v_i^-(\sigma_i) - v_i^+(\sigma_i)) \cos_{(\sigma_i, s_i)} \ , \tag{6.3.9}$$

where s_i is the direction on H_1 perpendicular to the helices.

Because of their orthogonality with the helices and the H_i being impervious, the velocities v_i^+ and v_i^- in the neighbourhood of the intersection of two reference surfaces can be approximated by the corresponding velocity components of a 2-dimensional potential flow in an angle $\tilde{\beta}$. The plane of this 2-dimensional flow is the flat plane W perpendicular to the common helix of the two considered reference surfaces. The angle $\tilde{\beta}$ is the dihedral angle, which is the angle enclosed at the angular point by the intersection lines of the plane W with the reference surfaces. The nearer to the common helix, the better the approximation for v_i^+ and v_i^-.

For instance, in the case of the intersection helix of the reference surface H_1 of the blade with the reference surface H_2 of a two-sided end plate, a dihedral angle $\tilde{\beta}$ is related to one of the two angles $\beta < \pi$ between the lifting lines l_1 and l_2 by

$$\tan \tilde{\beta} = (1 + a^2 r_1^2) \tan \beta \ , \qquad a = \omega / U \ . \tag{6.3.10}$$

For instance, we can take β between $l_{1,1}^-$ and $l_{2,1}^-$ (Figure 6.1.2). In the plane W we denote the magnitude of the velocities of the 2-dimensional flow inside $\tilde{\beta}$ and at the sides of $\tilde{\beta}$ by $w(R)$, where R is the distance of the point under consideration to the angular point. It follows from the flow within the angle $\tilde{\beta}$ that (for instance [4])

$$|v_1^-(R)| = |v_2^-(R)| = w(R) = \text{const.}\ R^p \ , \quad R \to 0 \tag{6.3.11}$$

$$p = \frac{\pi}{\tilde{\beta}} - 1 \ , \qquad 0 < p < \infty \ , \tag{6.3.12}$$

where const. is unknown. When we suppose that l_2 has no kink at the point of intersection, the other angle between $l_{1,1}^+$ and $l_{2,1}^-$ becomes $\pi - \beta$ for which we have an analogous result.

It is clear that at the junction of the reference surfaces H_1 of the blade and H_2 of the end plate the velocities (6.3.11) inside both angles tend to zero for $R \to 0$. This means that

$$\gamma_1(\sigma_1) \to 0 \ , \qquad \sigma_1 \to \sigma_{1,e} \ . \tag{6.3.13}$$

Then by (6.1.10) it follows that the circulation $\Gamma_1(\sigma_1)$ of the blade has a zero derivative at its junction with the end plate, $d\Gamma_1(\sigma_1)/d\sigma_1 = 0$. Also we can calculate by (6.3.11) in which way $\gamma_1(\sigma_1)$ tends to zero for $\sigma_1 \to \sigma_{1,e}$ which is useful for a numerical calculation.

The free vorticity at H_2 for $\sigma_2 = \sigma_{2,bl}$, hence at its junction with H_1, is continuous and is equal to the velocity of the fluid at H_2^+ at that place. The derivative of this free vorticity, with respect to σ_2, however can be discontinuous because of the behaviour of the velocities at H_2^- for $\sigma_2 \uparrow \sigma_{2,bl}$ and for $\sigma_2 \downarrow \sigma_{2,bl}$.

When we have a one-sided end plate, we have (Figure 6.1.1) for instance $x_2 = x_1$ and $r_2 = r_1$. In the (y, z) plane we then take as the angle β the smallest angle between the lifting lines l_1 and l_2, hence $0 < \beta < \pi$. Now we have to consider the 2-dimensional flow inside the dihedral angle $\tilde{\beta}$ and also the flow "around the corner" outside $\tilde{\beta}$. Inside $\tilde{\beta}$ we have again a stagnation point of the flow at the angular point. At the corner outside $\tilde{\beta}$, however, we have an infinite velocity. Hence also the free vorticity (see (6.3.9)) at H_1 and H_2 becomes infinite at their junction, we find

$$\gamma_1(\sigma_1) \approx \text{const.}\ (\sigma_{1,e} - \sigma_1)^p \ , \quad \gamma_2(\sigma_2) \approx \text{const.}\ (\sigma_2 - \sigma_{2,bl})^p \ , \tag{6.3.14}$$

$$p = \left(\frac{-\pi + \tilde{\beta}}{2\pi - \tilde{\beta}} \right) \ , \qquad (-\tfrac{1}{2} < p < 0) \ , \tag{6.3.15}$$

where the unknown constants in (6.3.14) are equal. Hence in this case the optimum free vorticity density becomes infinite at the junction. This means that by (6.1.10) the circulation $\Gamma_1(\sigma_1)$ of the blade for $\sigma_1 \to \sigma_{1,e}$ and $\Gamma_2(\sigma_2)$ of the end plate for $\sigma_2 \to \sigma_{2,bl}$, have an infinite derivative.

Analogous considerations can be given for the circulations of the blades in the neighbourhood of the hub, when the hub is represented by the lifting line l_3.

It goes without saying that at the tips of the end plates or at the tips of the blades when no end plates are present, we have a square root singularity of the free vorticity as follows from Section A.3.

Other cases can be treated in the same way, for instance when a two-sided end plate instead of being smooth at its junction with the blade has a kink at that place.

6.4. Numerical Method and Results, the Quality Number

The geometry of the reference surfaces can vary rather strongly, so the numerical method for the optimization problems needs to be flexible. For the vorticity $\gamma_i(\sigma_{i,n})$, $\sigma_{i,n,s} \le \sigma_{i,n} \le \sigma_{i,n,e}$ ($i = 1, 2, 3$; $n = 1, \ldots, N$), belonging to the boundary value problem (6.2.4)–(6.2.6), we can write down three coupled singular integral equations. As the relevant part of the reference surfaces we choose the part with parameters $\sigma_{i,1}$ ($i = 1, 2, 3$), see Figure 6.4.1.

At a point $\sigma_{i,1} = \tilde{\sigma}_{i,1}$ of the line $l_{i,1}$, the velocity component normal to the reference surface $H_{i,1}$, induced by a concentrated helicoidal vortex of unit strength cutting the (y, z) plane at the point $\sigma_{j,n}$ is denoted by

$$K_{i,j,n}(\tilde{\sigma}_{i,1}, \sigma_{j,n}) \ , \qquad (i, j = 1, 2, 3 \ ; \ n = 1, \ldots, N) \ , \tag{6.4.1}$$

which, because they are very complicated, will not be given in formulas. Using this notation, we find three coupled singular integral equations for the free vorticity $\gamma_i(\sigma_{i,n})$ shed by the N-bladed propeller.

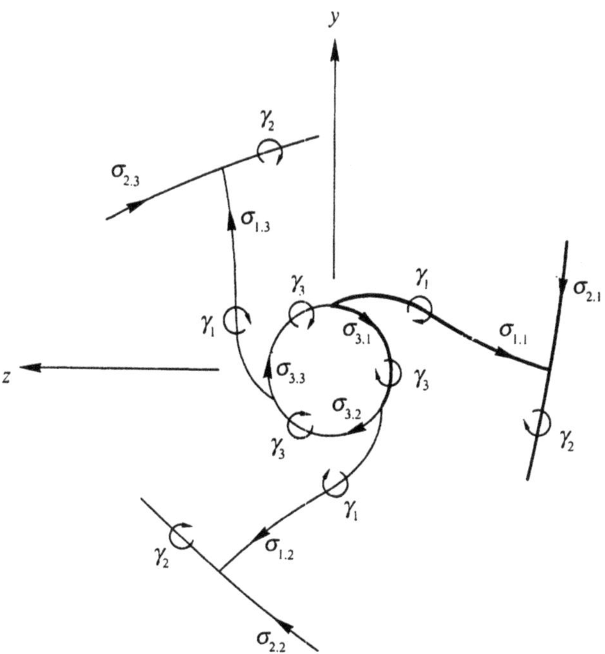

Fig. 6.4.1. (y, z) plane, length parameters $\sigma_{i,n}$ and free vorticity γ_i, ($i = 1, 2, 3$; $n = 1, \ldots, N$ (= 3)), used in integral equations (6.4.2).

$$\sum_{j=1}^{3}\sum_{n=1}^{N}\int_{\sigma_{j,n,s}}^{\sigma_{j,n,e}} K_{i,j,n}(\tilde{\sigma}_{i,1},\sigma_{j,n})\gamma_j(\sigma_{j,n})\,d\sigma_{j,n}$$

$$= (\vec{n}_{i,1}\cdot\vec{e}_x)(\tilde{\sigma}_{i,1})\ , \qquad \sigma_{i,1,s}\le\tilde{\sigma}_{i,1}\le\sigma_{i,1,e}\ ;i=1,2,3\ , \qquad (6.4.2)$$

where for $(j,n) = (i,1)$ we have to take the integrals as Cauchy principal value integrals.

Having solved these equations for some configuration, we can calculate the potential jump $[\tilde{\Phi}_2]_-^+$ from the free vorticity at the reference surfaces. Then by (6.2.19) follows the quality coefficient q of the optimum propeller corresponding to this configuration. Some results will be given in the remaining part of this section, where we take $r_h/r_p = 0.2$.

In Table 6.4.1 we assume the generator lines g_1 (6.3.5) and g_2 (6.3.6) to be straight, while g_2 is part of the line $y = r_1 = r_p$ (Figure 6.1.1), hence the planforms of the end plates are parts of the circular cylinder through the blade tips. Furthermore, we assume for the end plate ratio (6.3.7) $\xi = 1$.

The values of q in Table 6.4.1 are given in dependence on the number of blades N, the parameter $\mu = \omega r_p/U$, the ratio of covering k (6.1.9) and the rake angle α (6.3.5). It is seen that q increases with N and μ. Also q increases when k increases from 0 towards 1. For $k = 1$ it seems by numerical means that the partial derivative of q with respect to k is equal to zero. Further the influence of the rake angle α on q is very little for smaller values of $|\alpha|$ as was already predicted by an application of the symmetry property of optimum propellers (see below (6.3.7)). This is especially the case when end plates are present ($k > 0$).

It is of interest to consider the changes of q for the different configurations from the point of view of the statement given in Section 5.10 about the comparison by inspection of the efficiency of different propellers. The results of Table 6.4.1 are indeed in agreement with that section because also here it is found that the more the working region $r \le r_p$ is crowded with reference surfaces the higher is the value of the quality number q.

In Table 6.4.2 the influence of the asymmetry of the end plate reference surface is shown. Again the propeller has straight generator lines and the end plate reference surfaces are part of the circular cylinder through the blade tips, as in the preceding case. The difference is that now these reference surfaces can be asymmetric with respect to the blade tips, ($\xi \ne 1$, (6.3.7)). For screws with zero rake angle, the results are given for the ratios of covering $k = 0.25$, $k = 0.5$ and $k = 0.75$. It follows that the asymmetry has some positive influence on q, however the larger k, the less q is influenced by ξ. For $k = 1$, the values of q are independent of ξ because then the whole working region of the propeller is enclosed by the reference surfaces of the end plates, these values of q can be found in Table 6.4.1.

Because of the symmetry property (Section 6.3) in this case ($\alpha = 0°$) a screw propeller with end plate ratio ξ has the same quality number as one with end plate ratio ξ^{-1}, so Table 6.4.2 yields also the values of q for $\xi = 0.5$ and $\xi = 0$. From this we can expect that the asymmetry of the end plate reference surface cannot cause, in the neighbourhood of $\xi = 1$, substantial changes in q. This is analogous to the

Table 6.4.1. Quality number q, straight generator lines, end plate ratio $\xi = 1$.

k	0°			15°			30°			45°			α
	.450	.532	.582	.453	.535	.584	.463	.544	.592	.482	.561	.608	3
	.566	.646	.691	.569	.649	.694	.580	.658	.702	.602	.676	.717	4
0.00	.648	.722	.762	.652	.725	.764	.663	.733	.771	.685	.750	.784	5
	.708	.774	.809	.711	.776	.811	.722	.784	.817	.742	.799	.828	6
	.752	.811	.841	.755	.813	.844	.765	.820	.849	.783	.833	.859	7
	.587	.636	.665	.588	.636	.665	.588	.638	.667	.592	.643	.673	3
	.690	.736	.762	.691	.737	.763	.693	.739	.765	.700	.745	.770	4
0.25	.758	.800	.822	.759	.800	.822	.762	.803	.825	.769	.808	.829	5
	.804	.841	.861	.805	.842	.861	.808	.844	.863	.815	.849	.867	6
	.837	.870	.887	.838	.871	.888	.841	.873	.889	.847	.877	.893	7
	.679	.704	.719	.679	.704	.719	.679	.705	.720	,679	.706	.721	3
	.774	.796	.809	.774	.796	.809	.774	.797	.809	.775	.798	.811	4
0.50	.832	.851	.861	.832	.851	.861	.832	.852	.862	.834	.853	.863	5
	.869	.886	.895	.869	.886	.895	.870	.887	.895	.871	.888	.896	6
	.895	.909	.917	.895	.910	.917	.895	.910	.917	.897	.911	.918	7
	.732	.744	.750	.732	.744	.751	.733	.745	.751	.734	.746	.752	3
	.823	.831	.835	.823	.831	.835	.823	.831	.836	.824	.832	.836	4
0.75	.875	.881	.884	.875	.881	.884	.875	.881	.884	.876	.882	.885	5
	.907	.912	.914	.907	.912	.914	.907	.912	.914	.906	.912	.914	6
	.928	.932	.934	.928	.932	.934	.928	.932	.934	.928	.932	.934	7
	.750	.757	.761	.750	.757	.761	.751	.758	.762	.753	.760	.763	3
	.839	.842	.844	.839	.842	.844	.840	.843	.845	.841	.844	.845	4
1.00	.889	.891	.891	.889	.891	.891	.889	.891	.892	.890	.891	.892	5
	.919	.920	.920	.919	.920	.920	.919	.920	.921	.920	.920	.921	6
	.939	.939	.939	.939	.939	.939	.939	.939	.940	.939	.939	.940	7
N	2	3	4	2	3	4	2	3	4	2	3	4	μ

Table 6.4.2. Quality number q, straight generator lines, rake angle $\alpha = 0°$, $N = 3$.

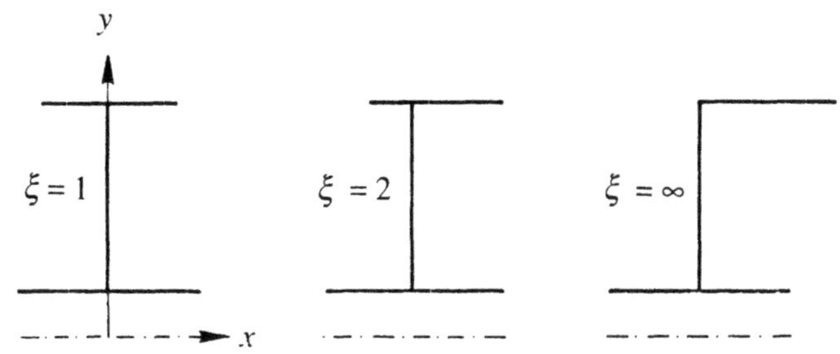

μ	0.25	0.5	0.75	0.25	0.5	0.75	0.25	0.5	0.75	k
3	.636	.704	.744	.636	.704	.744	.643	.709	.746	
4	.736	.796	.831	.737	.796	.831	.744	.802	.834	
5	.800	.851	.881	.800	.851	.881	.806	.857	.884	
		1			2			∞		ξ

Table 6.4.3. Quality number q, straight generator lines, rake angle $\alpha = 30°$, $N = 3$.

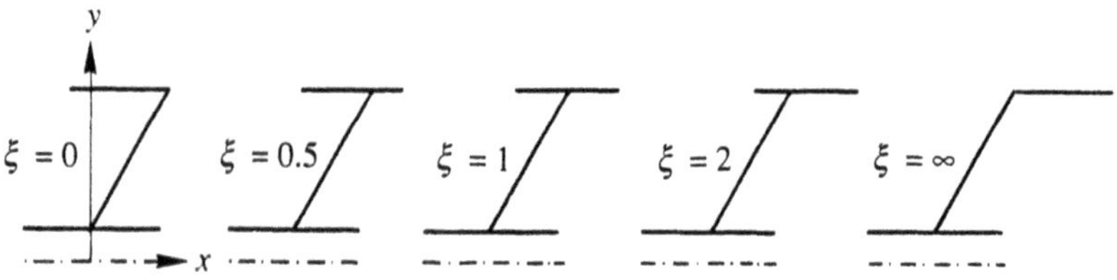

ξ	0			0.5			1			2			∞		
μ \ k	.25	.50	.75	.25	.50	.75	.25	.50	.75	.25	.50	.75	.25	.50	.75
3	.639	.707	.745	.637	.705	.715	.638	.705	.745	.640	.705	.744	.651	.712	.748
4	.739	.799	.832	.738	.797	.832	.739	.797	.831	.741	.797	.831	.752	.805	.835
5	.803	.853	.882	.802	.851	.881	.803	.852	.881	.805	.852	.881	.814	.859	.884

influence on q of the rake angle α in the neighbourhood of $\alpha = 0°$. The results of Table 6.4.2 confirm this expectation, even in the case of a one-sided end plate ($\xi = \infty$), for which q is maximum, the increase of q is small.

To illustrate the effect of $\xi \neq 1$ when the screw has a non-zero rake angle, results are given in Table 6.4.3 for analogous propellers, however, with a rake angle of 30°. Then we cannot apply the symmetry property of Section 6.3, hence we consider also the end plate ratios $\xi = 2^{-1} = 0.5$ and $\xi = \infty^{-1} = 0$.

The maximum values of q occur for the one-sided end plate with $\xi = \infty$. The increase of q is small with respect to the values of Table 6.4.2. Also here for the ratio of covering $k = 1$, the quality number q is independent of ξ, its values can be found in Table 6.4.1.

Next we consider in Table 6.4.4 the influence on q of a curvature of the generator line g_1 of the blade. The first screw of this table is of the same type as the first one of Table 6.4.2. The generator line of the second screw is obtained by interpolating with a cubic spline the points

$$(x, y) = (0, r_h), (0.1r_p, 0.4r_p), (0.15r_p, 0.6r_p), (0.1r_p, 0.8r_p), (0, r_p) \; . \quad (6.4.3)$$

By our symmetric property the quality number of this screw is the same as for the propeller obtained by using a similar spline through the points

$$(x, y) = (0, r_h), (-0.1r_p, 0.4r_p), (-0.15r_p, 0.6r_p),$$
$$(-0.1r_p, 0.8r_p), (0, r_p) \; . \quad (6.4.4)$$

The third screw has a rake of 30°. Next we apply a skew transformation, with an angle of 30°, to the points (6.4.3) and (6.4.4), for constant values of y. Then the generator lines of the fourth and the fifth screws are obtained by interpolating the resulting points with a cubic spline.

Table 6.4.4. Quality number q, end plate ratio $\xi = 1$, $N = 3$.

μ	0	0.5	0	0.5	0	0.5	0	0.5	0	0.5	k
3	.532	.704	.545	.704	.544	.705	.533	.706	.568	.705	
4	.645	.796	.660	.797	.658	.797	.646	.797	.684	.798	
5	.721	.851	.735	.852	.733	.852	.722	.852	.757	.853	

When the screw has no end plates ($k = 0$), a rake combined with curvature can have a favourable influence on q. Compare for $\mu = 3$ the first and the last screw but one. When end plates with $k = 0.5$ are applied, the influence of rake combined with curvature is negligible.

Finally, in Table 6.4.5 we look at the influence of the shape of the generator line g_2 of the end plates. The generator lines g_1 of the blades are straight and have zero rake except the last one which has a rake of 15°. All screws have end plates of which the range in radial direction is $r_p/20$, except the first one of which the end plates are part of the circular cylinder through the blade tips.

Again the first screw of this table is of the same type as the first one of Table 6.4.2. The third and fourth screw have parabolic generator lines g_2, while the second screw has a g_2 which is half straight and half parabolic. The generator lines g_2 of the end plates of the fifth and sixth screw are straight and inclined, they are connected to g_1 at their midpoint.

It follows from Table 6.4.5 that the highest quality number occurs for the screw of which the end plates are part of the circular cylinder through the blade tips.

Summarizing, it follows from the Tables 6.4.1–6.4.5 that it can be expected that from the hydrodynamical point of view a good propeller can be constructed on basis of straight generator lines and zero rake angle, while the reference surfaces of the end plates are on the circular cylinder through the blade tips. Hence we will consider in the following sections screw propellers of the type of Table 6.4.2. Then still a choice has to be made with respect to the ratio of covering k (6.1.9) and the end plate ratio ξ (6.3.8). From the point of view of potential theory we have (Table 6.4.1) to take $k = 1$. In that case, the choice of ξ is unimportant because, as we mentioned before, for $k = 1$ the value of ξ has no influence on q. However, the values of k and ξ have

Table 6.4.5. Quality number q, ratio of covering $k = 0.5$, $\xi = 1$, $N = 3$.

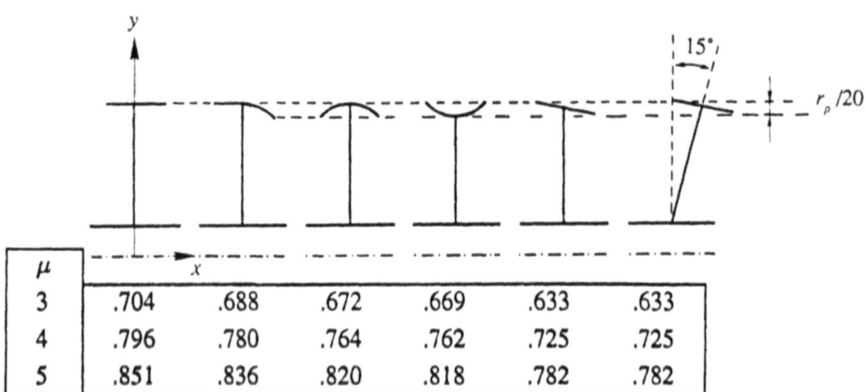

μ						
3	.704	.688	.672	.669	.633	.633
4	.796	.780	.764	.762	.725	.725
5	.851	.836	.820	.818	.782	.782

influence on the viscous resistance of the end plates in real water and, therefore, have influence on the efficiency of a propeller. This will be discussed in Section 6.6.

6.5. On the Shape of End Plates

There are three important items which have to be taken into account with respect to the choice of the shape of the planform of an end plate. They are: prevention of cavitation, low viscous resistance, and a good conveyance of the bound vorticity of the blade to the end plate. In this section we will discuss some aspects of these subjects.

Kinetic and viscous energy are in the following way related to the thrust T that has to be delivered. For a given type of propeller (fixed value of q), the kinetic energy loss E_2 is proportional to T^2, see (6.2.18) and (6.2.19)

$$E_2 = \frac{T^2}{2\rho\, qU\pi(r_p^2 - r_h^2)} \, . \tag{6.5.1}$$

For a propeller with given lifting lines, the total area of the screw blades and end plates depends on the chord lengths of the blades and end plates. In order to avoid cavitation, the chord length can be chosen to be about proportional to the circulation Γ_i of the lifting line at the place under consideration (proportionality assumption)

$$\text{chord length} = \text{const. } \Gamma_i(\sigma_i)(U^2 + \omega^2 r^2)^{1/2} \, . \tag{6.5.2}$$

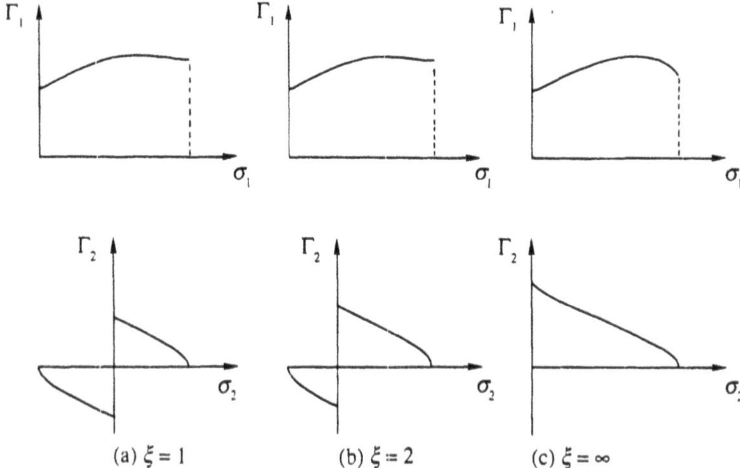

Fig. 6.5.1. Optimum circulation distributions on blade and end plate for mentioned screws, $k = 0.5$, $\mu = 4$, $N = 3$.

Note that strictly speaking, also the thickness distribution of the profiles influences the possible occurrence of cavitation which we will neglect here for the sake of simplicity. This effect is discussed in detail in [33]. The proportionality (6.5.2) can in general be used for regions not too close to the hub, because at the root of a blade we will need some minimum profile length in agreement with conditions of strength.

Because the optimum circulation distribution is proportional to T, it follows that the viscous energy loss will be more or less proportional to T. So the larger the prescribed thrust, the more important the kinetic energy loss will become while the viscous loss becomes relatively less important. The kinetic energy loss can be reduced by taking larger values of $q \leq 1$, hence larger values of the ratio of covering $k \leq 1$, hence larger values of the span of the end plates. This means that for a sufficiently large thrust it seems possible that a screw with "larger" end plates can be more efficient than a screw with "smaller" end plates, in spite of the higher viscous resistance of the larger end plates.

Now we discuss the influence of viscosity on the choice of the asymmetry ($\xi \neq 1$, (6.3.7)) of the end plate reference surface. As has been stated at the end of the previous section we will consider screws of the type of Table 6.4.2. More specific we take here $k = 0.5$, $\mu = \omega r_p/U = 4$ and $N = 3$. The calculated optimum circulation distributions along the span of screw blade and end plate for these screws are shown in Figure 6.5.1 for three values of ξ.

For the end plate the highest circulation appears at the junction of the end plate and the blade tip, while its circulation decreases monotonically to its tip. Let us now adopt the proportionality assumption (6.5.2), with the same factor (const.) for blade and end plate. Hence, because at the plate $r = r_p$, the area of an end plate and

Table 6.5.1. Relative areas of end plates (area $\xi = 1$ is equal to 1).

ξ	1	2	∞
area	1.00	1.06	1.54
q	.796	.796	.802

consequently its viscous resistance are roughly proportional to the surface integral of its circulation distribution when the same proportionality constant is used for the different configurations of Figure 6.5.1. Values of this area are given in Table 6.5.1, where the area for $\xi = 1$ is taken equal to 1. Also in this table the values q are given, which are taken from Table 6.4.2.

It follows from Table 6.5.1 that the area of the one-sided end plate ($\xi = \infty$) will be about 1.54 times as large as the area of the symmetric one ($\xi = 1$). This means that the viscous losses will also be about 1.54 times the viscous losses of the symmetric end plate.

The increase of q for the one-sided end plate however is small. The conclusion is that presumably the symmetric end plate reference surface, which belongs to Figure 6.5.1 (a), $\xi = 1$, is to be preferred even for larger thrusts T. From the last paragraph of Section 6.1 and from (6.3.4) we find that in this case the circulation Γ_1 of the blade is split into two parts of equal strength $\Gamma_2 = \pm\frac{1}{2}\,\Gamma_1$ each of which is conveyed to an end plate half.

Next we will give an impression how we can convey smoothly the bound vorticity from blade to end plate. Suppose that along the chord of the blade tip (of length c) the bound vorticity γ^*, which is the component of the vorticity perpendicular to the local relative flow direction, is distributed as drawn in Figure 6.5.2 (a). This graph is also, within a constant factor, a picture of the pressure jump in the chord-wise direction. Along the blade tip we have the length coordinate χ which starts at the leading edge. The bound vorticity increases linearly from zero at the leading edge upto $\chi = \alpha_1 c$, then keeps the constant value $\tilde{\gamma}$ upto $\beta_2 c$ and after this point decreases linearly to zero at the trailing edge $\chi = c$. The constant value $\tilde{\gamma}$ of γ^* for $\alpha_1 c \leq \chi \leq \beta_2 c$ has to be such that

$$\int_0^c \gamma^*(\chi)\, d\chi = \Gamma_1 \ , \tag{6.5.3}$$

where Γ_1 is the calculated optimum circulation at the blade tip.

This bound vorticity γ^* has to be divided into two parts of equal strength $\frac{1}{2}\,\Gamma_1$, each of which has to be conveyed to a side of the end plate. Such a division can be made for instance, as drawn in the Figures 6.5.2 (b) and (c), where the sum of γ_{I}^* and γ_{II}^* equals the distribution γ^* and the intervals are related by $(1+\alpha_1+\beta_2) = 2(\alpha_2+\beta_1)$. Here we have chosen for an overlap of the end plate halves between $\alpha_2 c$ and $\beta_1 c$ in order to let γ_{I}^* and γ_{II}^* be zero at $\chi = \beta_1 c$ and at $\chi = \alpha_2 c$, respectively. Then γ_{I}^* and

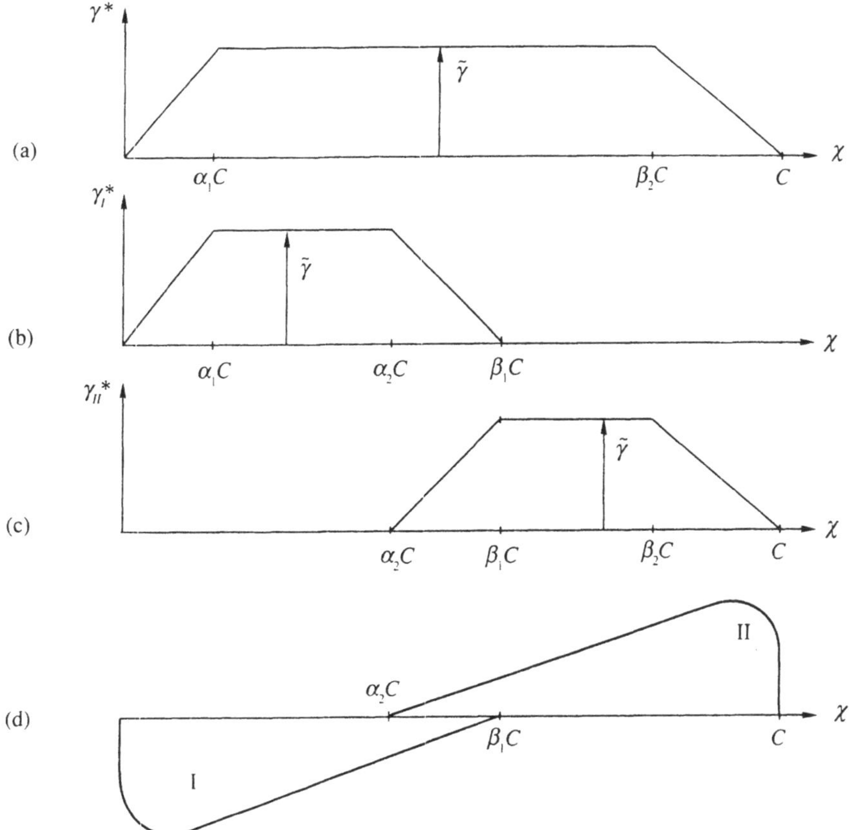

Fig. 6.5.2. (a) Bound vorticity distribution along chord of blade tip; (b) and (c) bound vorticity distributions along roots of end plate halves; (d) planforms of shifted halves of end plate; $0 < \alpha_1 < \alpha_2 < \beta_1 < \beta_2 < 1$.

γ_{II}^* are the bound vorticity distributions at the root of each half of the end plate. By this the halves I and II of the end plate become shifted with respect to each other, as is drawn in Figure 6.5.2 (d). The half I brings γ_{I}^* as trailing vorticity into the fluid and the half II does the same with γ_{II}^*. This has to be done in the optimum way, which means that the circulation distribution along the span of the end plate halves, has to be the calculated optimum one.

A valuable consequence of the splitting of the blade tip circulation as given in Figures 6.5.2 (b) and (c), is that no concentrated vortex is needed at the tip of the blade to keep the vorticity field free of divergence. Such a vortex would of course make the calculation of the profiles at the tip of the blade and at the roots of the end plate halves senseless, because it would induce "strongly" infinite velocities at the junction of blade and end plate.

However, another difficulty can remain. For instance, consider in the half planes

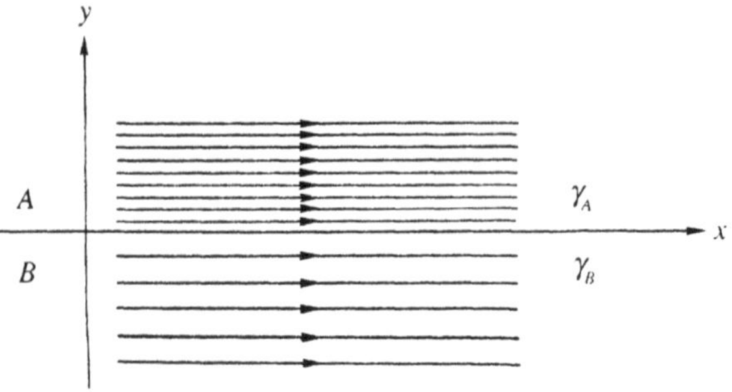

Fig. 6.5.3. Two half-planes A and B with vorticity sheets of constant strengths γ_A and γ_B, $\gamma_A \neq \gamma_B$.

A and B (Figure 6.5.3) two vortex sheets of constant strength γ_A and γ_B respectively with $\gamma_A \neq \gamma_B$. The vorticity is parallel to the x-axis at which the two sheets are connected. Then in the neighbourhood of the x-axis the induced velocity behaves as $O(\ln |y|)$ as can be derived easily from the law of Biot and Savart.

This phenomenon can, for instance, occur with the end plate half I (Figure 6.5.2 (d)) at its root for $0 \leq \chi < \alpha_2 c$, when the vorticity at I has a component parallel to the χ-axis, because at the opposite side of this axis there is no vorticity at all. It can also happen at the remaining part $\alpha_2 c \leq \chi \leq c$ of the χ-axis when the vorticity at both sides of the χ-axis has different components parallel to this axis. Finally, it can happen when the vorticity at the blade in the neighbourhood of the tip has a component parallel to the tip.

The type of shapes chosen of the end plate halves in Figure 6.5.2 (d) seem to be favourable. The reason is that then at $0 \leq \chi \leq \alpha_2 c$, the vorticity at I is nearly perpendicular to this part of its boundary, while for $\alpha_2 c \leq \chi \leq c$ the vorticity components parallel to the χ-axis are not too much unequal. With respect to the singular behaviour at the blade tip, it seems that a "straight" blade is useful, because then the vorticity ends more or less perpendicular to the tip chord.

Next we make a remark about the distribution of the chord lengths along the span of the end plate halves. In order to keep the danger of cavitation the same in span-wise direction, we could choose in principle the chord lengths of the end plates proportional to their calculated optimum circulation distribution. However, this will be done only approximately in general. A reason is that it can be desirable to change in span-wise direction, the bound vorticity distribution along the chords of the end plate halves, in order to avoid too steep an increase or a decrease of this vorticity at the leading edge and the trailing edge, respectively, when the chord length becomes smaller towards the end plate tips.

In Figure 6.5.4 we give the vorticity of an optimum propeller model with the following characteristics (Ae/Ao is the expanded blade area ratio)

$$r_p = 0.12 \text{ m} \ , \quad Ae/A_0 = \frac{N}{\pi r_p^2} \int_{r_h}^{r_p} c(r)\, dr = 0.52 \ , \quad r_h = 0.021 m \ ,$$

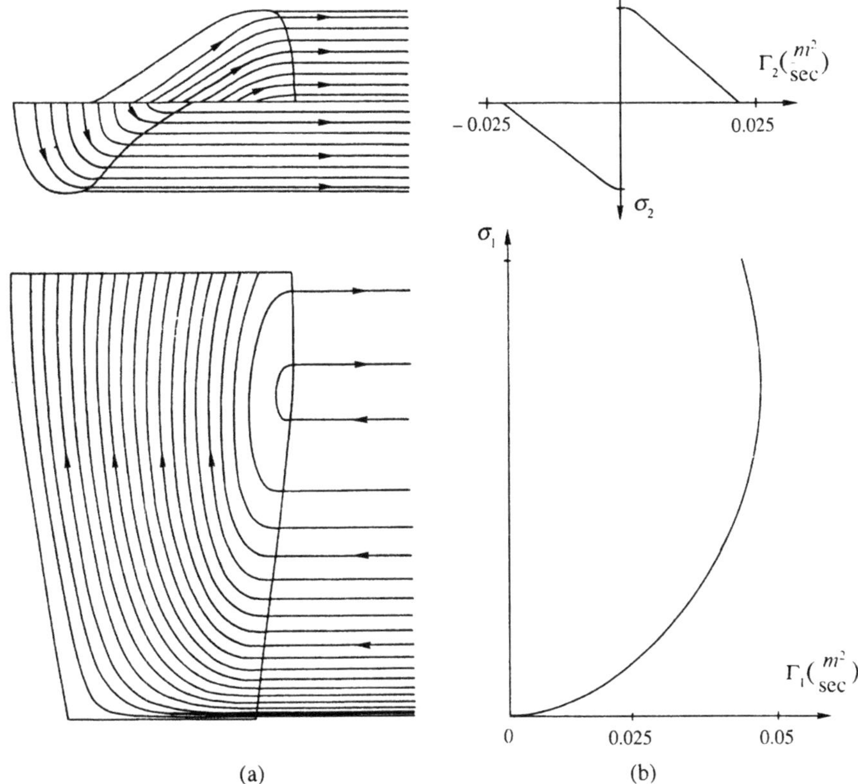

Fig. 6.5.4. Expanded planform of blade and developed planform of end plate with: (a) vorticity distributions, (b) optimum circulation distributions.

$$k = 0.65 \ , \quad N = 4 \ , \quad U = 0.73 \text{ m/sec} \ ,$$

$$\omega = 30.5 \text{ rad/sec} \ , \quad T = 35.8 \text{ Newton} \ . \tag{6.5.4}$$

We remark that the propeller of Figure 6.5.4 is calculated while its hub is neglected, so its blades turn around "freely" through the water at a distance $r_h = 0.021$ m from the axis. It follows that the circulation Γ_1 at the root of the blades has to be zero. However, this will not influence perceptibly the flow at the effective outer part of the blade and at the end plate. The vorticity lines have been calculated as we discussed at the end of Section 1.18.

It is seen from Figure 6.5.4 (a) that indeed by this choice of the shape of the blade and of the end plates, the vorticity is such that the undesirable configuration of Figure 6.5.3 is avoided to some extent.

6.6. Determination of Optimum Values of ω and k

In this section we will discuss how we can determine an optimum combination of the angular velocity ω of the propeller and the ratio of covering k of the end plates, in case we take into account the viscosity of water. If the viscosity is neglected, it is clear that an increase of ω and (or) of k causes an increase of the quality coefficient q, as also follows from Table 6.4.1. When viscosity is taken into account, this will no longer be true, because then larger values of ω and k, give rise to higher viscous losses. These losses can be more severe than the gain which follows from potential theory. So we can try to find an optimum combination of ω and k.

We suppose the propeller to be given in the sense of Figure 6.1.1. It is possible to omit the two-sided infinite hub. Then the previous theory is just as well applicable, only instead of the three coupled integral equations (6.4.2) where $i = 1$, 2 and 3, we have two of these equations $i = 1, 2$. For instance, we can put $\gamma_3 = 0$. In the latter case the circulation around the blade will tend to zero at the root of the blade as in Figure 6.5.4 and the Tables 6.4.1–6.4.5 will be changed slightly, while the trends remain the same.

We consider, in agreement with the previous section, propellers of the type of Table 6.4.2 and we take $\xi = 1$ (6.3.7). When the thrust T, the angular velocity and the ratio of covering are prescribed, we can calculate, for instance by the first method of Section 5.13, named "classical optimization in relation to viscosity" the efficiency (5.13.6)

$$\eta = \frac{TU}{TU + E_{\text{kin}} + E_{\text{visc}}} \ . \tag{6.6.1}$$

We now vary the angular velocity ω and the ratio of covering k. Then we can draw in the (ω, k) plane the level lines of the efficiency; an example is given in Figure 6.6.1.

For the calculation of the influence of the viscosity on the efficiency as given in Figure 6.6.1, some resistance law is chosen which we will not discuss here (several choices are possible, see for instance [30]). For all cases of this figure, a fixed relation between circulation and profile length is used. For a detailed discussion of such a relation we refer to [33]. From Figure 6.6.1 we can draw the following two conclusions.

First, for a moderately loaded propeller the increase of the efficiency that can be expected by the application of end plates is in the order of 2% with respect to a good designed classical screw propeller without end plates. Such an increase is confirmed by the Maritime Research Institute Netherlands, by means of open water tests carried out with an end plate propeller model based on the considerations of the preceding sections of this chapter. The end plate model was designed for the same diameter, thrust and velocity of advance as an already existing state-of-the-art conventional reference propeller model, without end plates which was used for comparison. The blade area ratio of the end plate model was such that it had the same safeguarding against cavitation as the reference propeller. By these experiments it was found that the efficiency of the end plate propeller was about 1.8% better than the efficiency of

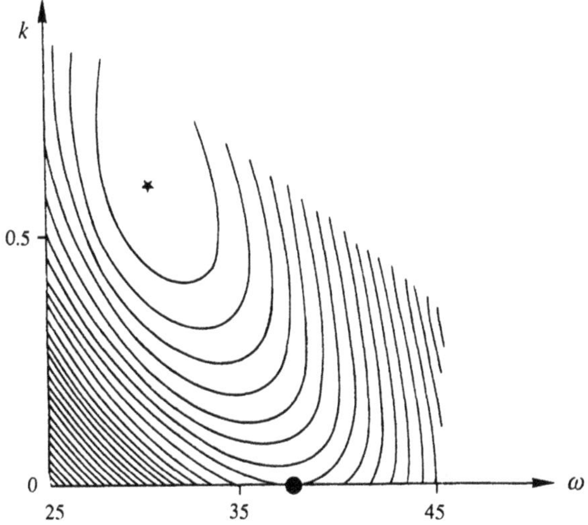

Fig. 6.6.1. Partially drawn level lines of efficiency η, influence of viscosity, for propeller model (6.5.4) with varying ω and k, $\Delta\eta = 0.2\%$; $*$ max efficiency $\eta = 57.16\%$, for $k = 0.60$, $\omega = 30$ rad/sec; • max efficiency $\eta = 55, 54\%$ for $k = 0$, $\omega = 37.5$ rad/sec.

the reference propeller.

Second, it follows from Figure 6.6.1 that the angular velocity of the optimum end plate propeller is lower than that of the optimum propeller without end plates ($k = 0$). Also this was confirmed by the mentioned experiments. There the angular velocity of the end plate model was 31.7 rad/sec and of the classical reference model this was 47.7 rad/sec, when both delivered the same demanded thrust. Because the angular velocity of the end plate model is much lower,its rotational kinetic energy left behind in the fluid will be larger. Hence more of this kinetic energy can be regained by the rudder or by a specially designed wing system which can take the swirl out of the fluid, than in the case of a classical propeller.

For the design of the optimum propellers of Figure 6.6.1 the non-linear correction (5.13.4) was used where the incoming flow U was increased by the induced amount w_x in order to determine more adequate surfaces on which the shed free vorticity can be placed.

The propellers of Figure 6.6.1 were optimized with respect to an undisturbed inflow. When the inhomogeneous inflow of a wake would be taken into account, hence when we use $\Gamma_{i,00}$ (6.2.3) as a part of the circulation distribution, then the total circulation distribution will obtain a different shape. It seems to be of interest to compare the resulting circulation distribution with the one belonging to a homogeneous inflow.

Finally we mention a type of propeller which is related to the screw propeller with one-sided endplates, namely the tip fin propeller. Here the blades are in the neighbourhood of the tip smoothly bent to the suction side. For a discussion of this propeller we refer to Andersen and Schwanecke [2].

6.7. On the Optimum Large Hub Screw Propeller

This is a simple case of the optimization of lifting surfaces which act in a fluid with disturbances of $O(\varepsilon^\circ)$ under the assumption that the free vorticity shed into the fluid is $O(\varepsilon)$. These considerations have some points in common with the discussion given in Section 4.7 and 4.8 with respect to the unsteady propulsion of Regimes (3) and (4) of Section 4.2. By this the shortcoming of the model of a two-sided infinitely long cylindrical hub, especially in optimization theory becomes apparent. This type of propeller is also discussed by Andrews and Cummings in [3] from a more practical point of view.

We assume that the hub C induces disturbance velocities of $O(\varepsilon^\circ)$ and that the prescribed thrust T is of $O(\varepsilon)$, to be delivered by the propeller as one propulsion unit, hence by the blades and by the induced pressures on the hub together. Also we assume that the fluid is inviscid.

For the sake of simplicity we first consider a one-bladed propeller of which the blade is assumed to be without thickness. The reference surface of the blade can be found as follows. Choose some, not necessarily straight, line g which is connected to the hub and which rotates and translates with it. Along g we have a parameter σ which is zero at the hub and has the value $\sigma = \sigma_e$ at the end, $0 \leq \sigma \leq \sigma_e$. Now we take some point A_1 with parameter value $\sigma(A_1)$ at g and consider the fluid particles which "pass through" this point while g moves with the hub. These particles lie at some line denoted by the dashed line A_1, A_2, A_3, A_4, A_5 of Figure 6.7.1. Doing this for all points of g, these lines form a more or less helicoidal surface H^*. This surface when considered as a rigid and impermeable surface does not disturb the flow around the hub C when it rotates and translates with C, hence it can be used as a reference surface for the blade. The projection of the blade on this reference surface is called again its planform W.

The reference line of a profile of the blade is the line A_2, A_3 along which the fluid particles, within the accuracy of $O(\varepsilon)$, pass along the blade. The points A_2 and A_3 lie at the boundary of W. We can consider a surface formed by short straight line segments, cutting the reference line A_2A_3 and which are perpendicular to H^*. Then a profile is the intersection of the blade with this surface.

The surface H^* behind the planform W is, in the neighbourhood of the hub C, influenced by the flow around C. However, far downstream, this surface will have assumed a periodic limit shape H. The vorticity of the blade will be in our semi-linear theory at the planform W and the shed free vorticity at H^* and ultimately at H. This free vorticity will lie at H along helicoidal lines of the type

$$\varphi - ax = b(r) \ , \qquad r = c \ , \qquad a = \frac{\omega}{U} \ , \tag{6.7.1}$$

where $b(r)$ has to be calculated from the given hub geometry.

Now we consider a lifting line (B_1, B_2) far behind the propeller at the limit surface H. This line has to be curved in one way or another in order that it lies on H. Along (B_1, B_2) we have also the parameter σ by assigning to the point A_5 the

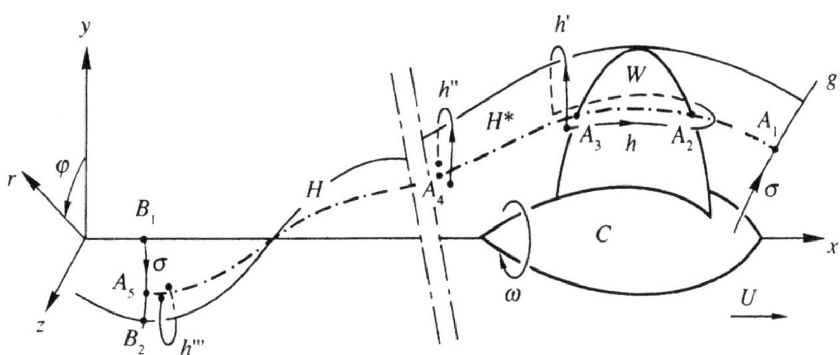

Fig. 6.7.1. One blade with vortex sheet of large hub propeller.

same parameter value as the value $\sigma(A_1)$ of A_1 on g. The lifting line (B_1, B_2) moves along H with the velocity U of the propeller in the positive x-direction. The bound vorticity $\Gamma(\sigma)$ of (B_1, B_2) is chosen such that behind (B_1, B_2) the free vorticity of the propeller is annihilated. It follows that within the accuracy of the linear theory the resultant thrust (of $O(\varepsilon)$) of the one-bladed propeller and the thrust of (B_1, B_2) are equal and opposite (Section 1.16). Hence the large hub propeller and the lifting line (B_1, B_2) but now provided with bound vorticity $\Gamma^*(\sigma) = -\Gamma(\sigma)$ are equivalent in the sense that both yield the same thrust at the cost of the same energy losses.

An exactly analogous reasoning can be given for the case of a propeller with more blades with or without end plates. Note, as we mentioned already, that with respect to the thrust of the propeller we have to take the resultant thrust of blades and hub together, because the induced pressures on the hub will have in general a non-zero resultant.

From the foregoing it follows that instead of the original large hub propeller we can also optimize, when we have N blades, a system $(B_1, B_2)_k$ $(k = 1, \ldots, N)$ of lifting lines moving far behind the propeller along the periodic surfaces H_k $(k = 1, \ldots, N)$ under the constraint of a prescribed thrust T. This optimization is the same as the problem discussed in Sections 6.1–6.4, be it that the reference surfaces stretch upto the x-axis and have to be calculated from the flow along the hub C in the way as we described. Hence the optimum free vorticity behind the lifting lines $(B_1, B_2)_k$ can be considered to be known. From this follows by a simple integration the optimum bound vorticity or circulation $\tilde{\Gamma}_k(\sigma_k)$ of the lifting lines $(B_1, B_2)_k$, where σ_k is the parameter along that line. This integration is related to (6.1.10), only we have to be careful with the meaning of the density of the free vorticity and of the parameter σ_k which is not a length parameter in general.

The optimum free vorticity at the H_k cannot have a concentrated vortex at the x-axis, otherwise the kinetic energy left behind per unit of length would be infinite and hence the efficiency would be zero. It follows that the optimum circulation distributions $\tilde{\Gamma}_k(\sigma_k)$ are zero at the x-axis, hence for $\sigma_k = 0$.

Next we consider the optimum circulation around the profiles of the blades W_k of the original propeller for a certain value of σ_k. This circulation can be defined for instance, along a contour h around the blade and close to the line along which the fluid particles are moving, or in other words the profile reference line $(A_2, A_3)_k$. This circulation equals the circulation $\tilde{\Gamma}_k(\sigma_k)$ as follows from deforming the contour $h \to h' \to h'' \to h'''$ (Figure 6.7.1) without cutting any vorticity.

Because $\tilde{\Gamma}_k(0) = 0$ it follows that the optimum circulation of the blades at their roots has to be zero, in contradistinction to the results for a two-sided infinitely long hub (Figure 6.5.1). The reason is that in the latter case the shed free concentrated vorticity flowing from the roots of the blades, never enters the fluid freely. It lies along the infinite hub and because it is "mirrored" in this surface, it cannot induce any velocity in the fluid. Hence from this point of view an infinitely long hub is not a good approximation of reality.

It is not impossible to make a more complicated model of a propeller with a hub of finite downstream length, which allows in the optimum case the circulation of the blades to be non-zero at their roots, while still the efficiency is not zero. In that case, we have to admit an infinitely long cavity with swirl around it behind the hub, see for instance Schmidt and Sparenberg [58].

When we make an assumption about the loading of the blade in chord-wise direction, hence about the distribution of the circulation $\tilde{\Gamma}_k(\sigma_k)$ as bound vorticity along the profiles, we know the vorticity at the blades as well as behind them. When we also make an assumption about the thickness distribution of the blades, we can, analogously as we have done in Chapter 3, calculate the shape of the profiles. However, there is one difference, here we have to take into account the hub more specifically. This can be done in two different ways.

The first one seems to be the more common way of approach. We calculate the velocities induced at the hub by the optimum vorticity and by the source distribution representing the thickness of the blades. Then we have to add unknown vorticity at the hub in order to compensate there the normal component of these velocities. Hence we have to solve an integral equation at the surface of the hub. Then we can by using the optimum vorticity, the source distribution at the blades and this latter vorticity, calculate the shape of the profiles of the blades.

The second method is, we calculate also the normal component of the induced velocities at the hub, but then change its shape so that a non-axisymmetric hub arises of which the surface follows the induced velocities. This new hub in our theory deviates from the original one by an amount of $O(\varepsilon)$. Although this approach is possibly more expensive in the process of manufacturing, a more streamlined shape of the propeller as a whole arises. Then the blades and the hub form one integrated unit.

For numerical calculations based on the theory presented in this section, we refer to Braam [7].

6.8. Optimum Large Amplitude Unsteady Propulsion, Wings of Finite Span

In Section 5.10 we considered, as an illustration of the comparison of efficiency by inspection, for the 2-dimensional case the values of the quality number q for a number of reference surfaces (Figure 5.10.1), along which "flexible" profiles had to move. It turned out that in the optimum case the interaction of the profiles was favourable. In this section we discuss some results with respect to possibly flexible optimum wings of finite span, which carry out a large-amplitude motion. In this case, the optimization problem is a 3-dimensional one. The propulsion systems deliver a mean value of thrust $\bar{T}$ and have a mean velocity of advance U in the positive x-direction.

The types of reference surfaces are indicated in Figure 6.8.1. First, one wing W of span h moving in the neighbourhood of the reference surface H (Figure 6.8.1 (a)). The projection of H on the (x, z) plane has the form

$$z = a\cos 2\pi x \ , \tag{6.8.1}$$

hence the width of the working region is $2a$ and the period of the motion in the x-direction is one. This chosen length of the period is not a restriction of generality because we can always take the length of the period as the unit of length. This holds also for the other two cases.

Second, two wings W_1 and W_2 each of span h moving in the neighbourhood of two reference surfaces H_1 and H_2, respectively (Figure 6.8.1 (b)). The projections of the intersecting H_1 and H_2 on the (x, z) plane are given by

$$z = \pm a\cos 2\pi x \ , \tag{6.8.2}$$

where the + sign refers to H_1 and the $-$ sign to H_2. As is obvious, the wings have to move in such a way that they do not collide with each other.

Third, two wings $\tilde{W}_1$ and $\tilde{W}_2$ moving "side by side" each in the neighbourhood of the non-intersecting reference surfaces $\tilde{H}_1$ and $\tilde{H}_2$, respectively (Figure 6.8.1 (c)). The projections of $\tilde{H}_1$ and $\tilde{H}_2$ on the (x, z) plane are given by

$$z = \pm\ \tfrac{1}{2}\{(a-\delta)\cos 4\pi x + (a+\delta)\} \ , \tag{6.8.3}$$

where the + sign refers to $\tilde{H}_1$ and the $-$ sign to $\tilde{H}_2$. Because the amplitude of each wing in this case is about half (when δ is small) of the amplitude in cases (a) and (b), we took the length of the period of the motion in the x-direction equal to a half. Then the maximum slope of the paths of the wings in the three cases is about the same.

In all three cases we have, as we discussed in Chapter 5, to translate the reference surfaces as impermeable rigid surfaces in the negative x-direction with some still unknown velocity λ, compare for instance (5.9.3) and (6.2.5). Then again the vorticity needed on the reference surfaces is the optimum free vorticity shed by the single wing or the two wings together.

The resulting boundary value problems are of the Neumann type which can be solved approximately by a vortex lattice method. Two checks on the method can be

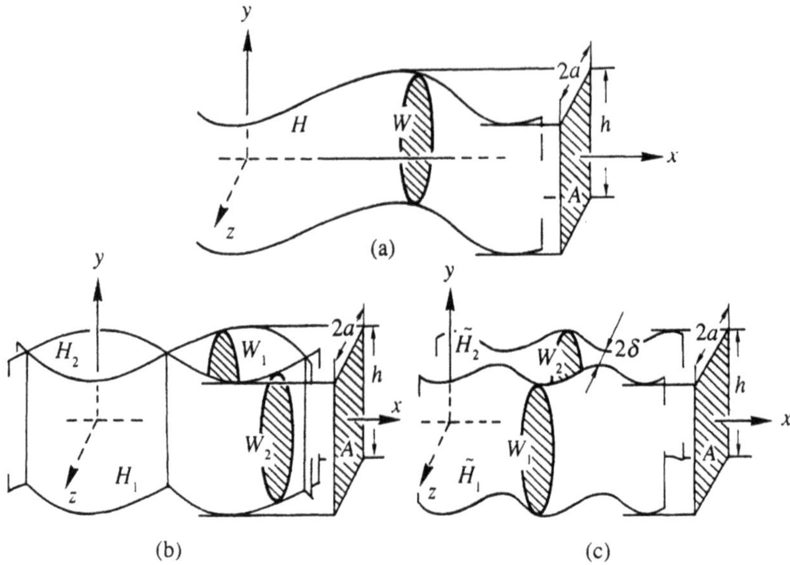

Fig. 6.8.1. Three types of reference surfaces.

made. For large values of the span h the results calculated by means of the vortex lattice method can be compared with those of the 2-dimensional case, of which some are given in Section 5.10. For small values of h the results obtained by slender body theory can be used for comparison. It turned out that in both checks the agreement was satisfactory. For a more detailed discussion of the method of computation and the mentioned checks we refer to Potze and Sparenberg [55].

For the calculation of the quality factor q we took in all cases, hence also in case (c) for $\delta \neq 0$, the working area A of the propulsion system equal to $2ah$. This differs from the choice we made in Section 5.10 with respect to Figure 5.10.1 where we took, in the notation of that section $A = A_1 + A_2$. There the slit between A_1 and A_2 did not belong to A. The reason is that here we want to compare three different types of propellers (a), (b) and (c), which have to propel the same ship and which are allowed to use the full breadth of the hull. However, when the wings W_1 and W_2 move side by side as we assumed in case (c), it will be necessary to keep some clearance 2δ between them when they approach each other. Hence this clearance can be looked upon as an unfavourable property of the device which has to be taken into account in the comparison.

First, we give in Table 6.8.1 the quality numbers q for case (a), hence for one wing. We repeat that this is with respect to the linearized case of Section 4.1. It would be of interest to consider also the semi-linear case of Regime (4) of Section 4.2 for one rigid wing of finite span and to calculate q by means of the periodic wake H which occurs far behind such a wing. However, these calculations have not been carried out up to now, so we have to content ourselves with the linearized case.

Table 6.8.1. Values of quality number q, case (a).

a	$h \ll 1$	0.05	0.2	0.4	0.8	1.6	3.2	∞	h
0.12	$0.782h$	0.0038	0.132	0.208	0.268	0.302	0.318	0.334	
0.25	$1.175h$	0.058	0.202	0.327	0.429	0.487	0.516	0.545	
0.38	$1.339h$	0.066	0.232	0.380	0.508	0.582	0.621	0.648	

Table 6.8.2. Values of quality number q, case (b).

a	$h \ll 1$	0.05	0.2	0.4	0.8	1.6	3.2	∞	h
0.12	$1.564h$	0.079	0.295	0.481	0.617	0.690	0.725	0.755	
0.25	$2.350h$	0.112	0.363	0.550	0.684	0.753	0.789	0.823	
0.38	$2.677h$	0.126	0.396	0.583	0.719	0.791	0.827	0.861	

In Table 6.8.1 and in the following tables, the column $h \ll 1$ follows from slender body theory. It is seen that for $h = 0.05$ slender body theory gives a reasonable approximation for q. The last column ∞ gives q for the 2-dimensional case.

In Table 6.8.2 we give the values of the quality number q for case (b), hence for two wings of which the reference surfaces intersect each other.

In case (c) we have the additional parameter δ which determines the smallest value of the distance between the two wings. Table 6.8.3 shows besides the dependence of q on a and on h also the dependence of q on δ.

From the numerical values of the tables we can draw the following conclusions. When we enlarge the span h of the wings or the amplitude of the motions or both at the same time, the optimum quality number q increases. When we increase the distance δ in between the two wings, which move side by side, (case (c)), the value of q decreases.

Comparing Tables 6.8.1 and 6.8.2 it is seen that for the amplitude $a = 0.12$ and say for $h \geq 0.2$, the quality number q for two wings is more than two times the value of q for one wing. This does not hold for larger values of the amplitude.

The maximum efficiency of a propulsion device as mentioned above, becomes by using (2.1.15)

$$\eta = \frac{U\bar{T}}{U\bar{T} + E} = \left(1 + \frac{E_i}{qU\bar{T}}\right)^{-1} = \left(1 + \frac{\bar{T}}{4\rho\, ahqU^2}\right)^{-1}, \tag{6.8.4}$$

where E_i is the kinetic energy lost per unit of time by the actuator disk with the constant load $\bar{T}/2ah$ and area $2ah$. We emphasize that this is the highest possible efficiency of this type of propellers in an unbounded inviscid fluid according to a linearized theory.

As we remarked just before, when the amplitude of the motion increases, also q increases and by (6.8.4) also η increases for the propeller acting in an inviscid fluid.

Table 6.8.3. Values of quality number q, case (c).

δ	a	$h \ll 1$	0.05	0.2	0.4	0.8	1.6	3.2	∞	h
	0.12	$1.564h$	0.081	0.299	0.466	0.574	0.627	0.653	0.670	
0	0.25	$2.351h$	0.116	0.379	0.554	0.668	0.723	0.753	0.774	
	0.38	$2.677h$	0.130	0.416	0.599	0.721	0.778	0.812	0.830	
	0.12	$1.137h$	0.055	0.170	0.237	0.276	0.295	0.305	0.314	
0.02	0.25	$2.090h$	0.100	0.300	0.410	0.476	0.507	0.523	0.537	
	0.38	$2.530h$	0.120	0.362	0.495	0.578	0.616	0.638	0.653	
	0.12	$0.759h$	0.036	0.107	0.145	0.167	0.178	0.183	0.188	
0.04	0.25	$1.832h$	0.087	0.256	0.344	0.396	0.420	0.433	0.444	
	0.38	$2.328h$	0.111	0.331	0.449	0.520	0.553	0.572	0.585	

However, when the amplitude becomes larger, also the velocities with respect to the water become higher. Hence because water is viscous, the viscous losses increase and the question arises whether η actually will become better. So also here we have to consider a second optimization problem analogous to the one of Section 6.6, where we discussed the optimum combination of the rotational velocity of a screw propeller and the span of its end plates with respect to the viscosity of water. For this second optimization problem we refer to [54].

6.9. Base Motion of Two Rigid Flat Profiles, 2-Dimensional

We will now consider as an example of the semi-linear theory (Regime (3), Section 4.2), the optimization of a simple model of the sculling propulsion of a ship by means of two wings. These wings are assumed to move sideways back and forth, each over the full breadth of the hull (Figure 6.9.1). One wing is in front of the other, so that they will not collide.

We approximate the configuration by assuming the wings to be two-sided infinitely long, hence our problem is 2-dimensional. The two identical profiles $(A_i - B_i)$ $(i = 1,2)$ are assumed to be flat and to have zero thickness. On each profile we have a pivotal point R_i around which the profile can oscillate. These points have a constant velocity U in the positive x-direction of a Cartesian coordinate system (x, y) and move along periodical paths L_i with period $2h$

$$L_i: \quad y = f_i(x) = f_i(x + 2h) \ , \qquad i = 1,2 \ . \tag{6.9.1}$$

On each profile we introduce a length parameter s with $s = l$ at the leading edge B_i and $s = -l$ at the trailing edge A_i. Both pivotal points R_i have the same parameter value $s = b$, the x-component of their distance is denoted by d.

The profiles have to deliver together a thrust of $O(\varepsilon)$ in the positive x-direction with a prescribed mean value $\bar{T}$ with respect to time, per unit of span. The angles

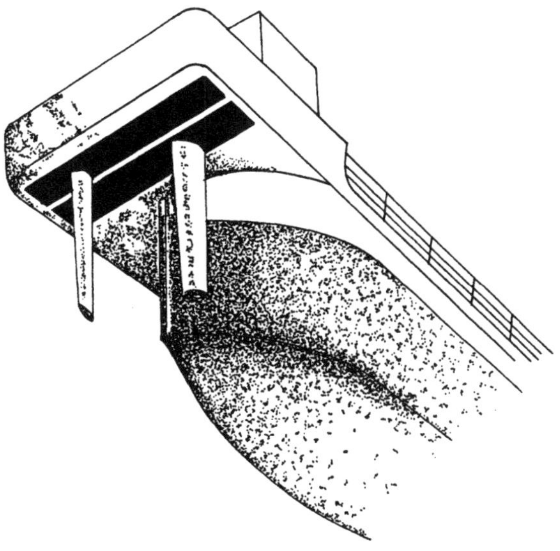

Fig. 6.9.1. Stern of ship from below, two propelling wings.

between the profiles and the tangent to the paths L_i have to be optimized so that this thrust is delivered at the cost of as little kinetic energy loss as possible.

We split the optimum angles into two parts $\alpha_i(t)$ and $\beta_i(t)$ (Figure 6.9.2). The first part $\alpha_i(t)$ is of $O(\varepsilon^\circ)$ and belongs to the base motion of the two profiles by which no vorticity is shed into the fluid. The second part $\beta_i(t)$ is of $O(\varepsilon)$, it belongs to the added motion which is responsible for the demanded thrust and by which vorticity is

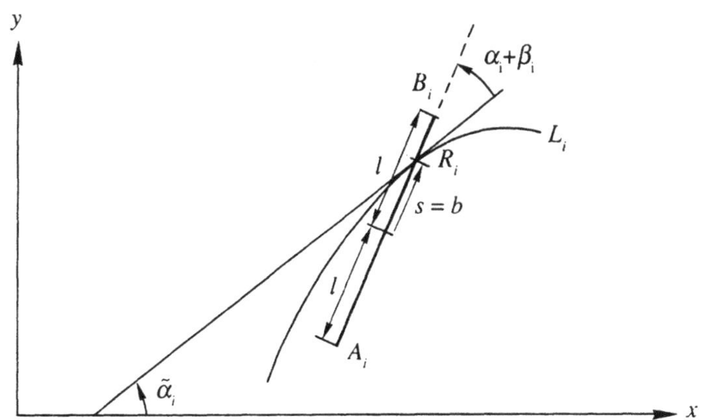

Fig. 6.9.2. Profile $(A_i - B_i)$ and its path L_i $(i = 1, 2)$.

shed into the fluid. Because we consider periodic motions we have

$$\alpha_i(t) = \alpha_i(t+\tau) \ , \quad \beta_i(t) = \beta_i(t+\tau) \ , \quad \tau = 2h/U \ . \tag{6.9.2}$$

In this section we consider the base motion of the two profiles, hence $\beta_i = 0$.

For the unit normal $\vec{n}_i$ on each profile we have

$$\vec{n}_i(t) = \left(-\sin\{\tilde{\alpha}_i(t) + \alpha_i(t)\} \ , \cos\{\tilde{\alpha}_i(t) + \alpha_i(t)\}\right) \ , \tag{6.9.3}$$

where $\tilde{\alpha}_i$ is the angle between the positive x-axis and the tangent to the path L_i at the pivotal point R_i at time t. The profiles $(A_i - B_i)$ have local velocity components in their normal direction, which are

$$v_{n,i}(s,t) = (s-b)\left\{\dot{\tilde{\alpha}}_i(t) + \dot{\alpha}_i(t)\right\} - V_i(t)\sin\alpha_i(t) \ , \quad |s| \le l \ , \tag{6.9.4}$$

where $V_i(t)$ is the velocity of R_i along L_i. The first part of (6.9.4) is caused by the rotation of the profile around R_i and the second part is caused by its translation along L_i.

For $\tilde{\alpha}_i$ and V_i we have

$$\tilde{\alpha}_i(t) = \arctan\left\{\frac{df_i}{dx}\left(x_{R_i}(t)\right)\right\} \ , \tag{6.9.5}$$

$$V_i(t) = \left[1 + \left\{\frac{df_i}{dx}\left(x_{R_i}(t)\right)\right\}^2\right]^{1/2} \cdot U \ , \tag{6.9.6}$$

where $x_{R_i}(t)$ is the x-coordinate of R_i at time t.

In order that the fluid flows along the profiles we need vorticity $\gamma_{bi}(s,t)$ $(i = 1, 2)$ on them. The index b denotes that this vorticity belongs to the base motion. Using the law of Biot and Savart these vorticities have to satisfy the two coupled integral equations

$$v_{ni}(s,t) = \frac{1}{2\pi}\oint_{-l}^{l}\frac{\gamma_{bi}(\sigma,t)\,d\sigma}{(s-\sigma)} + \frac{1}{2\pi}\int_{-l}^{l}\gamma_{bj}(\sigma,t)$$

$$\cdot\frac{[D_y(i,j,s,\sigma,t)\sin\{\alpha_j(t)+\tilde{\alpha}_j(t)\} + D_x(i,j,s,\sigma,t)\cos\{\alpha_j(t)+\tilde{\alpha}_j(t)\}]\,d\sigma}{D_x^2(i,j,s,\sigma,t) + D_y^2(i,j,s,\sigma,t)}$$

$$|s| < l \ , \tag{6.9.7}$$

$$j = 2 \quad \text{if} \quad i = 1 \ ; \qquad j = 1 \quad \text{if} \quad i = 2 \ , \tag{6.9.8}$$

$$\begin{aligned} D_x(i,j,\sigma,s,t) = {} & x_{R_i}(t) - x_{R_j}(t) + (s-b)\cos\{\alpha_i(t) + \tilde{\alpha}_i(t)\} \\ & - (\sigma - b)\cos\{\alpha_j(t) + \tilde{\alpha}_j(t)\} \ , \end{aligned} \tag{6.9.9}$$

$$D_y(i,j,\sigma,s,t) = f_i(x_{R_i}(t)) - f_j(x_{R_j}(t))$$
$$+ (s-b)\sin\{\alpha_i(t)+\tilde{\alpha}_i(t)\} - (\sigma-b)\sin\{\alpha_j(t)+\tilde{\alpha}_j(t)\} \ . \qquad (6.9.10)$$

The Kutta condition at the trailing edges ($s = -l$) of the profiles reads

$$\gamma_{bi}(-l,t) = 0 \ , \qquad i = 1,2 \ . \qquad (6.9.11)$$

The circulation Γ_{bi} around each of the profiles has to be constant, then the profiles do not shed free vorticity, which is required for the base motion. For these constants we take the value zero, because otherwise the profiles would yield unnecessary side forces, hence we demand

$$\Gamma_{bi} = \int_{-l}^{l} \gamma_{bi}(s,t)\,ds = 0 \ , \qquad i = 1,2 \ . \qquad (6.9.12)$$

In the case that the distance between the profiles is infinite, the integrand of the second integral at the right-hand side of (6.9.7) becomes zero, hence that integral is zero. Then each profile moves alone in the fluid and we obtain two simple uncoupled singular integral equations which can be solved explicitly by the methods of Appendix A, we find

$$\gamma_{bi}(s,t) = 2[\{\dot{\alpha}_i(t)+\dot{\tilde{\alpha}}_i(t)\}(b+l-s)$$
$$+ V_i(t)\sin\alpha_i(t)]\left(\frac{l+s}{l-s}\right)^{1/2} \ , \qquad (6.9.13)$$

which satisfies (6.9.11). Substituting (6.9.13) into (6.9.12) we find an ordinary differential equation for $\alpha_i(t)$

$$\{\dot{\alpha}_i(t)+\dot{\tilde{\alpha}}_i(t)\}(l+2b) + 2V_i(t)\sin\alpha_i(t) = 0 \ . \qquad (6.9.14)$$

It follows that if the pivotal point R_i happens to be the three-quarter point $b = -\frac{1}{2}\,l$, then

$$\alpha_i(t) = 0 \ , \qquad (6.9.15)$$

which is in agreement with the results of Section 4.7.

When the distance between the profiles is finite we have to take the second integral of (6.9.7) into account. To solve this equation in this case, it seems useful to write $\gamma_{bi}(s,t)$ in the form

$$\gamma_{bi}(s,t) = 2[\{\dot{\alpha}_i(t)+\dot{\tilde{\alpha}}_i(t)\}(b+l-s) + V_i(t)\sin\alpha_i(t)]\left(\frac{l+s}{l-s}\right)^{1/2}$$
$$+ \tilde{\gamma}_{bi}(s,t) \ , \qquad (6.9.16)$$

where by (6.9.13) $\tilde{\gamma}_{bi}(s,t)$ is the unknown, possibly small, disturbance vorticity caused by the nearness of the other profile.

Substituting (6.9.16) into (6.9.7), (6.9.11) and (6.9.12) we obtain together with the first equation of (6.9.2), four equations for the four unknown functions $\tilde{\gamma}_{bi}(s,t)$ and $\alpha_i(t)$ ($i = 1, 2$),

$$\oint_{-l}^{l} \frac{\tilde{\gamma}_{bi}(\sigma,t)\,d\sigma}{s-\sigma} = \int_{-l}^{l} \frac{\gamma_{bj}(\sigma,t)\{D_x \cdot \cos(\alpha_i + \tilde{\alpha}_i) + D_y \cdot \sin(\alpha_i + \tilde{\alpha}_i)\}}{D_x^2 + D_y^2}\,d\sigma \ ,$$

$$i = 1, 2 \ , \tag{6.9.17}$$

$$\tilde{\gamma}_{bi}(-l,t) = 0 \ , \tag{6.9.18}$$

$$\pi\{\dot{\alpha}_i(t) + \dot{\tilde{\alpha}}_i(t)\}\, l\,(l+2b) + 2\pi\, l\, V_i(t) \sin\alpha_i(t) + \int_{-l}^{l} \tilde{\gamma}_{bi}(\sigma,t)\,d\sigma = 0 \ ,$$

$$i = 1, 2 \ , \tag{6.9.19}$$

$$\alpha_i(t) = \alpha_i(t+\tau) \ . \tag{6.9.20}$$

These equations can be solved by an iteration method. Starting with some $\tilde{\gamma}_{bi}(\sigma,t)$ ($i = 1, 2$) we calculate from (6.9.19) by a shooting method an $\alpha_i(t)$ which satisfies (6.9.20). Next we substitute $\tilde{\gamma}_{bi}(\sigma,t)$, using (6.9.16), and the calculated $\alpha_i(t)$ into the right-hand side of (6.9.17). Then we solve (6.9.17) by which a new $\tilde{\gamma}_{bi}(\sigma,t)$ is found. With this new $\tilde{\gamma}_{bi}(\sigma,t)$ we can repeat the above-mentioned scheme. This can be carried out until the solutions change within the desired accuracy. A good start is $\tilde{\gamma}_{bi}(\sigma,t) = 0$.

6.10. Optimum Shed Vorticity, Quality Number and Added Motion

While carrying out their base motion, the profiles induce disturbance velocities of $O(\varepsilon^\circ)$ in the fluid. These velocities follow from the vorticity $\gamma_{bi}(s,t)$ ($|s| \leq l$) on the profiles as calculated in the preceding section. For the x-component $v_x(x,y,t)$ and the y-component $v_y(x,y,t)$ of this velocity field we find by the law of Biot and Savart

$$v_x(x,y,t) = -\frac{1}{2\pi}\sum_{i=1}^{2}\int_{-l}^{l} \frac{\gamma_{bi}(s,t)\{y - \eta_i(s,t)\}\,ds}{\{x - \xi_i(s,t)\}^2 + \{y - \eta_i(s,t)\}^2}$$

$$= \frac{d\tilde{x}}{dt}(x,y,t) \ , \tag{6.10.1}$$

$$v_y(x,y,t) = \frac{1}{2\pi}\sum_{i=1}^{2}\int_{-l}^{l}\frac{\gamma_{bi}(s,t)\{x-\xi_i(s,t)\}\,ds}{\{x-\xi_i(s,t)\}^2+\{y-\eta_i(s,t)\}^2}$$

$$= \frac{d\tilde{y}}{dt}(x,y,t)\ , \tag{6.10.2}$$

where

$$\xi_i(s,t) = x_{R_i}(t) + (s-b)\cos(\alpha_i(t)+\tilde{\alpha}_i(t))\ , \tag{6.10.3}$$

$$\eta_i(s,t) = f_i(x_{R_i}(t)) + (s-b)\sin(\alpha_i(t)+\tilde{\alpha}_i(t))\ , \tag{6.10.4}$$

and $(\tilde{x},\tilde{y})$ is the position of some fluid particle.

In the semi-linear theory the wake behind a profile was defined in Section 4.2 as the surface on which those fluid particles are situated, which have passed along the profile and have left it at the trailing edge, while the profile carries out the base motion. In the neighbourhood of the profiles the wakes are time-dependent and denoted by H_i^*. As the profiles move on, the wakes assume a fixed periodic shape which we denote by H_i. In order to calculate points of the wake we can determine for a number of moments t_n the coordinates $(\tilde{x}_0(t_n),\tilde{y}_0(t_n))$ of the trailing edge of the profile. By solving the coupled differential equations (6.10.1) and (6.10.2) with $\tilde{x}_0(t_n)$ and $\tilde{y}_0(t_n)$ as initial values, we know for $t > t_n$ the position of the fluid particle which has passed the trailing edge at time t_n. Doing this for a number of moments t_n, we can approximate the shape of the wake by interpolating the calculated points for instance with a cubic spline. Making no distinction between the exact and the approximate wake, we write in the neighbourhood of the profile

$$H_i^*:\quad y = g_i^*(x,t)\ ,\quad i = 1,2\ . \tag{6.10.5}$$

Far behind the profile we have the time-independent and periodic wake

$$H_i:\quad y = g_i(x) = g_i(x+2h)\ ,\quad i = 1,2\ . \tag{6.10.6}$$

As we discussed in Section 5.5, the optimum free vorticity infinitely far behind the profiles is characterized as follows. The wakes H_i (6.10.6) are translated as rigid and impermeable surfaces in the x-direction with some still unknown velocity λ. The vorticity needed at these surfaces in order to let the fluid flow along them is the optimum shed free vorticity, denoted by $\lambda\gamma_i(x)$, i = 1, 2. The unit normal at H_i has the form

$$\vec{n}_i = (-g_i'(x),1)\{1+g'(x)^2\}^{-1/2}\ , \tag{6.10.7}$$

hence the boundary condition at the translating H_i becomes

$$(v_x,v_y)\cdot\vec{n}_i = \lambda\cos_{(n_i,x)}(x) = -\lambda g'(x)\{1+g'(x)^2\}^{-1/2}\ . \tag{6.10.8}$$

Because the H_i are periodic in the x-direction with period $2h$, the vorticity $\lambda\gamma_i(x)$ will be periodic with the same period. Therefore, we consider a row of concentrated vortices of strength Γ, equally spaced in the x-direction with distance $2h$, at the points

$$x = \xi + 2mh \ , \qquad y = \eta \ , \qquad m = \dots, -2, -1, 0, 1, 2, \dots \ . \tag{6.10.9}$$

The velocity induced by this infinite row is [41]

$$(v_x(x,y), v_y(x,y)) = \frac{\Gamma}{4h} \frac{\{-\sinh\frac{\pi}{h}(y-\eta), \sin\frac{\pi}{h}(x-\xi)\}}{\{\cosh\frac{\pi}{h}(y-\eta) - \cos\frac{\pi}{h}(x-\xi)\}} \ . \tag{6.10.10}$$

When we take (ξ, η) on H_i ($i = 1, 2$) we can write down the equations which have to be satisfied by $\gamma_i(x)$. We obtain from (6.10.8)

$$\frac{1}{4h}\sum_{j=1}^{2} \oint_0^{2h} K_i(x,\xi)\, \gamma_j(\xi)\, d\xi = -g'(x) \ , \qquad i = 1, 2 \ , \tag{6.10.11}$$

where

$$K_i(x,\xi) = \frac{\{g_i'(x)\sinh\frac{\pi}{h}\left(g_i(x) - g_i(\xi)\right) + \sin\frac{\pi}{h}(x-\xi)\}}{\{\cosh\frac{\pi}{h}\left(g_i(x) - g_i(\xi)\right) - \cos\frac{\pi}{h}(x-\xi)\}} \cdot (1 + g'(\xi)^2)^{1/2} \ . \tag{6.10.12}$$

Because the mean value of the free shed vorticity is zero, (6.10.11) has to be solved under the condition

$$\int_0^{2h} \gamma_i(\xi)(1 + g'(\xi)^2)^{1/2}\, d\xi = 0 \ . \tag{6.10.13}$$

The scalar λ follows from the condition that the mean value with respect to time, of the thrust is $\bar{T}$. Then the equation from which λ follows is the balance of momentum

$$\tau \bar{T} = -\lambda\, \rho\, I(2h) \ , \tag{6.10.14}$$

where τ is the period of time (6.9.2) and $\rho I(2h)$ is the momentum of the fluid far behind the profiles over a period of length $2h$ in the x-direction. So we have to determine $I(2h)$.

We denote by $\Phi(x,y)$ the potential of the flow induced by $\gamma_i(x)$ far behind the profiles. Then because of (6.10.13) this potential is also periodic with period $2h$ in the x-direction. We have

$$\begin{aligned} I(2h) &= \int_0^{2h}\int_{-\infty}^{\infty} \frac{\partial\Phi}{\partial y}(x,y)\, dx\, dy \\ &= \sum_{i=1}^{2} \int_0^{2h} [\Phi]_-^+(x, g_i(x))\, g_i'(x)\, dx \ , \end{aligned} \tag{6.10.15}$$

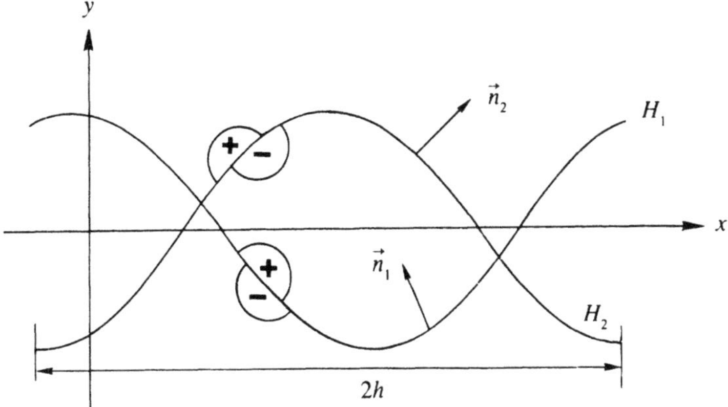

Fig. 6.10.1. The two wakes H_1 and H_2 behind the profiles.

where $[\Phi]_-^+(x, g_i(x))$ is the jump of Φ over H_i. The + side and the − side of H_i are denoted in Figure 6.10.1. Using the relation

$$[\Phi]_-^+(x, g_i(x)) = \int_0^x \gamma_i(\xi)\{1 + g_i'(\xi)^2\}^{1/2}\, d\xi \ , \tag{6.10.16}$$

and using the periodicity of Φ we can, by partial integration, rewrite (6.10.15) as

$$I(2h) = \sum_{i=1}^{2} \int_0^{2h} \gamma_i(x) g_i(x)\{1 + g_i'(x)^2\}^{1/2}\, dx \ . \tag{6.10.17}$$

Then (6.10.14) yields the Lagrange multiplier λ.

By an analogous reasoning we find for the kinetic energy E_{kin} lost by the propeller per unit of time

$$E_{\text{kin}} = \frac{\rho\,\lambda^2}{2\tau} \int_0^{2h} \int_{-\infty}^{\infty} |\text{grad}\ \Phi|^2\, dy\, dx = \frac{\rho\,\lambda^2}{2\tau} \int_0^{2h} \int_{-\infty}^{\infty} \text{div}\,(\Phi \text{grad}\ \Phi)\, dy\, dx$$

$$= -\frac{\rho\,\lambda^2}{2\tau}\, I(2h) = -\frac{h\,\bar{T}^2}{\rho\, U I(2h)} \ . \tag{6.10.18}$$

Next we calculate the quality number q of this propeller. We take as the working area A, per unit of span, the maximum width of the wake (see end of Section 5.8)

$$A = \max_{(i,j,\xi_1,\xi_2)} |g_i(\xi_1) - g_j(\xi_2)| \ . \tag{6.10.19}$$

Then the working region contains all the shed free vorticity, hence $q \leq 1$.

As follows from (2.1.15), the loss of kinetic energy E_{act} per unit of time of the actuator disk of area A and constant loading $\bar{T}/A$ has the value

$$E_{act} = \frac{\bar{T}^2}{2\rho\, A\, U} \quad . \tag{6.10.20}$$

Hence the quality number q becomes

$$q = \frac{E_{act}}{E_{kin}} = -\frac{I(2h)}{2hA} \leq 1 \ . \tag{6.10.21}$$

The efficiency becomes

$$\eta = \frac{\bar{T}U}{\bar{T}U + E_{kin}} = \left\{ 1 + \frac{\bar{T}}{2\rho\, U^2\, q\, A} \right\}^{-1} \quad . \tag{6.10.22}$$

Finally, we discuss shortly the added motion $\beta_i(t)(O(\varepsilon))$ which has to be superimposed on the base motion $\alpha_i(t)(O(\varepsilon^\circ))$. The added motion has to be calculated such that the desired mean value $\bar{T}(O(\varepsilon))$ of the thrust is delivered with as little kinetic energy losses as possible. In other words, the $\beta_i(t)$ have to be determined such that the vorticity far behind the propeller equals the optimum vorticity $\gamma_i(x)$ (6.10.11).

The shapes of the wakes H_i^* behind the profiles are known for each x and each t. In the neighbourhood of the profiles the optimum free shed vorticity denoted by $\lambda\, \gamma_i^*$, which is at the H_i^*, is also time-dependent. The vorticity γ_i^* can be considered to be known because it follows from γ_i by the conservation of vorticity

$$\gamma_i^*(x^*, t)\, ds_i^* = \gamma_i(x)\, ds_i \ , \quad i = 1, 2 \ , \tag{6.10.23}$$

where ds_i^* is an element of length along H_i^* and ds_i is the element of length along H_i, into which ds_i^* is transformed by the base motion and analogously for the value of $x^* \to x$.

The normal component of each moving profile is now given by

$$(s - a)(\dot{\alpha}_j + \dot{\tilde{\alpha}}_j + \dot{\beta}_j) - V_j(t)\sin(\alpha_j + \beta_j) \ , \quad j = 1, 2 \ . \tag{6.10.24}$$

Then we have to equate this velocity component for fixed j to the sum of the normal components of the velocities induced by the known free vorticity behind each profile and by the vorticities at the profiles $(\gamma_{bi}(s,t) + \gamma_{ai}(s,t))$ $(i = 1, 2)$, where $\gamma_{ai}(s,t)$ is the still unknown vorticity at $(A_i - B_i)$ needed for the added motion.

This yields two integral equations for the four unknown functions of $O(\varepsilon)$ namely $\beta_i(t)$ and $\gamma_{ai}(s,t)$ in which we can neglect terms of $O(\varepsilon^2)$. In these equations appear also terms of $O(\varepsilon^\circ)$ caused by $\alpha_i(t)$ and $\gamma_{bi}(s,t)$, however, these cancel each other by means of the equations (6.9.7) for the base motion. So two integral equations remain in which only occur terms of $O(\varepsilon)$.

We find another two equations by the condition that the total circulations around the profiles have to be such that

$$\frac{1}{W_i(t)}\, \frac{d}{dt} \int_{-l}^{l} \gamma_{ai}(s,t)\, ds = -\gamma_i^*\left(x(A_i), t\right) \ , \quad i = 1, 2 \ , \tag{6.10.25}$$

where $x(A_i)$ is the x-coordinate of the trailing edge A_i and $W_i(t)$ is the absolute value of the difference of the velocity of A_i and the velocity induced by the base motion.

Moreover, we have the Kutta condition for $\gamma_{ai}(s,t)$ which reads

$$\gamma_{ai}(-l,t) = \gamma_i^*(x(A_i),t) \ , \quad i = 1,2 \tag{6.10.26}$$

and the condition that the mean value with respect to time of the circulation around each profile is zero.

$$\int_0^\tau \int_{-l}^{l} \gamma_{ai}(s,t)\, ds\, dt = 0 \ , \quad i = 1,2 \ . \tag{6.10.27}$$

This latter equation prevents that a part of $O(\varepsilon)$ will occur in the vorticity $\gamma_{ai}(s,t)$ of the added motion, which would induce undesired mean values of $O(\varepsilon)$ of the side forces on the profiles.

For a possible numerical procedure to carry out the rather complicated calculations effectively we refer to Potze [53].

6.11. Numerical Results

We give some numerical results for the unsteady propeller discussed in the previous two sections, which is intended for a more or less realistic ship. The propeller is characterized by

Velocity of ship U = 8 m/sec,

Span of propeller wings = 8 m,

Chord length of wings, $2l$ = 2.2 m,

Pivotal point R_i, $s = b = \frac{1}{2}l$ = 0.55 m,

x-component of distance between R_i, d = 2.75 m,

Time period of wings, τ = 4.544 sec,

Stroke of motion of wings, 20 m,

Required mean value of thrust, $\bar{T} = 0.97\ 10^6 N$,

Density of sea water, ρ = 1023 kg/m^3. (6.11.1)

The wings move with their pivotal points R_i along the intersecting paths

$$y = f_1(x) = \left\{ 10 \cos \frac{2\pi x}{36.364} \right\} m \ , \quad i = 1 \ , \tag{6.11.2}$$

$$y = f_2(x) = \left\{ -10 \cos \frac{2\pi}{36.364}(x + 2.75) \right\} m \ , \quad i = 2 \ . \tag{6.11.3}$$

At the moment $t = 0$ we assume both wings to be in their extreme positions, for instance R_1 at $(x, y) = (0, 10)$ and R_2 at $(x, y) = (-2.75, -10)$. Hence wing $(A_1 - B_1)$ moves in front of wing $(A_2 - B_2)$.

By numerical calculations based on the previous 2-dimensional theory, it follows for this propeller that the quality coefficient q (6.10.21) and the efficiency η (6.10.22) assume the following values

$$q = 0.826 \ , \qquad \eta = 0.947 \ . \tag{6.11.4}$$

We can compare these results with results from the linear 3-dimensional theory of Section 6.8. There the shed free vorticity from the tips of the wings is taken into account, which is neglected in the 2-dimensional theory. We assume in order to obtain some insight that the free vorticity in the 3-dimensional theory is situated at the surfaces

$$y = \pm \left\{ 10 \cos \frac{2\pi x}{36.364} \right\} m \ . \tag{6.11.5}$$

When we assume that the free surface of the water has no influence, it follows from (6.11.1) that we have to take $|z| \leq 4m$, together with (6.11.5). Then we find

$$q = 0.396 \ , \qquad \eta = 0.895 \ . \tag{6.11.6}$$

When we assume that the free surface of the water is rigid, we have to mirror the wings with respect to this surface and we have to take $|z| \leq 8m$, together with (6.11.5). Then we find

$$q = 0.578 \ , \qquad \eta = 0.926 \ . \tag{6.11.7}$$

From the results (6.11.4), (6.11.6) and (6.11.7), it follows that for the relatively short wing (6.11.1), the influence of the tip vorticity on the quality coefficient q is important, however, this influence is much less with respect to the efficiency η. This means that the propeller is lightly loaded (see row $\eta \approx 1$ from Table 5.8.1 and the discussion of this table).

In Figures 6.11.1 and 6.11.2 we give some results calculated by means of the 2-dimensional theory on the motion of the profiles and on the forces exerted by the wings. The irregularities in the graphs at $t/\tau = 2.8$ and $t/\tau = 7.8$ are caused by the close passing of the profiles.

The mean values of the forces induced by the base motion in Figure 6.11.2, have to be zero. The maximum propulsive force in the x-direction is about twice the demanded mean value 0.97 $10^6 N$ of the thrust. Hence this type of propulsion is typically unsteady. The total side force in the y-direction remains small compared with the delivered mean thrust, because the forces on the separate wings in this direction nearly annihilate each other.

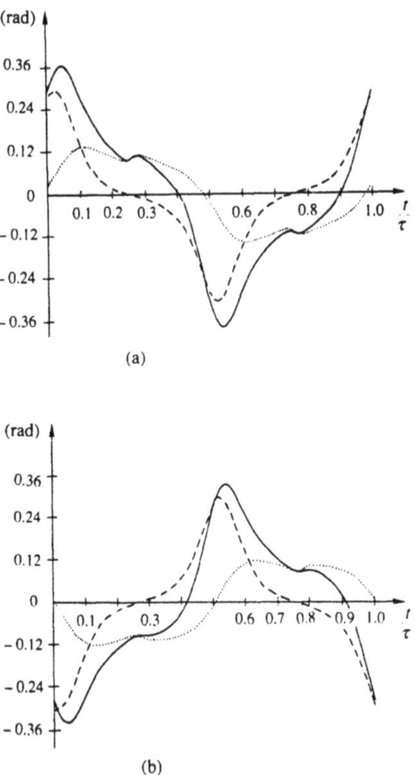

Fig. 6.11.1. (a) Angles of incidence of first wing: total motion $\alpha_1(t) + \beta_1(t)$ ——, base motion $\alpha_1(t)$ - - - -, added motion $\beta_1(t)$ $\cdots$; (b) Analogously for second wing.

We refer for further information about this way of approach to Potze [54] where a 3-dimensional theory for wings of finite span is developed. In this theory it is assumed that the free vorticity remains at the surfaces passed through by the trailing edges of the wings. The angles of incidence of the wings are according to the here discussed 2-dimensional theory, however the added motion is multiplied by a suitable constant such that the required mean value of the thrust is delivered.

Furthermore, we refer to Muijtjens [51] where experiments are described based on the theory given in [54] applied to optimum propulsion by one wing. It was found that only in a small number of cases a good agreement was obtained between theory and experiment with respect to the force actions and the efficiency. In a larger number of cases this does not happen, perhaps by the inaccuracy of the measurements. In [51] also a theoretical sensitivity analysis is given by using the computing program of [54]. From this the conclusion can be drawn that it is not necessary to follow very closely the calculated values of the angles, of incidence $(\alpha_i(t) + \beta_i(t))$ of the wings

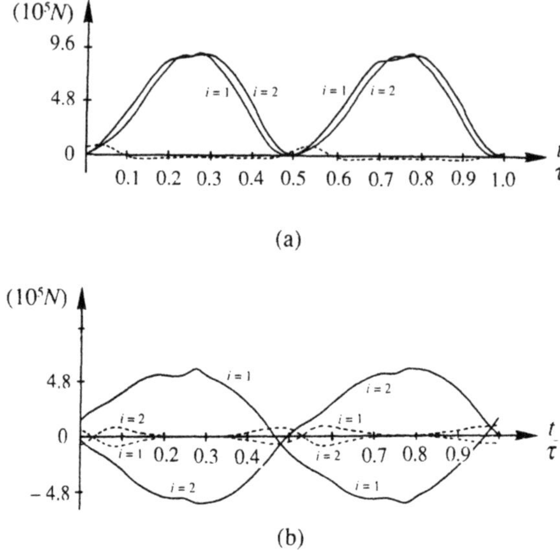

Fig. 6.11.2. (a) Force in x-direction acting on the wings; total motion, first wing ($i = 1$), second wing ($i = 2$) —— ; base motion both first and second wing - - - - . (b) Force in y-direction acting on the wings; total motion, first wing ($i = 1$), second wing ($i = 2$) —— ; base motion, first wing ($i = 1$) and second wing ($i = 2$) - - - - .

in order to obtain a good hydrodynamic propulsion efficiency.

Finally, we make some general remarks about unsteady propulsion:

1. The efficiency of an unsteady propeller can be made rather high because the amount of water that can be influenced by it is in general larger than the amount of water influenced by a screw propeller.
2. Two drawbacks are: the complicated machinery needed for the side ways motion of the wings, and in the seemingly natural configuration of Figure 6.9.1, the large bending moments which occur at the roots of the wings.

6.12. On the Optimum Voith-Schneider Propeller

In this section we discuss in short the optimization of a Voith-Schneider propeller (Section 4.9) which has a large, say countably infinite, number of very slender wings. The propeller is working in an undisturbed and inviscid fluid. In order to simulate the bottom of the ship which is just above the propeller, we mirror the blades (Figure 4.9.1) with respect to this bottom and consider a propeller of which the blades are twice as long as the real ones, but now working in an unbounded fluid. The working region of this new propeller is the cylinder with rectangular cross section denoted by G in Figure 6.12.1. This working region is densely crowded with

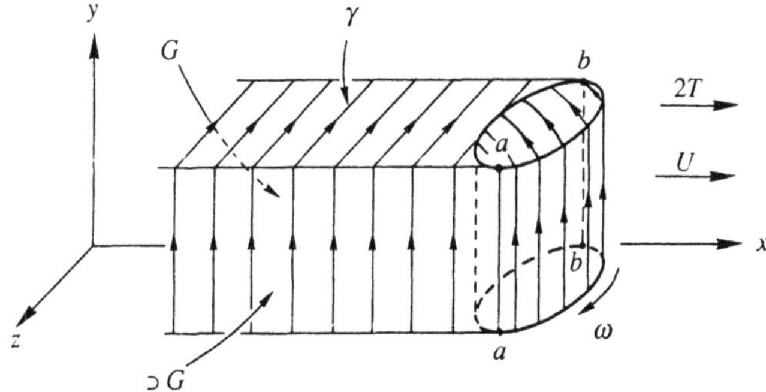

Fig. 6.12.1. The vorticity of an optimum Voith-Schneider propeller with "many" blades.

an infinite number of reference surfaces H_i, $i = 1, 2, 3, \ldots$, behind the blades.

We apply the reasoning which we used several times in the previous sections of this chapter and assume that the propeller started at $x = -\infty$ and has already moved to $x = \infty$. Hence we have to translate the H_i as rigid and impermeable surfaces in the x-direction with an unknown velocity λ, the vorticity needed on these surfaces is the free vorticity shed by the optimum Voith-Schneider propeller. The Lagrange multiplier λ follows from the demanded thrust.

Because the H_i are dense in G, it follows that the shed free vorticity γ in the optimum case is entirely at the boundary ∂G of G. This vorticity density is independent of the x-coordinate and consists of rectangular vortices perpendicular to the x-axis, they are denoted with a right-hand screw in Figure 6.12.1. It follows as we discussed in Section 5.9, that this model of the Voith-Schneider propeller yields an ideal propeller with a rectangular cross section of which the quality number $q = 1$.

Suppose that the bound vorticity of the blades per unit of length along the propeller circle is Γ at the front and $-\Gamma$ at the aft position of the blades. When this bound vorticity passes the line $(a$–$a)$, it changes from Γ to $-\Gamma$, hence it leaves behind free vorticity at the vertical part of ∂G, of strength

$$\gamma = \frac{2\Gamma\omega R}{U}\,, \tag{6.12.1}$$

per unit of length in the x-direction, where R is the radius of the propeller circle and ω the rotational velocity of the blades. The "inverse" happens at the line $(b$–$b)$.

The vorticity shed by the continuously distributed tips of the blades is also drawn in Figure 6.12.1. The free vorticity shed by the tips of the blades at the front position lies along cycloidal lines. It is a matter of simple analysis to find that the vorticity shed by the front and the aft tips combine behind the latter ones to vorticity of strength (6.12.1) at the lower and the upper boundary surfaces of G and is perpendicular to the x-axis.

Herewith the optimum bound vorticity Γ of the blades is found, which creates the optimum free vorticity γ demanded by the theory. By this the thrust is equally distributed over the front and aft position of the blades.

When h is the length of the original blades of the propeller and T is the prescribed thrust of it, the strength Γ follows from

$$2\int_0^{\pi} \rho\,\omega R\,\Gamma h \sin\varphi\, R\, d\varphi = T \ , \tag{6.12.2}$$

or

$$\Gamma = \frac{T}{4\rho\,\omega R^2 h} \ , \tag{6.12.3}$$

where in (6.12.2) φ is a suitably chosen angle which determines points at the propeller circle. It is not difficult to show that the momentum of the fluid in the wake caused by the free vorticity γ (6.12.1), with Γ from (6.12.3) is in agreement with the thrust T.

In [47] van Manen describes some results from experiments with Voith-Schneider propellers. Three cases were tested: (1) the thrust of the propeller was delivered only by the blades in the front position, (2) the thrust was delivered only by the blades in the aft position, and (3) the thrust was equally divided over the blades in the front position and the blades in the aft position. It turned out that the latter case had the highest efficiency. This corresponds with our result: bound vorticity Γ for the front position of the blades and $-\Gamma$ for their aft position.

When the propeller is underneath the aft of the hull, it is working in the boundary layer of the bottom. By this the incoming flow is inhomogeneous, its velocity in the x-direction depends on the y-coordinate. This means that the propeller works in a velocity field $\vec{v}_0^{\,*}$. Then the optimum circulation distribution Γ of the blades has to become an appropriate function of y, which can be determined by the theory of Section 5.5.

Finally, we remark that a more realistic way of describing the Voith-Schneider propeller would be, by using a finite number of lifting lines of finite span. Then we have a finite number of cycloidal reference strips to which we can apply our optimization theory, and we can calculate for different configurations the quality number. However, then we will not find such an elegant solution in closed form as before, which moreover gives essential information.

6.13. Optimization of the Sails of a Yacht

A somewhat different type of problem is the optimization of the sails of a yacht sailing close to wind. The difference with the previous problems is that here we do not prescribe the thrust of the sails, but we make the thrust as large as possible. Because the velocities of the air are small with respect to the velocity of sound, we can consider the air to be incompressible and also we assume the air to be inviscid.

Under sailing close to wind we will understand the situation that the wind relative to the yacht (apparent wind) makes a small angle of $O(\varepsilon)$ with the course of the

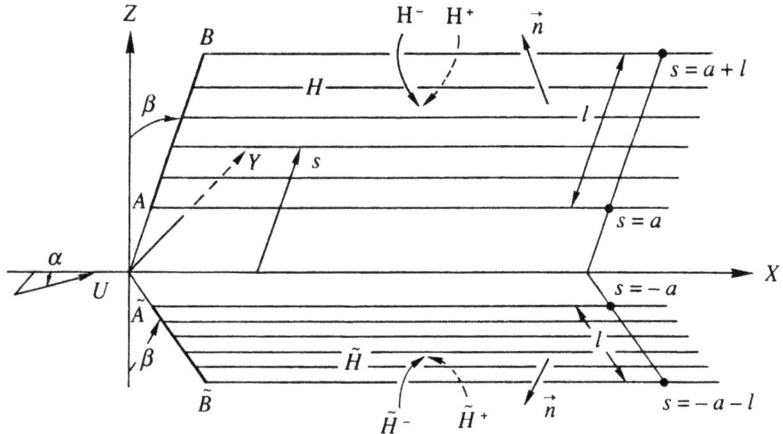

Fig. 6.13.1. Lifting lines $(A\text{–}B)$ and $(\tilde{A}\text{–}\tilde{B})$ and free vortex sheets.

yacht. Then the thrust T developed by the sails will be $O(\varepsilon^2)$. If desired, we can put constraints on the heeling force and heeling moment exerted on the hull by the sails.

In Figure 6.13.1 (X, Y, Z) represents a Cartesian coordinate system which is in rest with respect to the ship. The air is in the half-space region $Z > 0$, hence we neglect the existence of waves on the water surface. The relative velocity of the wind makes a small angle α with the X-axis and has the magnitude U. For the sake of simplicity, we omit the boundary layer of the wind at the water surface and take U independent of the Z-coordinate. Because we assume $\alpha = O(\varepsilon)$ the vortex sheets of foresail and mainsail are separated by a distance of $O(\varepsilon)$ and, within the accuracy of our theory, can be assumed to coincide. As we discussed already in the linearized optimization theory, lifting surfaces can be replaced without loosing generality, by lifting lines chosen at an arbitrary place at the vortex sheets. Hence the lifting lines representing foresail and mainsail can be assumed to coincide. So we replace the sails of the yacht by one lifting line $(A\text{–}B)$ through the origin O, in the (Y, Z) plane, with a heeling angle β, $0 \leq \beta \leq \frac{1}{2}\pi$. The free vortex sheet H stretches downstream from $(A\text{–}B)$. Because the undisturbed relative stream has only a component of $O(\varepsilon)$ perpendicular to the X-axis, H can be chosen within the accuracy of the theory to be parallel with the X-axis.

In order to simulate the boundary between air and water, we consider the image $(\tilde{A}\text{–}\tilde{B})$ of $(A\text{–}B)$ and the image $\tilde{H}$ of H, both reflections are with respect to the plane $Z = 0$. We define the $+$ ($-$ side) of H and $\tilde{H}$ to be oriented in the positive (negative) Y-direction and the unit normal vector on H and on $\tilde{H}$ from the $+$ side towards the $-$ side. On H and $\tilde{H}$ we introduce a Cartesian coordinate system (X, s) where on H the coordinate s is the distance from a point to the X-axis, while on $\tilde{H}$ this coordinate is minus this distance. The boundaries of H are denoted by $s = a$ and $s = a + l$ with $l > 0$, and of $\tilde{H}$ by $s = -a$ and $s = -a - l$, where l is the width of H and $\tilde{H}$ or the length of the lifting line. After this, we assume the whole space to be filled with air of which the incoming velocity is the same for $Z < 0$ as for $Z > 0$.

The driving force or thrust T of the sails, which is $O(\varepsilon^2)$, is defined as the component in the X-direction of the force acting on the bound vorticity of the lifting line (A–B), it is reckoned positive in the negative X-direction. It is our intention to determine within the realm of the linear theory the maximum thrust T_m.

First we make it plausible that an upperbound exists for T. Suppose we have at (A–B) and ($\tilde{A}$–$\tilde{B}$) some span-wise distribution $\nu\Gamma(s)$ where ν is some positive number. The bound vorticity $\Gamma(s)$ is reckoned positive with a right-hand screw in the positive s-direction. By symmetry, we have

$$\Gamma(s) = \Gamma(-s) \ . \tag{6.13.1}$$

We introduce the function $\nu w_n(s)$ which is the velocity induced in the direction of $\vec{n}$ at (A–B) by the free vortex sheets H and $\tilde{H}$. The straight lifting lines (A–B) and ($\tilde{A}$–$\tilde{B}$) do not contribute. Then the thrust T becomes

$$T = \rho \int_{a}^{a+l} \nu\Gamma(s)\{-\alpha U \cos\beta + \nu w_n(s)\}\, ds \ . \tag{6.13.2}$$

For sufficiently small values of ν it follows that the term in the integrand of (6.13.2) which is linear in ν dominates and that $\Gamma(s)$ has to be negative for positive values of T. It is reasonable that $w_n(s)$ is positive for negative values of $\Gamma(s)$, this follows by considering (A–B) as a wing of which the "down wash" is opposite to the direction of the "lift force". When ν increases from zero, the linear term increases and hence the same holds, in the first instance, for T. However, the induced velocity $\nu w_n(s)$ also increases and will more and more counteract the component $-\alpha U \cos\beta$ until for some value of ν the thrust has reached its maximum value and then starts to decrease with increasing values of ν. Hence because T depends quadratically on ν, one maximum value exists for the chosen $\Gamma(s)$. Note that this argumentation is valid because the two terms between brackets in the integrand of (6.13.2) are of the same order of magnitude, $\alpha = O(\varepsilon)$ and $w_n(s) = O(\varepsilon)$.

When we assume that an optimum distribution $\Gamma_m(s)$ exists, the same discussion holds and the thrust will have for this distribution a maximum value denoted by T_m.

We next define the heeling force F acting on (A–B) as the force perpendicular to H, reckoned positive in the negative $\vec{n}$-direction. The constraint that F has to have a prescribed value F_h which in general will be chosen smaller than the value which belongs to T_m, has the form

$$-\rho U \int_{a}^{a+l} \Gamma(s)\, ds = F_h \ . \tag{6.13.3}$$

Analogously, the constraint that the heeling moment has to have a prescribed value M_h, which is reckoned positive when it is connected with a right-hand screw to the negative X-direction is

$$-\rho U \int_{a}^{a+l} (s_0 + s)\, \Gamma(s)\, ds = M_h \ , \tag{6.13.4}$$

where s_0 is some constant depending on the line, parallel to the X-axis, with respect to which the moment is calculated. Also here the value M_h will be chosen to be smaller than the value which belongs to the optimization problem without constraints.

In this case of sailing close to wind, the heeling force on the sails is large with respect to the thrust. The first one is $O(\varepsilon)$ while the second one is $O(\varepsilon^2)$. This is analogous to the propulsion of Regimes (1.a) and (1.b) of Section 4.1.

We now change the formulation in such a way that we obtain a problem of energy extraction out of a slightly disturbed fluid. Consider the Cartesian coordinate system (x, y, z) which is related to the former system by

$$x = X - Ut \ , \qquad y = Y \ , \qquad z = Z \ . \tag{6.13.5}$$

The air is in rest with respect to (x, y, z), except for a small homogeneous flow of magnitude $U\alpha$, hence of $O(\varepsilon)$, in the y-direction and for a negligible flow of $O(\varepsilon^2)$ in the x-direction. Hence in our new coordinate system (x, y, z) we have a fluid disturbed by

$$v_0^* = (0, \alpha U, 0) \ . \tag{6.13.6}$$

The lifting lines $(A–B)$ and $(\tilde{A}–\tilde{B})$ (Figure 6.13.2) translate in the negative x-direction with velocity U, along two strips H and $\tilde{H}$ which stretch from $x = -\infty$ towards $x = \infty$. On these strips we use from now on the coordinates (x, s). The circulation $\Gamma(s)$ of these lines has to be such that under the constraints (6.13.3) and (6.13.4) the resulting force in the negative x-direction is maximum, hence the energy extraction out of the fluid has to be as large as possible.

As is shown in Section 5.2, we can replace v_0^* by v_0 which has the same normal component at H and $\tilde{H}$ and of which the vorticity $\vec{\gamma}_0(s)$ is confined to H and $\tilde{H}$. This vorticity is parallel to the x-axis and reckoned positive with a right-hand screw in the positive x-direction. Because $\vec{\gamma}_0(s)$ has only a component in the x-direction we will denote it simply by the scalar function $\gamma_0(s)$, which is the vorticity per unit of length in the s-direction. By symmetry it follows that $\gamma_0(s) = -\gamma_0(-s)$. As we discussed in Section 5.2, we have to take

$$\int_{a}^{a+l} \gamma_0(s)\, ds = 0 \ , \tag{6.13.7}$$

because a wing can only leave behind free vorticity for which (6.13.7) holds and inversely can annihilate only vorticity for which this condition is fulfilled. This total annihilation, or the total energy extraction out of the field $\vec{v}_0$, can occur when we

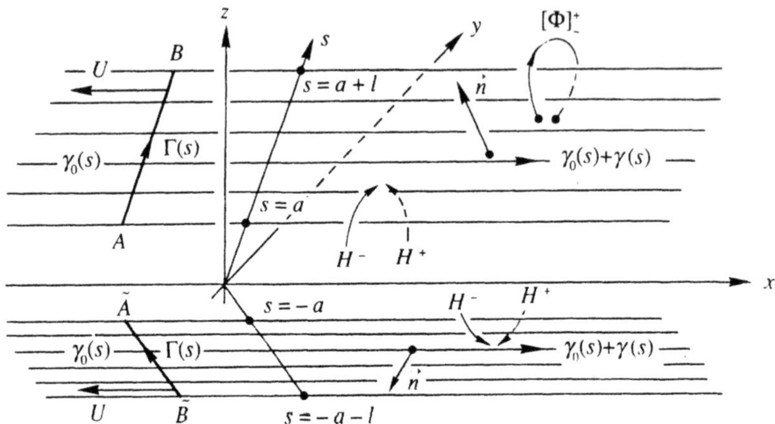

Fig. 6.13.2. The two-sided infinite free vortex sheets H and $\tilde{H}$.

omit the constraints (6.13.3) and (6.13.4). The potential Φ_0 of $\vec{v}_0$ satisfies at H and $\tilde{H}$

$$\frac{\partial \Phi_0}{\partial n} = -\alpha U \cos \beta \ . \tag{6.13.8}$$

The still unknown optimum bound vortices $\Gamma(s)$ shed free vorticity $\vec{\gamma}(s)$ which is also in the x-direction. We make the same agreement with respect to notation and sign for $\vec{\gamma}(s)$ as we did for $\vec{\gamma}_0(s)$. We have the relation

$$\gamma(s) = -\frac{d\Gamma(s)}{ds} \ . \tag{6.13.9}$$

When the bound vortices representing the sails and the mirrored ones are far away in the negative x-direction, the potential Φ of the velocity field induced by $\gamma(s)$ has become independent of time and of the x-coordinate hence $\Phi = \Phi(y, z)$. The jump of Φ across the strip H for some value of s, equals minus the amount of $\gamma(s)$ in the interval $(s, a + l)$. We obtain by (6.13.9)

$$[\Phi]_-^+(s) = -\int_s^{a+l} \gamma(s)\, ds = -\Gamma(s) \ , \tag{6.13.10}$$

and analogously for $\tilde{H}$.

The constraints (6.13.3) and (6.13.4) can now be written as

$$\rho U \int_{-a-l}^{a+l} [\Phi]_-^+(s)\, ds = 2F_h \ , \tag{6.13.11}$$

$$\rho U \int_{-a-l}^{a+l} [\Phi]_-^+(s)(s_0 + |s|)\, ds = 2M_h \ , \tag{6.13.12}$$

respectively, which have the form (5.4.9). The interval $-a < s < a$ gives no contribution to the integral because there $[\Phi]_-^+$ vanishes. Hence we arrived at a variational problem as treated in Section 5.5. We have, in order to obtain the maximum thrust, to minimize the kinetic energy E (5.4.1) (where we take $b = 1$) under the constraints (6.13.11) and (6.13.12). Note that in this problem we have no constraint on the thrust. Hence using (6.13.8), we obtain from (5.5.5)

$$\frac{\partial \Phi}{\partial n} = \alpha U \cos\beta + \lambda_1 + \lambda_2(s_0 + |s|) \ , \quad \text{on } H \text{ and } \tilde{H} \ , \tag{6.13.13}$$

where λ_1 and λ_2 are two Lagrange multipliers.

We introduce two potential functions $\varphi_1(y,z)$ and $\varphi_2(y,z)$, which satisfy

$$\frac{\partial \varphi_1}{\partial n} = 1 \ , \qquad \frac{\partial \varphi_2}{\partial n} = |s| \ , \quad \text{on } H \text{ and } \tilde{H} \ . \tag{6.13.14}$$

The potentials φ_1 and φ_2 can be determined by calculating their vorticity $\gamma_1(s)$ and $\gamma_2(s)$ respectively, reckoned positive with a right-hand screw in the positive x-direction. Using the law of Biot and Savart, it follows from (6.13.14) that $\gamma_1(s)$ and $\gamma_2(s)$ have to satisfy the following singular integral equations

$$-\frac{1}{2\pi} \oint_a^{a+l} \gamma_j(\sigma) \left\{ \frac{1}{(\sigma - s)} - \frac{(s + \sigma\cos 2\beta)}{(\sigma^2 + 2\sigma s \cos 2\beta + s^2)} \right\} d\sigma = g_j(s) \ ,$$

$$a \le s \le a + l \ , \qquad j = 1, 2 \ , \tag{6.13.15}$$

where $g_1(s) = 1$ and $g_2(s) = s$. The second term in the kernel arises from the vorticity at $\tilde{H}$. Because the total circulation of each strip H or $\tilde{H}$ has to be zero we have

$$\int_a^{a+l} \gamma_j(s)\, ds = 0 \ , \qquad j = 1, 2 \ , \tag{6.13.16}$$

by which the solutions for $j = 1$ and $j = 2$ of (6.13.5) become unique. These solutions have to be calculated by numerical means.

Integration of the vorticity yields

$$[\varphi_j]_-^+(s) = -\int_s^{a+l} \gamma_j(\sigma)\, d\sigma \ , \qquad j = 1, 2 \ , \tag{6.13.17}$$

hence these jumps $[\varphi_j]_-^+$, which are needed in the following, can be considered to be known. From (6.13.13) and (6.13.14) it follows that the potential Φ belonging to the shed vorticity in the optimum case can be written as

$$\Phi(y,z) = (\alpha U \cos\beta + \lambda_1 + \lambda_2 s_0)\,\varphi_1(y,z) + \lambda_2\varphi_2(y,z)\ . \tag{6.13.18}$$

The Lagrange multipliers λ_1 and λ_2 follow from (6.13.11) and (6.13.12).

Using (6.13.8) and (6.13.13) we find for the total velocity w_n in the direction $\vec{n}$ at the lifting line (A–B),

$$\begin{aligned} w_n(s) &= -\alpha U \cos\beta + \tfrac{1}{2}\,\frac{\partial\Phi}{\partial n}(s) \\ &= -\tfrac{1}{2}\,\alpha U \cos\beta + \tfrac{1}{2}\{\lambda_1 + \lambda_2(s_0 + |s|)\}\ , \end{aligned} \tag{6.13.19}$$

where the factor $1/2$ of $\partial\Phi/\partial n$ has to be introduced because behind (A–B) the shed vorticity $\gamma(s)$ is only "half infinite" of length. Hence the optimum thrust T has the value

$$\begin{aligned} T &= \rho \int_a^{a+l} \Gamma(s)\, w_n(s)\, ds = -\rho \int_a^{a+l} [\Phi]_-^+(s)\, w_n(s)\, ds \\ &= -\rho \int_a^{a+l} \{(\alpha U \cos\beta + \lambda_1 + \lambda_2 s_0)\,[\varphi_1]_-^+(s) + \lambda_2[\varphi_2]_-^+(s)\} \\ &\quad \cdot \{-\tfrac{1}{2}\,\alpha U \cos\beta + \tfrac{1}{2}(\lambda_1 + \lambda_2(s_0 + |s|))\}\, ds\ . \end{aligned} \tag{6.13.20}$$

In case we have no constraints, we can neglect equations (6.13.11) and (6.13.12) and put $\lambda_1 = \lambda_2 = 0$. Then we can calculate the values F_h^0 and M_h^0 for the heeling force and the heeling moment respectively, which belong to the optimum thrust in this case. We introduce $\nu_1 = F_h/F_h^0$ and $\nu_2 = M_h/M_h^0$, where F_h and M_h are the admitted heeling force and the admitted heeling moment. Hence for practical calculations we will take $0 \le \nu_1 \le 1$ and $0 \le \nu_2 \le 1$, because we have no interest in increasing unnecessarily the unprofitable heeling force and heeling moment.

6.14. Numerical Results

In this section we present some graphs which show numerical results of the previous theory. We assume that the constant s_0 (6.13.4) is zero. The value of a represents to a certain extent the gap between "sails and deck". We do not put a constraint on the heeling force but only on the heeling moment. This means that we neglect (6.13.11) and take $\lambda_1 = 0$. The results have been made dimensionless in an obvious way. Figure 6.14.1 (a) shows the maximum thrust T for the heeling angle $\beta = 0°$ as a function of the relative width of the gap between sails and deck and of the heeling

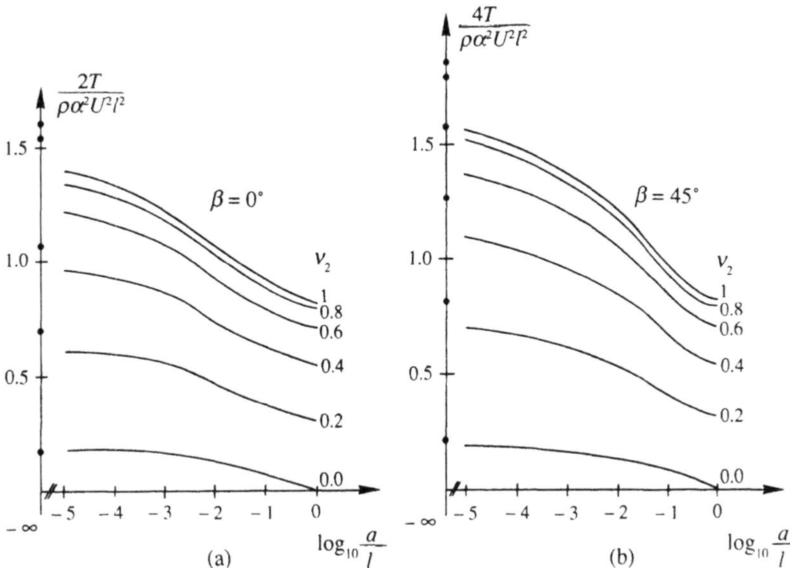

Fig. 6.14.1. Dependence of maximum thrust on a/l and on ν_2; (a) heeling angle $\beta = 0°$; (b) heeling angle $\beta = 45°$.

moment constraint coefficient ν_2. For $\nu_2 = 1$ we have no constraint on this moment and for $\nu_2 = 0$ this moment has to be zero. On the horizontal axis we have plotted a/l on a logarithmic scale, which is appropriate because of the sensitivity of the maximum thrust to narrow gaps. At the right-hand side of the lines the values of ν_2 are given. At the T-axis the values for $a/l = 0$ are denoted by means of points. It is seen that the relative width a/l of the gap affects strongly the optimum thrust. In case of no constraint $\nu_2 = 1$ and for $a/l = 10^{-4}$, still about 20% of the optimum thrust is lacking. In Figure 6.14.1 (b) we give the same results but now for a heeling angle $\beta = 45°$. It is seen that this heeling angle nearly halves the maximum possible thrust.

Figure 6.14.2 shows the values of the heeling moment. The lines for $\nu_2 \neq 1$ follow naturally from the lines for $\nu_2 = 1$ by multiplication by ν_2.

It turned out from the computations that the values given in Figures 6.14.1 and 6.14.2 for $\beta = 0°$ differ only by 6% at most from the corresponding values for $\beta = 30°$. This confirms a statement of Milgram [49] that mostly it will be sufficient to consider the case $\beta = 0$.

Figure 6.14.3 shows the optimum circulation distribution $\Gamma(\tilde{s})$ around the lifting line (A–B), hence around foresail and mainsail together, as a function of the dimensionless coordinate $\tilde{s} = s/l$ along the mast. Two cases are considered, $a/l = 0$ and $a/l = 0.01$, both for $\beta = 0$. The numbers inside the graphs denote the values of ν_2, $0 \leq \nu_2 \leq 1$, which is the admitted fraction of the unconstrained heeling moment

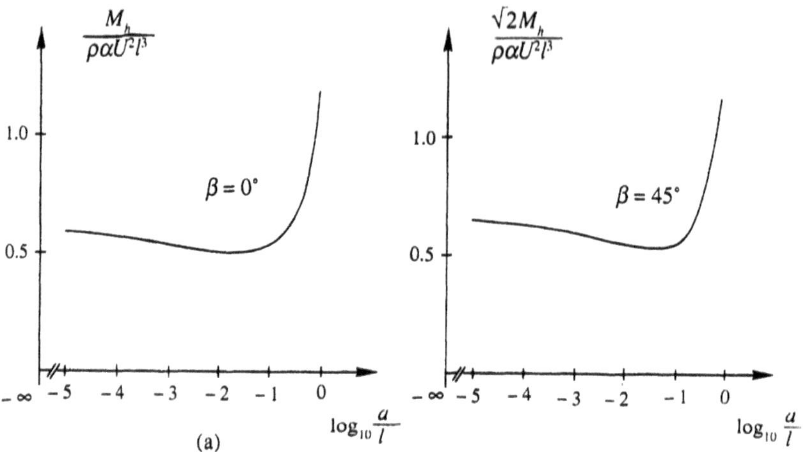

Fig. 6.14.2. Dependence of heeling moment on a/l for $\nu_2 = 1$; (a) heeling angle $\beta = 0°$; (b) heeling angle $\beta = 45°$.

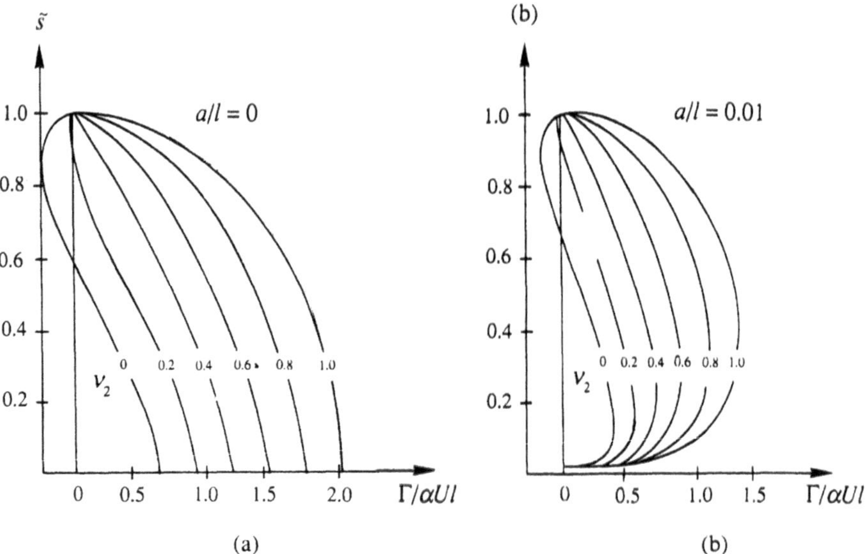

Fig. 6.14.3. Dependence on ν_2 of optimum circulation, around foresail and mainsail together, as a function of $\tilde{s} = s/l$, $\beta = 0°$; (a) without gap $a/l = 0$; (b) with gap $a/l = 0.01$.

($\nu_2 = 1$).

We next comment on some of the given results. From Figure 6.14.1 it is seen that with a moderate constraint on the heeling moment ($\nu_2 \approx 0.6$) not too much of the maximum thrust is lost. Hence it is even possible in connection with the transverse stability of the yacht, that by such a constraint the maximum thrust increases because the heeling angle β will become smaller.

The keel of a yacht considered as a lifting surface is needed to neutralize in combination with the hull the side force, which is the horizontal component of the heeling force, produced by the sails. By this force on the keel the heeling moment on the yacht is increased. The righting moment which occurs by the weight connected to the keel and the shape of the hull has to balance the heeling moment. It is clear that there is no fundamental difference between sail and keel. Both act as lifting surfaces, one in the air, the other in the water. In such a problem we have to optimize the circulation distribution of the sails as well as the circulation of the keel, from the point of view of their combined action and taking into account the righting moment. For this, we refer to the work of Wiersma [76].

When we consider a sail of span $l = 20$ m and a gap between sails and deck of only 1 cm, we have $a/l = 5 \cdot 10^{-4}$, hence $\log_{10} a/l = -3.3$. Then for a not too large heeling angle, it follows from Figure 6.14.1 (a) that a dramatic thrust reduction of about 25% would occur. The question arises if this happens in a real situation when the air is endowed with viscosity and turbulence. With respect to viscosity, this seems not too important for a gap of that width. The turbulence of the air, which has passed along the deck of the yacht, can induce the so-called eddy viscosity. However, the scale of the turbulence is possibly too large to have much influence. Hence it is not likely that the inviscid character of the flow has changed very much. It is possible, however, that the air which passes through the gap will not follow the lee side of the sail, but forms a 2-dimensional jet which partly stretches along the deck. Then possibly the effect is overestimated in our theory. Finally, we emphasize that the decline caused by the gap is calculated here for the maximum possible thrust. In general, a sail will not be shaped in such a way that it is able to produce this maximum thrust. Then of course it is not known what exactly the influence of the narrow gap will be. What really happens must be investigated by experiments.

For more practical information about the theory of sailing we refer to Marchaj [48].

Chapter 7

On the Existence of Optimum Propulsion

For most problems in hydrodynamics we have by experience sufficient information about the existence of a solution. In fact, hydrodynamic problems usually reduce to the solution of a boundary value problem or of an integral equation of some well-known type. In Section 5.7 however we discussed already the non-existence of an optimum propeller of a class of screw propellers for which the rotational velocity was left free. This phenomenon becomes more complicated in the case of unsteady propulsion where for a given base motion the added motion can have an infinite number of degrees of freedom. As we mentioned in the Preface of this book, the non-existence of an optimum motion for a class of propellers does not mean that there are no propulsive systems with a high efficiency in that class. It only states that it is not possible by using some algorithm to find a propulsive system of which the efficiency is better than or equal to the efficiency of any other system of the class.

We will discuss in detail three examples and give the results of a fourth one, all chosen from the unsteady propellers of Sections 4.1 and 4.2. The problems will be 2-dimensional and it is demanded that the profile delivers a prescribed mean value of thrust at the cost of lowest possible energy losses both per unit of span.

We start with two cases of non-existence. First, the small amplitude propulsion by means of a flexible profile (Section 4.1, Regime (1.a)). In this case the base motion is a uniform translational motion along the x-axis. The added motion, which yields the thrust $\bar{T} = O(\varepsilon^2)$, consists of small periodic wriggling motions of $O(\varepsilon)$. It will be shown that when we only put a constraint on the amplitude of the added motion, no optimum added motion need to exist.

Second, the semi-linear finite amplitude propulsion (Section 4.2, Regime (3)). Here the profile is assumed to be flat. For each fixed base motion there exists an infinite number of optimum added motions which yield the desired thrust $\bar{T} = O(\varepsilon)$ and which have the same maximum efficiency. The existence of these added motions follows from the existence of the solution of a Neumann problem by which the optimum shed free vorticity is calculated (6.10.8). This does not yield any essential difficulty from the point of view of applied mathematics. However, when we go one step further and also want to optimize these optimum propellers with respect to their base motions, a difficulty arises. It will be shown that when we only put a constraint on the amplitude of the base motion, no optimum base motion need to exist.

Third, we consider an existence proof. It discusses the small amplitude propulsion of which, as in the first case of non-existence, the amplitude of the added motion is bounded. But now the motion of the profile is much more restricted. The profile is assumed to be rigid and flat and only heaving motions are allowed.

Fourth, we give, without proof, some results with respect to the existence of an optimum small amplitude added motion of a flat profile when both heaving and pitching are allowed. Then besides the constraint on the amplitude still another constraint is needed in order that an optimum motion does exist.

7.1. Small Amplitude Flexible Profile

We consider the 2-dimensional version of the small amplitude propulsion of Section 4.1, Regime (1.a), Figure 4.1.1, hence the span of the wing W is infinite. In order to simplify the formulation we assume the profile to remain at its place and the fluid to come in with the velocity U as denoted in Figure 7.1.1. The profile has no thickness, its shape is given by

$$y = h(x,t) \ , \qquad -l \leq x \leq l \ , \tag{7.1.1}$$

where $h(x,t) = O(\varepsilon)$. The value of l is not yet fixed but will be discussed in the following. In first instance we only state that it has to be smaller than a given value $\tilde{l}, l \leq \tilde{l}$. The prescribed mean value of the thrust is given by $\bar{T} = O(\varepsilon^2)$. We assume the period of time $\tilde{\tau}$ of the motion to be given

$$h(x,t) = h(x, t+\tilde{\tau}) \ , \qquad -\infty < t < \infty \ . \tag{7.1.2}$$

We assume that the amplitude of the motion is bounded

$$|h(x,t)| \leq B = O(\varepsilon) \ . \tag{7.1.3}$$

The problem we consider is the following. Does an added motion $y = h(x,t)$ exist for prescribed values of $U, \bar{T}, \tilde{l}$ and $\tilde{\tau}$ which satisfies (7.1.3) and which leaves behind the least kinetic energy per unit of time? Note that because $h(x,t) = O(\varepsilon)$, this problem has no direct connection with the general optimization theory of Chapter 5.

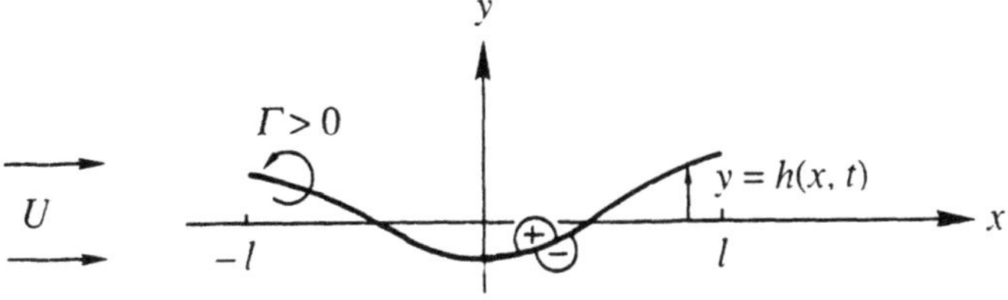

Fig. 7.1.1. The flexible profile $y = h(x,t)$, $-l \leq x \leq l$.

When we prescribe the period of time to be $\tilde{\tau}$, motions with period $\tilde{\tau}/j$, $j = 1, 2, 3, \ldots$ satisfy equation (7.1.2) and hence have also time period $\tilde{\tau}$. For given $\tilde{\tau}$, we can always find an integer $\tilde{j}$ and a length l such that

$$\frac{\tilde{\tau}}{\tilde{j}} = \frac{2l}{U} , \qquad l \leq \tilde{l} . \tag{7.1.4}$$

We choose $2l$ as the length of our profile and consider motions with time period $\tau = \tilde{\tau}/\tilde{j}$, which as we mentioned already, also have the desired period $\tilde{\tau}$. From now on l and τ are fixed. Instead of (7.1.2) we have

$$h(x,t) = h(x, t+\tau) , \qquad \tau = \frac{2l}{U} , \qquad -\infty < t < \infty . \tag{7.1.5}$$

Our first aim is to construct a minimizing sequence of profile motions $y = h_n(x,t)$ which satisfy our conditions and of which the shed kinetic energy E_n per unit of length in the x-direction, tends to zero for $n \to \infty$.

The force *exerted at the fluid* by a two-sided infinite bound vortex of strength Γ^* placed in a homogeneous flow with velocity U, is per unit of span

$$\rho\, U\, \Gamma^*(t) , \tag{7.1.6}$$

in the positive y-direction. When this vortex is placed at the point $x = \xi$ of the x-axis, it sheds a free vortex layer which at the place x with $x > \xi$, has the strength

$$-\frac{1}{U}\frac{\partial}{\partial t}\Gamma^*\left(t - \frac{(x-\xi)}{U}\right) \tag{7.1.7}$$

per unit of length in the x-direction.

We now consider a profile $y = h_n(x,t)$ which creates a pressure jump

$$\begin{aligned}[p(x,t)]_-^+ &= p(x,+0,t) - p(x,-0,t) \\ &= (-1)^n \rho\, U a \sin \pi n \frac{x}{l} \sin \pi n \frac{Ut}{l} , \qquad -l \leq x \leq l ,\end{aligned} \tag{7.1.8}$$

where a will be chosen later on and n can assume any of the values $1, 2, 3, \ldots$ This profile can be generated by elementary external forces acting at the fluid, of strength

$$[p(x,t)]_-^+ \, dx , \qquad -l \leq x \leq l , \tag{7.1.9}$$

in the positive y-direction. In this way we find by (7.1.6) and (7.1.7), for the total density of the vorticity at the profile

$$\Gamma(x,t) = \frac{1}{\rho\, U}\,[p(x,t)]_-^+ - \frac{1}{\rho\, U^2}\int_{-l}^{x} \frac{\partial}{\partial t}\left[p\left(\xi, t - \frac{(x-\xi)}{U}\right)\right]_-^+ d\xi ,$$

$$-l \leq x \leq l . \tag{7.1.10}$$

For $x > l$, that is behind the profile, the shed free vorticity density is denoted by $\gamma(x,t)$ and is calculated as

$$\gamma(x,t) = (-1)^n \pi a n \sin\left(\pi n \frac{(Ut-x)}{l}\right) , \qquad l < x . \tag{7.1.11}$$

It is easily checked that at the trailing edge

$$\begin{aligned}\gamma(l,t) &= -\frac{1}{U}\frac{d}{dt}\int_{-l}^{l} \Gamma(x,t)\, dx \\ &= -\frac{1}{\rho U^2}\int_{-l}^{l} \frac{\partial}{\partial t}\left[p\left(\xi, t - \frac{(l-\xi)}{U}\right)\right]_-^+ d\xi = \Gamma(l,t) ,\end{aligned} \tag{7.1.12}$$

which means that the Kutta condition is satisfied.

Next we estimate the velocity component in the y-direction $v_y(x,0,t)$, $-l \le x \le l$, induced by the vorticity (7.1.10) and (7.1.11).

$$v_y(x,0,t) = \frac{1}{2\pi}\oint_{-l}^{l} \frac{\Gamma(\xi,t)}{(x-\xi)}\, d\xi + \frac{1}{2\pi}\oint_{l}^{\infty} \frac{\gamma(\xi,t)}{(x-\xi)}\, d\xi , \tag{7.1.13}$$

where when necessary, one of the integrals is in the sense of Cauchy's principle value. An elementary but rather complicated calculation, which we do not reproduce here, shows that there exist constants C and N such that

$$|v_y(x,0,t)| \le C a\, n , \qquad -l \le x \le l , \qquad n \ge N , \tag{7.1.14}$$

where C is independent of a and n. In the following we will use C also for other such constants.

The shape $h_n(x,t)$ of a profile which creates the pressure difference (7.1.8) satisfies the equation

$$\frac{\partial}{\partial t} h_n(x,t) + U\frac{\partial}{\partial x} h_n(x,t) = v_y(x,0,t) , \tag{7.1.15}$$

where $v_y(x,0,t)$ is given by (7.1.13) and is estimated by (7.1.14). This partial differential equation can easily be solved, we find

$$\begin{aligned}h_n(x,t) &= \frac{1}{U}\int_{-l}^{x} v_y\left(\xi, 0, t + \frac{(\xi - x)}{U}\right) d\xi + g(x - Ut) \\ &\stackrel{\text{def}}{=} f_n(x,t) + g(x-Ut) .\end{aligned} \tag{7.1.16}$$

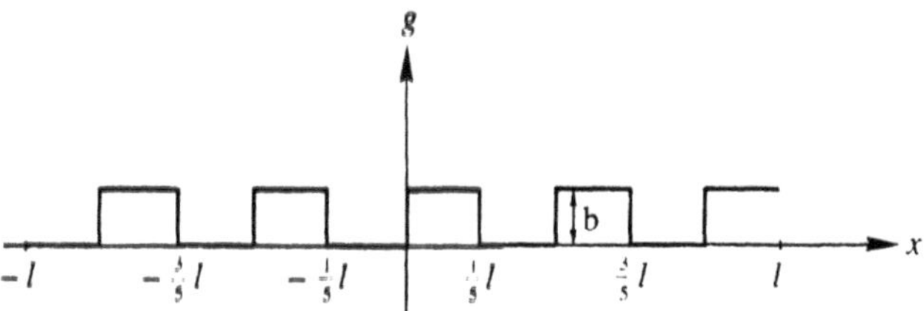

Fig. 7.1.2. The function $g_n(x - Ut)$, $n = 5$, $t = 0$.

The function $g(x - Ut)$ is an "arbitrary" function of $(x - Ut)$ and is the general solution of the homogeneous part of (7.1.15). By (7.1.14) we have

$$|h_n(x,t)| \leq \frac{C a\, n}{U}(l + x) + |g(x - Ut)| \ , \quad -l \leq x \leq l \ , \quad n \geq N \ . \tag{7.1.17}$$

Now we make a choice $g_n(x - Ut)$ for $g(x - Ut)$ which is essential for the non-existence proof. It will be a step function as shown in Figure 7.1.2, of which the horizontal parts have length l/n and the steps have the height $b = O(\varepsilon)$. Although we have chosen in this way a "strange function" with vertical parts (at the x-values of these parts the function is not uniquely defined), it will be clear that this is not essential. We could have taken instead, functions with steep parts, so that everywhere the slope would be finite. Then this steepness could be increased so fast with n that for $n \to \infty$ the new functions give results, which differ from those belonging to the corresponding original ones with vertical parts, by an amount which tends to zero. We also could have taken rounded edges so that the functions would have an infinite number of derivatives (C^∞ functions). These refinements however are not essential and would obscure the analysis. Hence we take $g_n(x - Ut)$ as shown in Figure 7.1.2.

The profile $g_n(x - Ut)$ is pushed to the right with velocity U, annihilated at $x = l$ and completed at $x = -l$ so that it preserves its character and stretches over the interval $-l \leq x \leq l$. Note that each moment the pressure difference (7.1.8) across the profile exerts a non-zero force to the left on the profile $g_n(x - Ut)$ except for those times where $[p]_-^+ = 0$. In this way the part $g_n(x - Ut)$ of the profile (7.1.16) is used to generate thrust without shedding additional vorticity with respect to the linearized theory. By this the efficiency increases. Perhaps the increase in efficiency of the propulsive action of a flexible profile by means of the function $g_n(x - Ut)$ has some relation with the fact that a swimming fish propagates waves along its body from head to tail [79]. A difference with the fish motion is that there the amplitude of the waves at the thicker forepart of the body is smaller than at the tail.

First, we estimate the mean value $\bar{T}_1$, of the thrust, delivered by the part $f_n(x, t)$ of the profile motion (7.1.16), reckoned positive in the negative x-direction. Using equations (7.1.15) and (7.1.16) we can write

$$\bar{T}_1 = -\frac{1}{\tau_n} \int_0^{\tau_n} \int_{-l}^{l} [p(x,t)]_-^+ \frac{\partial}{\partial x} f_n(x,t)\, dx\, dt$$

$$= -\frac{1}{\tau_n} \int_0^{\tau_n} \int_{-l}^{l} [p(x,t)]_-^+ \left\{ \frac{v_y(x,0,t)}{U} \right.$$

$$\left. - \frac{1}{U^2} \int_{-l}^{x} \frac{\partial}{\partial t} v_y \left(\xi, 0, t + \frac{(\xi - x)}{U} \right) d\xi \right\} dx\, dt \ , \tag{7.1.18}$$

where $\tau_n = 2l/nU = \tau/n$ is one period of the motion. It can be shown by (7.1.8) and (7.1.14) that

$$\bar{T}_1 \leq C a^2 n^2 \ , \tag{7.1.19}$$

where C is some constant independent of a and n. The factor n^2 arises in (7.1.19) because the velocity component v_y in (7.1.18) is differentiated with respect to time.

Second, we calculate the contribution $\bar{T}_2$ of the part $g_n(x - Ut)$ of the profile motion (7.1.16) to the mean value of the thrust. At $t = 0$ this profile is in the position as denoted in Figure 7.1.2, it is pushed to the right with velocity U, annihilated at $x = l$ and completed at $x = -l$ as we already mentioned. This profile does not disturb the fluid, however, it perceives the pressures (7.1.8) induced by $f_n(x,t)$. Hence

$$\bar{T}_2 = -\frac{2}{\tau_n} \int_0^{\tau_n/2} \int_{-l}^{l} [p(x,t)]_-^+ \frac{\partial}{\partial x} g_n(x - Ut)\, dx\, dt$$

$$= -\frac{2}{\tau_n} \int_0^{\tau_n/2} \int_{-l}^{l} [p(x,t)]_-^+ \sum_{j=0}^{2n-1} (-1)^{j+1}$$

$$\cdot\, b\, \delta \left(x - Ut + \frac{(n-j)l}{n} \right) dx\, dt \ , \tag{7.1.20}$$

where $\delta(x)$ is the delta function of Dirac. During the time interval $\tau_n/2$, the profile $g_n(x - Ut)$ moves to the right over a distance l/n. Substitution of the pressure jump (7.1.8) into (7.1.20) yields

$$\bar{T}_2 = \rho\, U a\, n\, b \ . \tag{7.1.21}$$

We now take

$$a\, n = A = \text{const.} \ , \tag{7.1.22}$$

where A is some constant which can be chosen at will, but once chosen remains fixed. Then it is possible to find for each n a value $b = b_n$ for the height of the steps of the $g_n(x - Ut)$, such that

$$\bar{T}_1 + \bar{T}_2 = \bar{T} \tag{7.1.23}$$

where $\bar{T}$ is the prescribed mean value of the thrust. It is easily seen that we have an upper bound b^* for the values $|b_n|$, for instance we can choose

$$b^* = \frac{CA^2 + \bar{T}}{\rho U A} \ . \tag{7.1.24}$$

From (7.1.17) and (7.1.24) it follows that when B (7.1.3) is sufficiently large the foregoing procedure is possible and the thrust $\bar{T}$ can be generated for each value of n.

The wasted kinetic energy per unit of length in the x-direction far behind the profile will be calculated now. We consider the free vorticity $\gamma(x,t)$ (7.1.11), with respect to a coordinate system which moves with the free stream. Then we have a vortex sheet of strength

$$\pi a \, n \sin \pi n \frac{x}{l} \ , \tag{7.1.25}$$

which can be assumed to extend from $x = -\infty$ towards $x = \infty$. The kinetic energy E_n of the fluid per unit of length in the x-direction induced by this sheet becomes

$$E_n = \frac{1}{8} \pi \rho \, l \, a^2 n = \frac{1}{8} \pi \rho \, l \, \frac{A^2}{n} \ . \tag{7.1.26}$$

Hence we have the result that for $n \to \infty$ this kinetic energy E_n tends to zero. This means that we have constructed a minimizing sequence of profile motions, each of which delivers a mean value of thrust $\bar{T}$ and for which the efficiency tends to one.

We note that also for flexible wings of finite span minimizing sequences in the above-mentioned sense can be constructed.

7.2. Non-Existence of Optimum Added Motion

We now show that the constraint on the amplitude of the profile motion (7.1.3) need not to be sufficient for the existence of an optimum motion. We first prove that when no free vorticity is shed behind a periodically moving profile, no mean value of thrust can occur. This has been proved in Section (1.16) with respect to the non-linear potential theory, however, it is not quite obvious in the approximative linearized small amplitude theory. It is even not true as we discussed in Section 4.1 (Figure 4.1.2) and in Section 4.10 (Figure 4.10.1) for a slotted profile.

Consider some profile (Figure 7.1.1) with total density of vorticity

$$\Gamma(x,t) \ , \qquad -l \leq x \leq l \ . \tag{7.2.1}$$

We use Bernoulli's theorem for unsteady flow (1.2.12), which reads

$$p = -\rho \left(U \frac{\partial \Phi}{\partial x} + \frac{\partial \Phi}{\partial t} \right) + \text{const.} \ , \tag{7.2.2}$$

where $\Phi(x, y, t)$ is the potential of the disturbance velocity field. Hence we can write

$$[p(x,t)]_{-}^{+} = -\rho \left[U\frac{\partial \Phi}{\partial x} + \frac{\partial \Phi}{\partial t} \right]_{-}^{+}$$

$$= \rho \left\{ U\Gamma(x,t) + \frac{\partial}{\partial t} \int_{-l}^{x} \Gamma(\xi,t)\, d\xi \right\} . \tag{7.2.3}$$

Then the mean value of the thrust becomes

$$\bar{T} = -\frac{1}{\tau} \int_{0}^{\tau} \int_{-l}^{l} [p(x,t)]_{-}^{+} \frac{\partial h}{\partial x}(x,t)\, dx\, dt$$

$$= -\frac{\rho}{\tau} \int_{0}^{\tau} \int_{-l}^{l} \left\{ U\Gamma(x,t) + \frac{\partial}{\partial t} \int_{-l}^{x} \Gamma(\xi,t)\, d\xi \right\} \frac{\partial h}{\partial x}(x,t)\, dx\, dt , \tag{7.2.4}$$

where τ is the period of the motion and we assume temporarily that no suction force occurs at the leading edge of the profile. Hence in other words we assume that

$$\lim_{x \to -l} \Gamma(x,t) \leq \frac{Q(t)}{(l+x)^{\alpha}} , \qquad 0 \leq \alpha < \tfrac{1}{2} . \tag{7.2.5}$$

When we suppose that no free vorticity is shed we have

$$\frac{d}{dt} \int_{-l}^{l} \Gamma(x,t)\, dx = 0 . \tag{7.2.6}$$

Integrating the second term of the integrand at the right-hand side of (7.2.4) first partially with respect to x, then with respect to t and using (7.2.6) and the periodicity of the motion we find

$$\bar{T} = -\frac{\rho}{\tau} \int_{0}^{\tau} \int_{-l}^{l} \Gamma(x,t) \left\{ U\frac{\partial h}{\partial x}(x,t) + \frac{\partial h}{\partial t}(x,t) \right\} dx\, dt$$

$$= -\frac{\rho}{\tau} \int_{0}^{\tau} \int_{-l}^{l} \Gamma(x,t) v_y(x,0,t)\, dx\, dt . \tag{7.2.7}$$

Again using the fact that there is no free vorticity for $x > l$, we can replace (7.2.7) by

$$\bar{T} = -\frac{\rho}{\tau} \int_{0}^{\tau} \int_{-l}^{l} \Gamma(x,t) \left\{ \oint_{-l}^{l} \frac{\Gamma(\xi,t)}{(x-\xi)}\, d\xi \right\} dx\, dt . \tag{7.2.8}$$

Because by (7.2.5) there is no serious singularity in the integrand of (7.2.8) for $x \to -l$ and $\xi \to -l$, this formula can be rewritten as

$$\bar{T} = -\frac{\rho}{\tau}\int_0^\tau \left\{\int_{-l}^{l}\int_{-l}^{l} \frac{\Gamma(x,t)\Gamma(\xi,t)}{(x-\xi)}\, d\xi\, dx \right\} dt \ , \tag{7.2.9}$$

where the integration with respect to ξ and x is as follows. Exclude from the region of integration a strip of width δ at each side of the diagonal $x = \xi$, perform the integration and take the limit $\delta \to 0$. By the antisymmetry of the integrand with respect to the diagonal it follows that

$$\bar{T} = 0 \ . \tag{7.2.10}$$

The case that we admit shapes $h(x,t)$ of the profile such that

$$\lim_{x \to -l} \Gamma(x,t) = \frac{Q(t)}{(l+x)^{1/2}} \ , \tag{7.2.11}$$

is slightly more complicated. In this case a suction force (1.21.13) arises at the leading edge of the profile which contributes to the thrust. We discuss this case by a limit procedure, which is analogous to the one mentioned at the end of Section 1.21. Consider instead of the correct $\Gamma(x,t)$ the following mutilated one

$$\Gamma^*(x,t) = 0 \ , \qquad -l \le x \le -l + \tilde{\varepsilon} \ , \tag{7.2.12}$$

$$\Gamma^*(x,t) = \Gamma(x,t) \ , \qquad -l + \tilde{\varepsilon} \le x \le l \ . \tag{7.2.13}$$

Now we can apply (7.2.9) to $\Gamma^*(x,t)$ and of course find again (7.2.10) for each $\tilde{\varepsilon}$ with $\tilde{\varepsilon} \to 0$. That by this procedure we have correctly taken into account the suction force follows from the consideration of

$$\bar{T}_\delta = -\frac{\rho}{\tau}\int_0^\tau \int_{-l}^{-l+\delta} \Gamma^*(x,t) \left\{ \oint_{-l}^{l} \frac{\Gamma^*(\xi,t)}{(x-\xi)}\, d\xi \right\} dx\, dt \ , \quad 0 < \delta \ , \tag{7.2.14}$$

which is the thrust exerted by the interval $(-l \le x \le -l+\delta)$ in the neighbourhood of the leading edge of the profile. When we consider the following limit, first, for fixed $\delta > 0$ we let $\tilde{\varepsilon} \to 0$ and second, $\delta \to 0$, we find that the integral (7.2.14) does not tend to zero but to the suction force at the leading edge. Hence it follows that also in the case of (7.2.11) we have $\bar{T} = 0$.

Note that such a limit procedure is a correct way to incorporate the suction force in (7.2.8). When for instance we take $\Gamma(x,t) = (l+x)^{-1/2}$, the successive integrals in (7.2.8) can be calculated analytically in closed form. It is found that the result equals minus the suction force, hence the integral is not zero. Addition of the suction force then gives the result $\bar{T} = 0$, which is the correct value. The reason is that the suction force is not represented in the integrand in the case of (7.2.11). When we consider the case given by (7.2.12) and (7.2.13), the suction force is, by the strongly curved shape of the profile in the neighbourhood of the leading edge, perceived by the integral.

The case for which the exponent α in (7.2.5) has a value with $1/2 < \alpha$ is of no interest. Then both the suction force as derived in Section 1.21 and the kinetic energy of the fluid in some neighbourhood of the leading edge become infinite. So we can restrict ourselves to vortex distributions $\Gamma(x,t)$ which satisfy (7.2.6) and

$$\lim_{x\to 0} \Gamma(x,t) \le \frac{Q(t)}{(l+x)^\alpha} \ , \qquad 0 \le \alpha \le \tfrac{1}{2} \ , \tag{7.2.15}$$

then it follows that $\bar{T} = 0$, also in the linearized theory, when no free vorticity is shed.

Inversely when we know of some vortex distribution $\Gamma(x,t)$ for which (7.2.15) holds, that it yields a thrust with a mean value $\bar{T} \neq 0$, then (7.2.6) cannot be true. Hence free vorticity of density of $O(\varepsilon)$ has to be shed and kinetic energy E of $O(\varepsilon^2)$ per unit of time is left behind, with

$$E > 0 \ . \tag{7.2.16}$$

This also applies to an optimum vortex distribution $\Gamma(x,t)$, if it exists.

In the previous section however we have constructed a minimizing series of profile motions $h_n(x,t)$ for which the lost kinetic energy E_n per unit of time has the property

$$E_n \to 0 \ , \qquad n \to \infty \ . \tag{7.2.17}$$

Hence no optimum profile motion can exist because its lost kinetic energy would be larger than the lost energy of these $h_n(x,t)$ with n larger than some suitable number N.

This result is valid when for given U, $\tilde{\tau}$, $\tilde{l}$, $\bar{T}$ the bound B (7.1.3) is sufficiently large. This does not mean that for small values of B an optimum motion will exist, only that the non-existence proof by means of the construction of the minimizing sequence of profile motions $h_n(x,t)$ is not possible.

The question can arise if the type of functions $g_n(x,-Ut)$ (Figure 7.1.2) can be used in a linear theory for larger values of n. However, this is not relevant for our problem where we only discussed if, within the framework of a linear theory, an algorithm exists for the calculation of an optimum profile motion which yields a mean thrust $\bar{T}$ under the only condition (7.1.3).

7.3. Large Amplitude Rigid Profile

The following is concerned with the existence or better with the non-existence of an optimum base motion (Section 4.7) in the case of the semi-linear theory for a rigid flat profile of zero thickness. The profile is drawn in Figure 7.3.1; it moves with its three quarter chord point Q along a periodic line L, while it remains tangent to L.

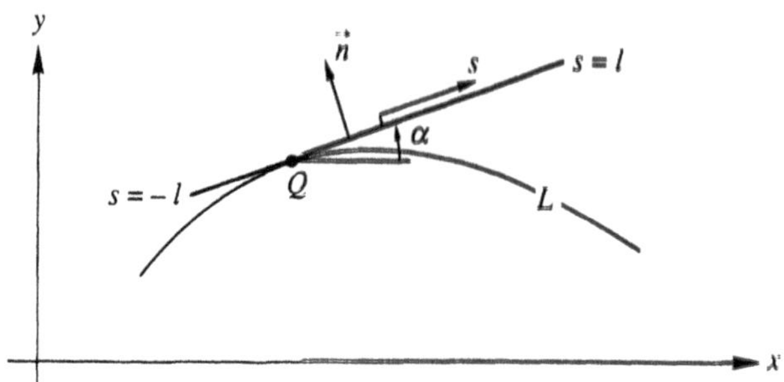

Fig. 7.3.1. Flat profile moving along L.

Its rotational velocity is $\omega = \dot{\alpha}(t)$, hence its vorticity $\Gamma(s,t)$ can by (4.7.3) (with $a = -\frac{1}{2}\,l$), be written as

$$\Gamma(s,t) = \dot{\alpha}(t)\,(l-2s)\left(\frac{l+s}{l-s}\right)^{1/2} , \qquad -l \le s \le l \;, \tag{7.3.1}$$

of which the total circulation is zero and where s is the length parameter along the profile.

We suppose the motion of the point Q to be prescribed by the functions

$$x_Q(t+\tau) = x_Q(t) + U\tau \;, \qquad y_Q(t) = y_Q(t+\tau) \;, \tag{7.3.2}$$

where U is the mean velocity of advance and τ is the time period. Then $\alpha(t)$ follows from

$$\tan\alpha(t) = \frac{\dot{y}_Q(t)}{\dot{x}_Q(t)} \;, \tag{7.3.3}$$

with

$$\alpha(t) = \alpha(t+\tau) \;. \tag{7.3.4}$$

In this way we have given the base motion of the profile because, as is discussed in Section 4.7, no vorticity is shed into the fluid.

Next we put constraints on the base motion of the profile. Its translational and its rotational velocity are assumed to be continuous functions of time and

$$|x_Q(t) - Ut| \le B_1 \;, \qquad |y_Q(t)| \le B_2 \;, \tag{7.3.5}$$

where B_1 and B_2 are prescribed quantities. Inequalities (7.3.5) mean that the point Q of the profile has to stay within a rectangle with sides $2B_1$ and $2B_2$, which translates in the positive x-direction with velocity U. Hence the motion of the pivotal point Q has a bounded amplitude and the profile cannot move backwards and forwards with respect to the propelled body over arbitrary large distances.

We now consider the set of all base motions with continuous $x_Q(t)$, $y_Q(t)$ and $\alpha(t)$, which satisfy (7.3.2), (7.3.4) and (7.3.5). To each base motion we assign an optimum added motion such that a prescribed mean value of the thrust $\bar{T}$ is delivered by the profile. This means that to each element of the set of admitted base motions an amount of kinetic energy loss per unit of time is "attached". As follows from Section 4.2 (Regime (3)) or Section 4.7 and from our optimization theory, its loss of kinetic energy does not depend on the chosen optimum added motion but only on the thrust $\bar{T}$. The question we will consider is: does there exist an optimum base motion for which this energy loss is a minimum (Sparenberg and Takens [65])?

7.4. The Wagging Motion

Because it is our intention to prove the non-existence of an optimum base motion we will, as we did in the previous non-existence problem, try to find a minimizing sequence but now of base motions instead of added motions. This sequence will be such that the lost kinetic energy of the corresponding optimum propellers, which deliver a prescribed thrust, tends to zero. Then it is clear that an optimum base motion cannot exist because any propeller which delivers a thrust has to shed vorticity (see below (1.16.8)).

We start with an important part of the base motion which we want to construct. We say that the profile carries out one wagging motion with amplitude 1, when $x_Q(t)$, $y_Q(t)$ and $\alpha(t)$ are the following functions of time

$$
\begin{aligned}
&t \in [0, 1/4]: \quad x_Q(t) = 0 \ ; \quad y_Q(t) \ , -1 \to 1 \ ; \quad \alpha(t) = \frac{\pi}{2} \ , \\
&t \in [1/4, 1/2]: \quad x_Q(t) = 0 \ ; \quad y_Q(t) = 1 \ ; \quad \alpha(t) \ , \frac{\pi}{2} \to -\frac{\pi}{2} \ , \\
&t \in [1/2, 3/4]: \quad x_Q(t) = 0 \ ; \quad y_Q(t) \ , 1 \to -1 \ ; \quad \alpha(t) = -\frac{\pi}{2} \ , \\
&t \in [3/4, 1]: \quad x_Q(t) = 0 \ ; \quad y_Q(t) = -1 \ ; \quad \alpha(t) \ , -\frac{\pi}{2} \to \frac{\pi}{2} \ .
\end{aligned} \tag{7.4.1}
$$

The line L along which the profile slides during the wagging motion is drawn in Figure 7.4.1, where we have to take the limit of zero deviation from the y-axis. Note that this wagging motion is only part of the total periodic base motion of the profile, which still has to be specified.

We want to investigate first how the fluid particles have moved after one wagging motion. That is, what can be said about the transformation M which is defined as follows; if $x(t), y(t)$ is a solution of (4.7.11) and (4.7.12) when the profile carries out a wagging motion $t = 0 \to t = 1$, then

$$M(x(0), y(0)) = (x(1), y(1)) \ , \tag{7.4.2}$$

where $(x(0), y(0))$ can be any point of the (x, y) plane. Hence M maps the (x, y) plane onto itself. It is clear that the intervals of time $[0, 1/4]$, $[1/4, 1/2]$, $[1/2, 3/4]$

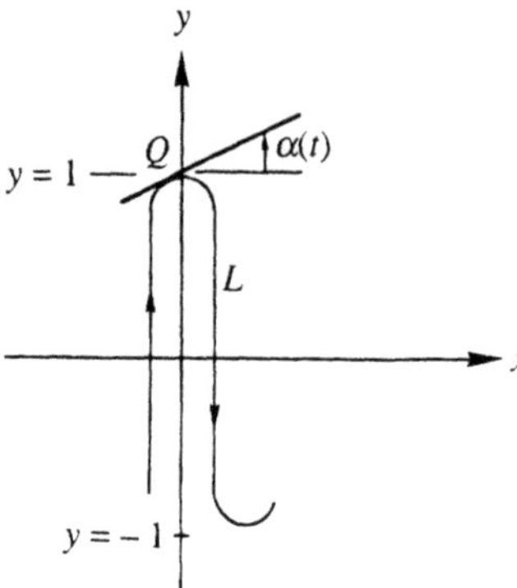

Fig. 7.4.1. Impression of one wagging motion.

and [3/4, 1], used in the definition (7.4.1) are not essential and can be changed into other intervals of time. The only thing that matters for the transformation M is that the profile moves from its starting position towards its ending position by passing through the indicated positions.

We define the displacement vector $\vec{D}$ belonging to M by

$$M(x(0), y(0)) = (x(1), y(1)) = (x(0), y(0)) + \vec{D}\,(x(0), y(0)) \ . \tag{7.4.3}$$

We shall prove that

$$\vec{D}(x, y) = \int_0^1 \{\dot{x}(x, y, t), \dot{y}(x, y, t)\}\, dt + O((x^2 + y^2)^{-5/2}) \ , \tag{7.4.4}$$

where we kept in the integrand the values of x and y unchanged. Here and in the following, we use $f(x, y) = O((x^2 + y^2)^{\beta})$ to denote that there exist C_1 and C_2, such that $|f(x, y)| \leq C_1(x^2 + y^2)^{\beta}$ for $(x^2 + y^2) \geq C_2$. From (4.7.11) and (4.7.12) and from the fact that the total circulation of the profile is zero (4.7.5) it follows that

$$\{\dot{x}(x, y, t), \dot{y}(x, y, t)\} = O((x^2 + y^2)^{-1}) \ , \tag{7.4.5}$$

and

$$\left\{\frac{\partial \dot{x}}{\partial x}(x, y, t), \frac{\partial \dot{x}}{\partial y}(x, y, t), \frac{\partial \dot{y}}{\partial x}(x, y, t), \frac{\partial \dot{y}}{\partial y}(x, y, t)\right\}$$

$$= O((x^2 + y^2)^{-3/2}) \ . \tag{7.4.6}$$

By this we have

$$\{x(0), y(0)\} - \{x(t), y(t)\} = O((x^2(0) + y^2(0))^{-1}) \ , \quad 0 \leq t \leq 1 \ , \tag{7.4.7}$$

because (7.4.6) tends to zero for $(x^2 + y^2) \to \infty$ more rapidly than (7.4.5). Hence

$$\vec{D}\,(x(0), y(0)) = \int_0^1 \{\dot{x}(x(t), y(t), t), \dot{y}(x(t), y(t), t)\}\, dt$$

$$= \int_0^1 \{\dot{x}(x(0), y(0), t), \dot{y}(x(0), y(0), t)\}\, dt$$

$$+ O\left((x(0)^2 + y(0)^2)^{-3/2}\right) \cdot O\left((x(0)^2 + y(0)^2)^{-1}\right) \ , \qquad (7.4.8)$$

which proves (7.4.4).

Next we show that the displacement vector $\vec{D}$ satisfies

$$\vec{D}(x, y) - 2C_3 \left\{ \frac{x^3 - 3xy^2}{(x^2 + y^2)^3} \ , \frac{3x^2y - y^3}{(x^2 + y^2)^3} \right\} = O((x^2 + y^2)^{-2}) \ . \qquad (7.4.9)$$

To this end we introduce the following functions of x and y

$$X(x, y) = -\frac{1}{2\pi} \int_{-\pi/2}^{\pi/2} \int_{-l}^{l}$$

$$\frac{(l - 2s)\left(\frac{l+s}{l-s}\right)^{1/2} \{y - (s + l/2)\sin\alpha\}\, ds\, d\alpha}{\{x - (s + l/2)\cos\alpha\}^2 + \{y - (s + l/2)\sin\alpha\}^2} \ , \qquad (7.4.10)$$

$$Y(x, y) = \frac{1}{2\pi} \int_{-\pi/2}^{\pi/2} \int_{-l}^{l}$$

$$\frac{(l - 2s)\left(\frac{l+s}{l-s}\right)^{1/2} \{x - (s + l/2)\cos\alpha\}\, ds\, d\alpha}{\{x - (s + l/2)\cos\alpha\}^2 + \{y - (s + l/2)\sin\alpha\}^2} \ . \qquad (7.4.11)$$

These functions arise from(4.7.11) and (4.7.12) by an integration with respect to time. The point (x, y) is assumed to be far away from the origin, in fact we assume $|x|$ and $|y|$ to be large with respect to l. The three quarter point (x_Q, y_Q) is chosen at the origin. Under the integral signs the small changes of x and y have been neglected. Hence the functions $X(x, y)$ and $Y(x, y)$ represent an approximation of the displacement of far away points when the profile carries out a rotation with its three quarter point at the origin from $\alpha = -\pi/2$ towards $\pi/2$.

Expanding the integrands in (7.4.10) and (7.4.11) for large values of $|x|$ and $|y|$ and using again that the total circulation of the profile is zero, it can be shown that there exists a positive constant C_4, such that

$$\left| \frac{\partial X}{\partial y}(x, y) - C_4 \left\{ \frac{x^3 - 3xy^2}{(x^2 + y^2)^3} \right\} \right| = O((x^2 + y^2)^{-2}) \ , \qquad (7.4.12)$$

$$\left| \frac{\partial Y}{\partial y}(x,y) - C_4 \left\{ \frac{3x^2y - y^3}{(x^2+y^2)^3} \right\} \right| = O((x^2+y^2)^{-2}) \ . \tag{7.4.13}$$

From the motion (7.4.1) of the profile and from (7.4.4) it follows that

$$\vec{D}(x,y) = \left(\int_{1/4}^{1/2} + \int_{3/4}^{1} \right) \{\dot{x}(x,y,t) + \dot{y}(x,y,t)\} \, dt$$
$$+ O((x^2+y^2)^{-5/2}) \ . \tag{7.4.14}$$

By (4.7.11), (4.7.12), (7.4.10) and (7.4.11)

$$\int_{1/4}^{1/2} \{\dot{x}(x,y,t), \dot{y}(x,y,t)\} \, dt = -\{X(x,y-1), Y(x,y-1)\} \ , \tag{7.4.15}$$

$$\int_{3/4}^{1} \{\dot{x}(x,y,t), \dot{y}(x,y,t)\} \, dt = \{X(x,y+1), Y(x,y+1)\} \ , \tag{7.4.16}$$

where the arguments $(y-1)$ and $(y+1)$ show that the rotations of the profile in (7.4.1) and hence in (7.4.4), take place at $(x_Q, y_Q) = (0,1)$ and at $(x_Q, y_Q) = (0,-1)$, respectively. Hence

$$\begin{aligned} \vec{D}(x,y) &= \{X(x,y+1), Y(x,y+1)\} \\ &\quad - \{X(x,y-1), Y(x,y-1)\} + O((x^2+y^2)^{-5/2}) \\ &= 2\left\{ \frac{\partial X}{\partial y}(x, y+\theta_1) \ , \frac{\partial Y}{\partial y}(x, y+\theta_2) \right\} \\ &\quad + O((x^2+y^2)^{-5/2}) \ , \end{aligned} \tag{7.4.17}$$

where θ_1 and θ_2 depend on (x,y) and are within the closed interval $[-1,1]$. By (7.4.12) and (7.4.13) we find from (7.4.17) that (7.4.9) is correct. In Figure 7.4.2 we have drawn a sketch of the directions of the motion of the fluid particles at a large distance from the origin, induced by one wagging motion of the profile.

We now introduce the set l_a, which consists of points (x,y) with $x \leq -4a$, $|y| = a$, $a > 0$ (Figure 7.4.3). A curve σ will be called of type a and will be denoted by $\sigma(a)$ if it contains a curve $\sigma^*(a)$ such that the points (x,y) of $\sigma^*(a)$ satisfy

$$x \leq -4a \ , \qquad |y| \leq a \ , \tag{7.4.18}$$

while $\sigma^*(a)$ connects the two components of l_a. We will prove the following statement:

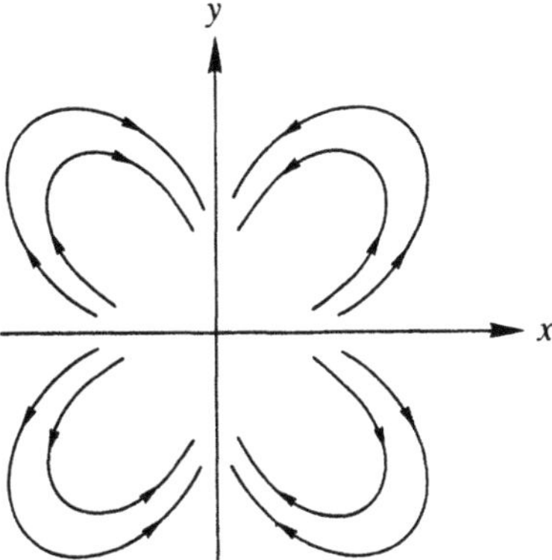

Fig. 7.4.2. Impression of the directions of the displacement vector $\vec{D}$ for one wagging motion.

For some $a_0 > 0$ holds, that for every curve $\sigma(a)$ with $a \geq a_0$ and any $b > a$ there is an $\tilde{N}$ such that for any $N \geq \tilde{N}$, $M^N(\sigma(a))$ is of type b.

By $M^N(\sigma(a))$ we mean the new curve that arises by applying N times the transformation M (7.4.2) to all points of $\sigma(a)$. This statement shows that it is possible to "blow up" the curve $\sigma(a)$ into a "larger" curve $\sigma(b)$ ($b > a$) by a suitable number of wagging motions of the profile (Figure 7.4.3).

The proof is as follows. For each point (x, y) of $\sigma^*(a)$ belonging to $\sigma(a)$ with a larger than some sufficiently large a_0, there holds by (7.4.9) for the x-component $D_1(x, y)$ of $\vec{D}(x, y)$

$$\frac{C_5}{(x^2+y^2)^{3/2}} \leq D_1(x,y) \leq \frac{C_6}{(x^2+y^2)^{3/2}} < 0 \ , \tag{7.4.19}$$

for some constants $C_5 < 0$ and $C_6 < 0$. The inequalities (7.4.19) follow from the fact that at $\sigma^*(a)$, $x \leq -4|y|$.

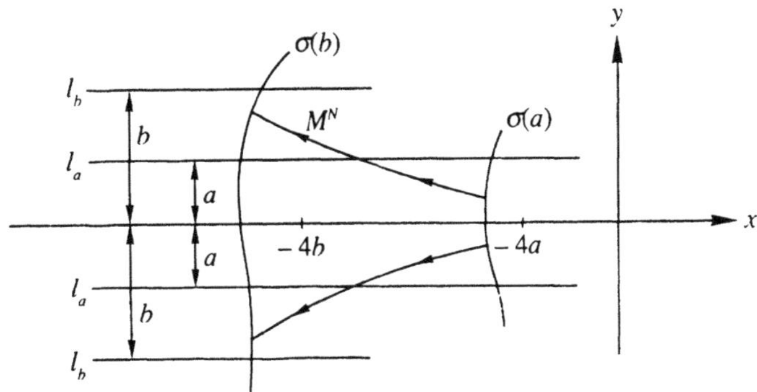

Fig. 7.4.3. Sets of points l_a and l_b , $b > a > 0$.

For any point $(x, y) \in l_a$ we have for the y-component $D_2(x, y)$ of $\vec{D}(x, y)$

$$\frac{C_7\, y}{(x^2+y^2)^2} \leqq (\geqq)\, D_2(x,y)\ , \quad y > 0\ , \quad (y < 0)\ , \tag{7.4.20}$$

for some constant $C_7 > 0$. The number a_0 being such that when we take larger values of a_0, the values of C_5, C_6 and C_7 do not change.

Now repeat the wagging motion a sufficient number N_1 of times such that the x-coordinates of

$$\left(M^{N_1}(\sigma(a))\right)^*(a) \tag{7.4.21}$$

are smaller than $-4b$. This is possible because of (7.4.19). Then repeat the wagging motion a sufficient number N_2 of times so that the points of (7.4.21) with $y = a$ and $y = -a$, pass the two components of l_b. This is possible by (7.4.20) and the first inequality of (7.4.19). Hence the curve $\sigma(a)$ is changed into a curve $\sigma(b)$ by a number $N = N_1 + N_2$ of wagging motions.

7.5. Non-Existence of Optimum Base Motion

In Figure 7.5.1 we have schematically drawn the line $L_N = (x_Q(t), y_Q(t))$ periodic in the x-direction with length period $U\tau$, along which the three quarter chord point Q of the profile moves, while the profile remains tangent to L_N. First the profile carries out one wagging motion with a "large amplitude" B_2 (7.3.5) at $x = -U\tau$, then it slides along the x-axis and in the neighbourhood of $x = 0$ it carries out N wagging motions with amplitude 1. Then a new period starts with a large amplitude wagging motion. Contrary to Figure 7.5.1, which is for stimulating the imagination, the wagging motions are assumed to happen exactly at $x = nU\tau$, $n = -1, 0, 1, 2, \ldots$

The reason for choosing the path L_N in this way is the following. In the first part of the period, for instance at $x = -U\tau$ the profile carries out the large wagging motion.

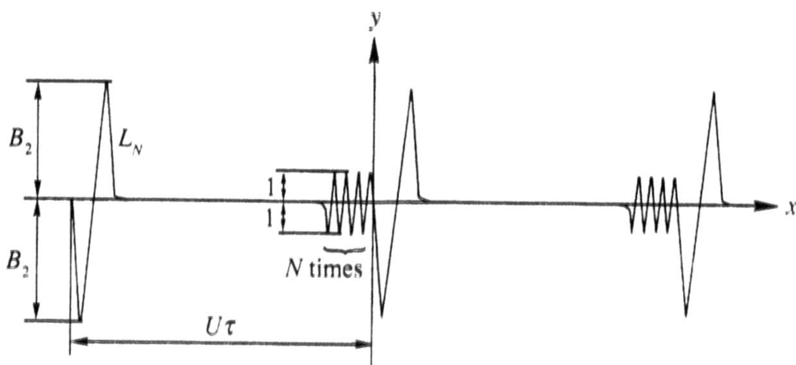

Fig. 7.5.1. Path L_n of the three quarter chord point Q.

Its wake (see definition below (4.7.13)) is a line which reaches large distances from the x-axis. In the second part of the period at $x = 0$, by the N wagging motions with amplitudes 1 this wake is 'blown up' to still further distances at both sides of the x-axis. Then the process is repeated.

We now determine the constants τ (7.3.2), B_1 and B_2 (7.3.5) and the number N of wagging motions of amplitude 1 so that we achieve our aim. B_1 can be chosen as any number larger than zero, in order to let Q follow the vertical parts of L with finite velocity. Next we choose B_2 and $U\tau$ such that the wake produced in the first part of the period is a line of type a_0, used in the last part of the previous section, with respect to the wagging motions of the second half of the period. It is easy to see that such B_2 and $U\tau$ exist. Then we consider the effect of the mth period, which starts at the position of the profile with $x_Q = (m-2)U\tau$, $y_Q = 0$. From (4.7.11) and (4.7.12) it follows that $\dot{x}$ and $\dot{y}$ are $O(\{(x-x_Q)^2+(y-y_Q)^2\}^{-1})$. Hence there exist constants $C > 0$ and $b_0 > 0$ such that, if σ is a curve of type $a > b_0$, due to the first part of the mth period σ is moved to a curve which is anyhow of type

$$a - C\{(m-2)U\tau + 4a\}^{-2} \ , \quad a > b_0 \ . \tag{7.5.1}$$

From the previous section it follows that if σ is a curve of type $a > a_0$ (a_0 as before), then σ is moved due to the second part of the mth period, to a curve which is again of type a.

Now we have to sum up the effects. Consider the curve σ_0, which is the wake produced in the first part of the first period and which is of type a_0. By the N wagging motions of the second part of the first period, this curve σ_0 is changed into σ_1, which will be of type a_1, with $a_1 \gg a_0$. Let σ_n be this curve after n periods then σ_n is a curve of type a_n with

$$a_n = a_{n-1} - C\{(n-2)U\tau + 4a_{n-1}\}^{-2} \ , \tag{7.5.2}$$

provided that $a_{n-1} > b_0$. When σ_n is of type a_n it is also of type $\tilde{a}_n$ with $\tilde{a}_n < a_n$, then the following calculation will be clear

$$a_2 = a_1 - C(4a_1)^{-2} \ , \tag{7.5.3}$$

$$a_3 = a_2 - C(U\tau + 4a_2)^{-2} \ , \quad a_3 > a_2 - C(U\tau)^{-2} \ , \tag{7.5.4}$$

$$a_4 = a_3 - C(2U\tau + 4a_3)^{-2} \ , \quad a_4 > a_2 - C\{(U\tau)^{-2} + (2U\tau)^{-2}\} \ , \tag{7.5.5}$$

$$\vdots$$

$$a_n > a_2 - C\sum_{j=1}^{n} (jU\tau)^{-2} \ , \tag{7.5.6}$$

as long as $a_n > b_0$ and $a_n > a_0$.

From (7.5.3) and (7.5.6), we find by estimating the summation in (7.5.6)

$$a_n > a_1 - C\left\{\frac{1}{(4a_1)^2} + \frac{2}{(U\tau)^2}\right\} \stackrel{\text{def}}{=} \beta \ . \tag{7.5.7}$$

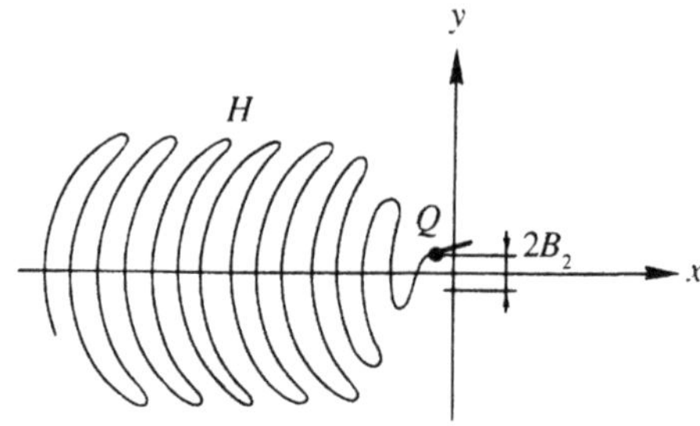

Fig. 7.5.2. Impression of a profile with an ultimate wide wake H.

Hence if we choose the number N of wagging motions of amplitude 1, in the second half of each period, so large that a_1 becomes so large that $\beta > a_0$ and $\beta > b_0$, the foregoing analysis is valid. Next if we want the wake to go out of the region $|y| \leq B$, hence to contain a curve of type B, we have to choose N so large that also $\beta > B$.

It follows from these considerations that for l, U, τ, B_1 and B_2 suitably chosen we can construct a periodic base motion of which the wake H becomes as wide as we want. Also it follows from (7.4.5) that when the profile moves on, the wake far behind it cannot become wide beyond all bounds, hence its width will tend to a finite value. Because the motion of the profile is periodic for $t > 0$, it is clear that also the wake becomes periodic with period $U\tau$ in the x-direction. Its shape will have the appearance of the wake drawn in Figure 7.5.2.

We now consider an optimum added motion of $O(\varepsilon)$ of the profile which has to be such that for a fixed line L_N (Figure 7.5.1) the mean value of the thrust becomes $\bar{T} = O(\varepsilon)$ at the cost of least energy losses. The shed free vorticity, which in this semi-linear theory is assumed to lie at the wake, follows from the shape of the ultimate wake H hence far behind the profile, in the way as we discussed explicitly in Section 6.10 be it that there we considered two profiles. Also in that section we discussed how to calculate the shed free vorticity just behind the profile and situated at the still deforming wake H^*. From this we can calculate an optimum added motion of the profile. This motion is not uniquely determined because the demanded optimum free vorticity just behind the profile can be created by infinitely many added motions.

We now consider a series of base motions determined by the lines $L_N = (x_Q(t), y_Q(t))$ with an increasing number N of wagging motions with amplitude 1 in the second half of the period. Then the periodic ultimate wakes H_N become wider and wider. By translating these surfaces H_N as rigid and impermeable surfaces in order to determine the optimum free vorticity on them, it is clear that this vorticity is concentrated relatively more and more at the outer edges of the wake. This means that over an arbitrarily wide neighbourhood of the x-axis a parallel flow by the free vorticity is induced of which the backwards directed momentum per length U in the x-direction equals the mean thrust $\bar{T}$ of the profile. Hence by increasing N the kinetic energy in the wake per unit of length tends to zero. It means that we have found a minimizing sequence of base motions for which the efficiency tends to one. Note that in Section 7.1 we constructed a minimizing sequence of added motions, the base motion in that section is purely translational.

It is possible to show that for any *fixed* given base motion the ultimate wake H of the profile will remain bounded. So this will also hold for the optimum base motion *if it exists*. However, in order to deliver a mean thrust $\bar{T}$ of $O(\varepsilon)$, it has in connection with the momentum theorem, to shed free vorticity of $O(\varepsilon)$ and hence it leaves behind kinetic energy $E_{\text{opt}} > 0$ which is of $O(\varepsilon^2)$. However, when l, U, τ, B_1 and B_2 are such that our previous theory about the creation of an arbitrary wide wake applies, we can always find a base motion L_N of our minimizing sequence which for the same $\bar{T}$ leaves behind less kinetic energy E_N

$$E_N < E_{\text{opt}} \ . \tag{7.5.8}$$

Hence we have shown that there are constraints for which no optimum base motion can be found.

Finally we return to the last three paragraphs of Section 5.8, where we made some remarks about the quality number q of a semi-linear propeller model. For the lines L_N $(N = 1, 2, \dots)$ of the minimizing sequence constructed in this section it is clear that for N larger than some value $\tilde{N}$ the quality number q, based on a working region which is the most narrow cylinder which can contain the moving profile, will become larger than 1. This is because the lost kinetic energy of the corresponding actuator surface in the linear theory has for a prescribed thrust $\bar{T}$, some value $E_{\text{ac}} > 0$, while for the same value $\bar{T}$ the lost kinetic energies E_N of the propellers belonging to L_N tend to zero for $N \to \infty$. Hence

$$q_N = E_{\text{ac}}/E_N \to \infty \ , N \to \infty \ . \tag{7.5.9}$$

7.6. Small Amplitude Heaving Motion

We consider the 2-dimensional problem of Section 4.1, Regime (1.a) for a purely heaving motion of a flat and infinitely thin rigid profile. The profile lies in the neighbourhood of the x-axis and stretches from $x = -l$ towards $x = l$, it is placed in a parallel flow of velocity U (Figure 7.6.1). Its periodic heaving motion is given by

$$y = h(t) = h(t + \tau) \ , \quad \tau = \frac{2\pi}{\omega} \ , \quad |h(t)| \le B = O(\varepsilon) \ , \tag{7.6.1}$$

where ω is the angular frequency of the motion.

In this case thrust is generated only by means of suction forces at the leading edge of the profile. Of course this type of thrust production is not very practical because it is liable to disturbances caused by flow separation at the leading edge. However, the analysis will show that when the wriggling type of motion of Figure 7.1.2 is prevented, the profile being rigid, we can prove that an optimum motion exists

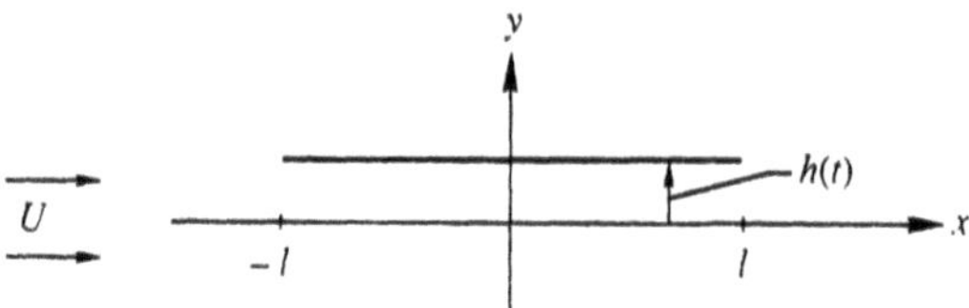

Fig. 7.6.1. Heaving motion of rigid flat profile.

(Sparenberg and Thomas [66]). The same holds for the more realistic type of motion of the profile when besides heaving also pitching is allowed. This more general and more complicated problem is discussed by Urbach in [72] and [73]. We will treat here the more simple heaving problem because it shows already to a certain extent the methods needed for the proof of the existence of an optimum motion. In Section 7.10 we will give results from [72] and [73] for the optimum motion with combined heaving and pitching.

We use the results of Sections 4.3 and 4.4. First we specialize to

$$h(x,t) = a\, e^{j\omega t} \ , \tag{7.6.2}$$

hence in formula (4.4.16) we have $h(x) = a$. Then by (4.3.12), (4.3.22) and (4.4.22)

$$V(x) = j\omega a \ , \quad \psi_1(x) = a\{-j\omega U + \omega^2(x+l)\} \ . \tag{7.6.3}$$

In the complex acceleration potential $f(z,t)$ (4.4.10) we have to substitute for A (4.4.28)

$$A = \frac{1}{\pi}\int_{-l}^{l} \frac{a\{-j\omega U + \omega^2(x+l)\}}{(l^2-\xi^2)^{1/2}}\, d\xi$$

$$+ \frac{a\, j\, \omega\, U}{\pi l}\int_{-l}^{l} \frac{1}{(l^2-\xi^2)^{1/2}}\left\{\frac{jlH_1^{(2)}(\alpha)+\xi H_0^{(2)}(\alpha)}{H_0^{(2)}(\alpha)-jH_1^{(2)}(\alpha)}\right\} d\xi \ , \tag{7.6.4}$$

where, as in Section 4.4, $\alpha = \omega l/U > 0$ is the reduced frequency. The Hankel functions we used above are related to Bessel functions by

$$H_\nu^{(2)}(x) = J_\nu(x) - jY_\nu(x) \ , \quad \nu = 0,1,2,\ldots \tag{7.6.5}$$

Hence we can split A into its real and imaginary part and write

$$(\omega a U)^{-1}A = A_1(\alpha) + iA_2(\alpha) \ , \tag{7.6.6}$$

with

$$A_1(\alpha) = -\left[\alpha + \frac{\{J_0(\alpha)J_1(\alpha)+Y_0(\alpha)Y_1(\alpha)\}}{\{J_0(\alpha)-Y_1(\alpha)\}^2+\{Y_0(\alpha)+J_1(\alpha)\}^2}\right] \ , \tag{7.6.7}$$

$$A_2(\alpha) = \frac{\{J_0^2(\alpha) + Y_0^2(\alpha) + 2/\pi\alpha\}}{\{J_0(\alpha) - Y_1(\alpha)\}^2 + \{Y_0(\alpha) + J_1(\alpha)\}^2} \ . \tag{7.6.8}$$

For the calculation of the power, which is a quadratic quantity, needed to maintain the motion we go one step further and assume $h(t)$ to be real

$$h(t) = a\cos\omega t \ , \tag{7.6.9}$$

in order to avoid the possible occurrence of $(j)^2 = -1$, which is real and hence can give wrong results. By (7.6.9) we have to take in (4.5.1) $h_1(x) = a$ and $h_2(x) = 0$. The mean value with respect to time of the power P becomes

$$P = \frac{\omega}{2\pi}\int_0^{2\pi/\omega}\left\{\int_{-l}^{l}[p]_-^+\frac{\partial h}{\partial t}\,dx\right\}dt = -a\,\omega\rho\int_{-l}^{l}\mathrm{Im}_j\varphi^+(x)\,dx \ . \tag{7.6.10}$$

After some reductions we find

$$P = \left(\frac{\pi\rho\,U^3a^2}{l}\right)\alpha^2\{1 - A_2(\alpha)\} \stackrel{\text{def}}{=} \left(\frac{\pi\rho\,U^3a^2}{l}\right)P^*(\alpha) \ , \tag{7.6.11}$$

where $P^*(\alpha)$ is dimensionless. The mean value $\bar{T}$ of the suction force at the leading edge is by (1.21.13)

$$\bar{T} = \frac{\omega}{2\pi}\int_0^{2\pi/\omega}\frac{\rho\,\pi}{4}\left\{\lim_{x\to -l}\varphi^+(t)\,(l+x)^{1/2}\right\}^2 dt \ , \tag{7.6.12}$$

where $\varphi^+(x,t) = \varphi(x,+0,t)$ is given in (4.5.4) with $h_1(x) = a$ and $h_2(x) = 0$. We obtain

$$\begin{aligned}\bar{T} &= \left(\frac{\pi\rho\,U^2a^2}{l}\right)\alpha^2\{(1 - A_2(\alpha))^2 + (\alpha + A_1(\alpha))^2\} \\ &\stackrel{\text{def}}{=} \left(\frac{\pi\rho\,U^2a^2}{l}\right)T^*(\alpha) \ ,\end{aligned} \tag{7.6.13}$$

where $T^*(\alpha)$ is dimensionless.

Note that when we had started in (7.6.9) from $h(t) = a\sin\omega t$ we had found the same mean values for the power and the thrust.

Next we consider some properties of the dimensionless $P^*(\alpha)$ and $T^*(\alpha)$. We will show that

$$P^*(\alpha) \ , \quad T^*(\alpha) \ , \quad P^*(\alpha)/T^*(\alpha) \ , \tag{7.6.14}$$

are positive and are increasing functions of α.

First we show that the clearly positive function

$$\alpha\{1 - A_2(\alpha)\} = \frac{\alpha^2\{J_1^2(\alpha) + Y_1^2(\alpha)\} + 2\alpha/\pi}{\alpha[\{J_0(\alpha) - Y_1(\alpha)\}^2 + \{Y_0(\alpha) + J_1(\alpha)\}^2]}$$

$$\stackrel{\text{def}}{=} \frac{\mathrm{I}(\alpha) + 2\alpha/\pi}{\mathrm{II}(\alpha)} , \tag{7.6.15}$$

where we introduced $\mathrm{I}(\alpha)$ and $\mathrm{II}(\alpha)$, is an increasing function of α. For the proof we use Nicholson's integral [74]

$$\{J_\nu^2(\alpha) + Y_\nu^2(\alpha)\} = \frac{8}{\pi^2} \int_0^\infty K_0(2\alpha \sinh\lambda) \cosh 2\nu\lambda \, d\lambda , \tag{7.6.16}$$

where $K_0(\alpha)$ is a modified Bessel function. Then with $\nu = 1$, the derivative of $\mathrm{I}(\alpha)$ can be written as

$$\frac{d}{d\alpha}\mathrm{I}(\alpha) = \frac{8}{\pi^2}\frac{d}{d\alpha}\alpha^2 \int_0^\infty K_0(2\alpha \sinh\lambda) \cosh 2\lambda \, d\lambda$$

$$= \frac{8\alpha}{\pi^2} \int_0^\infty K_0(2\alpha \sinh\lambda) \cosh 2\lambda$$

$$\cdot \left\{2 - \frac{1}{\cosh^2\lambda} - 2\tanh\lambda \tanh 2\lambda\right\} d\lambda . \tag{7.6.17}$$

It is easily seen that the expression between braces in the last integral of (7.6.17) is positive, hence $\mathrm{I}(\alpha)$ and by this the numerator of the right-hand side of (7.6.15) is an increasing function of α. Next we write

$$\mathrm{II}(\alpha) = \alpha\{J_0^2(\alpha) + Y_0^2(\alpha) + J_1^2(\alpha) + Y_1^2(\alpha)\} + 4\pi , \tag{7.6.18}$$

$$\frac{d}{d\alpha}\mathrm{II}(\alpha) = \frac{8}{\pi^2} \int_0^\infty K_0(2\alpha \sinh\lambda)$$

$$\cdot \left\{1 + \cosh 2\lambda - \frac{(1 + \cosh\lambda)}{\cosh^2\lambda} - 2\tanh\lambda \sinh 2\lambda\right\} d\lambda . \tag{7.6.19}$$

Now the expression between brackets in (7.6.19) is negative, hence $\mathrm{II}(\alpha)$ is a decreasing function of α. Then it follows that $\alpha\{1 - A_2(\alpha)\}$ (7.6.15) is an increasing function of α. Hence $P^*(\alpha)$ (7.6.11) is a positive and increasing function of α.

Next we show that

$$-\alpha\{\alpha + A_1(\alpha)\} = \frac{\alpha^2\{J_0(\alpha)J_1(\alpha) + Y_0(\alpha)Y_1(\alpha)\}}{\alpha[\{J_0(\alpha) - Y_1(\alpha)\}^2 + \{Y_0(\alpha) + J_1(\alpha)\}^2]}$$

$$= \frac{\mathrm{III}(\alpha)}{\mathrm{II}(\alpha)} , \tag{7.6.20}$$

where we introduced $\text{III}(\alpha)$, is positive and is an increasing function of α. That (7.6.20) is positive follows from

$$\{J_0(\alpha)J_1(\alpha) + Y_0(\alpha)Y_1(\alpha)\} = -\frac{d}{d\alpha}\{J_0^2(\alpha) + Y_0^2(\alpha)\} \ , \tag{7.6.21}$$

which by (7.6.16) turns out to be positive. Consider

$$\begin{aligned}\frac{d}{d\alpha}\text{III}(\alpha) &= -\frac{d}{d\alpha}\,\alpha^2\frac{d}{d\alpha}\{J_0^2(\alpha) + Y_0^2(\alpha)\} \\ &= \frac{8}{\pi^2}\int_0^\infty K_0(2\alpha\sinh\lambda)\left\{\frac{3\sinh^2\lambda}{\cosh^4\lambda}\right\}d\lambda \ .\end{aligned} \tag{7.6.22}$$

Because the integrand in (7.6.22) is positive we have the desired result.

Using this result and the fact that (7.6.15) is an increasing function of α we find that $T^*(\alpha)$ (7.6.13) is positive and is an increasing function of α.

Next we consider $P^*(\alpha)/T^*(\alpha)$ which is clearly positive. Instead of showing this function to be increasing with α, it is more easy to show that

$$\frac{T^*(\alpha)}{P^*(\alpha)} - 1 = \frac{T^*(\alpha) - P^*(\alpha)}{P^*(\alpha)} \tag{7.6.23}$$

is a decreasing function of α. Using the well known relation

$$J_0(\alpha)Y_1(\alpha) - J_1(\alpha)Y_0(\alpha) = -\frac{2}{\pi\alpha} \ , \tag{7.6.24}$$

we find after a straightforward calculation that (7.6.23) can be written as

$$-\frac{2}{\pi}\left[\alpha\{J_1^2(\alpha) + Y_1^2(\alpha) + \frac{2}{\pi}\}\right]^{-1} \ . \tag{7.6.25}$$

From (7.6.16) it follows in an analogous way as before that (7.6.25) is a decreasing function of α. Hence $P^*(\alpha)/T^*(\alpha)$ is increasing with α.

It follows from the asymptotic representations of Bessel functions that

$$\lim_{\alpha\to\infty}\frac{P^*(\alpha)}{\alpha^2} = \frac{1}{2} \ , \quad \lim_{\alpha\to\infty}\frac{T^*(\alpha)}{\alpha^2} = \frac{1}{4} \ , \quad \lim_{\alpha\to\infty}\frac{P^*(\alpha)}{T^*(\alpha)} = 2 \ . \tag{7.6.26}$$

7.7. The Optimization Problem

We assume without loosing generality, the time to be scaled such that the heaving motion of the profile $y = h(t)$, has period 2π and is given by

$$h(t) = \sum_{n=1}^{\infty}(a_n\sin nt + b_n\cos nt) \ ; \quad a_{2n} = b_{2n} = 0 \ , \quad n = 1,2,3,\ldots \tag{7.7.1}$$

From this representation it follows that $h(t+\pi) = -h(t)$, hence the profile carries out the same motion above and below its neutral position $y = 0$. We introduce the mean value P_n of the power and the mean value T_n of the thrust, which belong to a purely sinusoidal motion $h(t) = \sin(nt + \beta)$, where β is any phase angle. Then we can write the mean values of power and thrust belonging to a general motion of the type (7.7.1) as

$$P(h) = \sum_{n=1}^{\infty} P_n \cdot (a_n^2 + b_n^2) \ , \quad T(h) = \sum_{n=1}^{\infty} T_n \cdot (a_n^2 + b_n^2) \ . \tag{7.7.2}$$

From the previous section it follows

$$P_n > P_m \ , \quad T_n > T_m \ , \quad \frac{P_n}{T_n} > \frac{P_m}{T_m} \ ; \quad n > m \ , \tag{7.7.3}$$

$$\lim_{n\to\infty} \frac{P_n}{n^2} = \frac{\pi \rho\, U^3}{2l} > 0 \ , \quad \lim_{n\to\infty} \frac{T_n}{n^2} = \frac{\pi \rho\, U^2}{4l} > 0 \ . \tag{7.7.4}$$

The optimization problem is the following: minimize $P(h)$ under the condition that the prescribed mean value $\bar{T}$ of the thrust is generated while the amplitude of the motion is smaller than or equal to a given number B. In formula

$$\min P(h) \ ; \quad T(h) = \bar{T} \ , \quad |h(t)| \le B \ . \tag{7.7.5}$$

It follows from (7.7.4) that for each prescribed $\bar{T} > 0$ and $B > 0$ there exist motions, for instance

$$h(t) = \left(\frac{\bar{T}}{T_n}\right)^{1/2} \sin nt, \tag{7.7.6}$$

by choosing n sufficiently large for which the thrust $T(h) = \bar{T}$ and $|h(t)| \le B$. We emphasize again that the motion (7.7.6) has also period 2π.

We now establish a simple property of the power P. Consider two motions $h^{(j)}(t)$, $(j = 1, 2)$ with coefficients $a_n^{(j)}, b_n^{(j)}$, power $P^{(j)}$ and equal thrust $T^{(j)} = \bar{T}$. Suppose that the coefficients of $h^{(1)}(t)$ and $h^{(2)}(t)$ satisfy

$$(a_n^{(2)})^2 + (b_n^{(2)})^2 = (a_n^{(1)})^2 + (b_n^{(1)})^2 + \Delta_n \ , \tag{7.7.7}$$

$$\Delta_n \le 0 \ , \quad 1 \le n \le N_1 \ ; \quad \Delta_n \ge 0 \ , \quad N_1 < n \le N_2 \ , \tag{7.7.8}$$

for some numbers N_1 and N_2, while other coefficients are equal. Then because the thrust is the same

$$\sum_{n=1}^{N_1} T_n \Delta_n + \sum_{n=N_1+1}^{N_2} T_n \Delta_n = 0 \ . \tag{7.7.9}$$

For the difference in power we have

$$P^{(2)} - P^{(1)} = \sum_{n=1}^{N_1} P_n \Delta_n + \sum_{n=N_1+1}^{N_2} P_n \Delta_n \ . \tag{7.7.10}$$

Using (7.7.3) and (7.7.9) we find

$$\sum_{n=N_1+1}^{N_2} P_n \Delta_n \geq \frac{P_{N_1+1}}{T_{N_1+1}} \sum_{n=N_1+1}^{N_2} T_n \Delta_n = -\frac{P_{N_1+1}}{T_{N_1+1}} \sum_{n=1}^{N_1} T_n \Delta_n \ . \tag{7.7.11}$$

Hence we have shown, under the assumptions (7.7.7) and (7.7.8)

$$P^{(2)} - P^{(1)} \geq \sum_{n=1}^{N_1} T_n \left(\frac{P_n}{T_n} - \frac{P_{N_1+1}}{T_{N_1+1}} \right) \Delta_n \geq 0 \ , \tag{7.7.12}$$

where we have strict inequalities if there are $\Delta_n \neq 0$.

Stated somewhat loosely, *admitted propulsive motions yielding the prescribed thrust $\bar{T}$, need less power according to their dominant frequencies being lower.*

Now we consider the simple case that the bound B (7.7.5) on the amplitude of the motion is sufficiently large so that we can generate the prescribed thrust $T(h) = \bar{T}$ by means of the lowest frequency ($n = 1$)

$$h(t) = (a_1 \sin t + b_1 \cos t) \ , \quad (a_1^2 + b_1^2)^{1/2} = \left(\frac{\bar{T}}{T_1} \right)^{1/2} \leq B \ . \tag{7.7.13}$$

Then by our previous property of P this is an optimum motion.

If now we vary the prescribed thrust $\bar{T}$ such that

$$0 \leq \left(\frac{\bar{T}}{T_1} \right)^{1/2} \leq B \ , \tag{7.7.14}$$

then for this range all motions with the lowest frequency are allowed and are optimum and by (7.6.11) and (7.6.13) their efficiency

$$\eta = \frac{\bar{T} U}{P} = \text{const.} \tag{7.7.15}$$

When we have

$$B < \left(\frac{\bar{T}}{T_1} \right)^{1/2} \ , \tag{7.7.16}$$

then in order to generate the desired thrust $\bar{T}$, higher frequencies have to occur in the motion. It is clear from the property of P, that in this case the optimum motion of the profile, if it exists, has to touch its boundaries at least one time per period at $y = B$ and $y = -B$. Otherwise it is possible to enlarge amplitudes of lower frequencies and to reduce amplitudes of higher frequencies without changing $T(h)$. Then as we have shown, P decreases and hence our motion was not optimum.

In the next section we prove that also in the case (7.7.16) an optimum motion exists.

7.8. Existence of Optimum Added Motion

In the two previous cases of this chapter, non-existence could be carried out by the construction of a minimizing sequence of propellers of which the efficiency tends to one. These sequences could be found by using classical analysis. However, when we try to prove the existence of an optimum propeller it seems not well possible to develop a convincing theory without the use of the more abstract methods of functional analysis. For general considerations on this subject we refer to [46] and [69].

We introduce the Hilbert space S of real valued and absolutely continuous and periodic functions on the interval $0 \leq t \leq 2\pi$ by

$$S = \left\{ h : [0, 2\pi] \to \mathbb{R} \ ; \frac{dh}{dt} \in L^2 \ ; h(t) \equiv -h(t+\pi) \ , 0 \leq t \leq \pi \right\} \ , \tag{7.8.1}$$

where $\mathbb{R}$ represents the real numbers and L^2 is the space of quadratic integrable functions in the sense of Lebesque. The inner product of two elements $h^{(1)}$ and $h^{(2)}$ of S is defined by

$$(h^{(1)}, h^{(2)}) = \int_0^{2\pi} \frac{d}{dt} h^{(1)}(t) \frac{d}{dt} h^{(2)}(t) \, dt \ . \tag{7.8.2}$$

Hence the norm in this space is defined by

$$\|h\|^2 = \int_0^{2\pi} \left\{ \frac{d}{dt} h(t) \right\}^2 \, dt \ . \tag{7.8.3}$$

When $\|h\| = 0$ it follows that $h(t) =$ const., however, because of the demand $h(t) = -h(t+\pi)$ we find $h(t) \equiv 0$. Substitution of the representation (7.7.1) in (7.8.2) and in (7.8.3) yields

$$(h^{(1)}, h^{(2)}) = \pi \sum_{n=1}^{\infty} n^2 (a_n^{(1)} a_n^{(2)} + b_n^{(1)} b_n^{(2)}) \ , \tag{7.8.4}$$

$$\|h^2\| = \pi \sum_{n=1}^{\infty} n^2 (a_n^2 + b_n^2) \ . \tag{7.8.5}$$

Because $h(t)$ is a continuous function and $h(t) = -h(t+\pi)$, there is at least one point $t = t_0$ such that $h(t_0) = 0$, hence

$$h(t) = \int_{t_0}^{t} \frac{d}{d\xi} h(\xi) \, d\xi \ . \tag{7.8.6}$$

Using Schwarz' inequality we find

$$|h(t)|^2 \leq (t - t_0) \int_{t_0}^{t} \left\{ \frac{d}{d\xi} h(\xi) \right\}^2 d\xi \leq 2\pi \|h\|^2 \ . \tag{7.8.7}$$

It follows from this that the maximum norm

$$\|h\|_\infty = \max |h(t)| \leq (2\pi)^{1/2} \|h\| \ , \quad 0 \leq t \leq 2\pi \ , \tag{7.8.8}$$

is a continuous function on S. This means that when we have a sequence $\{h^{(n)}\}$ (n = 1, 2, ...) which converges to an element h, in other words $\|h^{(n)} - h\| \to 0$ for $n \to \infty$, then it follows that $\|h^{(n)}\|_\infty \to \|h\|_\infty$. Indeed by (7.8.7)

$$\|h^{(n)}\|_\infty = \|h^{(n)} - h + h\|_\infty \leq (2\pi)^{1/2} \|h^{(n)} - h\| + \|h\|_\infty \ , \tag{7.8.9}$$

$$\lim_{n \to \infty} \|h^{(n)}\|_\infty \leq \|h\|_\infty \ . \tag{7.8.10}$$

Then by interchanging $h^{(n)}$ and h we find the desired result.

The convergence $h^{(n)} \to h$ for which holds $\|h - h^{(n)}\| \to 0$ is called "strong" convergence. In the following we also need the concept "weak" convergence in our Hilbert space S. This is denoted by $h^{(n)} \rightharpoonup h$ and is defined as

$$h^{(n)} \rightharpoonup h : \lim_{n \to \infty} (h^{(n)}, \tilde{h}) = (h, \tilde{h}) \ , \quad \text{for all } \tilde{h} \in S \ . \tag{7.8.11}$$

It is clear that from the strong convergence of $\{h^{(n)}\}$ to h follows the weak convergence to the same element, because $|((h^{(n)} - h), \tilde{h})| \leq \|h^{(n)} - h\| \cdot \|\tilde{h}\|$. There are, however, sequences which converge weakly but not strongly.

On S we have two functionals (7.7.2), one representing the mean value $P(h)$ of the power and the other the mean value $T(h)$ of the thrust. From (7.7.4) it follows that there exist two constants C_1 and C_2 such that

$$C_1 \|h\|^2 \leq P(h) \leq C_2 \|h\|^2 \ . \tag{7.8.12}$$

For the constants we can take (7.7.2)

$$C_1 = \inf_n \left(\frac{P_n}{n^2} \right) \ , \quad C_2 = \sup_n \left(\frac{P_n}{n^2} \right) \ . \tag{7.8.13}$$

It follows from (7.8.12) that $P(h)$ is a continuous functional on S and because of the two inequalities $\{P(h)\}^{1/2}$ can be considered as an equivalent norm on S. This norm is related to the inner product

$$(h^{(1)}, h^{(2)})_P = \sum_{n=1}^{\infty} P_n \cdot (a_n^{(1)} a_n^{(2)} + b_n^{(1)} b_n^{(2)}) \ . \tag{7.8.14}$$

For $T(h)$ we have analogous formulas as for $P(h)$.

We introduce the set $G \subset S$ by

$$G = \{h \in S : T(h) \leq \bar{T} \ ; \|h\|_\infty \leq B\} \ . \tag{7.8.15}$$

This set is bounded because $\{T(h)\}^{1/2}$ just as $\{P(h)\}^{1/2}$, is an equivalent norm. Because both $T(h)$ and $\|h\|_\infty$ are continuous on S, the set G is closed, which can be seen as follows. Suppose $\{h^{(n)}\} \subset G$, $(n \geq 1)$ and $\|h^{(n)} - h\| \rightarrow 0$ for $n \rightarrow \infty$, then we have

$$\begin{aligned} \{T(h)\}^{1/2} &\leq \{T(h - h^{(n)})\}^{1/2} + \{T(h^{(n)})\}^{1/2} \\ &\leq C_2^{1/2} \|h - h^{(n)}\| + \bar{T}^{1/2} \ , \end{aligned} \tag{7.8.16}$$

hence $T(h) \leq \bar{T}$. In the same way we can show $\|h\|_\infty \leq B$. Then by definition (7.8.15) $h \in G$, hence G is closed.

G is a convex set, which means that for $h^{(1)} \in G$ and $h^{(2)} \in G$ also

$$h^{(3)} = (\nu h^{(1)} + (1 - \nu) h^{(2)}) \in G \ , \quad 0 \leq \nu \leq 1 \ . \tag{7.8.17}$$

This is correct because

$$\begin{aligned} \{T(h^{(3)})\}^{1/2} &= \{T(\nu h^{(1)} + (1 - \nu) h^{(2)})\}^{1/2} \\ &\leq \nu \{T(h^{(1)})\}^{1/2} + (1 - \nu) \{T(h^{(2)})\}^{1/2} \\ &\leq \bar{T}^{1/2} \ , \end{aligned} \tag{7.8.18}$$

hence $T(h^{(3)}) \leq \bar{T}$. In the same way we can prove $|h^{(3)}(t)| \leq B$ or $\|h^{(3)}\|_\infty \leq B$. It follows that $h^{(3)} \in G$ and G is convex.

Summarizing, the set G of the Hilbert space S is bounded, closed and convex. Then it is a standard result in functional analysis that G is weakly compact. In our case this implies that any sequence $\{h^{(n)}\}$ in G has a subsequence which converges weakly to an element of G.

In the following we need the notion of an extreme element of a set. An extreme element $h^{(1)}$ of a set G is defined by the property; when $h^{(2)}$ and $h^{(3)} \in G$ are such that

$$h^{(1)} = \tfrac{1}{2} \, (h^{(2)} + h^{(3)}) \tag{7.8.19}$$

then $h^{(1)} = h^{(2)} = h^{(3)}$. The set of extreme elements of G is denoted by $E(G)$.

We now consider the set $G_1 \subset G$

$$G_1 = \{h \in S : T(h) = \bar{T} \ , \|h\|_\infty \leq B\} \tag{7.8.20}$$

and show first that each extreme element of G belongs to G_1, or $E(G) \subset G_1$.

Suppose $h \in E(G)$ and $T(h) < \bar{T}$, otherwise $h \in G_1$ already. Because there exists a t_0 with $h(t_0) = 0$, there exists an interval with $|h(t)| < B$. Hence we can find a smooth function $\tilde{h} \in S$ with $\|\tilde{h}\| \neq 0$, such that

$$\|h \pm \tilde{h}\|_\infty \leq B \ , \quad T(h \pm \tilde{h}) < \bar{T} \ , \tag{7.8.21}$$

the latter inequality is possible because T is a continuous functional and $\|\tilde{h}\|$ can be chosen arbitrarily small. Then both functions $(h + \tilde{h})$ and $(h - \tilde{h}) \in G$ and $h + \tilde{h} \neq h - \tilde{h}$. However

$$h = \tfrac{1}{2}\{(h + \tilde{h}) + (h - \tilde{h})\} \ , \tag{7.8.22}$$

hence $h \notin E(G)$ and we arrive at a contradiction. Hence $T(h) = \bar{T}$ for any $h \in E(G)$ or $E(G) \subset G_1$.

Inversely, each element of G_1 is extreme in G, because each element $h \in G$ with $T(h) = \bar{T}$ is even an extreme element of the sphere S_T with

$$S_T = \{h \in S : T(h) \leq \bar{T}\} \ , \tag{7.8.23}$$

of which G is a subset. That such a h (with $T(h) = \bar{T}$) is an extreme element of S_T follows directly by the application of the parallelogram law

$$\|h^{(1)} + h^{(2)}\|_T^2 + \|h^{(1)} - h^{(2)}\|_T^2 = 2\|h^{(1)}\|_T^2 + 2\|h^{(2)}\|_T^2 \ , \tag{7.8.24}$$

for the Hilbert space but now with norm $\|h\|_T = \{T(h)\}^{1/2}$. Summing up we have found

$$G_1 = E(G) \ . \tag{7.8.25}$$

We now write

$$\lambda_n = \left(\lim_{m\to\infty} \frac{P_m}{T_m}\right) - \frac{P_n}{T_n} = C - \frac{P_n}{T_n} \ , \tag{7.8.26}$$

where by (7.6.11), (7.6.13) and (7.6.26), $C = 2U$. It follows from the monotony properties (7.6.14) that

$$\lambda_n > 0 \ , \quad \lim_{n\to 0} \lambda_n = 0 \ . \tag{7.8.27}$$

Next we introduce the operator K and the functional F by

$$\begin{aligned} P(h) &= CT(h) - \sum_{n=1}^{\infty} T_n \lambda_n (a_n^2 + b_n^2) = CT(h) - (Kh, h) \\ &= CT(h) - F(h) \ , \end{aligned} \tag{7.8.28}$$

where

$$Kh = \sum_{n=1}^{\infty} \mu_n(a_n \sin nt + b_n \cos nt) \ , \quad \mu_n = \frac{T_n \lambda_n}{\pi n^2} > 0 \ , \tag{7.8.29}$$

and $\lim_{n\to\infty} \mu_n = 0$.

We will show that K is a compact operator and take as a definition of such an operator: an operator is called compact when it transforms each bounded sequence $\{h^{(n)}\}$, hence with $\|h^{(n)}\| \leq$ const., into a sequence which has a converging subsequence.

We use the following property of compact operators. Suppose we have in S a sequence of compact operators $\{K_N\}$, $N = 1, 2, \ldots$, which converge in the operator norm $\| \ \|_O$ to an operator K, hence $\|K_N - K\|_O \to 0$ for $N \to \infty$, then K is also compact. The operator norm of a linear operator K is the smallest number $\|K\|_O$ for which

$$\|Kh\| \leq \|K\|_O \|h\| \ , \quad \text{for all } h \in S \ . \tag{7.8.30}$$

We choose in this case the sequence of compact operators $\{K_N\}$ defined by

$$K_N h = \sum_{n=1}^{N} \mu_n(a_n \sin nt + b_n \cos nt) \ . \tag{7.8.31}$$

Consider a bounded sequence $\{h^{(j)}\}$

$$h^{(j)} = \sum_{n=1}^{\infty} (a_n^{(j)} \sin nt + b_n^{(j)} \cos nt) \ ,$$

$$\|h^{(j)}\| \leq \text{ const.} \ , \quad j = 1, 2, \ldots \ , \tag{7.8.32}$$

then there exists a constant C_3 such that $a_n^{(j)} \leq C_3$ and $b_n^{(j)} \leq C_3$ for all n and j. For a fixed chosen N we consider the sequence

$$\{K_N h^{(j)}\} \ , \quad j = 1, 2, \ldots \tag{7.8.33}$$

Because each of its elements has only a finite number of N components unequal to zero it is easy to construct a convergent subsequence of (7.8.33). Hence K_N is compact for each N.

Now we consider

$$\|(K - K_N)h\| = \left\| \sum_{n=N+1}^{\infty} \mu_n(a_n \sin nt + b_n \cos nt) \right\|$$
$$\leq C(N)\|h\| \to \|K - K_N\|_O \leq C(N) \ , \tag{7.8.34}$$

where $C(N) \to 0$ for $N \to \infty$ because $\mu_{N+1} \to 0$. Then it follows

$$\lim_{N\to\infty} \|K - K_N\|_O = 0 \ . \tag{7.8.35}$$

Hence K is the limit in the operator norm, of the compact operators K_N and hence K is also compact.

We next show that the functional F is a convex functional. This means that we have to prove that

$$F(\nu h + (1-\nu)\tilde{h}) \leq \nu F(h) + (1-\nu)F(\tilde{h}) \ , \quad 0 \leq \nu \leq 1 \ , \tag{7.8.36}$$

or by the definition (7.8.28) of $F(h)$ that

$$\Big(K(\nu h + (1-\nu)\tilde{h}), (\nu h + (1-\nu)\tilde{h})\Big)$$
$$\leq \nu(Kh, h) + (1-\nu)(K\tilde{h}, \tilde{h}) \ , \quad 0 \leq \nu \leq 1 \ . \tag{7.8.37}$$

Because (7.8.31) $h = \sum_{n=1}^{\infty} h_n$, $h_n = (a_n \sin nt + b_n \cos nt)$, we will prove (7.8.37) for each term separately. Then by (7.8.31) we have to show

$$\Big((\nu h_n + (1-\nu)\tilde{h}_n) \ , (\nu h_n + (1-\nu)\tilde{h}_n)\Big)$$
$$\leq \nu(h_n, h_n) + (1-\nu)(\tilde{h}_n, \tilde{h}_n) \tag{7.8.38}$$

or after some simple reductions

$$-((h_n - \tilde{h}_n) \ , (h_n - \tilde{h}_n)) \leq 0 \ , \tag{7.8.39}$$

which is obviously true. So F is a convex functional.

Finally we show that F is weakly continuous or that

$$F(h^{(n)}) \to F(h) \quad \text{for} \quad h^{(n)} \rightharpoonup h \ . \tag{7.8.40}$$

Because K is a compact operator, the sequence $\{Kh^{(n)}\}$ converges strongly to $K(h)$ ([46], p. 146),

$$\|Kh^{(n)} - Kh\| \to 0 \quad \text{for} \quad h^{(n)} \rightharpoonup h \ . \tag{7.8.41}$$

However, then

$$|F(h^{(n)}) - F(h)|$$
$$\leq |(Kh^{(n)}, h^{(n)}) - (Kh, h^{(n)})| + |(Kh, h^{(n)}) - (Kh, h)|$$
$$= |(K(h^{(n)} - h), h^{(n)})| + |(Kh, (h^{(n)} - h))|$$
$$\leq \|K(h^{(n)} - h)\| \cdot \|h^{(n)}\| + |(Kh, (h^{(n)} - h))| \ . \tag{7.8.42}$$

The last but one term of (7.8.42) tends to zero because K is compact, and by this transforms the weakly convergent series $h^{(n)}$ into a strongly convergent one [46]. The last term of (7.8.42) tends to zero because $h^{(n)}$ converges weakly to h. Hence we find from (7.8.42)

$$|F(h^{(n)}) - F(h)| \to 0 \quad \text{for} \quad h^{(n)} \rightharpoonup h \tag{7.8.43}$$

by which F is weakly continuous.

Summarizing we have found the results:

- G is a convex set and weakly compact,
- $G_1 = E(G)$, the extremal points of G,
- F is a convex functional and weakly continuous.

Now we can apply Bauer's maximum principle ([11], p. 102) which states that a continuous convex functional defined on a compact and convex set assumes its maximum in at least one of the extreme points of the set. This holds for any locally convex topology hence also for the weak topology used above.

We have to minimize $P(h)$ (7.8.28)

$$P(h) = CT(h) - F(h) \ , \quad h \in G \ , \tag{7.8.44}$$

under the condition $T(h) = \bar{T}$, hence in fact we have to take $h \in G_1 = E(G) \subset G$. However, by Bauer's maximum principal and the results for G, G_1 and F we know that F assumes its maximum on G_1, hence

$$\min P(h) = C\bar{T} - \max F(h) \ , \quad h \in G_1 = E(G) \ . \tag{7.8.45}$$

So we know that a motion exists which satisfies the minimization problem as stated in (7.7.5).

7.9. Numerical Results for Optimum Heaving Motion

First we discuss by dimension analysis, on which dimensionless parameters the optimum heaving motion and its efficiency depend. We start with the optimum motion, which can be written as

$$h(t) = f_1(\bar{T}, U, \rho, B, l, \omega, t) \ , \quad |h(t)| \leq B \ , \tag{7.9.1}$$

where the mean thrust $\bar{T}$ is a force per unit of length. We find for the dimensions of the quantities in (7.9.1)

$$[h] = [l] \ , \quad [\bar{T}] = [m][t]^{-2} \ , \quad [U] = [l][t]^{-1} \ , \quad [\rho] = [m][l]^{-3} \ ,$$
$$[B] = [l] \ , \quad [l] = [l] \ , \quad [\omega] = [t]^{-1} \ , \quad [t] = [t] \ , \tag{7.9.2}$$

where some ambiguity exists in connection with the notation of l and t, which however is harmless. Then by Appendix E, it follows from (7.9.1) and (7.9.2) that we have to consider

$$[l] = ([m][t]^{-2})^{\beta_1} \cdot ([l][t]^{-1})^{\beta_2} \cdot ([m][l]^{-3})^{\beta_3} \cdot [l]^{\beta_4} \cdot [l]^{\beta_5} \cdot [t]^{-\beta_6} \cdot [t]^{\beta_7} . \tag{7.9.3}$$

Comparing both sides of (7.9.3) we find the following equations for $\beta_1, \ldots, \beta_7$

$$1 = \beta_2 - 3\beta_3 + \beta_4 + \beta_5 \ , \quad 0 = \beta_1 + \beta_2 \ ,$$
$$0 = -2\beta_1 - \beta_2 - \beta_6 + \beta_7 \ . \tag{7.9.4}$$

Solving (7.9.4) for β_1, β_2 and β_4 we find

$$h(t) = B\left(\frac{\rho U^2 B}{\bar{T}}\right)^{\beta_3} \cdot \left(\frac{l}{B}\right)^{\beta_5} \cdot \left(\frac{B\omega}{U}\right)^{\beta_6} \cdot \left(\frac{Ut}{B}\right)^{\beta_7} . \tag{7.9.5}$$

From this it follows after suitably combining the above quantities and because β_3, β_5, β_6 and β_7 are arbitrarily

$$\frac{h(t)}{l} = f_1\left(\frac{\bar{T}}{\rho U^2 l}, \frac{B}{l}, \frac{l\omega}{U}, \omega t\right) . \tag{7.9.6}$$

Hence the shape of the oscillating motion is determined by

$$\frac{\bar{T}}{\rho U^2 l}, \frac{B}{l}, \frac{l\omega}{U} = \alpha . \tag{7.9.7}$$

In the same way it can be shown that the dimensionless efficiency η can be written as

$$\eta = f_2\left(\frac{\bar{T}}{\rho U^2 l}, \frac{B}{l}, \frac{l\omega}{U}\right) , \tag{7.9.8}$$

and hence is determined also by the three quantities of (7.9.7).

In [72] and [73] is discussed the problem of the optimization of the heaving motion as a special case of the more general problem of the existence of an optimum heaving and pitching profile motion. Some numerical results are calculated by means of a method described in [18]. The figures are based on

$$\alpha = \frac{l\omega}{U} = 0.3 , \quad \frac{B}{l} = 0.35 , \tag{7.9.9}$$

hence by (7.9.7) we have still one free parameter, namely the dimensionless thrust $C_T = \bar{T}/\rho U^2 l$.

The mean thrust C_T for a purely sinusoidal motion can be calculated by (7.6.12) for the value $\alpha = 0.3$

$$C_T = 0.0164 . \tag{7.9.10}$$

Hence when we prescribe $0 \leq C_T \leq 0.0164$, the optimum motion consists of the lowest harmonic, see below (7.7.14). For this interval, the efficiency η is constant (7.7.15) and turns out to be 0.712. When we prescribe $C_T > 0.0164$, higher harmonics of the motion have to appear. For $0.0164 < C_T < 0.0550$ it is found numerically that the first harmonic is dominant, while for some range with $C_T > 0.0550$ the second harmonic is dominant. These phenomena are shown in Figure 7.9.1. For sufficiently large values of C_T, third or higher harmonics dominate.

At the transition value $C_T = 0.0550$, within the accuracy of the calculations, two different optimum motions are possible. One with a dominant first harmonic and one with a dominant second harmonic, both with the "same" efficiency. It is interesting

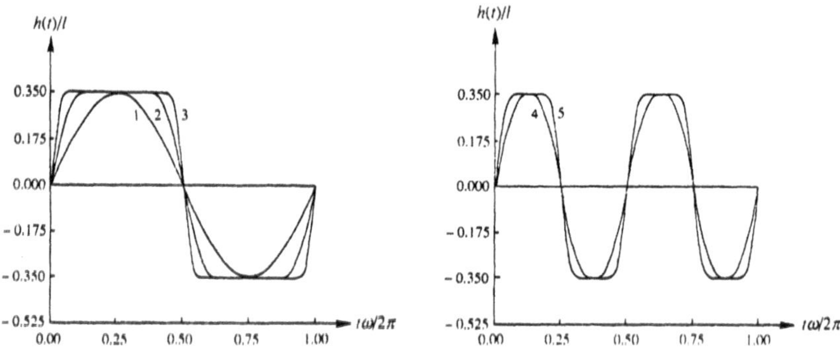

Fig. 7.9.1. The shape of $h(t)/l$ for $\alpha = l\omega/U = 0.3$, $B/l = 0.35$: (1) $C_T = 0.0164$, (2) $C_T = 0.0314$, (3) and (4) $C_T = 0.0550$, (5) $C_T = 0.0965$.

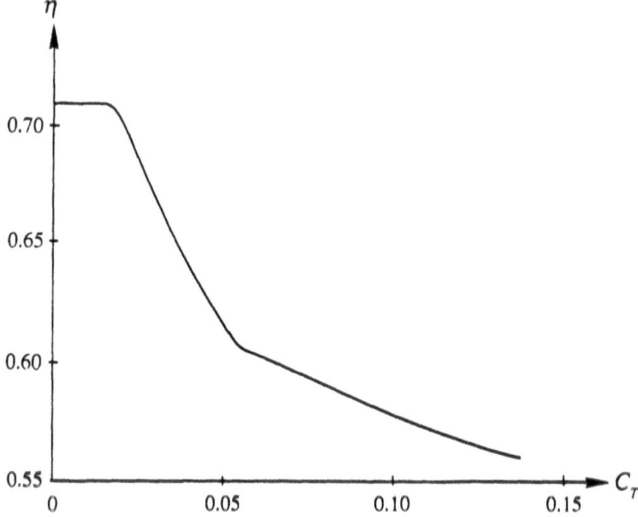

Fig. 7.9.2. Efficiency η of the heaving motion, $\alpha = l\omega/U = 0.3$, $B/l = 0.35$.

to see that for C_T slightly smaller than 0.0550, the profile is at rest most of the time at one of the two extreme positions. Still thrust can be produced in these positions because of the velocities induced by the shed free vorticity.

In Figure 7.9.2 we have drawn the efficiency η of the propulsion. As we mentioned already for $0 \leq C_T \leq 0.0164$ the efficiency is constant, $\eta = 0.712$, because the profile oscillates with its lowest harmonic and can adapt its amplitude to the demanded thrust. For $C_T > 0.0164$ the efficiency decreases smoothly with increasing C_T until this quantity becomes 0.0550. After this value a dominant second harmonic occurs and the efficiency curve obtains a different slope.

7.10. Results about Optimum Heaving and Pitching Motion

We will give now a number of results for a flat profile of zero thickness which carries out an optimum heaving and pitching motion (Figure 7.10.1). This problem is a generalization of the heaving motion discussed in Sections 7.6–7.9 and belongs also to Regime (1.a) of Section 4.1. For an exhaustive treatment of the elegant functional analytic theory we refer again to [72] and [73].

The profile stretches from $x = -l$ to $x = l$, its midpoint ($x = 0$) moves along the y-axis and carries out the motion $y = h(t)$. The pitching motion is given by $\beta(t)$. The whole motion has the angular frequency ω. The pivotal point Q, where the profile is thought to be controlled by means of forces and moments, has a fixed position $x = x_Q$ with respect to the profile.

The motion $g(t) = h(t) + x_Q\beta(t)$ of the pivotal point has to satisfy

$$|g(t)| = |h(t) + x_Q\beta(t)| \leq B \ , \tag{7.10.1}$$

where $B = O(\varepsilon)$, which means that the boundary conditions are satisfied at $-l < x < l, y = 0$. In the following we admit also symbolically the value $B = \infty$, or in words, no constraint is put on the amplitude. We denote the demanded mean value of the thrust by $\bar{T}$ and the mean value of that part of the thrust delivered by the suction forces at the leading edge $x = -l$, by $\bar{T}_s$. Then a second inequality constraint is introduced with respect to $\bar{T}_s$, namely

$$\bar{T}_s \leq r\bar{T} \ , \quad 0 \leq r < \infty \ , \tag{7.10.2}$$

for some chosen fixed value of r. No constraint is put on the pitching angle $\beta(t)$.

Using these assumptions and constraints it is proved in [72] and [73]:

(i) For every

$$0 < \omega < \infty \ , \quad 0 \leq \bar{T} < \infty \ , \quad 0 \leq r < \infty \ , \quad 0 < B \leq \infty \ , \tag{7.10.3}$$

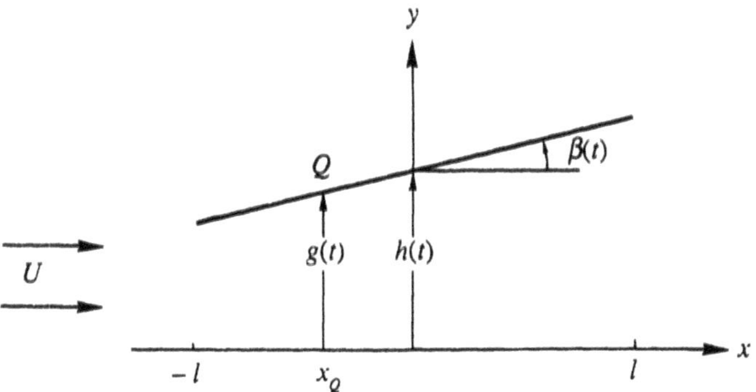

Fig. 7.10.1. Heaving ($h(t)$) and pitching ($\beta(t)$) motion of flat profile, $|h(t) + x_Q\beta(t)| \leq B$.

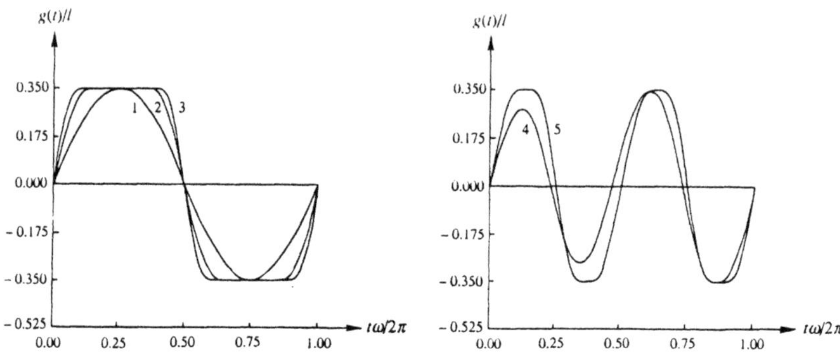

Fig. 7.10.2. The shape of $g(t)/l$ for $x_Q = -l/2$, $\alpha = l\omega/U = \pi/3$, $B/l = 0.35$, $r = 0.4$. (1) $C_T = 0.052$, (2) $C_T = 0.157$, (3) and (4) $C_T = 0.277$, (5) $C_T = 0.471$.

and an arbitrary position of Q, there exists an optimum periodic motion of the profile.

(ii) The case that $B = \infty$, hence when there is no constraint on the amplitude of Q, is special because in this case the optimum motion can be determined analytically. Then $h(t)$ and $\beta(t)$ are harmonic functions with the lowest permitted angular frequency ω. Also in this case of heaving and pitching, the efficiency is independent of the desired mean thrust $\bar{T}$, when the other parameters are kept constant.

(iii) When $0 < B \leq \infty$ and $r = 0$, the suction force has to be identically equal to zero because it is a positive quantity. Then every periodic solution which satisfies the other constraints has efficiency

$$\eta = \tfrac{1}{2} \ . \tag{7.10.4}$$

For $r > 1$ it follows by (7.10.2) that $\bar{T}_s$ is allowed to be larger than $\bar{T}$. If this happens, the pressure jumps over the profile have to cause a negative thrust in order to obtain the desired mean thrust $\bar{T}$. In case (ii), hence for the unconstrained amplitude of the pivotal point Q, it is proved that the efficiency is an increasing function of r. This means that for $B = \infty$ the mean suction force will obtain its maximum allowed value in the case of the optimum motion and for $r > 1$ the above mentioned negative thrust of the pressure jumps will occur. It follows that for $B = \infty$, no optimum motion will exist when the constraint (7.10.2) is omitted.

In [73] numerical calculations have been carried out for optimum heaving and pitching motions. For these the pivotal point Q is the 1/4-chord point ($x_Q = -l/2$) and the following values are taken

$$\alpha = \frac{\omega l}{U} = \frac{\pi}{3} \ , \quad \frac{B}{l} = 0.35 \ , \quad r = 0.4 \ . \tag{7.10.5}$$

It was found that for the considered values of $C_T = \bar{T}/\rho \, U^2 l$ also in this case the maximum allowed mean value of the suction force $\bar{T}_s = 0.4\bar{T}$ occurred.

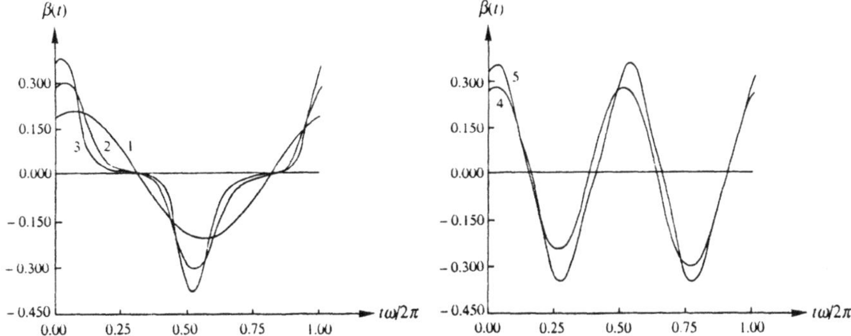

Fig. 7.10.3. The shape of $\beta(t)$, for same values of x_Q, α, B/l, r and C_T as in Figure 7.10.2.

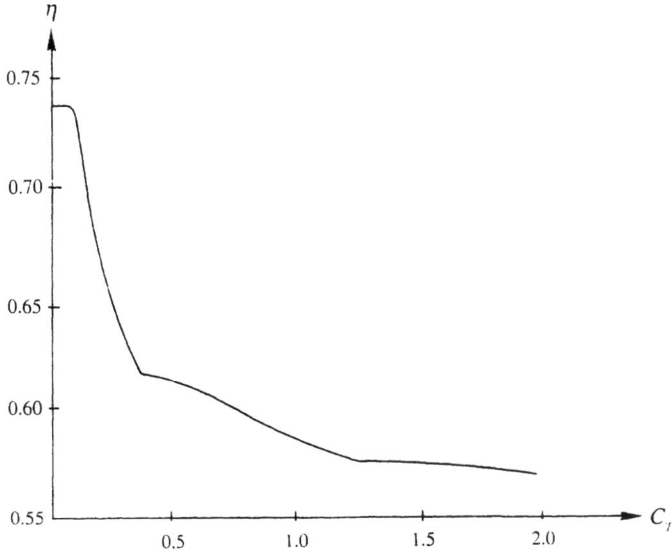

Fig. 7.10.4. Efficiency η of the optimum heaving and pitching motion for same values of x_Q, α, B/l and r as in Figure 7.10.2.

In Figure 7.10.2 the motion $y = g(t)$ of the pivotal point Q is given and in Figure 7.10.3 the pitching motion $\beta(t)$. When $C_T < 0.052$ the constraint $B/l = 0.35$ (7.10.1) is not perceived by the motion of Q and the optimum motion consists only of the lowest harmonic of $h(t)$ and of $\beta(t)$. For $0.052 < C_T < 0.277$ the lowest harmonic prevails and for some range with $0.277 < C_T$ the second harmonic prevails. For larger values of C_T higher harmonics become dominant.

In Figure 7.10.4 the efficiency is drawn, which for $C_T < 0.052$ is constant, because then only the lowest harmonic occurs. After this value the efficiency decreases smoothly with increasing values of C_T until 0.277 where a dominant second harmonic occurs and the efficiency curve obtains a different slope.

We remark that also other constraints can be chosen, for instance instead of (7.10.2) a constraint on $\beta(t)$ such as

$$|\beta(t)| \leq \text{const.} \tag{7.10.6}$$

But then the existence of an optimum motion has to be proved again, which has not been done up to now. Also it is possible to add (7.10.6) to the two constraints (7.10.1) and (7.10.2) by which different optimum motions can occur. In this way many other optimization problems can be posed.

Appendices

A. The Hilbert Problem

The formulas of Plemelj, the Hilbert problem and the singular integral equation are thoroughly discussed in [52]. For direct reference, however, we sketch the theory to an extent necessary for the understanding of the applications.

A.1. The Formulas of Plemelj

Let L be a smooth arc in the complex plane (Figure A.1.1), defined by

$$z = z(s) = x(s) + iy(s)\ , \quad s_a \le s \le s_b\ , \quad z(s_a) = a\ , \quad z(s_b) = b\ , \tag{A.1.1}$$

where s is a real parameter and $x(s)$ and $y(s)$ are real functions which have continuous first order derivatives which do not vanish simultaneously. Also we assume L to be simple, this means that $(x(s_1), y(s_1)) \neq (x(s_2), y(s_2))$ for $s_1 \neq s_2$.

When t denotes a point of L we consider a function $\varphi = \varphi(t)$ which satisfies the Hölder condition

$$|\varphi(t_2) - \varphi(t_1)| \le A\,|t_2 - t_1|^{\mu}\ , \tag{A.1.2}$$

where A and μ are positive constants. Then we form the Cauchy integral

$$\Phi(z) = \frac{1}{2\pi i}\int_L \frac{\varphi(t)}{(t-z)}\,dt\ , \quad z \notin L\ . \tag{A.1.3}$$

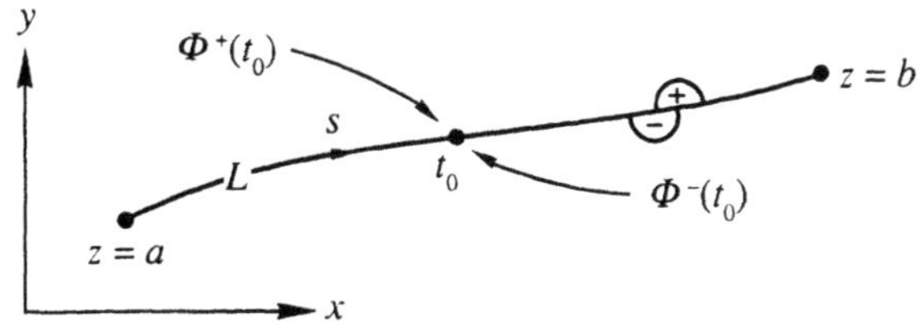

Fig. A.1.1. Arc of discontinuity for a sectionally holomorphic function.

Fig. A.1.2. Deformation of L when $\varphi(z)$ is analytic.

The resulting function $\Phi(z)$ is analytic in the entire complex plane minus the arc L. Such a function is called a sectionally holomorphic function. At L the values of $\Phi(z)$ exhibit a jump by passing from one side of L to the other. The left-hand side of L, with respect to the positive direction of s, is the + side and the right-hand side is the − side. The limiting values of $\Phi(z)$ are continuous up to L, possibly with the exception of the endpoints. In general we consider limiting values of Φ at the open arc hence at L without endpoints. The limiting values satisfy the following relations of Plemelj

$$\Phi^+(t_0) - \Phi^-(t_0) = \varphi(t_0) \ , \tag{A.1.4}$$

$$\Phi^+(t_0) + \Phi^-(t_0) = \frac{1}{\pi i} \oint_L \frac{\varphi(t)}{(t - t_0)} \, dt \ , \tag{A.1.5}$$

where the integration is in the sense of the Cauchy principal value.

Formulas (A.1.4) and (A.1.5) can be verified directly in the case that $\varphi(t)$ represents the values on L of a function $\varphi(z)$ which is analytic in a neighbourhood of L. In this case we may deform L slightly (Figure A.1.2) in order to calculate for instance the limit value

$$\Phi^+(t_0) = \frac{1}{2\pi i} \oint_L \frac{\varphi(t)}{(t - t_0)} \, dt + \frac{1}{2}\varphi(t_0) \ . \tag{A.1.6}$$

Analogously we find for the other side

$$\Phi^-(t_0) = \frac{1}{2\pi i} \oint_L \frac{\varphi(t)}{(t - t_0)} \, dt - \frac{1}{2}\,\varphi(t_0) \ . \tag{A.1.7}$$

Subtraction of (A.1.6) and (A.1.7) yields (A.1.4), addition yields (A.1.5).

A.2. The Hilbert Problem for an Arc

The Hilbert problem is to find a sectionally holomorphic function $\Phi(z)$ which satisfies the relation

$$\Phi^-(t) - G(t)\Phi^+(t) = g(t) \ , \quad t \in L \ , \tag{A.2.1}$$

where $G(t)$ and $g(t)$ are given functions, which satisfy the Hölder condition and $G(t) \neq 0$.

First we consider the homogeneous equation

$$\Psi^-(t) = G(t)\Psi^+(t) \ , \quad t \in L \ . \tag{A.2.2}$$

We try to find a sectionally holomorphic function $\Psi(z)$ without zeros in the whole complex plane. Then $\ln \Psi(z)$ is also sectionally holomorphic and satisfies

$$\{\ln \Psi(t)\}^+ - \{\ln \Psi(t)\}^- = -\ln G(t) \ , t \in L \ . \tag{A.2.3}$$

Comparing (A.2.3) and (A.1.4) we find by (A.1.3) as a solution of (A.2.2)

$$\Psi(z) = \exp\left\{ -\frac{1}{2\pi i} \int_L \frac{\ln G(t)}{(t-z)}\, dt \right\} \ , \quad z \notin L \ . \tag{A.2.4}$$

Multiplying this function by an arbitrary rational function $R_1(z)$, which is allowed to possess poles only at the ends of L, we do not disturb relation (A.2.2). Hence a more general solution of (A.2.2) is

$$\Psi(z) = \left[\exp\left\{ -\frac{1}{2\pi i} \int_L \frac{\ln G(t)}{(t-z)}\, dt \right\} \right] R_1(z) \ , \quad z \notin L \ . \tag{A.2.5}$$

In order to deal with the inhomogeneous equation we write (A.2.1) by using (A.2.2) in the form

$$\frac{\Phi^-(t)}{\Psi^-(t)} - \frac{\Phi^+(t)}{\Psi^+(t)} = \frac{g(t)}{\Psi^-(t)} \ . \tag{A.2.6}$$

Hence again by (A.1.4) and (A.1.3) we obtain

$$\Phi(z) = -\frac{\Psi(z)}{2\pi i} \left\{ \int_L \frac{g(t)}{\Psi^-(t)(t-z)}\, dt + R_2(z) \right\} \ , \quad z \notin L \ , \tag{A.2.7}$$

where it is assumed that $\Psi(z)$ is chosen in such a way that the integral exists and $R_2(z)$ is also an arbitrary rational function with the same restriction as $R_1(z)$. In [52] it is shown that (A.2.7) and (A.2.5) are the general solutions of the equations (A.2.1) and (A.2.2), respectively, when the behaviour at infinity is prescribed to be algebraic.

A.3. Singular Integral Equations

Next we discuss the relation between a simple type of singular integral equation and the Hilbert problem. Consider the equation

$$\frac{1}{\pi i} \oint_{-1}^{1} \frac{\varphi(t)}{(t-t_0)}\, dt = g(t_0) \ , \tag{A.3.1}$$

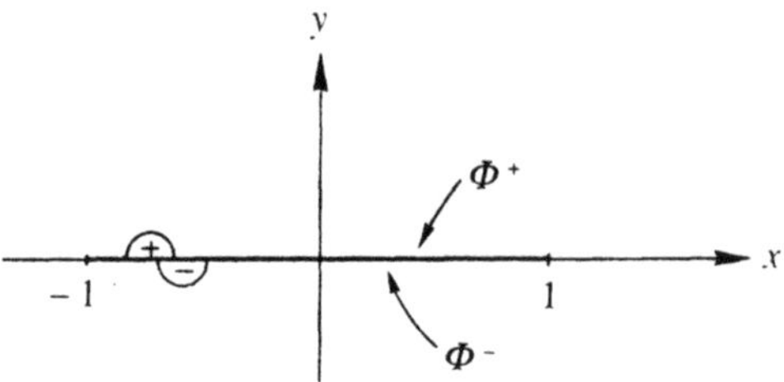

Fig. A.3.1. The real interval $[-1, 1]$.

where t and t_0 are real variables, $\varphi(t)$ is the unknown function and $g(t)$ is given. Introducing

$$\Phi(z) = \frac{1}{2\pi i} \int_{-1}^{1} \frac{\varphi(t)}{(t-z)}\, dt \ , \quad z \in L, \tag{A.3.2}$$

we find from (A.1.5)

$$\Phi^+(t) + \Phi^-(t) = g(t) \ , \quad -1 \leq t \leq 1 \ . \tag{A.3.3}$$

The arc L of Section A.1 is here the interval $[-1, 1]$ of the real axis, with its + side facing the upper half plane and its − side facing the lower half plane (Figure A.3.1). Substitution of $G(t) = -1$ in (A.2.5) yields for the solution $\Psi(z)$ of the homogeneous part of (A.3.3)

$$\Psi(z) = \left\{ \exp -\tfrac{1}{2} \int_{-1}^{1} \frac{dt}{(t-z)} \right\} R_1(z) = \left(\frac{z+1}{z-1} \right)^{1/2} R_1(z) \ , \tag{A.3.4}$$

where we took $\ln -1 = \pi i$. Choosing $R_1(z) = 1$, $R_1(z) = (z+1)^{-1}$ or $R_1(z) = (z-1)(z+1)^{-1}$ yields a number of functions $\Psi(z)$ namely

$$\Psi(z) = \left(\frac{z+1}{z-1} \right)^{1/2} , \quad \Psi(z) = (z^2-1)^{-1/2} , \quad \Psi(z) = \left(\frac{z-1}{z+1} \right)^{1/2} , \tag{A.3.5}$$

respectively. The choice we make is not essential, however in connection with the problem of Section 5.12 where we have to satisfy the Kutta condition at the trailing edge ($x = 1$) of the profile and where we tolerate only a square root singularity at the leading edge ($x = -1$), the third expression in (A.3.5) yields a more simple analysis. Of course we have to define the meaning of the square roots. We will take them real and positive for z real and $z > 1$. Then by analytic continuation the square root is defined in the whole complex plane with the exception of the real interval $[-1, 1]$.

In this way we find for the solution (A.3.3) by using (A.2.7)

$$\Phi(z) = -\frac{1}{2\pi i} \left(\frac{z-1}{z+1} \right)^{1/2} \left\{ \int_{-1}^{1} \frac{g(t)}{-i \left(\frac{1-t}{1+t} \right)^{1/2} (t-z)}\, dt + R_2(z) \right\} \ . \tag{A.3.6}$$

From (A.3.2) it follows that $\Phi(z)$ has to tend to zero for $|z| \to \infty$ and because of the admitted square root singularity at $z = -1$ and the Kutta condition at $z = 1$, the rational function $R_2(z)$ has to be identically zero, $R_2(z) \equiv 0$. By (A.1.6) and (A.1.7) for $-1 < t_0 < 1$, we have

$$\Phi^+(t_0) = -\frac{i}{2\pi}\left(\frac{1-t_0}{1+t_0}\right)^{1/2} \cdot \left\{ \oint_{-1}^{1} \left(\frac{1+t}{1-t}\right)^{1/2} \frac{g(t)}{(t-t_0)}\, dt + \pi i \left(\frac{1+t_0}{1-t_0}\right)^{1/2} g(t_0) \right\}, \tag{A.3.7}$$

$$\Phi^-(t_0) = \frac{i}{2\pi}\left(\frac{1-t_0}{1+t_0}\right)^{1/2} \cdot \left\{ \oint_{-1}^{1} \left(\frac{1+t}{1-t}\right)^{1/2} \frac{g(t)}{(t-t_0)}\, dt - \pi i \left(\frac{1+t_0}{1-t_0}\right)^{1/2} g(t_0) \right\}. \tag{A.3.8}$$

Then by (A.1.4) we find for the desired solution of (A.3.1)

$$\varphi(t_0) = -\frac{i}{2\pi}\left(\frac{1-t_0}{1+t_0}\right)^{1/2} \oint_{-1}^{1} \left(\frac{1+t}{1-t}\right)^{1/2} \frac{g(t)}{(t-t_0)}\, dt\ . \tag{A.3.9}$$

When we tolerate at each end of the interval, hence at $x = -1$ and at $x = 1$, a square root singularity, we can use in (A.3.4) $R_1(z) = (z+1)^{-1}$ hence $\Psi(z) = (z^2-1)^{-1/2}$. Then with (A.2.7)

$$\Phi(z) = -\frac{1}{2\pi i (z^2-1)^{1/2}} \left\{ \int_{-1}^{1} \frac{-i(1-t^2)^{1/2} g(t)}{(t-z)}\, dt + R_2(z) \right\}. \tag{A.3.10}$$

Demanding again that $\Phi(z)$ tends to zero for $|z| \to \infty$, we find that the rational function can only be a constant for which we take $-ic_2$. From (A.3.10) it follows

$$\Phi^+(t_0) = \frac{-i}{2\pi(1-t_0^2)^{1/2}} \cdot \left\{ \oint_{-1}^{1} \frac{(1-t^2)^{1/2} g(t)}{(t-t_0)}\, dt + \pi i (1-t_0^2)^{1/2} g(t_0) + c_2 \right\}, \tag{A.3.11}$$

$$\Phi^-(t_0) = \frac{i}{2\pi(1-t_0^2)^{1/2}} \cdot \left\{ \oint_{-1}^{1} \frac{(1-t^2)^{1/2} g(t)}{(t-t_0)}\, dt - \pi i (1-t_0^2)^{1/2} g(t_0) + c_2 \right\}. \tag{A.3.12}$$

Hence by (A.1.4)

$$\varphi(t_0) = \frac{-i}{\pi(1-t_0^2)^{1/2}} \left\{ \oint_{-1}^{1} \frac{(1-t^2)^{1/2} g(t)}{(t-t_0)} \, dt + c_2 \right\} . \tag{A.3.13}$$

We now suppose $g(t) = c_1 = \text{const.}$ and using (5.12.11) we obtain

$$\varphi(t_0) = \frac{i}{\pi(1-t_0^2)^{1/2}} \{\pi c_1 t_0 - c_2\} , \tag{A.3.14}$$

where c_2 is still arbitrary. In fact $(1-t_0^2)^{-1/2}$ is a solution of the homogeneous part of (A.3.1).

B. Curvilinear Coordinates

In this appendix we discuss some formulas used in Section 1.20 and which are needed for the calculation of operations such as grad, the Laplacian Δ, div and rot in curvilinear coordinates. These coordinates do not need to be locally orthogonal, which is important with respect to helicoidal coordinates which are useful in the theory of screw propellers. We do not enter into a discussion about the concept of a tensor. For the incorporation of the introduced subjects in a systematic theory we refer for instance to [63].

B.1. Concepts of Tensor Analysis

Suppose (ξ^1, ξ^2, ξ^3) and (η^1, η^2, η^3) are two curvilinear coordinate systems. Their relation is defined by

$$\xi^l = \xi^l(\eta^k) , \quad \eta^l = \eta^l(\xi^k) , \quad k, l = 1, 2, 3 . \tag{B.1.1}$$

In order that (B.1.1) is an acceptable transformation we demand that the Jacobian determinant

$$J = \left| \frac{\partial \xi^i}{\partial \eta^j} \right| \neq 0 . \tag{B.1.2}$$

A tensor in a 3-dimensional space is a set of 3^{r+s} functions

$$A^{j_1,\dots,j_s}_{i_1,\dots,i_r} \ (\xi_i), j_l = 1, 2, 3 \ , 1 \le l \le s \ ; \quad i_l = 1, 2, 3 \ , 1 \le l \le r \ , \tag{B.1.3}$$

defined in the ξ^i coordinate system, provided that the corresponding set of functions in the η^i coordinate system are given by

$$A^{j_1,\dots,j_s}_{i_1,\dots,i_r} \ (\eta^i) = \frac{\partial \xi^{\alpha_1}}{\partial \eta^{i_1}} \cdots \frac{\partial \xi^{\alpha_r}}{\partial \eta^{i_r}} \cdot \frac{\partial \eta^{j_1}}{\partial \xi^{\beta_1}} \cdots \frac{\partial \eta^{j_s}}{\partial \xi^{\beta_s}} A^{\beta_1,\dots,\beta_s}_{\alpha_1,\dots,\alpha_r} \ (\xi^i) . \tag{B.1.4}$$

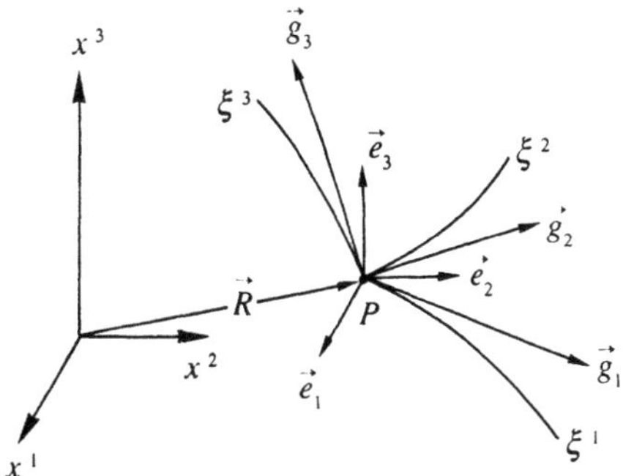

Fig. B.1.1. Base vectors $\vec{e}_j$ and $\vec{g}_j$ at P.

In (B.1.4) we assumed and in the following we will assume the summation convention. This convention states that: if in an expression an index occurs twice, once in a lower position and once in an upper position, then this expression is summed with respect to this index over its values 1, 2 and 3. When a summation sign $\sum$ is used this convention does not apply. Formula (B.1.4) defines a tensor, covariant of rank r and contravariant of rank s.

Now consider a Cartesian coordinate system x^i and a curvilinear system ξ^i, which are related by

$$x^l = x^l(\xi^k) \ , \quad \xi^l = \xi^l(x^k) \ , \quad k, l = 1, 2, 3 \ . \tag{B.1.5}$$

In the Cartesian system x^l we have a point P with vector of position $\vec{R}$ (Figure B.1.1). The base vectors $\vec{e}_l$ in the Cartesian system are of unit length. The curvilinear system ξ^l has the covariant base vectors

$$\vec{g}_i = \frac{\partial \vec{R}}{\partial \xi^i} = \frac{\partial}{\partial \xi^i} \left(x^1(\xi^k), x^2(\xi^k), x^3(\xi^k) \right) \ , \tag{B.1.6}$$

which are tangent to the ξ^i coordinate lines, along which the ξ^j with $j \neq i$ are constant. Equation (B.1.6) gives the components of the $\vec{g}_i$ with respect to the Cartesian system.

The contravariant base vectors $\vec{g}^{\,i}$, the first fundamental tensor g_{ij}, the second fundamental tensor g^{ij} and the determinant g are defined by

$$\vec{g}^{\,i} \cdot \vec{g}_j = \delta^i_j \ , \quad g_{ij} = \vec{g}_i \cdot \vec{g}_j \ , \quad g^{ij} = \vec{g}^{\,i} \cdot \vec{g}^{\,j} \ , \quad g = |g_{ij}| \ . \tag{B.1.7}$$

From (B.1.7) it follows that

$$g^{ij} = \frac{G(i,j)}{g} \ , \quad \vec{g}^{\,i} = g^{ij} \vec{g}_j \ , \quad \vec{g}_i = g_{ij} \vec{g}^{\,j} \ ,$$

$$\vec{g}^{\,i} \cdot \vec{g}_l = g^{ik} g_{kl} = \delta^i_l \ , \tag{B.1.8}$$

where $G(i,j)$ is the cofactor (including sign) of g_{ij} in the expansion of $g = |g_{kl}|$ and δ^i_l is the Kronecker symbol, $\delta^i_l = 1$ for $i = l$, $\delta^i_l = 0$ for $i \neq l$.

We now can write a vector $\vec{v}$ in the ξ^i system as

$$\vec{v} = v^i \vec{g}_i = v_i \vec{g}^{\,i} = {}^*v^i \vec{c}_i = {}^*v_i \vec{c}^{\,i} \ , \tag{B.1.9}$$

where v^i is a contravariant tensor of rank one, v_i is a covariant tensor of rank one and $\vec{c}_i$ and $\vec{c}^{\,i}$ are vectors of *unit length* in the direction of $\vec{g}_i$ and $\vec{g}^{\,i}$, respectively. Using (B.1.7) and (B.1.8) we have between the components of $\vec{v}$ the relations

$$v_i = g_{ij} v^j \ , \quad v^i = g^{ij} v_j \ , \quad {}^*v^i = (g_{ii})^{1/2} v^i \ , \quad {}^*v_i = (g^{ii})^{1/2} v_i \ , \tag{B.1.10}$$

where in agreement with the summation convention no summation occurs in the last two equalities. We remark that ${}^*v^i$ and ${}^*v_i$ are not tensors, they are called *the physical components* of $\vec{v}$.

For the special case of a locally orthogonal curvilinear coordinate system, we have $g_{ij} = g^{ij} = 0$, $(i \neq j)$, because then $\vec{g}_i \perp \vec{g}_j$, $(i \neq j)$ and $\vec{g}^{\,i} \perp \vec{g}^{\,j}$, $(i \neq j)$. From this it follows easily that $g^{ii} = (g_{ii})^{-1}$, then for fixed i,

$${}^*v^i = (g_{ii})^{1/2} v^i = (g_{ii})^{1/2} g^{ij} v_j = (g_{ii})^{1/2} g^{ii} v_i = (g^{ii})^{1/2} v_i = {}^*v_i \ , \tag{B.1.11}$$

which also follows directly from a geometrical observation.

The length $|\vec{v}|$ of a vector $\vec{v}$ in a general curvilinear system follows from

$$|\vec{v}|^2 = \vec{v} \cdot \vec{v} = v^i v^j g_{ij} = v_i v_j g^{ij} = v_i v^i \ . \tag{B.1.12}$$

Next we introduce the Christoffel symbols $\left\{ {k \atop i\ j} \right\}$ which are not tensors but which can be used to form tensors,

$$\left\{ {k \atop i\ j} \right\} = \frac{1}{2} g^{kl} \left(\frac{\partial g_{il}}{\partial \xi^j} + \frac{\partial g_{jl}}{\partial \xi^i} - \frac{\partial g_{ij}}{\partial \xi^l} \right) = \left\{ {k \atop j\ i} \right\} \ . \tag{B.1.13}$$

Herewith we define covariant differentiation of a general tensor. The covariant differentiation with respect to the coordinate ξ^l is denoted by "$|_l$", then

$$\begin{aligned} A^{j_1,\ldots,j_s}_{i_1,\ldots,i_r}(\xi^i)\big|_l = {} & \frac{\partial}{\partial \xi^l} A^{j_1,\ldots,j_s}_{i_1,\ldots,i_r} - \left\{ {\alpha \atop i_1\ l} \right\} A^{j_1,\ldots,j_s}_{\alpha,i_2,\ldots,i_r} \\ & - \left\{ {\alpha \atop i_2\ l} \right\} A^{j_1,\ldots,j_s}_{i_1,\alpha,\ldots,i_r} - \ldots - \left\{ {\alpha \atop i_r\ l} \right\} A^{j_1,\ldots,j_s}_{i_1,\ldots,\alpha} \\ & + \left\{ {j_1 \atop \alpha\ l} \right\} A^{\alpha,j_2,\ldots,j_s}_{i_1,\ldots,i_r} + \left\{ {j_2 \atop \alpha\ l} \right\} A^{j_1,\alpha,\ldots,j_s}_{i_1,\ldots,i_r} \\ & + \cdots + \left\{ {j_s \atop \alpha\ l} \right\} A^{j_1,\ldots,\alpha}_{i_1,\ldots,i_r} \ . \end{aligned} \tag{B.1.14}$$

This is again a tensor, but now covariant of rank $r + 1$ and still contravariant of rank s. From (B.1.14) it follows that covariant differentiation of a scalar function with respect to ξ^l, is ordinary partial differentiation with respect to ξ^l. The result is a covariant tensor of rank 1, in other words a covariant vector ($l = 1, 2, 3$).

When the transformation (B.1.5) is orthogonal, hence when (ξ^1, ξ^2, ξ^3) is again Cartesian, then the Christoffel symbols are zero and covariant differentiation becomes ordinary differentiation.

We remark that the first and the second fundamental tensor behave as constants for covariant differentiation. Hence $g_{ij}|_l = g^{ij}|_l = 0$ and these tensors can be brought outside the differentiation, for instance

$$(g_{ij}v^j)|_l = g_{ij}(v^j|_l) \ , \quad (g^{ij}v_j)|_l = g^{ij}(v_j|_l) \ . \tag{B.1.15}$$

Next we introduce the permutation tensors ε_{ijk} and ε^{ijk} by

$$\varepsilon_{ijk} = g^{1/2}e_{ijk} \ , \quad \varepsilon^{ijk} = g^{-1/2}e_{ijk} \ , \tag{B.1.16}$$

where the permutation symbols e_{ijk} are defined by $e_{123} = e_{231} = e_{312} = 1$, $e_{132} = e_{213} = e_{321} = -1$ and $e_{ijk} = 0$ otherwise.

By the above we can write down the following operations which are valid for a general curvilinear coordinate system

$$\text{grad}\,\Phi = \frac{\partial\Phi}{\partial\xi^i}\,\vec{g}^{\,i} = \frac{\partial\Phi}{\partial\xi^i}\,g^{ij}\vec{g}_j \ , \tag{B.1.17}$$

$$\begin{aligned}\Delta\Phi &= g^{ij}\left(\frac{\partial\Phi}{\partial\xi^i}\right)|_j = g^{ij}\left[\frac{\partial^2\Phi}{\partial\xi^i\partial\xi^j} - \left\{\begin{matrix}\alpha\\ i\,j\end{matrix}\right\}\frac{\partial\Phi}{\partial\xi^\alpha}\right]\\ &= g^{-1/2}\frac{\partial}{\partial\xi^i}\left(g^{1/2}g^{ij}\frac{\partial\Phi}{\partial\xi^j}\right) \ ,\end{aligned} \tag{B.1.18}$$

$$\text{div}\,\vec{v} = v^i|_i = \frac{\partial v^i}{\partial\xi^i} + \left\{\begin{matrix}i\\ \alpha\, i\end{matrix}\right\}v^\alpha = g^{-1/2}\frac{\partial}{\partial\xi^l}(g^{1/2}v^l) \ , \tag{B.1.19}$$

$$\begin{aligned}\text{rot}\,\vec{v} &= -\varepsilon^{ijk}(v_j|_k)\vec{g}_i = -\varepsilon^{ijk}\left[\frac{\partial v_j}{\partial\xi^k} - \left\{\begin{matrix}\alpha\\ j\,k\end{matrix}\right\}v_\alpha\right]\vec{g}_i\\ &= g^{-1/2}\left\{\left(\frac{\partial v_3}{\partial\xi^2} - \frac{\partial v_2}{\partial\xi^3}\right)\vec{g}_1\right.\\ &\quad \left. + \left(\frac{\partial v_1}{\partial\xi^3} - \frac{\partial v_3}{\partial\xi^1}\right)\vec{g}_2 + \left(\frac{\partial v_2}{\partial\xi^1} - \frac{\partial v_1}{\partial\xi^2}\right)\vec{g}_3\right\} \ .\end{aligned} \tag{B.1.20}$$

The last equalities in (B.1.18)–(B.1.20) are derived in [63] (pp. 249–251).

We remark that when the transformation (B.1.5) is Cartesian, then (B.1.17)–(B.1.20) assume their well-known form for a Cartesian coordinate system, which read for $\xi^1 = x$, $\xi^2 = y$, $\xi^3 = z$,

$$\text{grad}\,\Phi = \frac{\partial\Phi}{\partial x}\,\vec{e}_x + \frac{\partial\Phi}{\partial y}\,\vec{e}_y + \frac{\partial\Phi}{\partial z}\,\vec{e}_z \ , \tag{B.1.21}$$

$$\Delta\Phi = \frac{\partial^2\Phi}{\partial x^2} + \frac{\partial^2\Phi}{\partial y^2} + \frac{\partial^2\Phi}{\partial z^2} \ , \tag{B.1.22}$$

$$\mathrm{div}\, \vec{v} = \frac{\partial v_x}{\partial x} + \frac{\partial v_y}{\partial y} + \frac{\partial v_z}{\partial z} \ , \tag{B.1.23}$$

$$\mathrm{rot}\, \vec{v} = \left(\frac{\partial v_z}{\partial y} - \frac{\partial v_y}{\partial z}\right)\vec{e}_x + \left(\frac{\partial v_x}{\partial z} - \frac{\partial v_z}{\partial x}\right)\vec{e}_y + \left(\frac{\partial v_y}{\partial x} - \frac{\partial v_x}{\partial y}\right)\vec{e}_z \ , \tag{B.1.24}$$

where $\vec{e}_x$, $\vec{e}_y$ and $\vec{e}_z$ are the unit vectors along the x, y and z-axis, respectively. From this it follows that, because (B.1.17)–(B.1.20) are in tensor notation, that (B.1.17)–(B.1.20) are valid in all coordinate systems.

B.2. Cylindrical and Helicoidal Coordinate Systems

First we pass from the Cartesian system (x, y, z) to the cylindrical system (x, r, φ), which is a locally orthogonal curvilinear system. We use the index notation and give the relation between the two coordinate systems by

$$x^1 = x \ , x^2 = y \ , x^3 = z \ ; \quad \xi^1 = x \ , \xi^2 = r \ , \xi^3 = \varphi \tag{B.2.1}$$

$$x^1 = \xi^1 \ , \quad x^2 = \xi^2 \cos\xi^3 \ , x^3 = \xi^2 \sin\xi^3 \ . \tag{B.2.2}$$

Then we obtain by (B.1.6) and (B.1.7)

$$\vec{g}_1 = (1, 0, 0) \ , \quad \vec{g}_2 = (0, \cos\xi^3, \sin\xi^3) \ ,$$
$$\vec{g}_3 = (0, -\xi^2 \sin\xi^3, \xi^2 \cos\xi^3) \ , \tag{B.2.3}$$

$$\vec{g}^{\,1} = (1, 0, 0) \ , \quad \vec{g}^{\,2} = (0, \cos\xi^3, \sin\xi^3) \ ,$$
$$\vec{g}_3 = \left(0, -\frac{1}{\xi^2} \sin\xi^3, \frac{1}{\xi^2} \cos\xi^3\right) \ , \tag{B.2.4}$$

$$g_{11} = g_{22} = 1 \ , \quad g_{33} = (\xi^2)^2 \ , \quad g_{ij} = 0 \quad \text{for} \quad i \neq j \ , \quad g = (\xi^2)^2 \ , \tag{B.2.5}$$

$$g^{11} = g^{22} = 1 \ , \quad g^{33} = (\xi^2)^{-2} \ , \quad g^{ij} = 0 \quad \text{for} \quad i \neq j \ , \tag{B.2.6}$$

where the components of $\vec{g}_i$ and $\vec{g}^{\,i}$ are with respect to the Cartesian system.

We denote in the cylindrical system the physical components (B.1.10) of some vector $\vec{v}$ by v_x, v_r and v_φ. Then we find by (B.1.17)–(B.1.20) and using (B.1.9) and (B.1.10),

$$\mathrm{grad}\ \Phi = \frac{\partial\Phi}{\partial x}\vec{c}_1 + \frac{\partial\Phi}{\partial r}\vec{c}_2 + \frac{1}{r}\frac{\partial\Phi}{\partial\varphi}\vec{c}_3 \ , \tag{B.2.7}$$

$$\Delta\Phi = \frac{\partial^2\Phi}{\partial x^2} + \frac{1}{r}\frac{\partial}{\partial r} r \frac{\partial\Phi}{\partial r} + \frac{1}{r^2}\frac{\partial^2\Phi}{\partial\varphi^2} , \tag{B.2.8}$$

$$\operatorname{div}\vec{v} = \frac{\partial v_x}{\partial x} + \frac{1}{r}\frac{\partial r v_r}{\partial r} + \frac{1}{r}\frac{\partial v_\varphi}{\partial\varphi} , \tag{B.2.9}$$

$$\operatorname{rot}\vec{v} = \left\{ \left(\frac{1}{r}\frac{\partial r v_\varphi}{\partial r} - \frac{1}{r}\frac{\partial v_r}{\partial\varphi}\right)\vec{c}_1 + \left(\frac{1}{r}\frac{\partial v_x}{\partial\varphi} - \frac{\partial v_\varphi}{\partial x}\right)\vec{c}_2 + \left(\frac{\partial v_r}{\partial x} - \frac{\partial v_x}{\partial r}\right)\vec{c}_3 \right\} , \tag{B.2.10}$$

where the $\vec{c}_i$ are vectors of unit length in the direction of the $\vec{g}_i$.

Next we pass from the Cartesian system (x, y, z) to the helicoidal coordinates $(\tilde{\rho}, \zeta, \sigma)$. We use the index notation and give the relation between the coordinate systems by

$$x^1 = x \ , x^2 = y \ , x^3 = z \ ; \quad \xi^1 = \tilde{\rho} \ , \xi^2 = \zeta \ , \xi^3 = \sigma \ ; \tag{B.2.11}$$

$$x^1 = \frac{1}{a}(\xi^2 - \xi^3) \ , x^2 = \frac{\xi^1}{a}\cos\xi^3 \ ,$$

$$x^3 = \frac{\xi^1}{a}\sin\xi^3 \ ; \quad (\tilde{\rho} = a(y^2 + z^2)^{1/2}) \ , \tag{B.2.12}$$

where a is some positive number. By (B.1.6) and (B.1.7) we obtain

$$\vec{g}_1 = \left(0, \frac{1}{a}\cos\xi^3, \frac{1}{a}\sin\xi^3\right) \ , \quad \vec{g}_2 = \left(\frac{1}{a}, 0, 0\right) \ ,$$

$$\vec{g}_3 = \left(-\frac{1}{a}, -\frac{\xi^1}{a}\sin\xi^3, \frac{\xi^1}{a}\cos\xi^3\right) \ , \tag{B.2.13}$$

$$\vec{g}^{\,1} = (0, a\cos\xi^3, a\sin\xi^3) \ , \quad \vec{g}^{\,2} = \left(a, -\frac{a}{\xi^1}\sin\xi^3, \frac{a}{\xi^1}\cos\xi^3\right) \ ,$$

$$\vec{g}^{\,3} = \left(0, -\frac{a}{\xi_1}\sin\xi^3, \frac{a}{\xi^1}\cos\xi^3\right) \ , \tag{B.2.14}$$

$$g_{11} = g_{22} = -g_{23} = -g_{32} = \frac{1}{(a)^2} \ ,$$

$$g_{33} = \frac{1}{(a)^2}\left(1 + (\xi^1)^2\right) \ , \quad g = \frac{(\xi^1)^2}{(a)^6} \ , \tag{B.2.15}$$

$$g^{11} = (a)^2 \ , \quad g^{22} = \frac{(a)^2}{(\xi^1)^2}\left(1 + (\xi^1)^2\right) \ , \quad g^{23} = g^{32} = g^{33} = \frac{(a)^2}{(\xi^1)^2} \ , \tag{B.2.16}$$

where the components of $\vec{g}_i$ and $\vec{g}^{\,i}$ are with respect to the Cartesian system and the not mentioned g_{ij} and g^{ij} are zero.

Because $g_{23} \neq 0$ it follows that the helicoidal system is not locally orthogonal, hence (B.1.11) is not valid. This means that the formulas for grad Φ, div $\vec{v}$ or rot $\vec{v}$ will depend on the choice of the base vectors of unit length $\vec{c}_i$ or $\vec{c}^{\,i}$. When needed these expressions can be found from the general formulas (B.1.17)–(B.1.20). For the scalar $\Delta\Phi$ we obtain

$$\Delta\Phi = a^2 \left\{ \frac{1}{\tilde{\rho}} \frac{\partial}{\partial\tilde{\rho}} \tilde{\rho} \frac{\partial\Phi}{\partial\tilde{\rho}} + \left(1 + \frac{1}{\tilde{\rho}^2}\right) \frac{\partial^2\Phi}{\partial\zeta^2} + \frac{2}{\tilde{\rho}^2} \frac{\partial^2\Phi}{\partial\zeta\partial\sigma} + \frac{1}{\tilde{\rho}^2} \frac{\partial^2\Phi}{\partial\sigma^2} \right\} . \quad \text{(B.2.17)}$$

C. Some Identities

In this appendix we give a number of identities which can be useful in applied analysis. Their representation in curvilinear coordinates can be found by means of Appendix B, as is shown by way of example in (C.1) and (C.7)

$$\vec{a} \times \vec{b} = \varepsilon^{ijk} a_i b_j \vec{g}_k = \varepsilon_{ijk} a^i b^j \vec{g}^{\,k} \ , \quad \text{(C.1)}$$

$$\vec{a} \cdot (\vec{b} \times \vec{c}\,) = \vec{b} \cdot (\vec{c} \times \vec{a}\,) = \vec{c} \cdot (\vec{a} \times \vec{b}\,) \ , \quad \text{(C.2)}$$

$$\vec{a} \times (\vec{b} \times \vec{a}\,) = |\vec{a}|^2\, \vec{b} - (\vec{a} \cdot \vec{b}\,)\, \vec{a} \ , \quad \text{(C.3)}$$

$$\text{rot grad}\ \Phi = 0 \ , \quad \text{(C.4)}$$

$$\text{div rot}\ \Phi = 0 \ , \quad \text{(C.5)}$$

$$\Delta\Phi = \text{div grad}\ \Phi \ , \quad (\Delta \text{ is Laplacian}) \ , \quad \text{(C.6)}$$

$$\text{div}\ (\Phi\, \vec{a}\,) = \Phi\ \text{div}\ \vec{a} + (\text{grad}\ \Phi) \cdot \vec{a} = \Phi\, a^i|_i + \frac{\partial\Phi}{\partial\xi^i}\, a^i \ , \quad \text{(C.7)}$$

$$\text{rot}\ (\Phi\, \vec{a}\,) = \Phi\ \text{rot}\ \vec{a} + (\text{grad}\ \Phi) \times \vec{a} \ , \quad \text{(C.8)}$$

$$\text{div}\ (\vec{a} \times \vec{b}\,) = -\vec{a} \cdot \text{rot}\ \vec{b} + (\text{rot}\ \vec{a}\,) \cdot \vec{b} \ , \quad \text{(C.9)}$$

$$\text{rot}\ (\vec{a} \times \vec{b}\,) = -(\vec{a} \cdot \text{grad})\vec{b} + \vec{a}\ \text{div}\ \vec{b} + (\vec{b} \cdot \text{grad})\vec{a} - \vec{b}\ \text{div}\ \vec{a} \ , \quad \text{(C.10)}$$

$$\text{grad}\ (\vec{a} \cdot \vec{b}\,) = (\vec{a} \cdot \text{grad})\vec{b} + \vec{a} \times \text{rot}\ \vec{b} + (\vec{b} \cdot \text{grad})\vec{a} + \vec{b} \times \text{rot}\ \vec{a} \ , \quad \text{(C.11)}$$

$$\Delta\vec{a} = \text{grad div}\ \vec{a} - \text{rot rot}\ \vec{a}. \quad \text{(C.12)}$$

When $\vec{a}$ and $\vec{b}$ are independent of $\vec{x}$ then

$$\text{grad}\ \left\{\text{grad}\ \Phi \cdot (\vec{a} \times \vec{b}\,)\right\} = \Delta\Phi\ (\vec{a} \times \vec{b}\,) + (\vec{b} \cdot \text{grad})(\text{grad}\ \Phi \times \vec{a}\,)$$
$$-(\vec{a} \cdot \text{grad})(\text{grad}\ \Phi \times \vec{b}\,) \ . \quad \text{(C.13)}$$

D. On Linear Partial Differential Equations

We will shortly discuss the solution of linear partial differential equations by means of convolutions. The method we deal with is used in Section 1.4 for the solution of the linearized equations of motion. For more general considerations we refer to [70].

D.1. The Convolution

We start by stating the convolution condition for two sets $B_i \subset \mathbb{R}^3$ $(i = 1, 2)$. These sets satisfy the convolution condition when for each compact set $K \subset \mathbb{R}^3$, the set Q

$$Q = \{\vec{x}_1, \vec{x}_2 : \vec{x}_i \in B_i \ , i = 1, 2, \ (\vec{x}_1 + \vec{x}_2) \in K\} \ , \tag{D.1.1}$$

is a compact subset of $\mathbb{R}^6$. A sufficient condition that this will happen is that one of the B_i is compact.

The smallest closed region, outside which a function $f(\vec{x})$ is zero, is called the support of f (supp f). Now we say that two functions $f(\vec{x})$ and $g(\vec{x})$ satisfy the convolution condition if their supports satisfy the convolution condition.

The convolution of two functions $f(\vec{x})$ and $g(\vec{x})$, which satisfy the convolution condition, is defined by

$$(f * g)(\vec{x}) = \iiint\limits_{\mathbb{R}^3} g(\vec{\xi})f(\vec{x} - \vec{\xi}) \, d\xi \, d\eta \, d\zeta = (g * f)(\vec{x}) \ . \tag{D.1.2}$$

An important property of the convolution, which follows directly from (D.1.2), is

$$\frac{\partial}{\partial x}(f * g)(\vec{x}) = \left(\frac{\partial f}{\partial x} * g\right)(\vec{x}) = \left(f * \frac{\partial g}{\partial x}\right)(\vec{x}) \ , \tag{D.1.3}$$

and analogous for partial differentiations with respect to y and z, $(\vec{x} = (x, y, z))$.

Sometimes the convolution of a scalar function f and a vector function $\vec{g}$ is needed

$$\begin{aligned}(f * \vec{g})(\vec{x}) &= \iiint\limits_{\mathbb{R}^3} \vec{g}(\vec{\xi})f(\vec{x} - \vec{\xi}) \, d\xi \, d\eta \, d\zeta \\ &= \iiint\limits_{\mathbb{R}^3} f(\vec{\xi}) \, \vec{g}(\vec{x} - \vec{\xi}) \, d\xi \, d\eta \, d\zeta = (\vec{g} * f)(\vec{x}) \ ,\end{aligned} \tag{D.1.4}$$

where the products under the integral signs are component wise. Finally the convolution between two vector functions $\vec{f}$ and $\vec{g}$

$$(\vec{f} * \vec{g})(\vec{x}) = \iiint\limits_{\mathbb{R}^3} \vec{g}(\vec{\xi}) \cdot \vec{f}(\vec{x} - \vec{\xi}) \, d\xi \, d\eta \, d\zeta = (\vec{g} * \vec{f})(\vec{x}) \ , \tag{D.1.5}$$

where under the integral sign the inner product of the two vector functions is taken.

As an example of the use of (D.1.3) and (D.1.5) we can write

$$(f * \operatorname{div} \vec{g})(\vec{x}) = (\operatorname{grad} f * \vec{g})(\vec{x}) = \iiint_{\mathbb{R}^3} \vec{g}(\vec{\xi}) \cdot (\operatorname{grad} f)(\vec{x} - \vec{\xi}) \, d\xi \, d\eta \, d\zeta \; . \qquad \text{(D.1.6)}$$

D.2. Solution of Linear Partial Differential Equations

Consider a linear partial differential operator L, which differentiates with respect to the space variables x, y and z. We say that $E(\vec{x})$ is a fundamental solution of L when

$$L(E(\vec{x})) = \delta(\vec{x}) \; , \quad \vec{x} = (x, y, z) \; , \qquad \text{(D.2.1)}$$

where $\delta(\vec{x})$ is the delta function of Dirac. Then a solution of the inhomogeneous equation

$$L(f(\vec{x}, t)) = a(\vec{x}, t) \; , \qquad \text{(D.2.2)}$$

can be written as

$$f(\vec{x}, t) = (E * a)(\vec{x}, t) = \iiint_{\mathbb{R}^3} a(\vec{\xi}, t) E(\vec{x} - \vec{\xi}) \, d\xi \, d\eta \, d\zeta \; , \qquad \text{(D.2.3)}$$

in (D.2.2) and (D.2.3) the time t appears more or less as a parameter. That (D.2.3) is correct follows by repeated application of (D.1.3)

$$L(f(\vec{x}, t)) = L((E * a)(\vec{x}, t)) = ((LE) * a)(\vec{x}, t) = (\delta(\vec{x}) * a)(\vec{x}, t) = \vec{a}(\vec{x}, t) \; . \qquad \text{(D.2.4)}$$

For the important case of the Laplacian Δ, we have

$$\Delta E = \left(\frac{\partial^2}{\partial x^2} + \frac{\partial^2}{\partial y^2} + \frac{\partial^2}{\partial z^2} \right) E = \delta(\vec{x}) = \delta(x)\delta(y)\delta(z) \; , \qquad \text{(D.2.5)}$$

$$E(x, y, z) = -\frac{1}{4\pi R} \; , \quad R = (x^2 + y^2 + z^2)^{1/2} \; , \qquad \text{(D.2.6)}$$

which we call a "pole" or the potential of a "source" of unit strength at the origin. The velocity field of this source becomes grad E and the divergence of this velocity field is div grad $E = \Delta E = \delta(\vec{x})$.

When we consider minus the derivative of (D.2.6) in the direction of the unit vector $\vec{n} = (n_x, n_y, n_z)$ we find

$$D(\vec{x}) = -\vec{n} \cdot \operatorname{grad} E(\vec{x}) = -\frac{\vec{n} \cdot \vec{R}}{4\pi R^3} \; , \quad (\vec{R} = (x, y, z)) \; , \qquad \text{(D.2.7)}$$

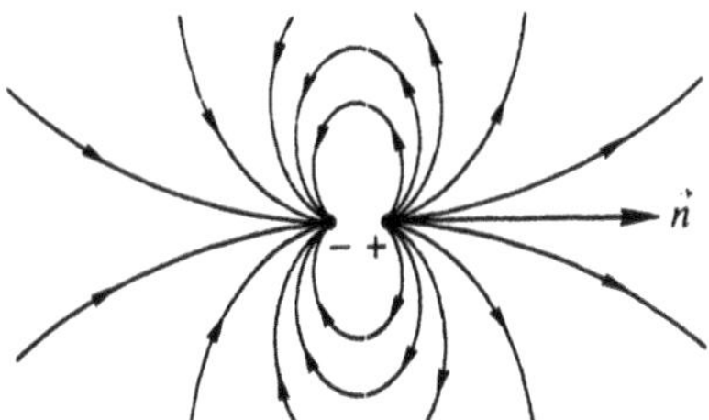

Fig. D.2.1. Impression of the flow caused by a classical dipole.

which we call a "dipole" of unit strength at the origin, with its axis along $\vec{n}$ (Figure D.2.1). The flow induced by this classical dipole, considered as a velocity dipole, is given by grad D and the divergence of this flow becomes

$$\begin{aligned} &\operatorname{div} \operatorname{grad} D(\vec{x}) = \Delta D \\ &= -(n_x\delta'(x)\delta(y)\delta(z) + n_y\delta(x)\delta'(y)\delta(z) + n_z\delta(x)\delta(y)\delta'(z)) \neq 0 \ . \end{aligned} \quad \text{(D.2.8)}$$

It follows from (D.2.8) that the velocity field of a classical dipole is *not free of divergence*. At the origin (D.2.8) has a complicated singular behaviour. In Section 1.5 is introduced a *divergenceless* dipole, which has outside the origin the same field of flow as the classical one. It can be considered as the building stone for linearized hydrodynamics of divergenceless flow fields.

E. Dimension Analysis

In this section we give a derivation of the π theorem of Buckingham. For further information we refer to [9] or [59].

We assume that we have three relevant dimensions, length denoted by $[l]$, mass by $[m]$ and time by $[t]$, which are sufficient for our applications. Consider a physical quantity A_1 which depends on a number of other physical quantities $A_2, A_3, \ldots, A_n$,

$$A_1 = f(A_2, A_3, \ldots, A_n) \ , \quad \text{(E.1)}$$

which equation is valid for any system of units for $[l]$, $[m]$ and $[t]$. The intention is to make combinations of the arguments of f, so that a smaller number of arguments occur which in addition are dimensionless. The dimensions $[A_j]$ of the A_j are denoted by

$$[A_j] = [l]^{\varepsilon_{1j}}[m]^{\varepsilon_{2j}}[t]^{\varepsilon_{3j}} \ , \quad j = 1, \ldots, n \ , \quad \text{(E.2)}$$

where ε_{1j}, ε_{2j} and ε_{3j} are known numbers.

Now we assume that the function f in (E.1) can be expanded as follows

$$A_1 = \sum_{k=1}^{\infty} \alpha_k (A_2^{\beta_{2k}} \cdot A_3^{\beta_{3k}} \cdot \ldots \cdot A_n^{\beta_{nk}}) \ , \quad \text{(E.3)}$$

where the α_k are dimensionless numbers. Then all terms at the right-hand side of (E.3) have to have the same dimension, which has to be equal to the dimension of A_1. Hence

$$[l]^{\varepsilon_{11}}[m]^{\varepsilon_{21}}[t]^{\varepsilon_{31}} = \prod_{j=2}^{n}([l]^{\varepsilon_{1j}}[m]^{\varepsilon_{2j}}[t]^{\varepsilon_{3j}})^{\beta_{jk}} , \quad k = 1, 2, \ldots . \tag{E.4}$$

This means that we have for the β_{jk} $(j = 2, \ldots, n)$ the following equations

$$[l] : \quad \varepsilon_{11} = \sum_{j=2}^{n} \varepsilon_{1j}\beta_{jk} , \quad k = 1, 2, \ldots , \tag{E.5}$$

$$[m] : \quad \varepsilon_{21} = \sum_{j=2}^{n} \varepsilon_{2j}\beta_{jk} , \quad k = 1, 2, \ldots , \tag{E.6}$$

$$[t] : \quad \varepsilon_{31} = \sum_{j=2}^{n} \varepsilon_{3j}\beta_{jk} , \quad k = 1, 2, \ldots \tag{E.7}$$

By means of these equations we can express, for any fixed value of k, the β_{jk} for three values of j into the other β_{jk} with different values of j. In these expressions occur also the known left-hand sides ε_{11}, ε_{21} and ε_{31} of the equations (E.5), (E.6) and (E.7). These ε_{11}, ε_{21} and ε_{31} are independent of k, hence those A_j $(j = 2, \ldots, n)$ which have such a number in their exponent can be brought "partly" as a factor before the summation sign in (E.3). The product of these factors which come from several of the A_j $(j = 2, \ldots, n)$, has the same dimension as A_1. The summation of the remaining products of A_j yield some function of which the arguments are now dimensionless combinations of the A_j.

We will elucidate the above reasoning by an example. Consider the thrust T delivered by a screw propeller under open water condition. The thrust depends on the diameter D of the propeller, the number N of revolutions per second, the velocity of advance U, the density of the water ρ, the viscosity coefficient μ of water, the acceleration of gravity g, the depth b of immersion and on form factors r_i. The form factors are dimensionless, they are for instance vectors of position of the material points of the screw propeller, divided by the diameter D. When we consider two similar propellers these form factors are the same for both propellers. First we leave them out of consideration and add them as arguments of the function we obtain at the end of our derivation. So we consider

$$T = f(D, N, U, \rho, \mu, g, b) . \tag{E.8}$$

The arguments of f have the following dimensions

$$[T] = [l]\,[m]\,[t]^{-2} ; \quad [D] = [l] ; \quad N = [t]^{-1} ; \quad U = [l]\,[t]^{-1} ;$$
$$[\rho] = [l]^{-3}[m] ; \quad [\mu] = [l]^{-1}[m]\,[t]^{-1} ;$$
$$[g] = [l]\,[t]^{-2} ; \quad [b] = [l] . \tag{E.9}$$

From the remark below (E.3) and from (E.4) it follows that

$$[l]\,[m]\,[t]^{-2} = ([l])^{\beta_{2k}} \cdot ([t]^{-1})^{\beta_{3k}} \cdot ([l]\,[t]^{-1})^{\beta_{4k}} \cdot ([l]^{-3}[m])^{\beta_{5k}} \cdot ([l]^{-1}[m]\,[t]^{-1})^{\beta_{6k}} \cdot ([l]\,[t]^{-2})^{\beta_{7k}} \cdot ([l])^{\beta_{8k}} \;, \quad k = 1,2,\ldots \qquad \text{(E.10)}$$

The three equations (E.5)–(E.7) become

$$[l]: \quad 1 = \beta_{2k} + \beta_{4k} - 3\beta_{5k} - \beta_{6k} + \beta_{7k} + \beta_{8k} \;, \quad k = 1,2,\ldots\;, \qquad \text{(E.11)}$$

$$[m]: \quad 1 = \beta_{5k} + \beta_{6k} \;, \quad k = 1,2,\ldots\;, \qquad \text{(E.12)}$$

$$[t]: \quad -2 = -\beta_{3k} - \beta_{4k} - \beta_{6k} - 2\beta_{7k} \;, \quad k = 1,2,\ldots\;. \qquad \text{(E.13)}$$

Now we solve for instance for β_{2k}, β_{3k} and β_{5k} then we obtain

$$\beta_{2k} = 4 - \beta_{4k} - 2\beta_{6k} - \beta_{7k} - \beta_{8k} \;; \quad \beta_{3k} = 2 - \beta_{4k} - \beta_{6k} - 2\beta_{7k} \;; \quad \beta_{5k} = 1 - \beta_{6k} \;, \quad k = 1,2,\ldots \qquad \text{(E.14)}$$

Hence D has 4, N has 2 and ρ has 1 as the parts of their exponents which are independent of k. Bringing the factor $D^4N^2\rho$, which has the dimension of T, outside the summation sign of (E.3), this equation yields

$$\begin{aligned} T &= D^4N^2\rho \sum_{k=1}^{\infty} \alpha_k \cdot \left\{ D^{(-\beta_{4k}-2\beta_{6k}-\beta_{7k}-\beta_{8k})} \cdot N^{(-\beta_{4k}-\beta_{6k}-2\beta_{7k})} \cdot U^{\beta_{4k}} \cdot \rho^{-\beta_{6k}} \cdot g^{\beta_{7k}} \cdot b^{\beta_{8k}} \right\} \\ &= D^4N^2\rho \sum_{k=1}^{\infty} \alpha_k \left\{ \left(\frac{U}{DN}\right)^{\beta_{4k}} \cdot \left(\frac{\mu}{D^2N\rho}\right)^{\beta_{6k}} \cdot \left(\frac{g}{DN^2}\right)^{\beta_{7k}} \cdot \left(\frac{b}{D}\right)^{\beta_{8k}} \right\} \\ &= D^4N^2\rho\, \tilde{h}\left(\frac{U}{DN}, \frac{\mu}{D^2N\rho}, \frac{g}{DN^2}, \frac{b}{D}\right) \;, \end{aligned} \qquad \text{(E.15)}$$

where $\tilde{h}$ is a function which now depends only on four dimensionless arguments. By dividing both sides of (E.15) by $D^4N^2\rho$ and using the remark above (E.8) about the form factors r_i, we obtain instead of (E.15)

$$K_T = \frac{T}{D^4N^2\rho} = h\left(\frac{U}{DN}, \frac{\mu}{D^2N\rho}, \frac{g}{DN^2}, \frac{b}{D}, r_i\right) \;, \qquad \text{(E.16)}$$

in which K_T is the dimensionless thrust coefficient. The argument U/DN is the advance ratio of the propeller often called J. The arguments $\mu/D^2N\rho$ and g/DN^2 are directly related to the Reynolds number and the Froude number respectively. In fact by combining them with U/DN such that N disappears we can replace them by $U\rho\, D/\mu$ and gD/U^2 which in their turn can be changed easily into the current

Reynolds number and Froude number. The argument b/D is the dimensionless depth of the propeller below the free water surface.

Finally we make a remark about the possibility to carry out dimension analysis in an abridged way. It is clear that for all values of $k = 1, 2, \ldots$, the relation between the β_{jk} ($j = 2, \ldots, n$) which follows from (E.5)–(E.7) or in the example from (E.11)–(E.13), is the same. So when we apply dimension analysis we will leave out, in the above described procedure, everywhere the α_k, the summation over k and k itself. Then we arrive for instance in the example of the screw propeller at

$$T = D^4 N^2 \rho \left\{ \left(\frac{U}{DN}\right)^{\beta_4} \cdot \left(\frac{\mu}{D^2 N \rho}\right)^{\beta_6} \cdot \left(\frac{g}{DN^2}\right)^{\beta_7} \cdot \left(\frac{b}{D}\right)^{\beta_8} \right\}, \quad \text{(E.17)}$$

where β_4, β_6, β_7 and β_8 are unknown. This means that the expression between braces represents a function of the four dimensionless arguments, in other words when the form factors r_i are added (E.16) arises again.

References

1. Adams, G. N., Propeller research at Canadian limited. *Cal/Usaav-Labs Symposium Proc., Aerodynamic Problems Associated with V-Stol Aircraft*, Vol. 1, 1966.
2. Andersen, P. and Schwanecke, H., Design and model tests of tip fin propellers. *The Royal Institution of Naval Architects, Spring Meetings*, 1992.
3. Andrews, J. B. and Cummings, D., A design procedure for large hub propellers. *Journal of Ship Research* **16**, No. 3, 1972.
4. Batchelor, G. K., *An Introduction to Fluid Dynamics*. Cambridge University Press, 1974.
5. Betz, A., Schraubenpropeller mit geringstem Energieverlust. *Kgl. Ges. d. Wiss., Nachrichten, Math.-Phys.*, Heft 2, 1919.
6. Birkhoff, G., Helmholtz and Taylor instability. *Proc. Symp. Appl. Math. Am. Math. Soc.* **8**, 1962.
7. Braam, H., Optimum screw propellers with a large hub of finite downstream length. *International Shipbuilding Progress* **31**, 1984.
8. Breslin, J. and Andersen, P., *Hydrodynamics of Ship Propellers*. Cambridge University Press, 1993.
9. Bridgman, P. W., *Dimensional Analysis*. Yale University Press, 1963.
10. Chopra, M. G., Hydrodynamics of lunate-tail swimming propulsion. *Journal of Fluid Mechanics*, **64**, part 2, 1972.
11. Choquet, G., *Lectures on Analysis*, Vol. II. Benjamin, 1969.
12. Chwang, A. T. and Wu, T. Y., A note on the helicoïdal movement of micro-organisms. *Proc. Roy. Soc. Lond. B* **178**, 1971.
13. Coney, W. B., *Optimum Circulation Distributions for a Class of Marine Propulsors*. Systems and Technologies Division, Bolt Beranck and Newmann, Inc., Cambridge, Massachusetts, 1991.
14. Couchet, G., *Les Profiles en Aerodynamique Instationnaire et la Condition de Joukowski*. Librairie Scientifique et Technique, Albart Blanchard, 1976.
15. Dickman, H. E., Schiffsantrieb mit instationären Vortriebsorganen. *Schiff und Hafen*, Heft 10, 1950.
16. Gelfand, J. M. and Fomin, S. V., *Calculus of Variations*. Prentice Hall, Inc., 1963.
17. Gent, W. van and Oossanen, P. van, Influence of wake on propeller loading and cavitation. *International Shipbuilding Progress* **20**, 1973.
18. Gill, P. E. and Murray, W., *Numerical Methods for Constraint Optimization*. Academic Press, 1974.
19. Goodman, T. R., Momentum theory of propeller in shear flow. *Journal of Ship Research* **23**, No. 4, 1979.
20. Graaf, R. de, On optimum fish tail propellers with two blades. Thesis, University of Groningen, The Netherlands, 1970.
21. Greenberg, M. D., Nonlinear actuator disk theory. *Zeitschrift für Flugwissenschaften*, Band 20, Heft 3, 1972.
22. Grimm, O., Propeller and vane wheel. *Journal of Ship Research* **24**, No. 4, 1980.
23. Gröbner, W. and Hofreiter, N., *Integraltafel I, II*. Springer-Verlag, 1973.
24. Hess, F., Boomerangs, aerodynamics and motion. Thesis, University of Groningen, The Netherlands, 1975.
25. Hochstadt, H., *Integral Equations*. John Wiley and Sons, Inc., 1973.
26. Hoeijmakers, H. W. M., Modelling and numerical simulation of vortex flow in aerodynamics. *Agard Conference Proceedings*, 494, 1990.
27. Horlock, J. H., *Actuator Disk Theory*. McGraw-Hill, Inc., 1978.
28. Isay, W. H., *Moderne Probleme der Propellertheory*. Springer-Verlag, 1970.
29. Isay, W. H., *Kavitation*. Schifffahrtsverlag "Hansa", C. Schroedter und Co., 1981.

30. I.T.T.C., *Proceedings of the 15th International Towing Tank Conference*, The Hague, The Netherlands, 1978.
31. I.T.T.C., *Proceedings of the 20th International Towing Tank Conference*, San Francisco, California, 1993.
32. Jacobs, W. R. and Tsakonas, S., Propeller induced velocity field due to thickness and loading effects. *Journal of Ship Research* **19**, No. 1, 1975.
33. Jong, K. de, On the optimization, including viscosity effects of ship screw propellers with optional end plates. *International Shipbuilding Progress* **38**, No. 414 and 415, 1991.
34. Jong, K. de and Sparenberg, J. A., On the influence of choice of generator lines on the optimum efficiency of screw propellers. *Journal of Ship Research* **34**, No. 2, 1990.
35. Katz, J. and Wheiss, D., Hydrodynamic propulsion by large amplitude propulsion of an airfoil with chordwise flexibility. *Journal of Fluid Mechanics* **88**, part 3, 1978.
36. Kellog, O. D., *Foundations of Potential Theory*. Verlag Julius Springer, 1929.
37. Kelly, H. R., Rentz, A. W. and Siekmann, J., Experimental studies on the motion of a flexible hydrofoil. *Journal of Fluid Mechanics* **19**, part 1, 1964.
38. Kerwin, J. E. and Leopold, R., Propeller incidence due to blade thickness. *Journal of Ship Research* **7**, No. 2, 1963.
39. Kinnas, S. A. and Coney, W. B., The generalized image model – An application to the design of ducted propellers. *Journal of Ship Research* **36**, No. 3, 1992.
40. Klaren, L., On the efficiency of a ducted non-stationary actuator disk. *Journal of Engineering Mathematics* **12**, No. 3, 1978.
41. Kotschin, N. J., Kibel, J. A. and Rose, N. W., *Theoretische Hydromechanik*, Band I. Akademie Verlag, 1954.
42. Krasny, R., Computation of vortex sheet roll-up in the Trefftz plane. *Journal of Fluid Mechanics* **184**, 123, 1976.
43. Kruppa, C., Practical aspects of high speed small propellers. *International Shipbuilding Progress* **23**, 1976.
44. Kuik, G. A. M. van, On the limitations of Froude's actuator disc concept. Thesis, Technical University Eindhoven, The Netherlands, 1991.
45. Lewis, E. V., *Principles of Naval Architecture*. The Society of Naval Architects and Marine Engineers, 1988.
46. Ljusternik, L. A. and Sobolev, W. I., *Elemente der Funktional Analysis*. Akademie Verlag, 1960.
47. Manen, J. D. van, Ergebnisse systematischer Versuche mit Propellern mit annäherend senkrecht stehende Achse. *Jahrbuch der Schiffbautechnische Gesellschaft*, Band 57, 1963.
48. Marchaj, C. A., *Aero-Hydrodynamics of Sailing*. Dodd, Mead and Company, 1980.
49. Milgram, J. H., *The Analytic Design of Yacht Sails*. The Society of Naval Architects and Marine Engineers, 1968.
50. Mueller, F., *Recent Developments in the Design and Application of the Vertical Axis Propeller*. The Society of Naval Architects and Marine Engineers, 1955.
51. Muijtjens, R. M. R., On the verification of a theory for sculling propulsion. Thesis, Technical University Eindhoven, The Netherlands, 1992.
52. Mushkhelishvily, N. I., *Singular Integral Equations*. P. Noordhoff N.V., 1953.
53. Potze, W., On optimum sculling propulsion. *Journal of Ship Research* **30**, No. 4, 1986.
54. Potze, W., On optimum large-amplitude sculling propulsion by wings of finite span. *Journal of Ship Research* **30**, No. 4, 1986.
55. Potze, W. and Sparenberg, J. A., On optimum large amplitude sculling propulsion, finite span. *International Shipbuilding Progress* **30**, No. 351, 1983.
56. Pullin, D. J., The large-scale structure of unsteady self-similar rolled-up vortex sheets. *Journal of Fluid Mechanics* **88**, part 3, 1978.
57. Schmidt, G. H. and Sparenberg, J. A., On the edge singularity of an actuator disk with large constant normal load. *Journal of Ship Research* **21**, No. 2, 1977.
58. Schmidt, G. H. and Sparenberg, J. A., On the linearized theory of hub cavity flow with swirl. *Twelfth Symposium Naval Hydrodynamics*, Office of Naval Research, 1978.
59. Sedov, L. J., *Similarity and Dimensional Analysis in Mechanics*. Infosearch Ltd., 1959.
60. Sijtsma, P., On the hydrodynamics of optimum sculling propulsion of ships and on the linearized

lifting surface theory. Thesis, University of Groningen, The Netherlands, 1992.
61. Sijtsma, P. and Sparenberg, J. A., On useful shapes of rigid wings for large amplitude sculling propulsion. *Journal of Ship Research* **36**, No. 3, 1992.
62. Sneddon, I. N., *The Use of Integral Transforms*. McGraw-Hill Publishing Company, Ltd., 1974.
63. Sokolnikoff, I. S., *Tensor Analysis*. John Wiley and Sons, 1960.
64. Sparenberg, J. A., Note on the streamfunction in curvilinear coordinates. *Fluid Dynamics Research* **5**, 1989.
65. Sparenberg, J. A. and Takens, F., On the optimum finite amplitude motion of a thrust producing profile. *Journal of Ship Research* **19**, No. 2, 1975.
66. Sparenberg, J. A. and Thomas, E. G. F., On the existence of small-amplitude optimum hydrofoil propulsion. *Mathematical Methods in the Applied Sciences* **3**, 1981.
67. Sparenberg, J. A. and Vries, J. de, On sculling propulsion by two elastically coupled profiles. *Journal of Ship Research* **27**, No. 2, 1983.
68. Szeless, A. G., Undulating plate type propeller: The two dimensional ideal case. *Journal of Ship Research* **13**, No. 3, 1969.
69. Taylor, A. E., *Functional Analysis*. John Wiley and Sons, Inc., 1958.
70. Trèves, F., *Basic Linear Partial Differential Equations*. Academic Press, 1975.
71. Uldrick, J. P. and Siekmann, J., On swimming of an elastic plate of arbitrary finite thickness. *Journal of Fluid Mechanics* **20**, part 1, 1964.
72. Urbach, H. P., Existence of optimum propulsion by means of periodic motion of rigid profile. *Studies in Applied Mathematics* **81**, 1989.
73. Urbach, H. P., On optimum propulsion by means of small periodic motions of a rigid profile, I. Properties of optimum motions, II. Optimization of the period and numerical results. *Studies in Applied Mathematics* **82**, 1990.
74. Watson, G. N., *A Treatise on the Theory of Bessel Functions*. Cambridge University Press, 1922.
75. Wheeden, R. L. and Zygmund, A., *Measure and Integral*. Marcel Dekker, 1977.
76. Wiersma, A. K., On the profit of optimizing the fin-keel of a yacht sailing close to wind. *Journal of Engineering Mathematics* **12**, 1978.
77. Wu, T. Y., Flow through a heavily loaded actuator disk. *Schiffstechnik*, Band 9, Heft 47, 1962.
78. Wu, T. Y., Hydromechanics of swimming propulsion. *Journal of Fluid Mechanics* **46**, parts 2 and 3, 1971.
79. Wu, T. Y., 'Hydromechanics of swimming of fishes and cetaceans', *Advances in Applied Mechanics* **11**, Academic Press, 1971.

Index

Mechanics

FLUID MECHANICS AND ITS APPLICATIONS

Series Editor: R. Moreau

Aims and Scope of the Series

The purpose of this series is to focus on subjects in which fluid mechanics plays a fundamental role. As well as the more traditional applications of aeronautics, hydraulics, heat and mass transfer etc., books will be published dealing with topics which are currently in a state of rapid development, such as turbulence, suspensions and multiphase fluids, super and hypersonic flows and numerical modelling techniques. It is a widely held view that it is the interdisciplinary subjects that will receive intense scientific attention, bringing them to the forefront of technological advancement. Fluids have the ability to transport matter and its properties as well as transmit force, therefore fluid mechanics is a subject that is particularly open to cross fertilisation with other sciences and disciplines of engineering. The subject of fluid mechanics will be highly relevant in domains such as chemical, metallurgical, biological and ecological engineering. This series is particularly open to such new multidisciplinary domains.

1. M. Lesieur: *Turbulence in Fluids.* 2nd rev. ed., 1990 ISBN 0-7923-0645-7
2. O. Métais and M. Lesieur (eds.): *Turbulence and Coherent Structures.* 1991 ISBN 0-7923-0646-5
3. R. Moreau: *Magnetohydrodynamics.* 1990 ISBN 0-7923-0937-5
4. E. Coustols (ed.): *Turbulence Control by Passive Means.* 1990 ISBN 0-7923-1020-9
5. A.A. Borissov (ed.): *Dynamic Structure of Detonation in Gaseous and Dispersed Media.* 1991 ISBN 0-7923-1340-2
6. K.-S. Choi (ed.): *Recent Developments in Turbulence Management.* 1991 ISBN 0-7923-1477-8
7. E.P. Evans and B. Coulbeck (eds.): *Pipeline Systems.* 1992 ISBN 0-7923-1668-1
8. B. Nau (ed.): *Fluid Sealing.* 1992 ISBN 0-7923-1669-X
9. T.K.S. Murthy (ed.): *Computational Methods in Hypersonic Aerodynamics.* 1992 ISBN 0-7923-1673-8
10. R. King (ed.): *Fluid Mechanics of Mixing.* Modelling, Operations and Experimental Techniques. 1992 ISBN 0-7923-1720-3
11. Z. Han and X. Yin: *Shock Dynamics.* 1993 ISBN 0-7923-1746-7
12. L. Svarovsky and M.T. Thew (eds.): *Hydroclones.* Analysis and Applications. 1992 ISBN 0-7923-1876-5
13. A. Lichtarowicz (ed.): *Jet Cutting Technology.* 1992 ISBN 0-7923-1979-6
14. F.T.M. Nieuwstadt (ed.): *Flow Visualization and Image Analysis.* 1993 ISBN 0-7923-1994-X
15. A.J. Saul (ed.): *Floods and Flood Management.* 1992 ISBN 0-7923-2078-6
16. D.E. Ashpis, T.B. Gatski and R. Hirsh (eds.): *Instabilities and Turbulence in Engineering Flows.* 1993 ISBN 0-7923-2161-8
17. R.S. Azad: *The Atmospheric Boundary Layer for Engineers.* 1993 ISBN 0-7923-2187-1
18. F.T.M. Nieuwstadt (ed.): *Advances in Turbulence IV.* 1993 ISBN 0-7923-2282-7
19. K.K. Prasad (ed.): *Further Developments in Turbulence Management.* 1993 ISBN 0-7923-2291-6
20. Y.A. Tatarchenko: *Shaped Crystal Growth.* 1993 ISBN 0-7923-2419-6

Kluwer Academic Publishers – Dordrecht / Boston / London

Mechanics

FLUID MECHANICS AND ITS APPLICATIONS

Series Editor: R. Moreau

21. J.P. Bonnet and M.N. Glauser (eds.): *Eddy Structure Identification in Free Turbulent Shear Flows.* 1993 ISBN 0-7923-2449-8
22. R.S. Srivastava: *Interaction of Shock Waves.* 1994 ISBN 0-7923-2920-1
23. J.R. Blake, J.M. Boulton-Stone and N.H. Thomas (eds.): *Bubble Dynamics and Interface Phenomena.* 1994 ISBN 0-7923-3008-0
24. R. Benzi (ed.): *Advances in Turbulence V.* 1994 ISBN 0-7923-3032-3
25. B.I. Rabinovich, V.G. Lebedev and A.I. Mytarev: *Vortex Processes and Solid Body Dynamics.* The Dynamic Problems of Spacecrafts and Magnetic Levitation Systems. 1994 ISBN 0-7923-3092-7
26. P.R. Voke, L. Kleiser and J.-P. Chollet (eds.): *Direct and Large-Eddy Simulation I.* Selected papers from the First ERCOFTAC Workshop on Direct and Large-Eddy Simulation. 1994 ISBN 0-7923-3106-0
27. J.A. Sparenberg: *Hydrodynamic Propulsion and its Optimization.* Analytic Theory. 1995 ISBN 0-7923-3201-6

Kluwer Academic Publishers – Dordrecht / Boston / London

Mechanics

SOLID MECHANICS AND ITS APPLICATIONS

Series Editor: G.M.L. Gladwell

Aims and Scope of the Series

The fundamental questions arising in mechanics are: *Why?*, *How?*, and *How much?* The aim of this series is to provide lucid accounts written by authoritative researchers giving vision and insight in answering these questions on the subject of mechanics as it relates to solids. The scope of the series covers the entire spectrum of solid mechanics. Thus it includes the foundation of mechanics; variational formulations; computational mechanics; statics, kinematics and dynamics of rigid and elastic bodies; vibrations of solids and structures; dynamical systems and chaos; the theories of elasticity, plasticity and viscoelasticity; composite materials; rods, beams, shells and membranes; structural control and stability; soils, rocks and geomechanics; fracture; tribology; experimental mechanics; biomechanics and machine design.

1. R.T. Haftka, Z. Gürdal and M.P. Kamat: *Elements of Structural Optimization.* 2nd rev.ed., 1990 ISBN 0-7923-0608-2
2. J.J. Kalker: *Three-Dimensional Elastic Bodies in Rolling Contact.* 1990 ISBN 0-7923-0712-7
3. P. Karasudhi: *Foundations of Solid Mechanics.* 1991 ISBN 0-7923-0772-0
4. N. Kikuchi: *Computational Methods in Contact Mechanics.* (forthcoming) ISBN 0-7923-0773-9
5. *Not published.*
6. J.F. Doyle: *Static and Dynamic Analysis of Structures.* With an Emphasis on Mechanics and Computer Matrix Methods. 1991 ISBN 0-7923-1124-8; Pb 0-7923-1208-2
7. O.O. Ochoa and J.N. Reddy: *Finite Element Analysis of Composite Laminates.* ISBN 0-7923-1125-6
8. M.H. Aliabadi and D.P. Rooke: *Numerical Fracture Mechanics.* ISBN 0-7923-1175-2
9. J. Angeles and C.S. López-Cajún: *Optimization of Cam Mechanisms.* 1991 ISBN 0-7923-1355-0
10. D.E. Grierson, A. Franchi and P. Riva: *Progress in Structural Engineering.* 1991 ISBN 0-7923-1396-8
11. R.T. Haftka and Z. Gürdal: *Elements of Structural Optimization.* 3rd rev. and exp. ed. 1992 ISBN 0-7923-1504-9; Pb 0-7923-1505-7
12. J.R. Barber: *Elasticity.* 1992 ISBN 0-7923-1609-6; Pb 0-7923-1610-X
13. H.S. Tzou and G.L. Anderson (eds.): *Intelligent Structural Systems.* 1992 ISBN 0-7923-1920-6
14. E.E. Gdoutos: *Fracture Mechanics.* An Introduction. 1993 ISBN 0-7923-1932-X
15. J.P. Ward: *Solid Mechanics.* An Introduction. 1992 ISBN 0-7923-1949-4
16. M. Farshad: *Design and Analysis of Shell Structures.* 1992 ISBN 0-7923-1950-8
17. H.S. Tzou and T. Fukuda (eds.): *Precision Sensors, Actuators and Systems.* 1992 ISBN 0-7923-2015-8
18. J.R. Vinson: *The Behavior of Shells Composed of Isotropic and Composite Materials.* 1993 ISBN 0-7923-2113-8

Kluwer Academic Publishers – Dordrecht / Boston / London

Mechanics

SOLID MECHANICS AND ITS APPLICATIONS

Series Editor: G.M.L. Gladwell

19. H.S. Tzou: *Piezoelectric Shells.* Distributed Sensing and Control of Continua. 1993 ISBN 0-7923-2186-3
20. W. Schiehlen: *Advanced Multibody System Dynamics.* Simulation and Software Tools. 1993 ISBN 0-7923-2192-8
21. C.-W. Lee: *Vibration Analysis of Rotors.* 1993 ISBN 0-7923-2300-9
22. D.R. Smith: *An Introduction to Continuum Mechanics.* 1993 ISBN 0-7923-2454-4
23. G.M.L. Gladwell: *Inverse Problems in Scattering.* An Introduction. 1993 ISBN 0-7923-2478-1
24. G. Prathap: *The Finite Element Method in Structural Mechanics.* 1993 ISBN 0-7923-2492-7
25. J. Herskovits (ed.): *Structural Optimization '93.* 1993 ISBN 0-7923-2510-9
26. M.A. González-Palacios and J. Angeles: *Cam Synthesis.* 1993 ISBN 0-7923-2536-2
27. W.S. Hall: *The Boundary Element Method.* 1993 ISBN 0-7923-2580-X
28. J. Angeles, G. Hommel and P. Kovács (eds.): *Computational Kinematics.* 1993 ISBN 0-7923-2585-0
29. A. Curnier: *Computational Methods in Solid Mechanics.* 1994 ISBN 0-7923-2761-6
30. D.A. Hills and D. Nowell: *Mechanics of Fretting Fatigue.* 1994 ISBN 0-7923-2866-3
31. B. Tabarrok and F.P.J. Rimrott: *Variational Methods and Complementary Formulations in Dynamics.* 1994 ISBN 0-7923-2923-6
32. E.H. Dowell, E.F. Crawley, H.C. Curtiss Jr., D.A. Peters, R. H. Scanlan and F. Sisto (eds.): *A Modern Course in Aeroelasticity.* Third Revised and Enlarged Edition. (forthcoming) ISBN 0-7923-2788-8; Pb: 0-7923-2789-6
33. A. Preumont: *Random Vibration and Spectral Analysis.* 1994 ISBN 0-7923-3036-6
34. J.N. Reddy (ed.): *Mechanics of Composite Materials.* Selected works of Nicholas J. Pagano. 1994 ISBN 0-7923-3041-2

Kluwer Academic Publishers – Dordrecht / Boston / London

Mechanics

From 1990, books on the subject of *mechanics* will be published under two series:

***FLUID* MECHANICS AND ITS APPLICATIONS**
Series Editor: R.J. Moreau

***SOLID* MECHANICS AND ITS APPLICATIONS**
Series Editor: G.M.L. Gladwell

Prior to 1990, the books listed below were published in the respective series indicated below.

MECHANICS: DYNAMICAL SYSTEMS
Editors: L. Meirovitch and G.Æ. Oravas

1. E.H. Dowell: *Aeroelasticity of Plates and Shells.* 1975 ISBN 90-286-0404-9
2. D.G.B. Edelen: *Lagrangian Mechanics of Nonconservative Nonholonomic Systems.* 1977 ISBN 90-286-0077-9
3. J.L. Junkins: *An Introduction to Optimal Estimation of Dynamical Systems.* 1978 ISBN 90-286-0067-1
4. E.H. Dowell (ed.), H.C. Curtiss Jr., R.H. Scanlan and F. Sisto: *A Modern Course in Aeroelasticity.* *Revised and enlarged edition see under Volume 11*
5. L. Meirovitch: *Computational Methods in Structural Dynamics.* 1980 ISBN 90-286-0580-0
6. B. Skalmierski and A. Tylikowski: *Stochastic Processes in Dynamics.* Revised and enlarged translation. 1982 ISBN 90-247-2686-7
7. P.C. Müller and W.O. Schiehlen: *Linear Vibrations.* A Theoretical Treatment of Multi-degree-of-freedom Vibrating Systems. 1985 ISBN 90-247-2983-1
8. Gh. Buzdugan, E. Mihăilescu and M. Radeş: *Vibration Measurement.* 1986 ISBN 90-247-3111-9
9. G.M.L. Gladwell: *Inverse Problems in Vibration.* 1987 ISBN 90-247-3408-8
10. G.I. Schuëller and M. Shinozuka: *Stochastic Methods in Structural Dynamics.* 1987 ISBN 90-247-3611-0
11. E.H. Dowell (ed.), H.C. Curtiss Jr., R.H. Scanlan and F. Sisto: *A Modern Course in Aeroelasticity.* Second revised and enlarged edition (of Volume 4). 1989 ISBN Hb 0-7923-0062-9; Pb 0-7923-0185-4
12. W. Szemplińska-Stupnicka: *The Behavior of Nonlinear Vibrating Systems.* Volume I: Fundamental Concepts and Methods: Applications to Single-Degree-of-Freedom Systems. 1990 ISBN 0-7923-0368-7
13. W. Szemplińska-Stupnicka: *The Behavior of Nonlinear Vibrating Systems.* Volume II: Advanced Concepts and Applications to Multi-Degree-of-Freedom Systems. 1990 ISBN 0-7923-0369-5
Set ISBN (Vols. 12–13) 0-7923-0370-9

MECHANICS OF STRUCTURAL SYSTEMS
Editors: J.S. Przemieniecki and G.Æ. Oravas

1. L. Frýba: *Vibration of Solids and Structures under Moving Loads.* 1970 ISBN 90-01-32420-2
2. K. Marguerre and K. Wölfel: *Mechanics of Vibration.* 1979 ISBN 90-286-0086-8

Mechanics

3. E.B. Magrab: *Vibrations of Elastic Structural Members*. 1979 ISBN 90-286-0207-0
4. R.T. Haftka and M.P. Kamat: *Elements of Structural Optimization*. 1985
   Revised and enlarged edition see under Solid Mechanics and Its Applications, Volume 1
5. J.R. Vinson and R.L. Sierakowski: *The Behavior of Structures Composed of Composite Materials*. 1986 ISBN Hb 90-247-3125-9; Pb 90-247-3578-5
6. B.E. Gatewood: *Virtual Principles in Aircraft Structures*. Volume 1: Analysis. 1989
   ISBN 90-247-3754-0
7. B.E. Gatewood: *Virtual Principles in Aircraft Structures*. Volume 2: Design, Plates, Finite Elements. 1989 ISBN 90-247-3755-9
   Set (Gatewood 1 + 2) ISBN 90-247-3753-2

MECHANICS OF ELASTIC AND INELASTIC SOLIDS

Editors: S. Nemat-Nasser and G.Æ. Oravas

---

1. G.M.L. Gladwell: *Contact Problems in the Classical Theory of Elasticity*. 1980
   ISBN Hb 90-286-0440-5; Pb 90-286-0760-9
2. G. Wempner: *Mechanics of Solids with Applications to Thin Bodies*. 1981
   ISBN 90-286-0880-X
3. T. Mura: *Micromechanics of Defects in Solids*. 2nd revised edition, 1987
   ISBN 90-247-3343-X
4. R.G. Payton: *Elastic Wave Propagation in Transversely Isotropic Media*. 1983
   ISBN 90-247-2843-6
5. S. Nemat-Nasser, H. Abé and S. Hirakawa (eds.): *Hydraulic Fracturing and Geothermal Energy*. 1983 ISBN 90-247-2855-X
6. S. Nemat-Nasser, R.J. Asaro and G.A. Hegemier (eds.): *Theoretical Foundation for Large-scale Computations of Nonlinear Material Behavior*. 1984 ISBN 90-247-3092-9
7. N. Cristescu: *Rock Rheology*. 1988 ISBN 90-247-3660-9
8. G.I.N. Rozvany: *Structural Design via Optimality Criteria*. The Prager Approach to Structural Optimization. 1989 ISBN 90-247-3613-7

MECHANICS OF SURFACE STRUCTURES

Editors: W.A. Nash and G.Æ. Oravas

---

1. P. Seide: *Small Elastic Deformations of Thin Shells*. 1975 ISBN 90-286-0064-7
2. V. Panc: *Theories of Elastic Plates*. 1975 ISBN 90-286-0104-X
3. J.L. Nowinski: *Theory of Thermoelasticity with Applications*. 1978
   ISBN 90-286-0457-X
4. S. Łukasiewicz: *Local Loads in Plates and Shells*. 1979 ISBN 90-286-0047-7
5. C. Fiřt: *Statics, Formfinding and Dynamics of Air-supported Membrane Structures*. 1983 ISBN 90-247-2672-7
6. Y. Kai-yuan (ed.): *Progress in Applied Mechanics*. The Chien Wei-zang Anniversary Volume. 1987 ISBN 90-247-3249-2
7. R. Negruţiu: *Elastic Analysis of Slab Structures*. 1987 ISBN 90-247-3367-7
8. J.R. Vinson: *The Behavior of Thin Walled Structures*. Beams, Plates, and Shells. 1988
   ISBN Hb 90-247-3663-3; Pb 90-247-3664-1

Mechanics

MECHANICS OF FLUIDS AND TRANSPORT PROCESSES
Editors: R.J. Moreau and G.Æ. Oravas

1. J. Happel and H. Brenner: *Low Reynolds Number Hydrodynamics.* With Special Applications to Particular Media. 1983 ISBN Hb 90-01-37115-9; Pb 90-247-2877-0
2. S. Zahorski: *Mechanics of Viscoelastic Fluids.* 1982 ISBN 90-247-2687-5
3. J.A. Sparenberg: *Elements of Hydrodynamics Propulsion.* 1984 ISBN 90-247-2871-1
4. B.K. Shivamoggi: *Theoretical Fluid Dynamics.* 1984 ISBN 90-247-2999-8
5. R. Timman, A.J. Hermans and G.C. Hsiao: *Water Waves and Ship Hydrodynamics.* An Introduction. 1985 ISBN 90-247-3218-2
6. M. Lesieur: *Turbulence in Fluids.* Stochastic and Numerical Modelling. 1987 ISBN 90-247-3470-3
7. L.A. Lliboutry: *Very Slow Flows of Solids.* Basics of Modeling in Geodynamics and Glaciology. 1987 ISBN 90-247-3482-7
8. B.K. Shivamoggi: *Introduction to Nonlinear Fluid-Plasma Waves.* 1988 ISBN 90-247-3662-5
9. V. Bojarevičs, Ya. Freibergs, E.I. Shilova and E.V. Shcherbinin: *Electrically Induced Vortical Flows.* 1989 ISBN 90-247-3712-5
10. J. Lielpeteris and R. Moreau (eds.): *Liquid Metal Magnetohydrodynamics.* 1989 ISBN 0-7923-0344-X

MECHANICS OF ELASTIC STABILITY
Editors: H. Leipholz and G.Æ. Oravas

1. H. Leipholz: *Theory of Elasticity.* 1974 ISBN 90-286-0193-7
2. L. Librescu: *Elastostatics and Kinetics of Aniosotropic and Heterogeneous Shell-type Structures.* 1975 ISBN 90-286-0035-3
3. C.L. Dym: *Stability Theory and Its Applications to Structural Mechanics.* 1974 ISBN 90-286-0094-9
4. K. Huseyin: *Nonlinear Theory of Elastic Stability.* 1975 ISBN 90-286-0344-1
5. H. Leipholz: *Direct Variational Methods and Eigenvalue Problems in Engineering.* 1977 ISBN 90-286-0106-6
6. K. Huseyin: *Vibrations and Stability of Multiple Parameter Systems.* 1978 ISBN 90-286-0136-8
7. H. Leipholz: *Stability of Elastic Systems.* 1980 ISBN 90-286-0050-7
8. V.V. Bolotin: *Random Vibrations of Elastic Systems.* 1984 ISBN 90-247-2981-5
9. D. Bushnell: *Computerized Buckling Analysis of Shells.* 1985 ISBN 90-247-3099-6
10. L.M. Kachanov: *Introduction to Continuum Damage Mechanics.* 1986 ISBN 90-247-3319-7
11. H.H.E. Leipholz and M. Abdel-Rohman: *Control of Structures.* 1986 ISBN 90-247-3321-9
12. H.E. Lindberg and A.L. Florence: *Dynamic Pulse Buckling.* Theory and Experiment. 1987 ISBN 90-247-3566-1
13. A. Gajewski and M. Zyczkowski: *Optimal Structural Design under Stability Constraints.* 1988 ISBN 90-247-3612-9

Mechanics

MECHANICS: ANALYSIS

Editors: V.J. Mizel and G.Æ. Oravas

1. M.A. Krasnoselskii, P.P. Zabreiko, E.I. Pustylnik and P.E. Sbolevskii: *Integral Operators in Spaces of Summable Functions.* 1976 ISBN 90-286-0294-1
2. V.V. Ivanov: *The Theory of Approximate Methods and Their Application to the Numerical Solution of Singular Integral Equations.* 1976 ISBN 90-286-0036-1
3. A. Kufner, O. John and S. Pučík: *Function Spaces.* 1977 ISBN 90-286-0015-9
4. S.G. Mikhlin: *Approximation on a Rectangular Grid.* With Application to Finite Element Methods and Other Problems. 1979 ISBN 90-286-0008-6
5. D.G.B. Edelen: *Isovector Methods for Equations of Balance.* With Programs for Computer Assistance in Operator Calculations and an Exposition of Practical Topics of the Exterior Calculus. 1980 ISBN 90-286-0420-0
6. R.S. Anderssen, F.R. de Hoog and M.A. Lukas (eds.): *The Application and Numerical Solution of Integral Equations.* 1980 ISBN 90-286-0450-2
7. R.Z. Has'minskiĭ: *Stochastic Stability of Differential Equations.* 1980 ISBN 90-286-0100-7
8. A.I. Vol'pert and S.I. Hudjaev: *Analysis in Classes of Discontinuous Functions and Equations of Mathematical Physics.* 1985 ISBN 90-247-3109-7
9. A. Georgescu: *Hydrodynamic Stability Theory.* 1985 ISBN 90-247-3120-8
10. W. Noll: *Finite-dimensional Spaces.* Algebra, Geometry and Analysis. Volume I. 1987 ISBN Hb 90-247-3581-5; Pb 90-247-3582-3

MECHANICS: COMPUTATIONAL MECHANICS

Editors: M. Stern and G.Æ. Oravas

1. T.A. Cruse: *Boundary Element Analysis in Computational Fracture Mechanics.* 1988 ISBN 90-247-3614-5

MECHANICS: GENESIS AND METHOD

Editor: G.Æ. Oravas

1. P.-M.-M. Duhem: *The Evolution of Mechanics.* 1980 ISBN 90-286-0688-2

MECHANICS OF CONTINUA

Editors: W.O. Williams and G.Æ. Oravas

1. C.-C. Wang and C. Truesdell: *Introduction to Rational Elasticity.* 1973 ISBN 90-01-93710-1
2. P.J. Chen: *Selected Topics in Wave Propagation.* 1976 ISBN 90-286-0515-0
3. P. Villaggio: *Qualitative Methods in Elasticity.* 1977 ISBN 90-286-0007-8